Teach Yourself Electricity and Electronics

Other books by Stan Gibilisco

The Illustrated Dictionary of Electronics, 7/e

TAB Encyclopedia of Electronics for Technicians and Hobbyists

McGraw-Hill Encyclopedia of Personal Computing

The McGraw-Hill Encyclopedia of Robotics and Artificial Intelligence

Teach Yourself Electricity and Electronics
Second Edition

Stan Gibilisco

McGraw-Hill

New York San Francisco Washington, D.C. Auckland Bogotá
Caracas Lisbon London Madrid Mexico City Milan
Montreal New Delhi San Juan Singapore
Sydney Tokyo Toronto

Gibilisco, Stan.
  Teach yourself electricity and electronics / Stan Gibilisco.—
2nd ed.
      p.   cm.
    Includes index.
    ISBN 0-07-024578-9 (hard edition).—ISBN 0-07-024579-7
(paper edition)
    1. Electronics   2. Electricity.   I. Title.
TK7819.G38 1997
621.3—dc21   97-16153
                                                   CIP

McGraw-Hill

A Division of The McGraw·Hill Companies

ISBN 0-07-024578-9 (HC)
3 4 5 6 7 8 9 10 FGR/FGR 9 9 8

ISBN 0-07-024579-7 (Pbk)
 5 6 7 8 9 10 FGR/FGR 9 9

The sponsoring editor for this book was Scott Grillo, the editing supervisor was
Ruth W. Mannino, and the production supervisor was Tina Cameron. It was set
in ITC Century Light by McGraw-Hill's Hightstown, N.J., Professional Book Group
composition unit.

McGraw-Hill books are available at special quantity discounts to use as premiums and
sales promotions, or for use in corporate training programs. For more information,
please write to the Director of Special Sales, McGraw-Hill, 11 West 19th Street,
New York, NY 10011. Or contact your local bookstore.

   This book is printed on recycled, acid-free paper containing a minimum of
50% recycled, de-inked fiber.

To Tony and Tim
from Uncle Stan

Contents

Preface

This book is for people who want to learn basic electricity, electronics, and communications concepts without taking a formal course. It can also serve as a classroom text. This second edition contains new material covering wireless communications, personal computers, and the Internet.

I recommend you start at the beginning of this book and go straight through. There are hundreds of quiz and test questions to fortify your knowledge and help you check your progress as you work your way along.

There is a short multiple-choice quiz at the end of every chapter. You may (and should) refer to the chapter texts when taking these quizzes. When you think you're ready, take the quiz, write down your answers, and then give your list of answers to a friend. Have the friend tell you your score, but not which questions you got wrong. The answers are listed in the back of the book. Stick with a chapter until you get most of the answers correct. Because you're allowed to look at the text during quizzes, the questions are written so that you really have to think before you write down an answer. Some are rather difficult, but there are no trick questions.

This book is divided into three major sections: direct current, alternating current, and basic electronics. At the end of each section is a multiple-choice test. Take these tests when you're done with the respective sections and have taken all the chapter quizzes. Don't look back at the text when taking these tests. A satisfactory score is 37 correct answers. Again, answers are in the back of the book.

There is a final exam at the end of the book. The questions are practical, mostly nonmathematical, and somewhat easier than those in the quizzes. The final exam contains questions drawn from all the chapters. Take this exam when you have finished all three sections, all three section tests, and all of the chapter quizzes. A satisfactory score is at least 75 percent correct answers.

With the section tests and final exam, as with the quizzes, have a friend tell you your score without letting you know which questions you missed. That way, you can't subconsciously memorize the answers. You might want to take a test two or three times. When you have gotten a score that makes you happy, you can check to see where your knowledge is strong and where it's weak.

It is not necessary to have a mathematical or scientific background in order to use this do-it-yourself course. Junior high school algebra, geometry, and physical science will suffice. I've tried to gradually introduce standard symbols and notations so that it will be evident what they mean as you go along. By the time you get near the end of this book, assuming you've followed it all along, you should be familiar with most of the symbols used in schematic diagrams.

I'd recommend doing a chapter a week. An hour daily ought to be more than enough time for this. That way, in less than eight months, you'll complete the course. You can then use this book, with its comprehensive index, as a permanent reference.

Suggestions for future editions are welcome.

Stan Gibilisco

Teach Yourself Electricity and Electronics

1
PART

Direct current

1
CHAPTER

Basic physical concepts

IT IS IMPORTANT TO UNDERSTAND SOME SIMPLE, GENERAL PHYSICS PRINCIPLES in order to have a full grasp of electricity and electronics. It is not necessary to know high-level mathematics.

In science, you can talk about *qualitative* things or about *quantitative* things, the "what" versus the "how much." For now, you need only be concerned about the "what." The "how much" will come later.

Atoms

All matter is made up of countless tiny particles whizzing around. These particles are extremely dense; matter is mostly empty space. Matter seems continuous because the particles are so small, and they move incredibly fast.

Even people of ancient times suspected that matter is made of invisible particles. They deduced this from observing things like water, rocks, and metals. These substances are much different from each other. But any given material—copper, for example—is the same wherever it is found. Even without doing any complicated experiments, early physicists felt that substances could only have these consistent behaviors if they were made of unique types, or arrangements, of particles. It took centuries before people knew just how this complicated business works. And even today, there are certain things that scientists don't really know. For example, is there a smallest possible material particle?

There were some scientists who refused to believe the atomic theory, even around the year of 1900. Today, practically everyone accepts the theory. It explains the behavior of matter better than any other scheme.

Eventually, scientists identified 92 different kinds of fundamental substances in nature, and called them *elements*. Later, a few more elements were artificially made. Each element has its own unique type of particle, known as its *atom*. Atoms of different elements are always different.

The slightest change in an atom can make a tremendous difference in its behavior. You can live by breathing pure oxygen, but you can't live off of pure nitrogen. Oxygen will cause metal to corrode, but nitrogen will not. Wood will burn furiously in an atmosphere of pure oxygen, but will not even ignite in pure nitrogen. Yet both are gases at room temperature and pressure; both are colorless, both are odorless, and both are just about of equal weight. These substances are so different because oxygen has eight *protons*, while nitrogen has only seven.

There are many other examples in nature where a tiny change in atomic structure makes a major difference in the way a substance behaves.

Protons, neutrons, and the atomic number

The part of an atom that gives an element its identity is the *nucleus*. It is made up of two kinds of particles, the *proton* and the *neutron*. These are extremely dense. A teaspoonful of either of these particles, packed tightly together, would weigh tons. Protons and neutrons have just about the same mass, but the proton has an electric charge while the neutron does not.

The simplest element, hydrogen, has a nucleus made up of only one proton; there are usually no neutrons. This is the most common element in the universe. Sometimes a nucleus of hydrogen has a neutron or two along with the proton, but this does not occur very often. These "mutant" forms of hydrogen do, nonetheless, play significant roles in atomic physics.

The second most abundant element is helium. Usually, this atom has a nucleus with two protons and two neutrons. Hydrogen is changed into helium inside the sun, and in the process, energy is given off. This makes the sun shine. The process, called *fusion*, is also responsible for the terrific explosive force of a hydrogen bomb.

Every proton in the universe is just like every other. Neutrons are all alike, too. The number of protons in an element's nucleus, the *atomic number*, gives that element its identity. The element with three protons is lithium, a light metal that reacts easily with gases such as oxygen or chlorine. The element with four protons is beryllium, also a metal. In general, as the number of protons in an element's nucleus increases, the number of neutrons also increases. Elements with high atomic numbers, like lead, are therefore much denser than elements with low atomic numbers, like carbon. Perhaps you've compared a lead sinker with a piece of coal of similar size, and noticed this difference.

Isotopes and atomic weights

For a given element, such as oxygen, the number of neutrons can vary. But no matter what the number of neutrons, the element keeps its identity, based on the atomic number. Differing numbers of neutrons result in various *isotopes* for a given element.

Each element has one particular isotope that is most often found in nature. But all elements have numerous isotopes. Changing the number of neutrons in an element's nucleus results in a difference in the weight, and also a difference in the density, of the element. Thus, hydrogen containing a neutron or two in the nucleus, along with the proton, is called *heavy hydrogen.*

The *atomic weight* of an element is approximately equal to the sum of the number of protons and the number of neutrons in the nucleus. Common carbon has an atomic weight of about 12, and is called carbon 12 or C12. But sometimes it has an atomic weight of about 14, and is known as carbon 14 or C14.

Table 1-1 lists all the known elements in alphabetical order, with atomic numbers in one column, and atomic weights of the most common isotopes in another column. The standard abbreviations are also shown.

Electrons

Surrounding the nucleus of an atom are particles having opposite electric charge from the protons. These are the *electrons.* Physicists arbitrarily call the electrons' charge *negative,* and the protons' charge *positive.* An electron has exactly the same charge quantity as a proton, but with opposite polarity. The charge on a single electron or proton is the smallest possible electric charge. All charges, no matter how great, are multiples of this unit charge.

One of the earliest ideas about the atom pictured the electrons embedded in the nucleus, like raisins in a cake. Later, the electrons were seen as orbiting the nucleus, making the atom like a miniature solar system with the electrons as the planets (Fig. 1-1). Still later, this view was modified further. Today, the electrons are seen as so fast-moving, with patterns so complex, that it is not even possible to pinpoint them at any given instant of time. All that can be done is to say that an electron will just as likely be inside a certain sphere as outside. These spheres are known as electron *shells.* Their centers correspond to the position of the atomic nucleus. The farther away from the nucleus the *shell,* the more energy the electon has (Fig. 1-2).

Electrons can move rather easily from one atom to another in some materials. In other substances, it is difficult to get electrons to move. But in any case, it is far easier to move electrons than it is to move protons. Electricity almost always results, in some way, from the motion of electrons in a material.

Electrons are much lighter than protons or neutrons. In fact, compared to the nucleus of an atom, the electrons weigh practically nothing.

Generally, the number of electrons in an atom is the same as the number of protons. The negative charges therefore exactly cancel out the positive ones, and the atom is electrically neutral. But under some conditions, there can be an excess or shortage of electrons. High levels of radiant energy, extreme heat, or the presence of an electric field (discussed later) can "knock" or "throw" electrons loose from atoms, upsetting the balance.

Ions

If an atom has more or less electrons than neutrons, that atom acquires an electrical charge. A shortage of electrons results in positive charge; an excess of electrons gives a negative charge. The element's identity remains the same, no matter how great the excess or shortage of electrons. In the extreme case, all the electrons might be removed

Table 1-1.   Atomic numbers and weights.

Element name	Abbreviation	Atomic number	Atomic weight*
Actinium	Ac	89	227
Aluminum	Al	13	27
Americium**	Am	95	243
Antimony	Sb	51	121
Argon	Ar	18	40
Arsenic	As	33	75
Astatine	At	85	210
Barium	Ba	56	138
Berkelium**	Bk	97	247
Beryllium	Be	4	9
Bismuth	Bi	83	209
Boron	B	5	11
Bromine	Br	35	79
Cadmium	Cd	48	114
Calcium	Ca	20	40
Californium**	Cf	98	251
Carbon	C	6	12
Cerium	Ce	58	140
Cesium	Cs	55	133
Chlorine	Cl	17	35
Chromium	Cr	24	52
Cobalt	Co	27	59
Copper	Cu	29	63
Curium**	Cm	96	247
Dysprosium	Dy	66	164
Einsteinium**	Es	99	254
Erbium	Er	68	166
Europium	Eu	63	153
Fermium	Fm	100	257
Fluorine	F	9	19
Francium	Fr	87	223
Gadolinium	Gd	64	158
Gallium	Ga	31	69
Germanium	Ge	32	74
Gold	Au	79	197
Hafnium	Hf	72	180
Helium	He	2	4
Holmium	Ho	67	165
Hydrogen	H	1	1
Indium	In	49	115
Iodine	I	53	127
Iridium	Ir	77	193
Iron	Fe	26	56
Krypton	Kr	36	84
Lanthanum	La	57	139
Lawrencium**	Lr or Lw	103	257

Table 1-1. Continued.

Element name	Abbreviation	Atomic number	Atomic weight*
Lead	Pb	82	208
Lithium	Li	3	7
Lutetium	Lu	71	175
Magnesium	Mg	12	24
Manganese	Mn	25	55
Mendelevium**	Md	101	256
Mercury	Hg	80	202
Molybdenum	Mo	42	98
Neodymium	Nd	60	142
Neon	Ne	10	20
Neptunium**	Np	93	237
Nickel	Ni	28	58
Niobium	Nb	41	93
Nitrogen	N	7	14
Nobelium**	No	102	254
Osmium	Os	76	192
Oxygen	O	8	16
Palladium	Pd	46	108
Phosphorus	P	15	31
Platinum	Pt	78	195
Plutonium**	Pu	94	242
Polonium	Po	84	209
Potassium	K	19	39
Praseodymium	Pr	59	141
Promethium	Pm	61	145
Proactinium	Pa	91	231
Radium	Ra	88	226
Radon	Rn	86	222
Rhenium	Re	75	187
Rhodium	Rh	45	103
Rubidium	Rb	37	85
Ruthenium	Ru	44	102
Samarium	Sm	62	152
Scandium	Sc	21	45
Selenium	Se	34	80
Silicon	Si	14	28
Silver	Ag	47	107
Sodium	Na	11	23
Strontium	Sr	38	88
Sulfur	S	16	32
Tantalum	Ta	73	181
Technetium	Tc	43	99
Tellurium	Te	52	130
Terbium	Tb	65	159
Thallium	Tl	81	205
Thorium	Th	90	232
Thulium	Tm	69	169

Table 1-1.   Continued.

Element name	Abbreviation	Atomic number	Atomic weight*
Tin	Sn	50	120
Titanium	Ti	22	48
Tungsten	W	74	184
Unnilhexium**	Unh	106	—
Unnilpentium**	Unp	105	—
Unnilquadium**	Unq	104	—
Uranium	U	92	238
Vanadium	V	23	51
Xenon	Xe	54	132
Ytterbium	Yb	70	174
Yttrium	Y	39	89
Zinc	Zn	30	64
Zirconium	Zr	40	90

*Most common isotope. The sum of the number of protons and the number of neutrons in the nucleus. Most elements have other isotopes with different atomic weights.

**These elements (atomic numbers 93 or larger) are not found in nature, but are human-made.

Nucleus

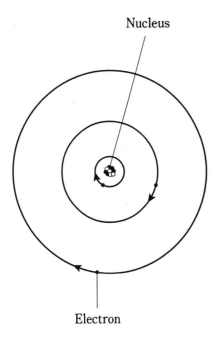

Electron

1-1   An early model of the atom, developed about the year 1900, rendered electrons like planets and the nucleus like the sun in a miniature solar system. Electric charge attraction kept the electrons from flying away.

from an atom, leaving only the nucleus. However it would still represent the same element as it would if it had all its electrons.

A charged atom is called an *ion*. When a substance contains many ions, the material is said to be *ionized*.

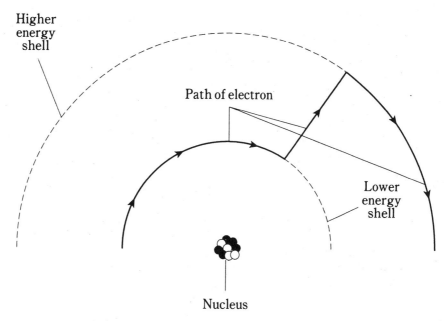

1-2   Electrons move around the nucleus of an atom at defined levels corresponding to different energy states. This is a simplified drawing, depicting an electron gaining energy.

A good example of an ionized substance is the atmosphere of the earth at high altitudes. The ultraviolet radiation from the sun, as well as high-speed subatomic particles from space, result in the gases' atoms being stripped of electrons. The ionized gases tend to be found in layers at certain altitudes. These layers are responsible for long-distance radio communications at some frequencies.

Ionized materials generally conduct electricity quite well, even if the substance is normally not a good conductor. Ionized air makes it possible for a lightning stroke to take place, for example. The ionization, caused by a powerful electric field, occurs along a jagged, narrow channel, as you have surely seen. After the lightning flash, the nuclei of the atoms quickly attract stray electrons back, and the air becomes electrically neutral again.

An element might be both an ion and an isotope different from the usual isotope. For example, an atom of carbon might have eight neutrons rather than the usual six, thus being the isotope C14, and it might have been stripped of an electron, giving it a positive unit electric charge and making it an ion.

Compounds

Different elements can join together to share electrons. When this happens, the result is a chemical *compound*. One of the most common compounds is water, the result of two hydrogen atoms joining with an atom of oxygen. There are literally thousands of different chemical compounds that occur in nature.

A compound is different than a simple mixture of elements. If hydrogen and oxygen are mixed, the result is a colorless, odorless gas, just like either element is a gas separately. A spark, however, will cause the molecules to join together; this will liberate energy in the form of light and heat. Under the right conditions, there will be a violent explosion, because the two elements join eagerly. Water is chemically illustrated in Fig. 1-3.

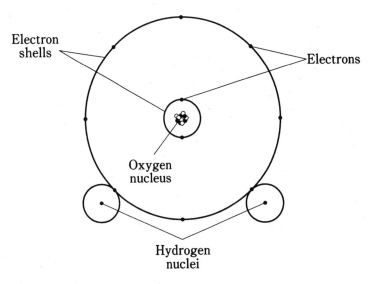

1-3 Simplified diagram of a water molecule.

Compounds often, but not always, appear greatly different from any of the elements that make them up. At room temperature and pressure, both hydrogen and oxygen are gases. But water under the same conditions is a liquid. If it gets a few tens of degrees colder, water turns solid at standard pressure. If it gets hot enough, water becomes a gas, odorless and colorless, just like hydrogen or oxygen.

Another common example of a compound is rust. This forms when iron joins with oxygen. While iron is a dull gray solid and oxygen is a gas, rust is a maroon-red or brownish powder, completely unlike either of the elements from which it is formed.

Molecules

When atoms of elements join together to form a compound, the resulting particles are *molecules*. Figure 1-3 is an example of a molecule of water, consisting of three atoms put together.

The natural form of an element is also known as its molecule. Oxygen tends to occur in pairs most of the time in the earth's atmosphere. Thus, an oxygen molecule is sometimes denoted by the symbol O_2: The "O" represents oxygen, and the subscript 2 indicates that there are two atoms per molecule. The water molecule is symbolized H_2O, because there are two atoms of hydrogen and one atom of oxygen in each molecule.

Sometimes oxygen atoms are by themselves; then we denote the molecule simply as O. Sometimes there are three atoms of oxygen grouped together. This is the gas called *ozone*, that has received much attention lately in environmental news. It is written O_3.

All matter, whether it is solid, liquid, or gas, is made of molecules. These particles are always moving. The speed with which they move depends on the temperature. The hotter the temperature, the more rapidly the molecules move around. In a solid, the molecules are interlocked in a sort of rigid pattern, although they vibrate continuously (Fig. 1-4A). In a liquid, they slither and slide around (Fig. 1-4B). In a gas, they are literally whizzing all over the place, bumping into each other and into solids and liquids adjacent to the gas (Fig. 1-4C).

Conductors

In some materials, electrons move easily from atom to atom. In others, the electrons move with difficulty. And in some materials, it is almost impossible to get them to move. An electrical *conductor* is a substance in which the electrons are mobile.

The best conductor at room temperature is pure elemental silver. Copper and aluminum are also excellent electrical conductors. Iron, steel, and various other metals are fair to good conductors of electricity.

In most electrical circuits and systems, copper or aluminum wire is used. Silver is impractical because of its high cost.

Some liquids are good electrical conductors. Mercury is one example. Salt water is a fair conductor.

Gases are, in general, poor conductors of electricity. This is because the atoms or molecules are usually too far apart to allow a free exchange of electrons. But if a gas becomes ionized, it is a fair conductor of electricity.

Electrons in a conductor do not move in a steady stream, like molecules of water through a garden hose. Instead, they are passed from one atom to another right next to it (Fig. 1-5). This happens to countless atoms all the time. As a result, literally trillions of electrons pass a given point each second in a typical electrical circuit.

You might imagine a long line of people, each one constantly passing a ball to the neighbor on the right. If there are plenty of balls all along the line, and if everyone keeps passing balls along as they come, the result will be a steady stream of balls moving along the line. This represents a good conductor.

If the people become tired or lazy, and do not feel much like passing the balls along, the rate of flow will decrease. The conductor is no longer very good.

Insulators

If the people refuse to pass balls along the line in the previous example, the line represents an electrical *insulator*. Such substances prevent electrical currents from flowing, except possibly in very small amounts.

Most gases are good electrical insulators. Glass, dry wood, paper, and plastics are other examples. Pure water is a good electrical insulator, although it conducts some current with even the slightest impurity. Metal oxides can be good insulators, even though the metal in pure form is a good conductor.

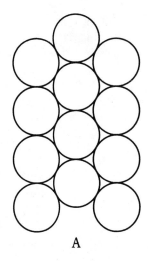

A

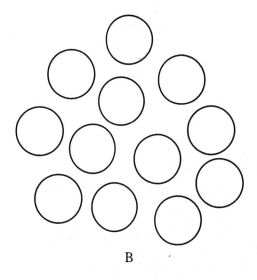

B

1-4   At A, simplified rendition of molecules in a solid; at B, in a liquid; at C, in a gas. The molecules don't shrink in the gas. They are shown smaller because of the much larger spaces between them.

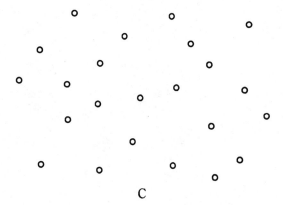

C

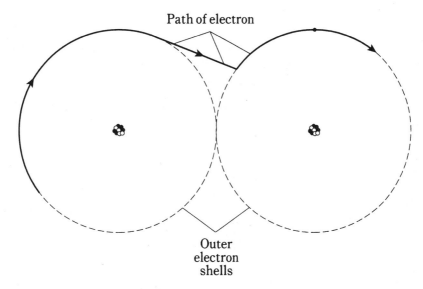

Path of electron

Outer
electron
shells

1-5 In a conductor, electrons are passed from atom to atom.

Electrical insulators can be forced to carry current. Ionization can take place; when electrons are stripped away from their atoms, they have no choice but to move along. Sometimes an insulating material gets charred, or melts down, or gets perforated by a spark. Then its insulating properties are lost, and some electrons flow.

An insulating material is sometimes called a *dielectric*. This term arises from the fact that it keeps electrical charges apart, preventing the flow of electrons that would equalize a charge difference between two places. Excellent insulating materials can be used to advantage in certain electrical components such as capacitors, where it is important that electrons not flow.

Porcelain or glass can be used in electrical systems to keep short circuits from occurring. These devices, called insulators, come in various shapes and sizes for different applications. You can see them on high-voltage utility poles and towers. They hold the wire up without running the risk of a short circuit with the tower or a slow discharge through a wet wooden pole.

Resistors

Some substances, such as carbon, conduct electricity fairly well but not really well. The conductivity can be changed by adding impurities like clay to a carbon paste, or by winding a thin wire into a coil. Electrical components made in this way are called *resistors*. They are important in electronic circuits because they allow for the control of current flow.

Resistors can be manufactured to have exact characteristics. Imagine telling each person in the line that they must pass a certain number of balls per minute. This is analogous to creating a resistor with a certain value of electrical *resistance*.

The better a resistor conducts, the lower its resistance; the worse it conducts, the higher the resistance.

Electrical resistance is measured in units called *ohms*. The higher the value in ohms, the greater the resistance, and the more difficult it becomes for current to flow. For wires, the resistance is sometimes specified in terms of *ohms per foot* or *ohms per kilometer*. In an electrical system, it is usually desirable to have as low a resistance, or ohmic value, as possible. This is because resistance converts electrical energy into heat. Thick wires and high voltages reduce this resistance loss in long-distance electrical lines. This is why such gigantic towers, with dangerous voltages, are necessary in large utility systems.

Semiconductors

In a *semiconductor*, electrons flow, but not as well as they do in a conductor. You might imagine the people in the line being lazy and not too eager to pass the balls along. Some semiconductors carry electrons almost as well as good electrical conductors like copper or aluminum; others are almost as bad as insulating materials. The people might be just a little sluggish, or they might be almost asleep.

Semiconductors are not exactly the same as resistors. In a semiconductor, the material is treated so that it has very special properties.

The semiconductors include certain substances, such as silicon, selenium, or gallium, that have been "doped" by the addition of impurities like indium or antimony. Perhaps you have heard of such things as *gallium arsenide, metal oxides,* or *silicon rectifiers.* Electrical conduction in these materials is always a result of the motion of electrons. However, this can be a quite peculiar movement, and sometimes engineers speak of the movement of *holes* rather than electrons. A hole is a shortage of an electron—you might think of it as a positive ion—and it moves along in a direction opposite to the flow of electrons (Fig. 1-6).

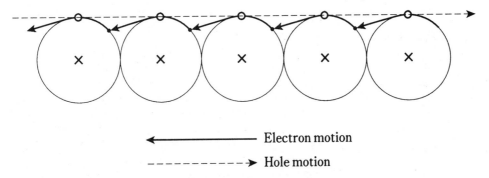

1-6   Holes move in the opposite direction from electrons in a semiconducting material.

When most of the *charge carriers* are electrons, the semiconductor is called *N-type*, because electrons are negatively charged. When most of the charge carriers are holes, the semiconducting material is known as *P-type* because holes have a positive electric charge. But P-type material does pass some electrons, and N-type material carries some holes. In a semiconductor, the more abundant type of charge carrier is called the *majority carrier*. The less abundant kind is known as the *minority carrier*.

Semiconductors are used in diodes, transistors, and integrated circuits in almost limitless variety. These substances are what make it possible for you to have a computer in a briefcase. That notebook computer, if it used vacuum tubes, would occupy a skyscraper, because it has billions of electronic components. It would also need its own power plant, and would cost thousands of dollars in electric bills every day. But the circuits are etched microscopically onto semiconducting wafers, greatly reducing the size and power requirements.

Current

Whenever there is movement of charge carriers in a substance, there is an electric *current*. Current is measured in terms of the number of electrons or holes passing a single point in one second.

Usually, a great many charge carriers go past any given point in one second, even if the current is small. In a household electric circuit, a 100-watt light bulb draws a current of about *six quintillion* (6 followed by 18 zeroes) charge carriers per second. Even the smallest mini-bulb carries *quadrillions* (numbers followed by 15 zeroes) of charge carriers every second. It is ridiculous to speak of a current in terms of charge carriers per second, so usually it is measured in *coulombs per second* instead. A coulomb is equal to approximately 6,240,000,000,000,000,000 electrons or holes. A current of one coulomb per second is called an *ampere*, and this is the standard unit of electric current. A 100-watt bulb in your desk lamp draws about one ampere of current.

When a current flows through a resistance—and this is always the case because even the best conductors have resistance—heat is generated. Sometimes light and other forms of energy are emitted as well. A light bulb is deliberately designed so that the resistance causes visible light to be generated. Even the best incandescent lamp is inefficient, creating more heat than light energy. Fluorescent lamps are better. They produce more light for a given amount of current. Or, to put it another way, they need less current to give off a certain amount of light.

Electric current flows very fast through any conductor, resistor, or semiconductor. In fact, for most practical purposes you can consider the speed of current to be the same as the speed of light: 186,000 miles per second. Actually, it is a little less.

Static electricity

Charge carriers, particularly electrons, can build up, or become deficient, on things without flowing anywhere. You've probably experienced this when walking on a carpeted floor during the winter, or in a place where the humidity was very low. An excess or shortage of electrons is created on and in your body. You acquire a *charge* of *static electricity.* It's called "static" because it doesn't go anywhere. You don't feel this until you touch some metallic object that is connected to earth ground or to some large fixture; but then there is a *discharge*, accompanied by a spark that might well startle you. It is the current, during this discharge, that causes the sensation that might make you jump.

If you were to become much more charged, your hair would stand on end, because every hair would repel every other. Like charges are caused either by an excess or a deficiency of electrons; they repel. The spark might jump an inch, two inches, or even six inches. Then it would more than startle you; you could get hurt. This doesn't happen with ordinary carpet and shoes, fortunately. But a device called a *Van de Graaff generator*, found in some high school physics labs, can cause a spark this large (Fig. 1-7). You have to be careful when using this device for physics experiments.

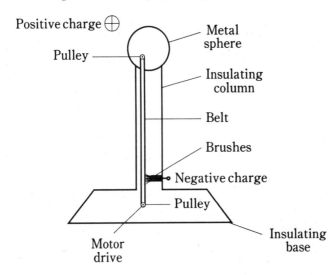

Positive charge ⊕

Pulley

Metal sphere

Insulating column

Belt

Brushes

Negative charge

Pulley

Motor drive

Insulating base

1-7   Simple diagram of a Van de Graaff generator for creating large static charges.

In the extreme, lightning occurs between clouds, and between clouds and ground in the earth's atmosphere. This spark is just a greatly magnified version of the little spark you get after shuffling around on a carpet. Until the spark occurs, there is a static charge in the clouds, between different clouds or parts of a cloud, and the ground. In Fig. 1-8, cloud-to-cloud (A) and cloud-to-ground (B) static buildups are shown. In the case at B, the positive charge in the earth follows along beneath the thunderstorm cloud like a shadow as the storm is blown along by the prevailing winds.

The current in a lightning stroke is usually several tens of thousands, or hundreds of thousands, of amperes. But it takes place only for a fraction of a second. Still, many coulombs of charge are displaced in a single bolt of lightning.

Electromotive force

Current can only flow if it gets a "push." This might be caused by a buildup of static electric charges, as in the case of a lightning stroke. When the charge builds up, with positive

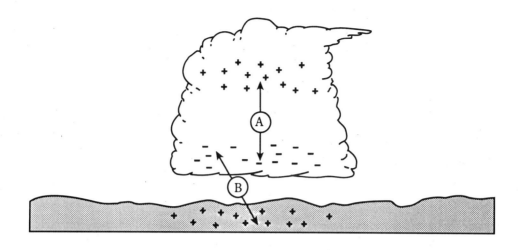

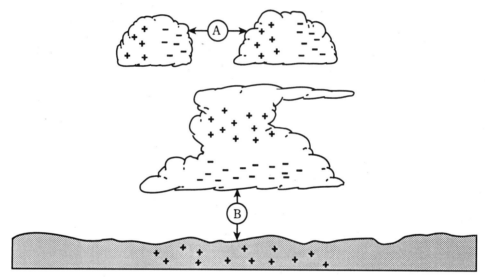

1-8 Cloud-to-cloud (A) and cloud-to-ground (B) charge buildup can both occur in a single thunderstorm.

polarity (shortage of electrons) in one place and negative polarity (excess of electrons) in another place, a powerful *electromotive force* exists. It is often abbreviated EMF. This force is measured in units called *volts*.

Ordinary household electricity has an effective voltage of between 110 and 130; usually it is about 117. A car battery has an EMF of 12 volts (six volts in some older systems). The static charge that you acquire when walking on a carpet with hard-soled shoes is often several thousand volts. Before a discharge of lightning, many millions of volts exist.

An EMF of one volt, across a resistance of one ohm, will cause a current of one ampere to flow. This is a classic relationship in electricity, and is stated generally as *Ohm's*

Law. If the EMF is doubled, the current is doubled. If the resistance is doubled, the current is cut in half. This important law of electrical circuit behavior is covered in detail a little later in this book.

It is possible to have an EMF without having any current. This is the case just before a lightning bolt occurs, and before you touch that radiator after walking on the carpet. It is also true between the two wires of an electric lamp when the switch is turned off. It is true of a dry cell when there is nothing connected to it. There is no current, but a current is possible given a conductive path between the two points. Voltage, or EMF, is sometimes called *potential* or *potential difference* for this reason.

Even a very large EMF might not drive much current through a conductor or resistance. A good example is your body after walking around on the carpet. Although the voltage seems deadly in terms of numbers (thousands), there are not that many coulombs of charge that can accumulate on an object the size of your body. Therefore in relative terms, not that many electrons flow through your finger when you touch a radiator so you don't get a severe shock.

Conversely, if there are plenty of coulombs available, a small voltage, such as 117 volts (or even less), can result in a lethal flow of current. This is why it is so dangerous to repair an electrical device with the power on. The power plant will pump an unlimited number of coulombs of charge through your body if you are foolish enough to get caught in that kind of situation.

Nonelectrical energy

In electricity and electronics, there are many kinds of phenomena that involve other forms of energy besides electrical energy.

Visible light is an example. A light bulb converts electricity into radiant energy that you can see. This was one of the major motivations for people like Thomas Edison to work with electricity. Visible light can also be converted into electric current or voltage. A *photovoltaic cell* does this.

Light bulbs always give off some heat, as well as visible light. Incandescent lamps actually give off more energy as heat than as light. And you are certainly acquainted with electric heaters, designed for the purpose of changing electricity into heat energy. This "heat" is actually a form of radiant energy called *infrared*. It is similar to visible light, except that the waves are longer and you can't see them.

Electricity can be converted into other radiant-energy forms, such as radio waves, ultraviolet, and X rays. This is done by things like radio transmitters, sunlamps, and X-ray tubes.

Fast-moving protons, neutrons, electrons, and atomic nuclei are an important form of energy, especially in deep space where they are known as *cosmic radiation*. The energy from these particles is sometimes sufficient to split atoms apart. This effect makes it possible to build an atomic reactor whose energy can be used to generate electricity. Unfortunately, this form of energy, called *nuclear energy*, creates dangerous by-products that are hard to dispose of.

When a conductor is moved in a magnetic field, electric current flows in that conductor. In this way, mechanical energy is converted into electricity. This is how a generator works. Generators can also work backwards. Then you have a motor that changes electricity into useful mechanical energy.

A magnetic field contains energy of a unique kind. The science of *magnetism* is closely related to electricity. Magnetic phenomena are of great significance in electronics. The oldest and most universal source of magnetism is the flux field surrounding the earth, caused by alignment of iron atoms in the core of the planet.

A changing magnetic field creates a fluctuating electric field, and a fluctuating electric field produces a changing magnetic field. This phenomenon, called *electromagnetism*, makes it possible to send radio signals over long distances. The electric and magnetic fields keep producing one another over and over again through space.

Chemical energy is converted into electricity in all dry cells, wet cells, and batteries. Your car battery is an excellent example. The acid reacts with the metal electrodes to generate an electromotive force. When the two poles of the batteries are connected, current results. The chemical reaction continues, keeping the current going for awhile. But the battery can only store a certain amount of chemical energy. Then it "runs out of juice," and the supply of chemical energy must be restored by *charging*. Some cells and batteries, such as lead-acid car batteries, can be recharged by driving current through them, and others, such as most flashlight and transistor-radio batteries, cannot.

Quiz

Refer to the text in this chapter if necessary. A good score is at least 18 correct answers out of these 20 questions. The answers are listed in the back of this book.

1. The atomic number of an element is determined by:
   A. The number of neutrons.
   B. The number of protons.
   C. The number of neutrons plus the number of protons.
   D. The number of electrons.

2. The atomic weight of an element is approximately determined by:
   A. The number of neutrons.
   B. The number of protons.
   C. The number of neutrons plus the number of protons.
   D. The number of electrons.

3. Suppose there is an atom of oxygen, containing eight protons and eight neutrons in the nucleus, and two neutrons are added to the nucleus. The resulting atomic weight is about:
   A. 8.
   B. 10.
   C. 16.
   D. 18.

4. An ion:
   A. Is electrically neutral.
   B. Has positive electric charge.
   C. Has negative electric charge.
   D. Might have either a positive or negative charge.

5. An isotope:
   A. Is electrically neutral.
   B. Has positive electric charge.
   C. Has negative electric charge.
   D. Might have either a positive or negative charge.

6. A molecule:
   A. Might consist of just a single atom of an element.
   B. Must always contain two or more elements.
   C. Always has two or more atoms.
   D. Is always electrically charged.

7. In a compound:
   A. There can be just a single atom of an element.
   B. There must always be two or more elements.
   C. The atoms are mixed in with each other but not joined.
   D. There is always a shortage of electrons.

8. An electrical insulator can be made a conductor:
   A. By heating.
   B. By cooling.
   C. By ionizing.
   D. By oxidizing.

9. Of the following substances, the worst conductor is:
   A. Air.
   B. Copper.
   C. Iron.
   D. Salt water.

10. Of the following substances, the best conductor is:
    A. Air.
    B. Copper.
    C. Iron.
    D. Salt water.

11. Movement of holes in a semiconductor:

    A. Is like a flow of electrons in the same direction.

    B. Is possible only if the current is high enough.

    C. Results in a certain amount of electric current.

    D. Causes the material to stop conducting.

12. If a material has low resistance:

    A. It is a good conductor.

    B. It is a poor conductor.

    C. The current flows mainly in the form of holes.

    D. Current can flow only in one direction.

13. A coulomb:

    A. Represents a current of one ampere.

    B. Flows through a 100-watt light bulb.

    C. Is one ampere per second.

    D. Is an extremely large number of charge carriers.

14. A stroke of lightning:

    A. Is caused by a movement of holes in an insulator.

    B. Has a very low current.

    C. Is a discharge of static electricity.

    D. Builds up between clouds.

15. The volt is the standard unit of:

    A. Current.

    B. Charge.

    C. Electromotive force.

    D. Resistance.

16. If an EMF of one volt is placed across a resistance of two ohms, then the current is:

    A. Half an ampere.

    B. One ampere.

    C. Two amperes.

    D. One ohm.

17. A backwards-working electric motor is best described as:

    A. An inefficient, energy-wasting device.

    B. A motor with the voltage connected the wrong way.

    C. An electric generator.

    D. A magnetic-field generator.

18. In some batteries, chemical energy can be replenished by:
    A. Connecting it to a light bulb.
    B. Charging it.
    C. Discharging it.
    D. No means known; when a battery is dead, you have to throw it away.

19. A changing magnetic field:
    A. Produces an electric current in an insulator.
    B. Magnetizes the earth.
    C. Produces a fluctuating electric field.
    D. Results from a steady electric current.

20. Light is converted into electricity:
    A. In a dry cell.
    B. In a wet cell.
    C. In an incandescent bulb.
    D. In a photovoltaic cell.

2
CHAPTER

Electrical units

THIS CHAPTER EXPLAINS SOME MORE ABOUT UNITS THAT QUANTIFY THE behavior of direct-current circuits. Many of these rules apply to utility alternating-current circuits also. Utility current is, in many respects, just like direct current because the frequency of alternation is low (60 complete cycles per second).

The volt

In chapter 1, you learned a little about the volt, the standard unit of electromotive force (EMF) or *potential difference*.

An accumulation of static electric charge, such as an excess or shortage of electrons, is always associated with a voltage. There are other situations in which voltages exist. Voltage is generated at a power plant, and produced in an electrochemical reaction, and caused by light falling on a special semiconductor chip. It can be produced when an object is moved in a magnetic field, or is placed in a fluctuating magnetic field.

A potential difference between two points produces an *electric field*, represented by *electric lines of flux* (Fig. 2-1). There is always a pole that is relatively positive, with fewer electrons, and one that is relatively negative, with more electrons. The positive pole does not necessarily have a deficiency of electrons compared with neutral objects, and the negative pole might not actually have a surplus of electrons with respect to neutral things. But there's always a difference in charge between the two poles. The negative pole always has more electrons than the positive pole.

The abbreviation for volt is V. Sometimes, smaller units are used. The *millivolt* (mV) is equal to a thousandth (0.001) of a volt. The *microvolt* (uV) is equal to a millionth (0.000001) of a volt. And it is sometimes necessary to use units much larger than one volt. One *kilovolt* (kV) is equal to one thousand volts (1,000). One *megavolt* (MV) is equal to one million volts (1,000,000) or one thousand kilovolts.

In a dry cell, the EMF is usually between 1.2 and 1.7 V; in a car battery, it is most

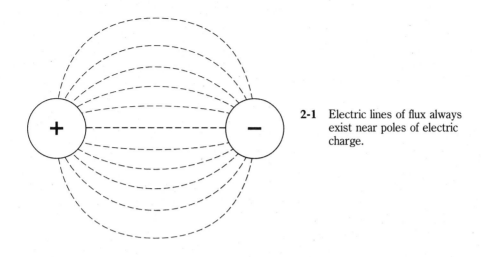

2-1 Electric lines of flux always exist near poles of electric charge.

often 12 V to 14 V. In household utility wiring, it is a low-frequency alternating current of about 117 V for electric lights and most appliances, and 234 V for a washing machine, dryer, oven, or stove. In television sets, transformers convert 117 V to around 450 V for the operation of the picture tube. In some broadcast transmitters, kilovolts are used. The largest voltages on Earth occur between clouds, or between clouds and the ground, in thundershowers; this potential difference is on the order of tens of megavolts.

In every case, voltage, EMF, or potential difference represents the fact that charge carriers will flow between two points if a conductive path is provided. The number of charge carriers might be small even if the voltage is huge, or very large even if the voltage is tiny. Voltage represents the pressure or driving force that impels the charge carriers to move. In general, for a given number of charge carriers, higher voltages will produce a faster flow, and therefore a larger current. It's something like water pressure. The amount of water that will flow through a hose is proportional to the water pressure, all other things being equal.

Current flow

If a conducting or semiconducting path is provided between two poles having a potential difference, charge carriers will flow in an attempt to equalize the charge between the poles. This flow of electric *current* will continue as long as the path is provided, and as long as there is a charge difference between the poles.

Sometimes the charge difference is equalized after a short while. This is the case, for example, when you touch a radiator after shuffling around on the carpet in your hard-soled shoes. It is also true in a lightning stroke. In these instances, the charge is equalized in a fraction of a second.

The charge might take longer to be used up. This will happen if you short-circuit a dry cell. Within a few minutes, or maybe up to an hour, the cell will "run out of juice" if you put a wire between the positive and negative terminals. If you put a bulb across the cell, say with a flashlight, it takes an hour or two for the charge difference to drop to zero.

In household electric circuits, the charge difference will essentially never equalize, unless there's a power failure. Of course, if you short-circuit an outlet (don't!), the fuse or breaker will blow or trip, and the charge difference will immediately drop to zero. But if you put a 100-watt bulb at the outlet, the charge difference will be maintained as the current flows. The power plant can keep a potential difference across a lot of light bulbs indefinitely.

You might have heard that "It's the current, not the voltage, that kills," concerning the danger in an electric circuit. This is a literal truth, but it plays on semantics. It's like saying "It's the heat, not the fire, that burns you." Naturally! But there can only be a deadly current if there is enough voltage to drive it through your body. You don't have to worry when handling flashlight cells, but you'd better be extremely careful around household utility circuits. A voltage of 1.2 to 1.7 V can't normally pump a dangerous current through you, but a voltage of 117 V almost always can.

Through an electric circuit with constant conductivity, the current is directly proportional to the applied voltage. That is, if you double the voltage, you double the current; if the voltage is cut in half, the current is cut in half too. Figure 2-2 shows this relationship as a graph in general terms. But it assumes that the power supply can provide the necessary number of charge carriers. This rule holds only within reasonable limits.

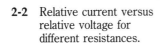 **2-2** Relative current versus relative voltage for different resistances.

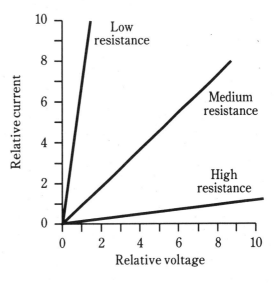

When you are charged up with static electricity, there aren't very many charge carriers. A dry cell runs short of energy after awhile, and can no longer deliver as much current. All power supplies have their limitations in terms of the current they can provide. A power plant, or a power supply that works off of the utility mains, or a very large electrochemical battery, has a large capacity. You can then say that if you cut the resistance by a factor of 100, you'll get 100 times as much current. Or perhaps even 1000 or 10,000 times the current, if the resistance is cut to 0.001 or 0.0001 its former value.

The ampere

Current is a measure of the rate at which charge carriers flow. The standard unit is the *ampere*. This represents one coulomb (6,240,000,000,000,000,000) of charge carriers per second past a given point.

An ampere is a comparatively large amount of current. The abbreviation is A. Often, current is specified in terms of *milliamperes*, abbreviated mA, where 1 mA = 0.001 A or a thousandth of an ampere. You will also sometimes hear of *microamperes* (uA), where 1 uA = 0.000001 A = 0.001 mA, a millionth of an ampere. And it is increasingly common to hear about *nanoamperes* (nA), where 1 nA = 0.001 uA = 0.000000001 A (a billionth of an ampere). Rarely will you hear of *kiloamperes* (kA), where 1 kA = 1000 A.

A current of a few milliamperes will give you a startling shock. About 50 mA will jolt you severely, and 100 mA can cause death if it flows through your chest cavity.

An ordinary 100-watt light bulb draws about 1 A of current. An electric iron draws approximately 10 A; an entire household normally uses between 10 A and 50 A, depending on the size of the house and the kinds of appliances it has, and also on the time of day, week or year.

The amount of current that will flow in an electrical circuit depends on the voltage, and also on the resistance. There are some circuits in which extremely large currents, say 1000 A, flow; this might happen through a metal bar placed directly at the output of a massive electric generator. The resistance is extremely low in this case, and the generator is capable of driving huge amounts of charge. In some semiconductor electronic devices, such as microcomputers, a few nanoamperes will suffice for many complicated processes. Some electronic clocks draw so little current that their batteries last as long as they would if left on the shelf without being put to any use at all.

Resistance and the ohm

Resistance is a measure of the opposition that a circuit offers to the flow of electric current. You might compare it to the diameter of a hose. In fact, for metal wire, this is an excellent analogy: small-diameter wire has high resistance (a lot of opposition to current flow), and large-diameter wire has low resistance (not much opposition to electric currents). Of course, the type of metal makes a difference too. Iron wire has higher resistance for a given diameter than copper wire. Nichrome wire has still more resistance.

The standard unit of resistance is the ohm. This is sometimes abbreviated by the upper-case Greek letter omega, resembling an upside-down capital letter U (Ω). In this book, we'll just write it out as "ohm" or "ohms."

You'll sometimes hear about *kilohms*, where 1 kilohm = 1,000 ohms, or about *megohms*, where 1 megohm = 1,000 kilohms = 1,000,000 ohms.

Electrical wire is sometimes rated for *resistivity*. The standard unit for this purpose is the *ohm per foot (ohm/ft)* or the *ohm per meter (ohm/m)*. You might also come across the unit *ohm per kilometer (ohm/km)*. Table 2-1 shows the resistivity for various common sizes of wire.

When 1 V is placed across 1 ohm of resistance, assuming that the power supply can

Table 2-1. Resistivity for copper wire, in terms of the size in American Wire Gauge (AWG).

Wire size, AWG	Resistivity, ohms/km
2	0.52
4	0.83
6	1.3
8	2.7
10	3.3
12	5.3
14	8.4
16	13
18	21
20	34
22	54
24	86
26	140
28	220
30	350

deliver an unlimited number of charge carriers, there will be a current of 1 A. If the resistance is doubled, the current is cut in half. If the resistance is cut in half, the current doubles. Therefore, the current flow, for a constant voltage, is *inversely proportional* to the resistance. Figure 2-3 is a graph that shows various currents, through various resistances, given a constant voltage of 1 V across the whole resistance.

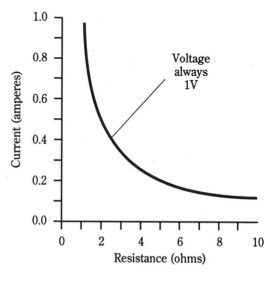

2-3 Current versus resistance through an electric device when the voltage is constant at 1 V.

Resistance has another property in an electric circuit. If there is a current flowing through a resistive material, there will always be a potential difference across the resistive object. This is shown in Fig. 2-4. The larger the current through the *resistor*, the greater the EMF across the resistor. In general, this EMF is directly proportional to the current through the resistor. This behavior of resistors is extremely useful in the design of electronic circuits, as you will learn later in this book.

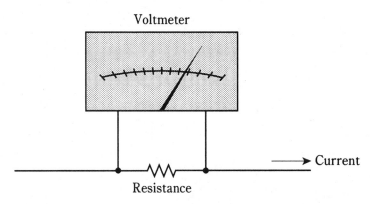

2-4   Whenever a resistance carries a current, there is a voltage across it.

Electrical circuits always have some resistance. There is no such thing as a perfect conductor. When some metals are chilled to extremely low temperatures, they lose practically all of their resistance, but they never become absolutely perfect, resistance-free conductors. This phenomenon, about which you might have heard, is called *superconductivity*. In recent years, special metals have been found that behave this way even at fairly moderate temperatures. Researchers are trying to concoct substances that will superconduct even at room temperature. Superconductivity is an active field in physics right now.

Just as there is no such thing as a perfectly resistance-free substance, there isn't a truly infinite resistance, either. Even air conducts to some extent, although the effect is usually so small that it can be ignored. In some electronic applications, materials are selected on the basis of how nearly infinite their resistance is. These materials make good electric insulators, and good dielectrics for *capacitors*, devices that store electric charge.

In electronics, the resistance of a component often varies, depending on the conditions under which it is operated. A transistor, for example, might have extremely high resistance some of the time, and very low resistance at other times. This high/low fluctuation can be made to take place thousands, millions or billions of times each second. In this way, oscillators, amplifiers and digital electronic devices function in radio receivers and transmitters, telephone networks, digital computers and satellite links (to name just a few applications).

Conductance and the siemens

The better a substance conducts, the less its resistance; the worse it conducts, the higher its resistance. Electricians and electrical engineers sometimes prefer to speak about the *conductance* of a material, rather than about its resistance. The standard unit of conductance is the *siemens*, abbreviated S. When a component has a conductance of 1 S, its resistance is 1 ohm. If the resistance is doubled, the conductance is cut in half, and vice-versa. Therefore, conductance is the reciprocal of resistance.

If you know the resistance in ohms, you can get the conductance in siemens by taking the quotient of 1 over the resistance. Also, if you know the conductance in siemens, you can get the resistance in ohms by taking 1 over the conductance. The relation can be written as:

$$\text{siemens} = 1/\text{ohms, or}$$
$$\text{ohms} = 1/\text{siemens}$$

Smaller units of conductance are often necessary. A resistance of one kilohm is equal to one *millisiemens*. If the resistance is a megohm, the conductance is one *microsiemens*. You'll also hear about *kilosiemens* or *megasiemens*, representing resistances of 0.001 ohm and 0.000001 ohm (a thousandth of an ohm and a millionth of an ohm) respectively. Short lengths of heavy wire have conductance values in the range of kilosiemens. Heavy metal rods might sometimes have conductances in the megasiemens range.

As an example, suppose a component has a resistance of 50 ohms. Then its conductance, in siemens, is ⅟₅₀, or 0.02 S. You might say that this is 20 mS. Or imagine a piece of wire with a conductance of 20 S. Its resistance is ⅟₂₀, or 0.05, ohm. Not often will you hear the term "milliohm"; engineers do not, for some reason, speak of subohmic units very much. But you could say that this wire has a resistance of 50 milliohms, and you would be technically right.

Conductivity is a little trickier. If wire has a resistivity of, say, 10 ohms per kilometer, you can't just say that it has a conductivity of ⅟₁₀, or 0.1, siemens per kilometer. It is true that a kilometer of such wire will have a conductance of 0.1 S; but 2 km of the wire will have a resistance of 20 ohms (because there is twice as much wire), and this is not twice the conductance, but half. If you say that the conductivity of the wire is 0.1 S/km, then you might be tempted to say that 2 km of the wire has 0.2 S of conductance. Wrong! Conductance decreases, rather than increasing, with wire length.

When dealing with wire conductivity for various lengths of wire, it's best to convert to resistivity values, and then convert back to the final conductance when you're all done calculating. Then there won't be any problems with mathematical semantics.

Figure 2-5 illustrates the resistance and conductance values for various lengths of wire having a resistivity of 10 ohms per kilometer.

Power and the watt

Whenever current flows through a resistance, heat results. This is inevitable. The heat can be measured in *watts*, abbreviated W, and represents electrical *power*. Power can be manifested in many other ways, such as in the form of mechanical motion, or radio waves, or visible light, or noise. In fact, there are dozens of different ways that power can be dissipated. But heat is always present, in addition to any other form of power in an electrical or electronic device. This is because no equipment is 100-percent efficient. Some power always goes to waste, and this waste is almost all in the form of heat.

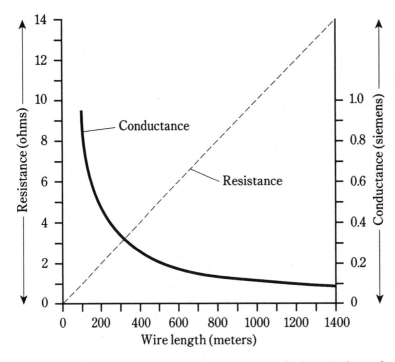

2-5   Total resistances and conductances for a wire having 10 ohms of resistivity per kilometer.

Look again at the diagram of Fig. 2-4. There is a certain voltage across the resistor, not specifically given in the diagram. There's also a current flowing through the resistance, not quantified in the diagram, either. Suppose we call the voltage E and the current I, in volts and amperes, respectively. Then the power in watts dissipated by the resistance, call it P, is the product E × I. That is: P = EI.

This power might all be heat. Or it might exist in several forms, such as heat, light and infrared. This would be the state of affairs if the resistor were an incandescent light bulb, for example. If it were a motor, some of the power would exist in the form of mechanical work.

If the voltage across the resistance is caused by two flashlight cells in series, giving 3 V, and if the current through the resistance (a light bulb, perhaps) is 0.1 A, then E = 3 and I = 0.1, and we can calculate the power P, in watts, as:

$$(watts) = EI = 3 \times 0.1 = 0.3 \text{ W}$$

Suppose the voltage is 117 V, and the current is 855 mA. To calculate the power, we must convert the current into amperes; 855 mA = 855/1000 = 0.855 A. Then

$$P \text{ (watts)} = 117 \times 0.855 = 100 \text{ W}$$

You will often hear about *milliwatts (mW)*, *microwatts (uW)*, *kilowatts (kW)* and *megawatts (MW)*. You should, by now, be able to tell from the prefixes what these units represent. But in case you haven't gotten the idea yet, you can refer to Table 2-2. This table gives the most commonly used prefix multipliers in electricity and electronics, and the fractions that they represent. Thus, 1 mW = 0.001 W; 1 uW = 0.001 mW = 0.000001 W; 1 kW = 1,000 W; and 1 MW = 1,000 kW = 1,000,000 W.

Table 2-2. Common prefix multipliers.

Prefix	Fraction
pico-	0.000000000001 (one-trillionth)
nano-	0.000000001 (one-billionth)
micro-	0.000001 (one-millionth)
milli-	0.001 (one-thousandth)
kilo-	1000
mega-	1,000,000
giga-	1,000,000,000 (one billion)
tera-	1,000,000,000,000 (one trillion)

Sometimes you need to use the power equation to find currents or voltages. Then you should use I = P/E to find current, or E = P/I to find power. It's easiest to remember that P = EI (watts equal volt-amperes), and derive the other equations from this by dividing through either by E (to get I) or by I (to get E).

Energy and the watt hour

There is an important difference between *energy* and power. You've probably heard the two terms used interchangeably, as if they mean the same thing. But they don't. Energy is power dissipated over a length of time. Power is the rate at which energy is expended.

Physicists measure energy in *joules*. One joule is the equivalent of one watt of power, dissipated for one second of time. In electricity, you'll more often encounter the *watt hour* or the *kilowatt hour*. As their names imply, a watt hour, abbreviated Wh, is the equivalent of 1 W dissipated for an hour (1 h), and 1 kilowatt hour (kWh) is the equivalent of 1 kW of power dissipated for 1 h.

An energy of 1 Wh can be dissipated in an infinite number of different ways. A 60-watt bulb will burn 60 Wh in an hour, or 1 Wh per minute. A 100-W bulb would burn 1 Wh in $\frac{1}{100}$ hour, or 36 seconds. A 6-watt Christmas tree bulb would require 10 minutes ($\frac{1}{6}$ hour) to burn 1 Wh. And the rate of power dissipation need not be constant; it could be constantly changing.

Figure 2-6 illustrates two hypothetical devices that burn up 1 Wh of energy. Device A uses its power at a constant rate of 60 watts, so it consumes 1 Wh in a minute. The power consumption rate of device B varies, starting at zero and ending up at quite a lot more than 60 W. How do you know that this second device really burns up 1 Wh of energy? You determine the area under the graph. This example has been chosen because figuring out this area is rather easy. Remember that the area of a triangle is equal to half the product of the base length and the height. This second device is on for 72 seconds, or 1.2 minute; this is 1.2/60 = 0.02 hour. Then the area under the "curve" is ½ × 100 × 0.02 = 1 Wh.

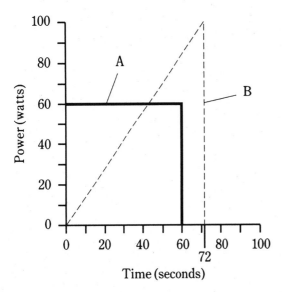

2-6   Two devices that burn 1 Wh of energy. Device A dissipates a constant power; device B dissipates a changing amount of power.

When calculating energy values, you must always remember the units you're using. In this case the unit is the watt hour, so you must multiply watts by hours. If you multiply watts by minutes, or watts by seconds, you'll get the wrong kind of units in your answer. That means a wrong answer!

Sometimes, the curves in graphs like these are complicated. In fact, they usually are. Consider the graph of power consumption in your home, versus time, for a whole day. It might look something like the curve in Fig. 2-7. Finding the area under this curve is no easy task, if you have only this graph to go by. But there is another way to determine the total energy burned by your household in a day, or in a week, or most often, in a month. That is by means of the electric meter. It measures electrical energy in kilowatt hours. Every month, without fail, the power company sends its representative to read that meter. This person takes down the number of kilowatt hours displayed, subtracts the number from the previous month, and a few days later you get a bill. This meter automatically keeps track of total consumed energy, without anybody having to do sophisticated integral calculus to find the areas under irregular curves such as the graph of Fig. 2-7.

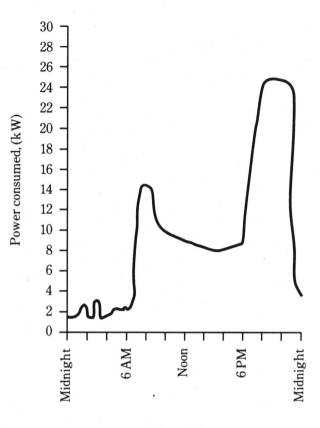

2-7 Hypothetical graph showing the power consumed by a typical household, as a function of the time of day.

Other energy units

As said before, physicists prefer to use the joule, or watt second, as their energy unit. This is the standard unit for scientific purposes.

Another unit is the *erg*, equivalent to one ten-millionth (0.0000001) of a joule. This is said to be roughly the amount of energy needed by a mosquito to take off after it has bitten you (not including the energy needed for the bite itself). The erg is used in lab experiments involving small amounts of expended energy.

You have probably heard of the *British thermal unit (Btu)*, equivalent to 1055 joules. This is the energy unit most often used to indicate the cooling or heating capacity of air-conditioning equipment. To cool your room from 85 to 78 degrees needs a certain amount of energy, perhaps best specified in Btu. If you are getting an air conditioner or furnace installed in your home, an expert will come look at your situation, and determine the size of air conditioning/heating unit, in Btu, that best suits your needs. It doesn't make any sense to get one that is way too big; you'll be wasting your money. But you want to be sure that it's big enough—or you'll also waste money because of inefficiency and possibly also because of frequent repair calls.

Physicists also use, in addition to the joule, a unit of energy called the *electron volt* (eV). This is an extremely tiny unit of energy, equal to just 0.00000000000000000016 joule (there are 18 zeroes after the decimal point and before the 1). The physicists writes 1.6×10^{-19} to represent this. It is the energy gained by a single electron in an electric field of 1 V. Atom smashers are rated by millions of electron volts (MeV) or billions of electron volts (GeV) of energy capacity. In the future you might even hear of a huge linear accelerator, built on some vast prairie, and capable of delivering trillions of electron volts (TeV).

Another energy unit, employed to denote work, is the *foot pound (ft-lb)*. This is the work needed to raise a weight of one pound by a distance of one foot, not including any friction. It's equal to 1.356 joules.

All of these units, and conversion factors, are given in Table 2-3. Kilowatt hours and watt hours are also included in this table. You don't really need to worry about the exponential notation, called *scientific notation*, here. In electricity and electronics, you need to be concerned only with the watt hour and the kilowatt hour for most purposes, and the conversions hardly ever involve numbers so huge or so miniscule that you'll need scientific notation.

Table 2-3.   Energy units.

Unit	To convert to joules, multiply by	Conversely, multiply by
Btu	1055	0.000948 or 9.48×10^{-4}
eV	1.6×10^{-19}	6.2×10^{18}
erg	0.0000001 or 10^{-7}	10,000,000 or 10^7
ft-lb	1.356	0.738
Wh	3600	0.000278 or 2.78×10^{-4}
kWh	3,600,000 or 3.6×10^6	0.000000278 or 2.78×10^{-7}

ac Waves and the hertz

This chapter, and this whole first section, is concerned with *direct current* (dc), that is, current that always flows in the same direction, and that does not change in intensity (at least not too rapidly) with time. But household utility current is not of this kind. It reverses direction periodically, exactly once every $\frac{1}{120}$ second. It goes through a complete *cycle* every $\frac{1}{60}$ second. Every repetition is identical to every other. This is *alternating current* (ac). In some countries, the direction reverses every $\frac{1}{100}$ second, and the cycle is completed every $\frac{1}{50}$ second.

Figure 2-8 shows the characteristic wave of alternating current, as a graph of voltage versus time. Notice that the maximum positive and negative voltages are not 117 V, as you've heard about household electricity, but close to 165 V. There is a reason for this difference. The *effective voltage* for an ac wave is never the same as the *instantaneous maximum*, or *peak*, voltage. In fact, for the common waveshape shown in Fig. 2-8, the effective value is 0.707 times the peak value. Conversely, the peak value is 1.414 times the effective value.

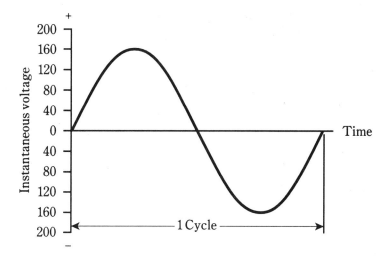

2-8 One cycle of utility alternating current. The peak voltage is about 165 V.

Because the whole cycle repeats itself every $1/60$ second, the *frequency* of the utility ac wave is said to be 60 *Hertz*, abbreviated 60 Hz. The word "Hertz" literally translates to "cycles per second." In the U.S., this is the standard frequency for ac. In some places it is 50 Hz. (Some remote places even use dc, but they are definitely the exception, not the rule.)

In radio practice, higher frequencies are common, and you'll hear about *kilohertz (kHz), megahertz (MHz) and gigahertz (GHz)*. You should know right away the size of these units, but in case you're still not sure about the way the prefixes work, the relationships are as follows:

$$1 \text{ kHz} = 1000 \text{ Hz}$$
$$1 \text{ MHz} = 1000 \text{ kHz} = 1,000,000 \text{ Hz}$$
$$1 \text{ GHz} = 1000 \text{ MHz} = 1,000,000 \text{ kHz}$$
$$= 1,000,000,000 \text{ Hz}$$

Usually, but not always, the waveshapes are of the type shown in Fig. 2-8. This waveform is known as a *sine wave* or a *sinusoidal* waveform.

Rectification and fluctuating direct current

Batteries and other sources of direct current (dc) produce a constant voltage. This can be represented by a straight, horizontal line on a graph of voltage versus time (Fig. 2-9). The peak and effective values are the same for pure dc. But sometimes the value of dc voltage fluctuates rapidly with time, in a manner similar to the changes in an ac wave. This might happen if the waveform in Fig. 2-8 is *rectified*.

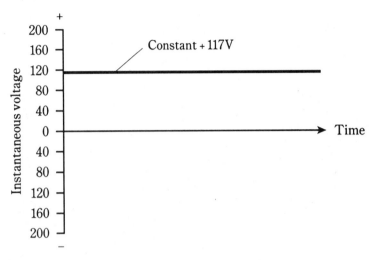

2-9   A representation of pure dc.

Rectification is a process in which ac is changed to dc. The most common method of doing this uses a device called the *diode*. Right now, you need not be concerned with how the rectifier circuit is put together. The point is that part of the ac wave is either cut off, or turned around upside-down, so that the *polarity* is always the same, either positive or negative.

Figure 2-10 illustrates two different waveforms of fluctuating, or pulsating, dc. In the waveform at A, the negative (bottom) part has simply been chopped off. At B, the negative portion of the wave has been turned around and made positive, a mirror image of its former self. The situation at A is known as *half-wave rectification*, because it makes use of only half the wave. At B, the wave has been subjected to *full-wave rectification*, because all of the original current still flows, even though the alternating nature has been changed so that the current never reverses.

The effective value, compared with the peak value, for pulsating dc depends on whether half-wave or full-wave rectification has been used. In the figure, effective voltage is shown as a dotted line, and the *instantaneous voltage* is shown as a solid line. Notice that the instantaneous voltage changes all the time, from instant to instant. This is how it gets this name! The peak voltage is the maximum instantanous voltage. Instantaneous voltage is never, ever any greater than the peak.

In Fig. 2-10B, the effective value is 0.707 times the peak value, just as is the case with ordinary ac. The direction of current flow, for many kinds of devices, doesn't make any difference. But in Fig. 2-10A, half of the wave has been lost. This cuts the effective value in half, so that it's just 0.354 times the peak value.

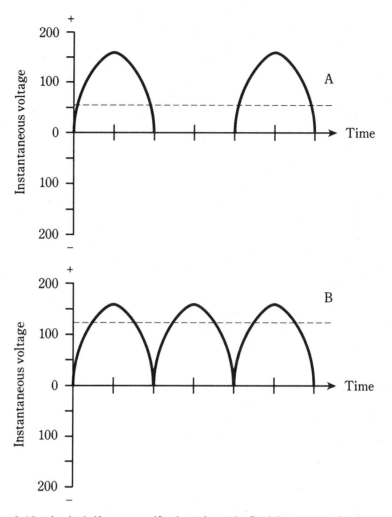

2-10 At A, half-wave rectification of ac. At B, full-wave rectification. Effective values are shown by dotted lines.

Using household ac as an example, the peak value is generally about 165 V; the effective value is 117V. If full-wave rectification is used (Fig. 2-10B), the effective value is still 117 V. If half-wave rectification is used, as in Fig. 2-10A, the effective value is about 58.5 V.

Safety considerations in electrical work

For your purposes here, one rule applies concerning safety around electrical apparatus. If you are in the slightest doubt about whether or not something is safe, *leave it to a professional electrician.*

Household voltage, normally about 117 V but sometimes twice that, or about 234 V, is more than sufficient to kill you if it appears across your chest cavity. Certain devices, such as automotive spark coils, can produce lethal currents even from the low voltage (12 V to 14 V) in a car battery.

Consult the American Red Cross or your electrician concerning what kinds of circuits, procedures and devices are safe, and which kinds aren't.

Magnetism

Electric currents and magnetic fields are closely related.

Whenever an electric current flows—that is, when charge carriers move—a magnetic field accompanies the current. In a straight wire, the magnetic *lines of flux* surround the wire in circles, with the wire at the center (Fig. 2-11). Actually, these aren't really lines or circles; this is just a convenient way to represent the magnetic field. You might sometimes hear of a certain number of flux lines per unit cross-sectional area, such as 100 lines per square centimeter. This is a relative way of talking about the intensity of the magnetic field.

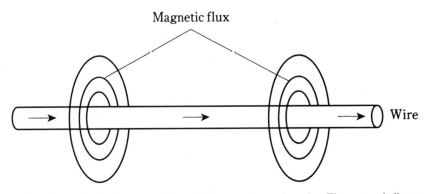

2-11 Magnetic flux lines around a straight, current-carrying wire. The arrows indicate current flow.

Magnetic fields can be produced when the atoms of certain materials align themselves. Iron is the most common metal that has this property. The iron in the core of the earth has become aligned to some extent; this is a complex interaction caused by the rotation of our planet and its motion with respect to the magnetic field of the sun. The magnetic field surrounding the earth is responsible for various effects, such as the concentration of charged particles that you see as the *aurora borealis* just after a solar eruption.

When a wire is coiled up, the resulting magnetic flux takes a shape similar to the flux field surrounding the earth, or the flux field around a bar magnet. Two well-defined *magnetic poles* develop, as shown in Fig. 2-12.

The intensity of a magnetic field can be greatly increased by placing a special core inside of a coil. The core should be of iron or some other material that can be readily

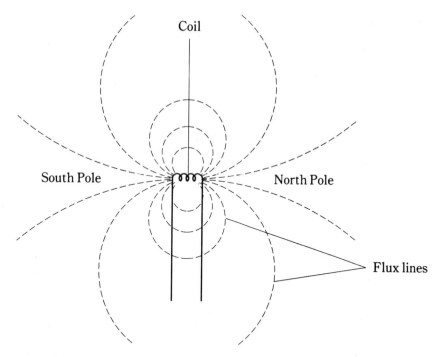

2-12 Magnetic flux lines around a coil of wire. The lines converge at the magnetic poles.

magnetized. Such substances are called *ferromagnetic*. A core of this kind cannot actually increase the total quantity of magnetism in and around a coil, but it will cause the lines of flux to be much closer together inside the material. This is the principle by which an *electromagnet* works. It also makes possible the operation of electrical transformers for utility current.

Magnetic lines of flux are said to emerge from the magnetic north pole, and to run inward toward the magnetic south pole. But this is just a semantical thing, about which theoretical physicists might speak. It doesn't need to concern you for ordinary electrical and electronics applications.

Magnetic units

The size of a magnetic field is measured in units called *webers*, abbreviated Wb. One weber is mathematically equivalent to one volt-second. For weaker magnetic fields, a smaller unit, called the *maxwell*, is sometimes used. One maxwell is equal to 0.00000001 (one hundred-millionth) of a weber, or 0.01 microvolt-second.

The *flux density* of a magnetic field is given in terms of webers or maxwells per square meter or per square centimeter. A flux density of one weber per square meter (1 Wb/m^2) is called one *tesla*. One *gauss* is equal to 0.0001 weber, or one maxwell per square centimeter.

In general, the greater the electric current through a wire, the greater the flux density near the wire. A coiled wire will produce a greater flux density than a single, straight wire. And, the more turns in the coil, the stronger the magnetic field will be.

Sometimes, magnetic field strength is specified in terms of *ampere-turns* (At). This is actually a unit of *magnetomotive force*. A one-turn wire loop, carrying 1 A of current, produces a field of 1 At. Doubling the number of turns, or the current, will double the number of ampere-turns. Therefore, if you have 10 A flowing in a 10-turn coil, the magnetomotive force is 10×10, or 100 At. Or, if you have 100 mA flowing in a 100-turn coil, the magnetomotive force is 0.1×100, or, again, 10 At. (Remember that 100 mA = 0.1 A.)

Another unit of magnetomotive force is the *gilbert*. This unit is equal to 0.796 At.

Quiz

Refer to the text in this chapter if necessary. A good score is at least 18 correct answers. The answers are listed in the back of this book.

1. A positive electric pole:
   A. Has a deficiency of electrons.
   B. Has fewer electrons than the negative pole.
   C. Has an excess of electrons.
   D. Has more electrons than the negative pole.

2. An EMF of one volt:
   A. Cannot drive much current through a circuit.
   B. Represents a low resistance.
   C. Can sometimes produce a large current.
   D. Drops to zero in a short time.

3. A potentially lethal electric current is on the order of:
   A. 0.01 mA.
   B. 0.1 mA.
   C. 1 mA.
   D. 0.1 A.

4. A current of 25 A is most likely drawn by:
   A. A flashlight bulb.
   B. A typical household.
   C. A power plant.
   D. A clock radio.

5. A piece of wire has a conductance of 20 siemens. Its resistance is:
   A. 20 Ω.

B. 0.5 Ω.

C. 0.05 Ω.

D. 0.02 Ω.

6. A resistor has a value of 300 ohms. Its conductance is:

A. 3.33 millisiemens.

B. 33.3 millisiemens.

C. 333 microsiemens.

D. 0.333 siemens.

7. A mile of wire has a conductance of 0.6 siemens. Then three miles of the same wire has a conductance of:

A. 1.8 siemens.

B. 1.8 Ω.

C. 0.2 siemens.

D. Not enough information has been given to answer this.

8. A 2-kW generator will deliver approximately how much current, reliably, at 117 V?

A. 17 mA.

B. 234 mA.

C. 17 A.

D. 234 A.

9. A circuit breaker is rated for 15 A at 117 V. This represents approximately how many kilowatts?

A. 1.76.

B. 1760.

C. 7.8.

D. 0.0078.

10. You are told that a certain air conditioner is rated at 500 Btu. What is this in kWh?

A. 147.

B. 14.7.

C. 1.47.

D. 0.147.

11. Of the following energy units, the one most often used to define electrical energy is:

A. The Btu.

B. The erg.

    C. The foot pound.

    D. The kilowatt hour.

12. The frequency of common household ac in the U.S. is:

    A. 60 Hz.

    B. 120 Hz.

    C. 50 Hz.

    D. 100 Hz.

13. Half-wave rectification means that:

    A. Half of the ac wave is inverted.

    B. Half of the ac wave is chopped off.

    C. The whole wave is inverted.

    D. The effective value is half the peak value.

14. In the output of a half-wave rectifier:

    A. Half of the wave is inverted.

    B. The effective value is less than that of the original ac wave.

    C. The effective value is the same as that of the original ac wave.

    D. The effective value is more than that of the original ac wave.

15. In the output of a full-wave rectifier:

    A. The whole wave is inverted.

    B. The effective value is less than that of the original ac wave.

    C. The effective value is the same as that of the original ac wave.

    D. The effective value is more than that of the original ac wave.

16. A low voltage, such as 12 V:

    A. Is never dangerous.

    B. Is always dangerous.

    C. Is dangerous if it is ac, but not if it is dc.

    D. Can be dangerous under certain conditions.

17. Which of these can represent magnetomotive force?

    A. The volt-turn.

    B. The ampere-turn.

    C. The gauss.

    D. The gauss-turn.

18. Which of the following units can represent magnetic flux density?

    A. The volt-turn.

    B. The ampere-turn.

C. The gauss.

D. The gauss-turn.

19. A ferromagnetic material:

    A. Concentrates magnetic flux lines within itself.

    B. Increases the total magnetomotive force around a current-carrying wire.

    C. Causes an increase in the current in a wire.

    D. Increases the number of ampere-turns in a wire.

20. A coil has 500 turns and carries 75 mA of current. The magnetomotive force will be:

    A. 37,500 At.

    B. 375 At.

    C. 37.5 At.

    D. 3.75 At.

3
CHAPTER

Measuring devices

NOW THAT YOU'RE FAMILIAR WITH THE PRIMARY UNITS COMMON IN ELECTRICITY and electronics, let's look at the instruments that are employed to measure these quantities.

Many measuring devices work because electric and magnetic fields produce forces proportional to the intensity of the field. By using a tension spring against which the electric or magnetic force can pull or push, a movable needle can be constructed. The needle can then be placed in front of a calibrated scale, allowing a direct reading of the quantity to be measured. These meters work by means of *electromagnetic deflection* or *electrostatic deflection*.

Sometimes, electric current is measured by the extent of heat it produces in a resistance. Such meters work by *thermal heating* principles.

Some meters work by means of small motors whose speed depends on the measured quantity. The rotation rate, or the number of rotations in a given time, can be measured or counted. These are forms of *rate meters*.

Still other kinds of meters actually count electronic pulses, sometimes in thousands, millions or billions. These are *electronic counters*. There are also various other metering methods.

Electromagnetic deflection

Early experimenters with electricity and magnetism noticed that an electric current produces a magnetic field. This discovery was probably an accident, but it was an accident that, given the curiosity of the scientist, was bound to happen. When a magnetic compass is placed near a wire carrying a direct electric current, the compass doesn't point toward magnetic north. The needle is displaced. The extent of the error depends on how close the compass is brought to the wire, and also on how much current the wire is carrying.

Scientific experimenters are like children. They like to play around with things. Most likely, when this effect was first observed, the scientist tried different arrangements to see how much the compass needle could be displaced, and how small a current could be detected. An attempt was made to obtain the greatest possible current-detecting sensitivity. Wrapping the wire in a coil around the compass resulted in a device that would indicate a tiny electric current (Fig. 3-1). This effect is known as *galvanism*, and the meter so devised was called a *galvanometer*.

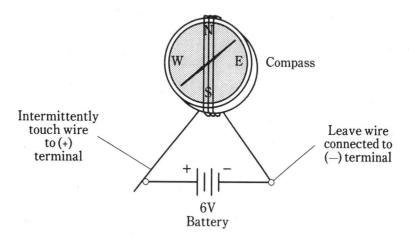

3-1   A simple galvanometer. The compass must lie flat.

Once this device was made, the scientist saw that the extent of the needle displacement increased with increasing current. Aha—a device for measuring current! Then, the only challenge was to calibrate the galvanometer somehow, and to set up some kind of standard so that a universal meter could be engineered.

You can easily make your own galvanometer. Just buy a cheap compass, about two feet of insulated bell wire, and a six-volt lantern battery. Set it up as shown in Fig. 3-1. Wrap the wire around the compass four or five times, and align the compass so that the needle points right along the wire turns while the wire is disconnected from the battery. Connect one end of the wire to the minus (−) terminal of the battery. Touch the other end to the plus (+) terminal, intermittently, and watch the compass needle. Don't leave the wire connected to the battery for any length of time unless you want to drain the battery in a hurry.

You can buy a *resistor* and a *potentiometer* at a place like Radio Shack, and set up an experiment that shows how galvanometers measure current. For a 6-V lantern battery, the fixed resistor should have a value of at least 330 Ω at 1/4 watt, and the potentiometer should have a value of 10 KΩ (10,000 Ω) maximum. Connect the resistor and potentiometer in series between one end of the bell wire and one terminal of the battery, as shown in Fig. 3-2. The center contact of the potentiometer should be short-circuited to one of the end contacts, and the resulting two terminals used in the circuit. When you adjust the potentiometer, the compass needle should deflect more or less, depending on the current through the wire. Early experimenters calibrated their meters by referring to the degree scale around the perimeter of the compass.

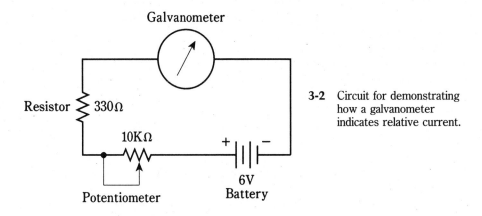

3-2 Circuit for demonstrating how a galvanometer indicates relative current.

Electrostatic deflection

Electric fields produce forces, just as do magnetic fields. You have probably noticed this when your hair feels like it's standing on end in very dry or cold weather. You've probably heard that people's hair really does stand straight out just before a lightning bolt hits nearby; this is no myth. Maybe you performed experiments in science classes to observe this effect.

The most common device for demonstrating electrostatic forces is the *electroscope*. It consists of two foil leaves, attached to a conducting rod, and placed in a sealed container so that air currents will not move the foil leaves (Fig. 3-3). When a charged object is brought near, or touched to, the contact at the top of the rod, the leaves stand apart from each other. This is because the two leaves become charged with like electric poles— either an excess or a deficiency of electrons—and like poles always repel.

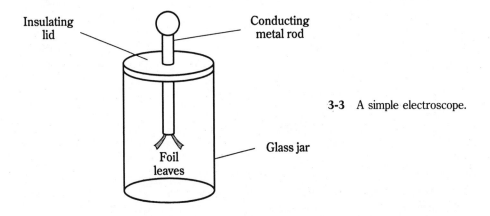

3-3 A simple electroscope.

The extent to which the leaves stand apart depends on the amount of electric charge. It is somewhat difficult to actually measure this deflection and correlate it with charge quantity; electroscopes do not make very good meters. But variations on this theme can be employed, so that electrostatic forces can operate against tension springs or magnets, and in this way, electrostatic meters can be made.

An electrostatic device has the ability to measure alternating electric charges as well as steady charges. This gives electrostatic meters an advantage over electromagnetic meters (galvanometers). If you connect ac to the coil of the galvanometer device in Fig. 3-1 or Fig. 3-2, the compass needle might vibrate, but will not give a clear deflection. This is because current in one direction pulls the meter needle one way, and current in the other direction will deflect the needle the opposite way. But if an alternating electric field is connected to an electrostatic meter, the plates will repel whether the charge is positive or negative. The deflection will be steady, therefore, with ac as well as with dc.

Most electroscopes aren't sensitive enough to show much deflection with ordinary 117-V utility voltage. Don't try connecting 117 V to an electroscope anyway; it might not deflect the foil leaves, but it can certainly present a danger to your body if you bring it out to points where you can readily come into physical contact with it.

An electrostatic meter has another property that is sometimes an advantage in electrical or electronic work. This is the fact that the device does not draw any current, except a tiny amount at first, needed to put a charge on the plates. Sometimes, an engineer or experimenter doesn't want the measuring device to draw current, because this affects the behavior of the circuit under test. Galvanometers, by contrast, always need at least a little bit of current in order to operate. You can observe this effect by charging up a laboratory electroscope, say with a glass rod that has been rubbed against a cloth. When the rod is pulled away from the electroscope, the foil leaves will remain standing apart. The charge just sits there. If the electroscope drew any current, the leaves would immediately fall back together again, just as the galvanometer compass needle returns to magnetic north the instant you take the wire from the battery.

Thermal heating

Another phenomenon, sometimes useful in the measurement of electric currents, is the fact that whenever current flows through a conductor having any resistance, that conductor is heated. All conductors have some resistance; none are perfect. The extent of this heating is proportional to the amount of current being carried by the wire.

By choosing just the right metal or alloy, and by making the wire a certain length and diameter, and by employing a sensitive thermometer, and by putting the entire assembly inside a thermally insulating package, a *hot-wire* meter can be made. The hot-wire meter can measure ac as well as dc, because the current-heating phenomenon does not depend on the direction of current flow.

A variation of the hot-wire principle can be used by placing two different metals into contact with each other. If the right metals are chosen, the junction will heat up when a current flows through it. This is called the *thermocouple principle*. As with the hot-wire meter, a thermometer can be used to measure the extent of the heating.

But there is also another effect. A thermocouple, when it gets warm, generates a direct current. This current can be measured by a more conventional, dc type meter. This method is useful when it is necessary to have a faster meter response time. The hot-wire and thermocouple effects are used occasionally to measure current at radio frequencies, in the range of hundreds of kilohertz up to tens of gigahertz.

Ammeters

Getting back to electromagnetic deflection, and the workings of the galvanometer, you might have thought by now that a magnetic compass doesn't make a very convenient type of meter. It has to be lying flat, and the coil has to be aligned with the compass needle when there is no current. But of course, electrical and electronic devices aren't all turned in just the right way, so as to be aligned with the north geomagnetic pole. That would not only be a great bother, but it would be ridiculous. Imagine a bunch of scientists running around, turning radios and other apparatus so the meters are all lying flat and are all lined up with the earth's magnetic field! In the early days of electricity and electronics, when the phenomena were confined to scientific labs, this was indeed pretty much how things were.

Then someone thought that the magnetic field could be provided by a permanent magnet right inside the meter, instead of by the earth. This would supply a stronger magnetic force, and would therefore make it possible to detect much weaker currents. It would let the meter be turned in any direction and the operation would not be affected. The coil could be attached right to the meter pointer, and suspended by means of a spring in the field of the magnet. This kind of meter, called a *D'Arsonval movement*, is still extensively used today. The assembly is shown in Fig. 3-4. This is the basic principle of the *ammeter*.

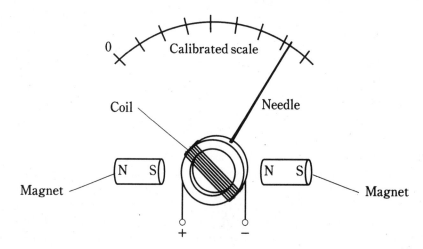

3-4   The D'Arsonval meter movement. The spring bearing is not shown.

A variation of this is the attachment of the meter needle to a permanent magnet, and the winding of the coil in a fixed form around the magnet. Current in the coil produces a magnetic field, and this in turn generates a force if the coil and magnet are aligned correctly with respect to each other. This meter movement is also sometimes called a D'Arsonval movement. This method will work, but the inertial mass of the permanent magnet causes a slower needle response. This kind of meter is also more prone to *overshoot* than the true D'Arsonval movement; the inertia of the magnet's mass, once overcome by the magnetic force, causes the needle to fly past the actual current level before finally coming to rest at the correct reading.

It is possible to use an electromagnet in place of the permanent magnet in the meter assembly. This electromagnet can be operated by the same current that flows in the coil attached to the meter needle. This gets rid of the need for a massive, permanent magnet inside the meter. It also eliminates the possibility that the meter sensitivity will change in case the strength of the permanent magnet deteriorates (such as might be caused by heat, or by severe mechanical vibration). The electromagnet can be either in series with, or in parallel with, the meter movement coil.

The sensitivity of the D'Arsonval meter, and of its cousins, depends on several factors. First is the strength of the permanent magnet, if the meter uses a permanent magnet. Second is the number of turns in the coil. The stronger the magnet, and the larger the number of turns in the coil, the less current is needed in order to produce a given magnetic force. If the meter is of the electromagnet type, the combined number of coil turns affects the sensitivity. Remember that the strength of a magnetomotive force is given in terms of ampere turns. For a given current (number of amperes), the force increases in direct proportion to the number of coil turns. The more force in a meter, the greater the needle deflection, and the smaller the amount of current that is needed to cause a certain amount of needle movement.

The most sensitive ammeters can detect currents of just a microampere or two. The amount of current for *full scale deflection* (the needle goes all the way up without banging against the stop pin) can be as little as about 50 uA in commonly available meters. Thus you might see a *microammeter,* or a *milliammeter,* quite often in electronic work. Meters that measure large currents are not a problem to make; it's easy to make an *in*sensitive device.

Sometimes, it is desirable to have an ammeter that will allow for a wide range of current measurements. The full-scale deflection of a meter assembly cannot easily be changed, since this would mean changing the number of coil turns and/or the strength of the magnet. But all ammeters have a certain amount of *internal resistance.* If a resistor, having the same internal resistance as the meter, is connected in parallel with the meter, the resistor will take half the current. Then it will take twice the current through the assembly to deflect the meter to full scale, as compared with the meter alone. By choosing a resistor of just the right value, the full-scale deflection of an ammeter can be increased by a factor of 10, or 100, or even 1000. This resistor must be capable of carrying the current without burning up. It might have to take practically all of the current flowing through the assembly, leaving the meter to carry only 1/10, or 1/100, or 1/1000 of the current. This is called a *shunt resistance* or *meter shunt* (Fig. 3-5).

Meter shunts are frequently used when it is necessary to measure very large currents, such as hundreds of amperes. They allow microammeters or milliammeters to be used in a versatile *multimeter,* with many current ranges.

Voltmeters

Current is a flow of charge carriers. Voltage, or electromotive force (EMF), or potential difference, is the "pressure" that makes a current possible. Given a circuit whose resistance is constant, the current that will flow in the circuit is directly proportional to the

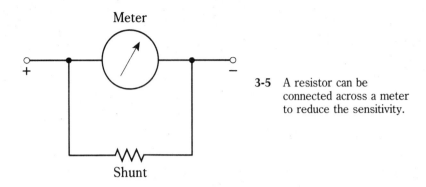

Meter

+ —

Shunt

3-5 A resistor can be connected across a meter to reduce the sensitivity.

voltage placed across it. Early electrical experimenters recognized that an ammeter could be used to measure voltage, since an ammeter is a form of constant-resistance circuit.

If you connect an ammeter directly across a source of voltage—a battery, say—the meter needle will deflect. In fact, a milliammeter needle will probably be "pinned" if you do this with it, and a microammeter might well be wrecked by the force of the needle striking the pin at the top of the scale. For this reason, you should never connect milliammeters or microammeters directly across voltage sources. An ammeter, perhaps with a range of 0-10 A, might not deflect to full scale if it is placed across a battery, but it's still a bad idea to do this, because it will rapidly drain the battery. Some batteries, such as automotive lead-acid cells, can explode under these conditions. This is because all ammeters have low internal resistance. They are designed that way deliberately. They are meant to be connected in series with other parts of a circuit, not right across the power supply.

But if you place a large resistor in series with an ammeter, and then connect the ammeter across a battery or other type of power supply, you no longer have a short circuit. The ammeter will give an indication that is directly proportional to the voltage of the supply. The smaller the full-scale reading of the ammeter, the larger the resistance to get a meaningful indication on the meter. Using a microammeter and a very large value of resistor in series, a voltmeter can be devised that will draw only a little current from the source.

A voltmeter can be made to have different ranges for the full-scale reading, by switching different values of resistance in series with the microammeter (Fig. 3-6). The internal resistance of the meter is large because the values of the resistors are large. The greater the supply voltage, the larger the internal resistance of the meter, because the necessary series resistance increases as the voltage increases.

It's always good when a voltmeter has a high internal resistance. The reason for this is that you don't want the meter to draw much current from the power source. This current should go, as much as possible, towards working whatever circuit is hooked up to the supply, and not into just getting a reading of the voltage. Also, you might not want, or need, to have the voltmeter constantly connected in the circuit; you might need the voltmeter for testing many different circuits. You don't want the behavior of the circuit to be affected the instant you connect the voltmeter to the supply. The less current a voltmeter draws, the less it will affect the behavior of anything that is working from the power supply.

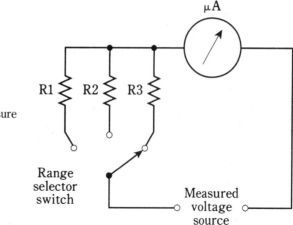

3-6 Circuit for using a microammeter to measure voltage.

Another type of voltmeter uses the effect of electrostatic deflection, rather than electromagnetic deflection. You remember that electric fields produce forces, just as do magnetic fields. Therefore, a pair of plates will attract or repel each other if they are charged. The *electrostatic voltmeter* makes use of this effect, taking advantage of the attractive force between two plates having opposite electric charge, or having a large potential difference. Figure 3-7 is a simplified drawing of the mechanics of an electrostatic voltmeter.

The electrostatic meter draws almost no current from the power supply. The only thing between the plates is air, and air is a nearly perfect insulator. The electrostatic meter will indicate ac as well as dc. The construction tends to be rather delicate, however, and mechanical vibration influences the reading.

Ohmmeters

You remember that the current through a circuit depends on the resistance. This principle can be used to manufacture a voltmeter using an ammeter and a resistor. The larger the value of the resistance in series with the meter, the more voltage is needed to produce a reading of full scale. This has a converse, or a "flip side": Given a constant voltage, the current through the meter will vary if the resistance varies. This provides a means for measuring resistances.

An *ohmmeter* is almost always constructed by means of a milliammeter or microammeter in series with a set of fixed, switchable resistances and a battery that provides a known, constant voltage (Fig. 3-8). By selecting the resistances appropriately, the meter will give indications in ohms over any desired range. Usually, zero on the meter is assigned the value of *infinity ohms*, meaning a perfect insulator. The full-scale value is set at a certain minimum, such as 1 Ω, 100 Ω, or 10 KΩ (10,000 Ω).

Ohmmeters must be precalibrated at the factory where they are made. A slight error in the values of the series resistors can cause gigantic errors in measured resistance. Therefore, precise tolerances are needed for these resistors. It is also necessary that the battery be exactly the right kind, and that it be reasonably fresh so that it will provide the

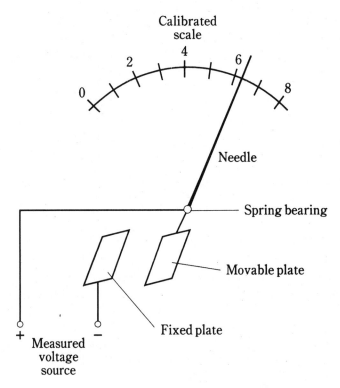

3-7 Simplified drawing of an electrostatic voltmeter.

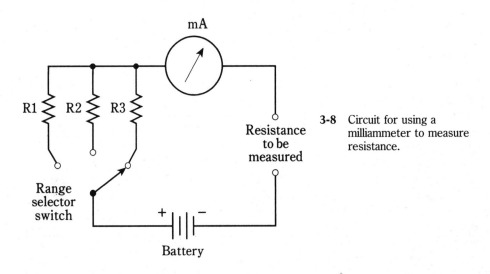

3-8 Circuit for using a milliammeter to measure resistance.

appropriate voltage. The smallest deviation from the required voltage can cause a big error in the meter indication.

The scale of an ohmmeter is nonlinear. That is, the graduations are not the same everywhere. Values tend to be squashed together towards the "infinity" end of the scale.

It can be difficult to interpolate for high values of resistance, unless the right scale is selected. Engineers and technicians usually connect an ohmmeter in a circuit with the meter set for the highest resistance range first; then they switch the range till the meter is in a part of the scale that is easy to read. Finally, the reading is taken, and is multiplied (or divided) by the appropriate amount as indicated on the range switch. Figure 3-9 shows an ohmmeter reading. The meter itself says 4.7, but the range switch says 1 KΩ. This indicates a resistance of 4.7 KΩ, or 4700 Ω.

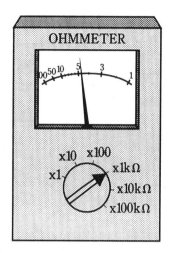

3-9 An example of an ohmmeter reading. This device shows about 4.7 × 1 K = 4.7 K = 4700 ohms.

Ohmmeters will give inaccurate readings if there is a voltage between the points where the meter is connected. This is because such a voltage either adds to, or subtracts from, the ohmmeter battery voltage. This in effect changes the battery voltage, and the meter reading is thrown way off. Sometimes the meter might even read "more than infinity" ohms; the needle will hit the pin at the left end of the scale. Therefore, when using an ohmmeter to measure resistance, you need to be sure that there is no voltage between the points under test. The best way to do this is to switch off the equipment in question.

Multimeters

In the electronics lab, a common piece of test equipment is the *multimeter,* in which different kinds of meters are combined into a single unit. The *volt-ohm-milliammeter (VOM)* is the most often used. As its name implies, it combines voltage, resistance and current measuring capabilities.

You should not have too much trouble envisioning how a single milliammeter can be used for measuring voltage, current and resistance. The preceding discussions for measurements of these quantities have all included methods in which a current meter can be used to measure the intended quantity.

Commercially available multimeters have certain limits in the values they can measure. The maximum voltage is around 1000 V; larger voltages require special leads and heavily insulated wires, as well as other safety precautions. The maximum current that a common VOM can measure is about 1 A. The maximum resistance is on the order of several megohms or tens of megohms. The lower limit of resistance indication is about an ohm.

FET and vacuum-tube voltmeters

It was mentioned that a good voltmeter will disturb the circuit under test as little as possible, and this requires that the meter have a high internal resistance. Besides the electrostatic type voltmeter, there is another way to get an extremely high internal resistance. This is to sample a tiny, tiny current, far too small for any meter to directly indicate, and then amplify this current so that a meter will show it. When a miniscule amount of current is drawn from a circuit, the equivalent resistance is always extremely high.

The most effective way to accomplish the amplification, while making sure that the current drawn really is tiny, is to use either a *vacuum tube* or a *field-effect transistor (FET)*. You needn't worry about how such amplifiers work right now; that subject will come much later in this book. A voltmeter that uses a vacuum tube amplifier to minimize the current drain is known as a *vacuum-tube voltmeter (VTVM)*. If an FET is used, the meter is called a *FET voltmeter (FETVM)*. Either of these devices provide an extremely high input resistance along with good sensitivity and amplification. And they allow measurement of lower voltages, in general, than electrostatic voltmeters.

Wattmeters

The measurement of electrical power requires that voltage and current both be measured simultaneously. Remember that power is the product of the voltage and current. That is, watts (P) equals volts (E) times amperes (I), written as $P = EI$. In fact, watts are sometimes called volt-amperes in a dc circuit.

You might think that you can just connect a voltmeter in parallel with a circuit, thereby getting a reading of the voltage across it, and also hook up an ammeter in series to get a reading of the current through the circuit, and then multiply volts times amperes to get watts consumed by the circuit. And in fact, for practically all dc circuits, this is an excellent way to measure power (Fig. 3-10).

Quite often, however, it's simpler than that. In many cases, the voltage from the power supply is constant and predictable. Utility power is a good example. The effective voltage is always very close to 117 V. Although it's ac, and not dc, power can be measured in the same way as with dc: by means of an ammeter connected in series with the circuit, and calibrated so that the multiplication (times 117) has already been done. Then, rather than 1 A, the meter would show a reading of 117 W, because $P = EI = 117 \times 1 = 117$ W. If the meter reading were 300 W, the current would be 300/117 = 2.56 A.

An electric iron might consume 1000 W, or a current of 1000/117 = 8.55 A. And a large heating unit might gobble up 2000 W, requiring a current of 2000/117 = 17.1 A. This might blow a fuse or breaker, since these devices are often rated for only 15 A. You've probably had an experience where you hooked up too many appliances to a single circuit, blowing the fuse or breaker. The reason was that the appliances, combined, drew too

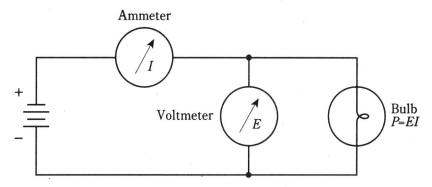

Ammeter

Voltmeter

Bulb
P=EI

3-10 Power can be measured with a voltmeter and an ammeter.

much current for the house wiring to safely handle, and the fuse or breaker, detecting the excess current, opened the circuit.

Specialized wattmeters are necessary for the measurement of radio-frequency (RF) power, or for peak audio power in a high-fidelity amplifier, or for certain other specialized applications. But almost all of these meters, whatever the associated circuitry, use simple ammeters as their indicating devices.

Watt-hour meters

The utility company is not too interested in how much power you're using with one appliance, or even how much power a single household is drawing, at any given time. By far the greater concern is the total energy that is used over a day, a week, a month or a year. Electrical energy is measured in watt hours, or, more commonly for utility purposes, in kilowatt hours (kWh). The device that indicates this is the *watt-hour meter* or *kilowatt-hour meter.*

The most often-used means of measuring electrical energy is by using a small electric motor device, whose speed depends on the current, and thereby on the power at a constant voltage. The number of turns of the motor shaft, in a given length of time, is directly proportional to the number of kilowatt hours consumed. The motor is placed at the point where the utility wires enter the house, apartment or building. This is usually at a point where the voltage is 234 V. This is split into some circuits with 234 V, for heavy-duty appliances such as the oven, washer and dryer, and the general household lines for lamps, clock radios and television sets.

You've surely seen the little disk in the utility meter going around and around, sometimes fast, other times slowly. Its speed depends on the power you're using. The total number of turns of this little disk, every month, determines the size of the bill you will get—as a function also, of course, of the cost per kilowatt hour for electricity.

Kilowatt-hour meters count the number of disk turns by means of geared, rotary drums or pointers. The drum type meter gives a direct digital readout. The pointer type has several scales calibrated from 0 to 9 in circles, some going clockwise and others going counterclockwise.

Reading a pointer type utility meter is a little tricky, because you must think in whatever direction (clockwise or counterclockwise) the scale goes. An example of a pointer type utility meter is shown in Fig. 3-11. Read from left to right. For each little meter, take down the number that the pointer has most recently passed. Write down the result as you go. The meter in the figure reads 3875 kWh. If you want to be really precise, you can say it reads 3875-1/2 kWh.

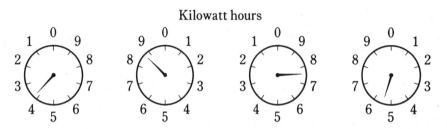

Kilowatt hours

3-11   An example of a utility meter. The reading is a little more than 3875 kWh.

Digital readout meters

Increasingly, metering devices are being designed so that they provide a direct readout, and there's no need (or possibility) for interpolation. The number on the meter is the indication. It's that simple. Such a meter is called a *digital meter*.

The advantage of a digital meter is that it's easy for anybody to read, and there is no chance for interpolation errors. This is ideal for utility meters, clocks, and some kinds of ammeters, voltmeters and wattmeters. It works very well when the value of the quantity does not change very often or very fast.

But there are some situations in which a digital meter is a disadvantage. One good example is the signal-strength indicator in a radio receiver. This meter bounces up and down as signals fade, or as you tune the radio, or sometimes even as the signal modulates. A digital meter would show nothing but a constantly changing, meaningless set of numerals. Digital meters require a certain length of time to "lock in" to the current, voltage, power or other quantity being measured. If this quantity never settles at any one value for a long enough time, the meter can never lock in.

Meters with a scale and pointer are known as *analog meters*. Their main advantages are that they allow interpolation, they give the operator a sense of the quantity relative to other possible values, and they follow along when a quantity changes. Some engineers and technicians prefer the "feel" of an analog meter, even in situations where a digital meter would work just as well.

One problem you might have with digital meters is being certain of where the decimal point goes. If you're off by one decimal place, the error will be by a factor of 10. Also, you need to be sure you know what the units are; for example, a frequency indicator might be reading out in megahertz, and you might forget and think it is giving you a reading in kilohertz. That's a mistake by a factor of 1000. Of course this latter type of error can happen with an analog meter, too.

Frequency counters

The measurement of energy used by your home is an application to which digital metering is well suited. It's easier to read the drum type, digital kilowatt-hour meter than to read the pointer type meter. When measuring frequencies of signals, digital metering is not only more convenient, but far more accurate.

The *frequency counter* measures by actually counting pulses, in a manner similar to the way the utility meter counts the number of turns of a motor. But the frequency counter works electronically, without any moving parts. It can keep track of thousands, millions or even billions of pulses per second, and it shows the rate on a digital display that is as easy to read as a digital watch. It measures frequency directly by tallying up the number of pulses in an oscillating wave, even when the number of pulses per second is huge.

The accuracy of the frequency counter is a function of the lock-in time. Lock-in is usually done in 0.1 second, 1 second or 10 seconds. Increasing the lock-in time by a factor of 10 will cause the accuracy to be good by one additional digit. Modern frequency counters are good to six, seven or eight digits; sophisticated lab devices will show frequency to nine or ten digits.

Other specialized meter types

The following are some less common types of meters that you might come across in electrical and electronic work.

VU and decibel meters

In high-fidelity equipment, especially the more sophisticated amplifiers ("amps"), loudness meters are sometimes used. These are calibrated in *decibels*, a unit that you will sometimes encounter in reference to electronic signal levels. A decibel is an increase or decrease in sound or signal level that you can just barely detect, if you are expecting the change.

Audio loudness is given in *volume units (VU)*, and the meter that indicates it is called a *VU meter*. Usually, such meters have a zero marker with a red line to the right and a black line to the left, and they are calibrated in decibels (dB) above and below this zero marker (Fig. 3-12). The meter might also be calibrated in *watts rms*, an expression for audio power.

As music is played through the system, or as a voice comes over it, the VU meter needle will kick up. The amplifier volume should be kept down so that the meter doesn't go past the zero mark and into the red range. If the meter does kick up into the red scale, it means that distortion is probably taking place within the amplifier circuit.

Sound level in general can be measured by means of a *sound-level meter*, calibrated in decibels (dB) and connected to the output of a precision amplifier with a microphone of known, standardized sensitivity (Fig. 3-13). You have perhaps heard that a vacuum cleaner will produce 80 dB of sound, and a large truck going by might subject your ears to 90 dB. These figures are determined by a sound-level meter. A VU meter is a special form of sound-level meter.

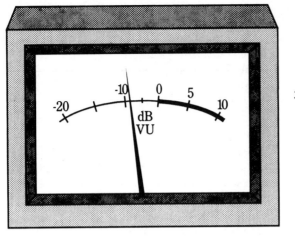

3-12   A VU meter. The heavy scale is usually red, indicating high risk of audio distortion.

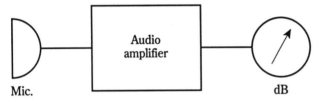

3-13   A sound-level meter.

Light meters

Light intensity is measured by means of a *light meter* or *illumination meter*. You might think that it's easy to make this kind of meter by connecting a milliammeter to a solar (photovoltaic) cell. And this is, in fact, a good way to construct an inexpensive light meter (Fig. 3-14). More sophisticated devices might use dc amplifiers to enhance sensitivity and to allow for several different ranges of readings.

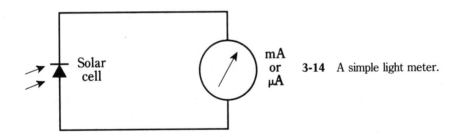

3-14   A simple light meter.

One problem with this design is that solar cells are not sensitive to light at exactly the same wavelengths as human eyes. This can be overcome by placing a colored filter in front of the solar cell, so that the solar cell becomes sensitive to the same wavelengths, in the same proportions, as human eyes. Another problem is calibrating the meter. This must usually be done at the factory, in units such as *lumens* or *candela*. It's not important that you know the precise definitions of these units in electricity and electronics.

Sometimes, meters such as the one in Fig. 3-14 are used to measure infrared or ultraviolet intensity. Different types of photovoltaic cells have peak sensitivity at different wavelengths. Filters can be used to block out wavelengths that you don't want the meter to detect.

Pen recorders

A meter movement can be equipped with a marking device, usually a pen, to keep a graphic record of the level of some quantity with respect to time. Such a device is called a *pen recorder*. The paper, with a calibrated scale, is taped to a rotating drum. The drum, driven by a clock motor, turns at a slow rate, such as one revolution per hour or one revolution in 24 hours. A simplified drawing of a pen recorder is shown in Fig. 3-15.

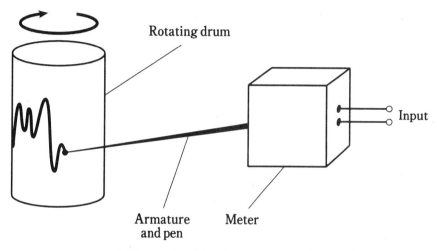

3-15 Simplified drawing of a pen recorder.

A device of this kind, along with a wattmeter, might be employed to get a reading of the power consumed by your household at various times during the day. In this way you might tell when you use the most power, and at what particular times you might be using too much.

Oscilloscopes

Another graphic meter is the *oscilloscope*. This measures and records quantities that vary rapidly, at rates of hundreds, thousands, or millions of times per second. It creates a "graph" by throwing a beam of electrons at a phosphor screen. A *cathode-ray tube*, similar to the kind in a television set, is employed.

Oscilloscopes are useful for looking at the shapes of signal waveforms, and also for measuring peak signal levels (rather than just the effective levels). An oscilloscope can also be used to approximately measure the frequency of a waveform. The horizontal scale of an oscilloscope shows time, and the vertical scale shows instantaneous voltage. An oscilloscope can indirectly measure power or current, by using a known value of resistance across the input terminals.

Technicians and engineers develop a sense of what a signal waveform should look like, and then they can often tell, by observing the oscilloscope display, whether or not the circuit under test is behaving the way it should. This is a subjective kind of "measurement," since it is qualitative as well as quantitative. If a wave shape "looks wrong," it might indicate distortion in a circuit, or possibly even betray a burned-out component someplace.

Bar-graph meters

A cheap, simple kind of meter can be made using a string of light-emitting diodes (LEDs) or a liquid-crystal display (LCD) along with a digital scale, to indicate approximate levels of current, voltage or power. This type of meter has no moving parts to break, just like a digital meter. But it also offers the relative-reading feeling you get with an analog meter. Figure 3-16 is an example of a bar-graph meter that is used to show the power output, in kilowatts, for a radio transmitter. It indicates 0.8 kW or 800 watts, approximately.

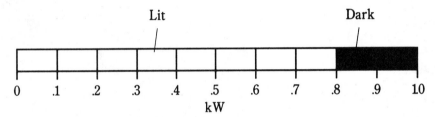

3-16 A bar-graph meter. This device shows a power level of about 0.8 kW or 800 W.

The chief disadvantage of the bar-graph meter is that it isn't very accurate. For this reason it is not generally used in laboratory testing. The LED or LCD devices sometimes also flicker when the level is "between" two values given by the bars. This can be annoying to some people.

Quiz

Refer to the text in this chapter if necessary. A good score is 18 out of 20 correct. Answers are in the back of the book.

1. The force between two electrically charged objects is called:
    A. Electromagnetic deflection.
    B. Electrostatic force.
    C. Magnetic force.
    D. Electroscopic force.

2. The change in the direction of a compass needle, when a current-carrying wire is brought near, is:
    A. Electromagnetic deflection.
    B. Electrostatic force.

C. Magnetic force.

D. Electroscopic force.

3. Suppose a certain current in a galvanometer causes the needle to deflect 20 degrees, and then this current is doubled. The needle deflection:

　A. Will decrease.

　B. Will stay the same.

　C. Will increase.

　D. Will reverse direction.

4. One important advantage of an electrostatic meter is that:

　A. It measures very small currents.

　B. It will handle large currents.

　C. It can detect ac voltages.

　D. It draws a large current from the source.

5. A thermocouple:

　A. Gets warm when current flows through it.

　B. Is a thin, straight, special wire.

　C. Generates dc when exposed to light.

　D. Generates ac when heated.

6. One advantage of an electromagnet meter over a permanent-magnet meter is that:

　A. The electromagnet meter costs much less.

　B. The electromagnet meter need not be aligned with the earth's magnetic field.

　C. The permanent-magnet meter has a more sluggish coil.

　D. The electromagnet meter is more rugged.

7. An ammeter shunt is useful because:

　A. It increases meter sensitivity.

　B. It makes a meter more physically rugged.

　C. It allows for measurement of a wide range of currents.

　D. It prevents overheating of the meter.

8. Voltmeters should generally have:

　A. Large internal resistance.

　B. Low internal resistance.

　C. Maximum possible sensitivity.

　D. Ability to withstand large currents.

9. To measure power-supply voltage being used by a circuit, a voltmeter:

　A. Is placed in series with the circuit that works from the supply.

B. Is placed between the negative pole of the supply and the circuit working from the supply.

C. Is placed between the positive pole of the supply and the circuit working from the supply.

D. Is placed in parallel with the circuit that works from the supply.

10. Which of the following will *not* cause a major error in an ohmmeter reading?

A. A small voltage between points under test.

B. A slight change in switchable internal resistance.

C. A small change in the resistance to be measured.

D. A slight error in range switch selection.

11. The ohmmeter in Fig. 3-17 shows a reading of about:

A. 33,000 Ω.

B. 3.3 KΩ.

C. 330 Ω.

D. 33 Ω.

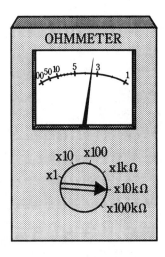

3-17 Illustration for quiz question 11.

12. The main advantage of a FETVM over a conventional voltmeter is the fact that the FETVM:

A. Can measure lower voltages.

B. Draws less current from the circuit under test.

C. Can withstand higher voltages safely.

D. Is sensitive to ac as well as to dc.

13. Which of the following is *not* a function of a fuse?

A. To be sure there is enough current available for an appliance to work right.

B. To make it impossible to use appliances that are too large for a given circuit.

C. To limit the amount of power that a circuit can deliver.

D. To make sure the current is within safe limits.

14. A utility meter's motor speed works directly from:

A. The number of ampere hours being used at the time.

B. The number of watt hours being used at the time.

C. The number of watts being used at the time.

D. The number of kilowatt hours being used at the time.

15. A utility meter's readout indicates:

A. Voltage.

B. Power.

C. Current.

D. Energy.

16. A typical frequency counter:

A. Has an analog readout.

B. Is usually accurate to six digits or more.

C. Works by indirectly measuring current.

D. Works by indirectly measuring voltage.

17. A VU meter is *never* used for measurement of:

A. Sound.

B. Decibels.

C. Power.

D. Energy.

18. The meter movement in an illumination meter measures:

A. Current.

B. Voltage.

C. Power.

D. Energy.

19. An oscilloscope *cannot* be used to indicate:

A. Frequency.

B. Wave shape.

C. Energy.

D. Peak signal voltage.

20. The display in Fig. 3-18 could be caused by a voltage of:

A. 6.0 V.

B. 6.6 V.

C. 7.0 V.

D. No way to tell; the meter is malfunctioning.

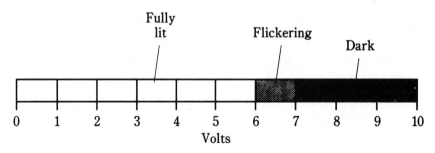

3-18 Illustration for quiz question 20.

4
CHAPTER

Basic dc circuits

YOU'VE ALREADY SEEN SOME SIMPLE ELECTRICAL CIRCUIT DIAGRAMS. SOME OF these are the same kinds of diagrams, using the same symbols, that professional technicians and engineers use. In this chapter, you'll get more acquainted with this type of diagram. You'll also learn more about how current, voltage, resistance, and power are related in direct-current (dc) and low-frequency alternating-current (ac) circuits.

Schematic symbols

In this course, the plan is to familiarize you with schematic symbols mainly by getting you to read and use them "in action," rather than by dryly drilling you with them. But it's a good idea now to check Appendix B and look over the various symbols. Some of the more common ones are mentioned here.

The simplest schematic symbol is the one representing a wire or electrical conductor: a straight, solid line. Sometimes dotted lines are used to represent conductors, but usually, dotted lines are drawn to partition diagrams into constituent circuits, or to indicate that certain components interact with each other or operate in step with each other. Conductor lines are almost always drawn either horizontally across, or vertically up and down the page, so that the imaginary charge carriers are forced to march in formation like soldiers. This keeps the diagram neat and easy to read.

When two conductor lines cross, they aren't connected at the crossing point unless a heavy, black dot is placed where the two lines meet. The dot should always be clearly visible wherever conductors are to be connected, no matter how many of them meet at the junction.

A resistor is indicated by a zig-zaggy line. A variable resistor, or potentiometer, is indicated by a zig-zaggy line with an arrow through it, or by a zig-zaggy line with an arrow pointing at it. These symbols are shown in Fig. 4-1.

A cell is shown by two parallel lines, one longer than the other. The longer line represents the plus terminal. A battery, or combination of cells in series, is indicated by

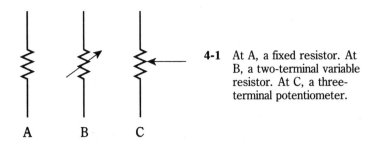

4-1 At A, a fixed resistor. At B, a two-terminal variable resistor. At C, a three-terminal potentiometer.

four parallel lines, long-short-long-short. It's not necessary to use more than four lines for any battery, even though sometimes you'll see six or eight lines. The symbols for a cell and a battery are shown in Fig. 4-2.

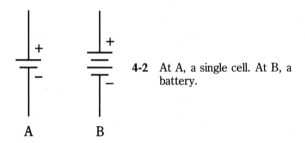

4-2 At A, a single cell. At B, a battery.

Meters are indicated as circles. Sometimes the circle has an arrow inside it, and the meter type, such as mA (milliammeter) or V (voltmeter) are written alongside the circle, as shown in Fig. 4-3A. Sometimes the meter type is indicated inside the circle, and there is no arrow (Fig. 4-3B). It doesn't matter which way it's done, as long as you're consistent everywhere in a schematic diagram.

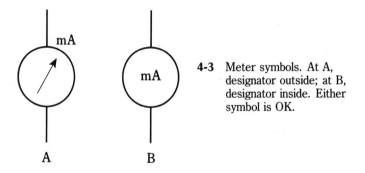

4-3 Meter symbols. At A, designator outside; at B, designator inside. Either symbol is OK.

Some other common symbols include the lamp, the capacitor, the air-core coil, the iron-core coil, the chassis ground, the earth ground, the alternating-current source, the set of terminals, and the "black box," a rectangle with the designator written inside. These are shown in Fig. 4-4.

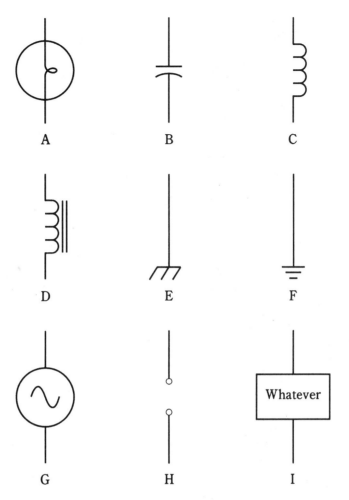

4-4 Nine common schematic symbols. A: Incandescent lamp. B: Capacitor. C: Air-core coil. D: Iron-core coil. E: Chassis ground. F: Earth ground. G: Source of alternating current (ac). H: Pair of terminals. J: Specialized component or device.

Schematic diagrams

Look back through the earlier chapters of this book and observe the *schematic diagrams*. These are all simple examples of how professionals would draw the circuits. There is no inscrutable gobbledygook to put in to make them into the sorts of circuit maps that the most brilliant engineer would need. The diagrams you have worked with are exactly like the ones that the engineer would use to depict these circuits.

Wiring diagrams

The difference between a schematic diagram and a *wiring diagram* is the amount of detail included. In a schematic diagram, the interconnection of the components is shown, but the actual values of the components are not necessarily indicated.

You might see a diagram of a two-transistor audio amplifier, for example, with resistors and capacitors and coils and transistors, but without any data concerning the values or ratings of the components. This is a schematic diagram, but not a true wiring diagram. It gives the *scheme* for the circuit, but you can't *wire* the circuit and make it work, because there isn't enough information.

Suppose you want to build the circuit. You go to an electronics store to get the parts. What sizes of resistors should you buy? How about capacitors? What type of transistor will work best? Do you need to wind the coils yourself, or can you get them ready-made? Are there test points or other special terminals that should be installed for the benefit of the technicians who might have to repair the amplifier? How many watts should the potentiometers be able to handle? All these things are indicated in a wiring diagram, a jazzed-up schematic. You might have seen this kind of diagram in the back of the instruction manual for a hi-fi amp or an FM stereo tuner or a television set. Wiring diagrams are especially useful and necessary when you must service or repair an electronic device.

Voltage/current/resistance circuit

Most dc circuits can be ultimately boiled down to three major components: a voltage source, a set of conductors, and a resistance. This is shown in the schematic diagram of Fig. 4-5. The voltage or EMF source is called E; the current in the conductor is called I; the resistance is called R. The standard units for these components are the volt, the ampere, and the ohm respectively.

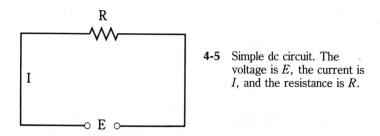

4-5 Simple dc circuit. The voltage is E, the current is I, and the resistance is R.

You already know that there is a relationship among these three quantities. If one of them changes, then one or both of the others will also change. If you make the resistance smaller, the current will get larger. If you make the EMF source smaller, the current will decrease. If the current in the circuit increases, the voltage across the resistor will increase. There is a simple arithmetic relationship between these three quantities.

Ohm's Law

The interdependence between current, voltage, and resistance is one of the most fundamental rules, or *laws*, in electrical circuits. It is called Ohm's Law, named after the scientist who supposedly first expressed it. Three formulas denote this law:

$$E = IR$$
$$I = E/R$$
$$R = E/I$$

You need only to remember the first one in order to derive the others. The easiest way to remember it is to learn the abbreviations E for EMF or voltage, I for current and R for resistance, and then remember that they appear in alphabetical order with the equals sign after the E.

Sometimes the three symbols are written in a triangle, as in Fig. 4-6. To find the value of one, you cover it up and read the positions of the others.

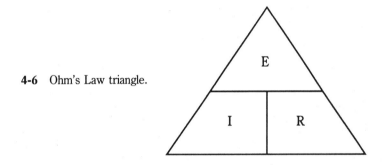

4-6   Ohm's Law triangle.

It's important to remember that you must use units of volts, amperes, and ohms in order for Ohm's Law to work right. If you use volts, milliamperes, and ohms or kilovolts, microamperes, and megohms you cannot expect to get the right answers.

If the initial quantities are given in units other than volts, amperes, and ohms, you must convert to these units, then calculate. After that, you can convert the units back again to whatever you like. For example, if you get 13,500,000 ohms as a calculated resistance, you might prefer to say that it's 13.5 megohms.

Current calculations

The first way to use Ohm's Law is to find current values in dc circuits. In order to find the current, you must know the voltage and the resistance, or be able to deduce them.

Refer to the schematic diagram of Fig. 4-7. It consists of a variable dc generator, a voltmeter, some wire, an ammeter, and a calibrated, wide-range potentiometer. Component values have been left out of this diagram, so it's not a wiring diagram. But values can be assigned for the purpose of creating sample Ohm's Law problems. While calculating the current in the following problems, it is necessary to mentally "cover up" the meter.

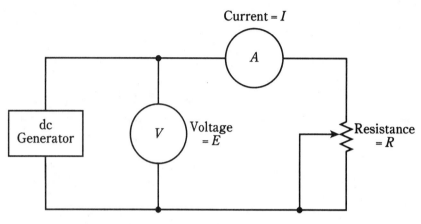

4-7   Circuit for working Ohm's Law problems.

Problem 4-1

Suppose that the dc generator (Fig. 4-7) produces 10 V, and that the potentiometer is set to a value of 10 Ω. Then what is the current?

This is easily solved by the formula $I = E/R$. Just plug in the values for E and R; they are both 10, because the units were given in volts and ohms. Then $I = 10/10 = 1$ A.

Problem 4-2

The dc generator (Fig. 4-7) produces 100 V and the potentiometer is set to 10 KΩ. What is the current?

First, convert the resistance to ohms: 10 KΩ = 10,000 Ω. Then plug the values in: $I = 100/10,000 = 0.01$ A. This might better be expressed as 10 mA.

Engineers and technicians prefer to keep the numbers within reason when specifying quantities. Although it's perfectly all right to say that a current is 0.01 A, it's best if the numbers can be kept at 1 or more, but less than 1,000. It is a little silly to talk about a current of 0.003 A, or a resistance of 107,000 Ω, when you can say 3 mA or 107 KΩ.

Problem 4-3

The dc generator (Fig. 4-7) is set to provide 88.5 V, and the potentiometer is set to 477 MΩ. What is the current?

This problem involves numbers that aren't exactly round, and one of them is huge. But you can use a calculator. The resistance is first changed to ohms, giving 477,000,000 Ω. Then you plug into the Ohm's Law formula: $I = E/R = 88.5/477,000,000 = 0.000000186$ A = 0.186 uA. This value is less than 1, but there isn't much you can do about it unless you are willing to use units of *nanoamperes (nA)*, or billionths of an ampere. Then you can say that the current is 186 nA.

Voltage calculations

The second use of Ohm's Law is to find unknown voltages when the current and the resistance are known. For the following problems, uncover the ammeter and cover the voltmeter scale instead in your mind.

Problem 4-4

Suppose the potentiometer (Fig. 4-7) is set to 100 ohms, and the measured current is 10 mA. What is the dc voltage?

Use the formula $E = IR$. First, convert the current to amperes: 10 mA = 0.01 A. Then multiply: $E = 0.01 \times 100 = 1$ V. That's a low, safe voltage, a little less than what is produced by a flashlight cell.

Problem 4-5

Adjust the potentiometer (Fig. 4-7) to a value of 157 KΩ, and let the current reading be 17 mA. What is the voltage of the source?

Now you have to convert both the resistance and the current values to their proper units. A resistance of 157 KΩ is 157,000 Ω; a current of 17 mA is 0.017 A. Then $E = IR$ = $0.017 \times 157,000 = 2669$ V = 2.669 kV. You might want to round this off to 2.67 kV. This is a dangerous voltage. If you touch the terminals you'll get clobbered.

Problem 4-6

You set the potentiometer (Fig. 4-7) so that the meter reads 1.445 A, and you observe that the potentiometer scale shows 99 ohms. What is the voltage?

These units are both in their proper form. Therefore, you can plug them right in and use your calculator: $E = IR = 1.445 \times 99 = 143.055$ V. This can, and should, be rounded off to 143 V. A purist would go further and round it to the nearest 10 volts, to 140 V.

It's never a good idea to specify your answer to a problem with more significant figures than you're given. The best engineers and scientists go by the *rule of significant figures*: keep to the *least* number of digits given in the data. If you follow this rule in Problem 4-6, you must round off the answer to two significant figures, getting 140 V, because the resistance specified (99 Ω) is only accurate to two digits.

Resistance calculations

Ohms' Law can be used to find a resistance between two points in a dc circuit, when the voltage and the current are known. For the following problems, imagine that both the voltmeter and ammeter scales in Fig. 4-7 are visible, but that the potentiometer is uncalibrated.

Problem 4-7

If the voltmeter reads 24 V and the ammeter shows 3.0 A, what is the value of the potentiometer?

Use the formula $R = E/I$, and plug in the values directly, because they are expressed in volts and amperes: $R = 24/3.0 = 8.0 \ \Omega$.

Note that you can specify this value to two significant figures, the eight and the zero, rather than saying simply 8 Ω. This is because you are given both the voltage and the current to two significant figures. If the ammeter reading had been given as 3 A (meaning some value between 2-½ A and 3-½ A), you would only be entitled to express the answer as 8 Ω (somewhere between 7-½ and 8-½ Ω). A zero can be a significant figure, just as well as the digits 1 through 9.

Problem 4-8

What is the value of the resistance if the current is 18 mA and the voltage is 229 mV?

First, convert these values to amperes and volts. This gives $I = 0.018$ A and $E = 0.229$ V. Then plug into the equation $R = E/I = 0.229/0.018 = 13 \ \Omega$. You're justified in giving your answer to two significant figures, because the current is only given to that many digits.

Problem 4-9

Suppose the ammeter reads 52 uA and the voltmeter indicates 2.33 kV. What is the resistance?

Convert to amperes and volts, getting $I = 0.000052$ A and $E = 2330$ V. Then plug into the formula: $R = 2330/0.000052 = 45,000,000 \ \Omega = 45 \ M\Omega$.

Power calculations

You can calculate the power, in watts, in a dc circuit such as that shown in Fig. 4-7, by the formula $P = EI$, or the product of the voltage in volts and the current in amperes. You might not be given the voltage directly, but can calculate it if you know the current and the resistance.

Remember the Ohm's Law formula for obtaining voltage: $E = IR$. If you know I and R, but don't know E, you can get the power P by means of the formula $P = (IR)I = I^2R$. That is, you take the current in amperes, multiply this figure by itself, and then multiply the result by the resistance in ohms.

You can also get the power if you aren't given the current directly. Suppose you're given only the voltage and the resistance. Remember the Ohm's Law formula for obtaining current: $I = E/R$. Therefore, $P = E(E/R) = E^2/R$. Take the voltage, multiply it by itself, and divide by the resistance.

Stated all together, these power formulas are:

$$P = EI = I^2R = E^2/R$$

Now you're ready to do some problems in power calculations. Refer once again to Fig. 4-7.

Problem 4-10

Suppose that the voltmeter reads 12 V and the ammeter shows 50 mA. What is the power dissipated by the potentiometer?

Use the formula $P = EI$. First, convert the current to amperes, getting $I = 0.050$ A. (Note that the zero counts as a significant digit.) Then $P = EI = 12 \times 0.050 = 0.60$ W.

You might say that this is 600 mW, although that is to three significant figures. It's not easy to specify the number 600 to two significant digits without using a means of writing numbers called *scientific notation*. That subject is beyond the scope of this discussion, so for now, you might want to say "600 milliwatts, accurate to two significant figures." (You can probably get away with "600 milliwatts" and nobody will call you on the number of significant digits.)

Problem 4-11

If the resistance in the circuit of Fig. 4-7 is 999 Ω and the voltage source delivers 3 V, what is the dissipated power?

Use the formula $P = E^2/R = 3 \times 3/999 = 9/999 = 0.009$ W $= 9$ mW. You are justified in going to only one significant figure here.

Problem 4-12

Suppose the resistance is 47 KΩ and the current is 680 mA. What is the power dissipated by the potentiometer?

Use the formula $P = I^2R$, after converting to ohms and amperes. Then $P = 0.680 \times 0.680 \times 47,000 = 22,000$ W $= 22$ kW.

This is a ridiculous state of affairs. An ordinary potentiometer, such as the one you would get at an electronics store, dissipating 22 kW, several times more than a typical household. The voltage must be phenomenal. It's not too hard to figure out that such a voltage would burn out the potentiometer so fast that it would be ruined before the little "Pow!" could even begin to register.

Problem 4-13

Just from curiosity, what is the voltage that would cause so much current to be driven through such a large resistance?

Use Ohm's Law to find the current: $E = IR = 0.680 \times 47,000 = 32,000$ V $= 32$ kV. That's the sort of voltage you'd expect to find only in certain industrial/commercial applications. The resistance capable of drawing 680 mA from such a voltage would surely not be a potentiometer, but perhaps something like an amplifier tube in a radio broadcast transmitter.

Resistances in series

When you place resistances in series, their ohmic values simply add together to get the total resistance. This is easy to see intuitively, and it's quite simple to remember.

Problem 4-14

Suppose the following resistances are hooked up in series with each other: 112 Ω, 470 Ω, and 680 Ω. What is the total resistance of the series combination (Fig. 4-8)?

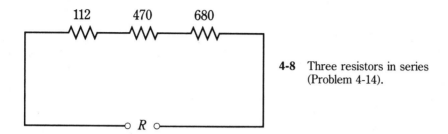

4-8   Three resistors in series (Problem 4-14).

Just add the values, getting a total of 112 + 470 + 680 = 1262 Ω. You might round this off to 1260 Ω. It depends on the tolerances of the components—how precise their actual values are to the ones specified by the manufacturer.

Resistances in parallel

When resistances are placed in parallel, they behave differently than they do in series. In general, if you have a resistor of a certain value and you place other resistors in parallel with it, the overall resistance will decrease.

One way to look at resistances in parallel is to consider them as conductances instead. In parallel, conductances add, just as resistances add in series. If you change all the ohmic values to siemens, you can add these figures up and convert the final answer back to ohms.

The symbol for conductance is G. This figure, in siemens, is related to the resistance R, in ohms, by the formulas:

$$G = 1/R, \text{ and}$$
$$R = 1/G$$

Problem 4-15

Consider five resistors in parallel. Call them R1 through R5, and call the total resistance R as shown in the diagram Fig. 4-9. Let R1 = 100 Ω, R2 = 200 Ω, R3 = 300 Ω, R4 = 400 Ω. and R5 = 500 Ω respectively. What is the total resistance, R, of this parallel combination?

Converting the resistances to conductance values, you get G1 = 1/100 = 0.01 siemens, G2 = 1/200 = 0.005 siemens, G3 = 1/300 = 0.00333 siemens, G4 = 1/400 = 0.0025 siemens, and G5 = 1/500 = 0.002 siemens. Adding these gives G = 0.01 + 0.005 + 0.00333 + 0.0025 + 0.002 = 0.0228 siemens. The total resistance is therefore R = 1/G = 1/0.0228 = 43.8 Ω.

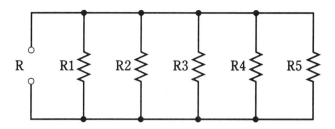

4-9 Five resistors in parallel, R1 through R5, give a total resistance *R*. See Problems 4-15 and 4-16.

When you have resistances in parallel and their values are all equal, the total resistance is equal to the resistance of any one component, divided by the number of components.

Problem 4-16

Suppose there are five resistors R1 through R5 in parallel, as shown in Fig. 4-9, all having a value of 4.7K Ω. What is the total resistance, *R*?

You can probably guess that the total is a little less than 1 KΩ or 1000 Ω. So you can convert the value of the single resistor to 4,700 Ω and divide by 5, getting a total resistance of 940 Ω. This is accurate to two significant figures, the 9 and the 4; engineers won't usually be worried about the semantics, and you can just say "940 Ω."

Division of power

When combinations of resistances are hooked up to a source of voltage, they will draw current. You can easily figure out how much current they will take by calculating the total resistance of the combination and then considering the network as a single resistor.

If the resistances in the network all have the same ohmic value, the power from the source will be evenly distributed among the resistances, whether they are hooked up in series or in parallel. If there are eight identical resistors in series with a battery, the network will consume a certain amount of power, each resistor bearing ⅛ of the load. If you rearrange the circuit so that the resistors are in parallel, the circuit will dissipate a certain amount of power (a lot more than when the resistors were in series), but again, each resistor will handle ⅛ of the total power load.

If the resistances in the network do not all have identical ohmic values, they divide up the power unevenly. Situations like this are discussed in the next chapter.

Resistances in series-parallel

Sets of resistors, all having identical ohmic values, can be connected together in parallel sets of series networks, or in series sets of parallel networks. By doing this, the total power handling capacity of the resistance can be greatly increased over that of a single resistor.

Sometimes, the total resistance of a series-parallel network is the same as the value of any one of the resistors. This is always true if the components are identical, and are in a network called an *n-by-n matrix*. That means, when n is a whole number, there are n parallel sets of n resistors in series (A of Fig. 4-10), or else there are n series sets of n resistors in parallel (B of Fig. 4-10). Either arrangement will give exactly the same results in practice.

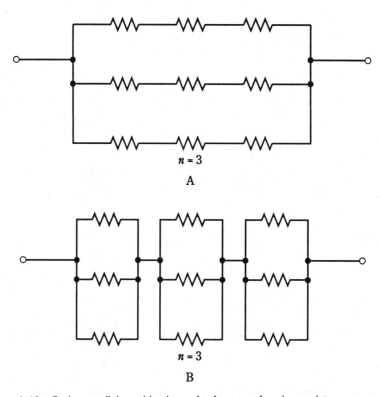

4-10 Series-parallel combinations. At A, sets of series resistors are connected in parallel. At B, sets of parallel resistors are in series.

Engineers and technicians sometimes use this to their advantage to get resistors with large power-handling capacity. Each resistor should have the same rating, say 1 W. Then the combination of n by n resistors will have n^2 times that of a single resistor. A 3 × 3 series-parallel matrix of 2-W resistors can handle $3^2 × 2 = 9 × 2 = 18$ W, for example. A 10 × 10 array of 1-W resistors can take 100 W. Another way to look at this is to see that the total power-handling capacity is multiplied by the total number of individual resistors in the matrix. But this is only true if all the resistors have the same ohmic values, and the same power-dissipation ratings.

It is unwise to build series-parallel arrays from resistors with different ohmic values or power ratings. If the resistors have values and/or ratings that are even a little nonuniform, one of them might be subjected to more current than it can withstand, and it will burn out. Then the current distribution in the network can change so a second component fails, and then a third. It's hard to predict the current and power distribution in an array when its resistor values are all different. So it's hard to know whether any of the components in such a matrix are going to burn out.

If you need a resistance with a certain power-handling capacity, you must be sure the network can handle at least that much power. If a 50-W rating is required, and a certain combination will handle 75 W, that's alright. But it isn't good enough to build a circuit that will handle only 48 W. Some extra tolerance, say 10 percent over the minimum rating needed, is good, but it's silly to make a 500-W network using far more resistors than necessary, unless that's the only convenient combination given the parts available.

Nonsymmetrical series-parallel networks, made up from identical resistors, will increase the power-handling capability. But in these cases, the total resistance will not be the same as the value of the single resistors. The overall power-handling capacity will always be multiplied by the total number of resistors, whether the network is symmetrical or not, provided all the resistors are the same. In engineering work, cases sometimes do arise where nonsymmetrical networks fit the need just right.

Resistive loads in general

The circuits you've seen here are good for illustrating the principles of dc. But some of the circuits shown here have essentially no practicality. You'll never find a resistor connected across a battery, along with a couple of meters, as shown in Fig. 4-7, for example. The resistor will get warm, maybe even hot, and it will eventually drain the battery in an unspectacular way. Aside from its educational value, the circuit does nothing of any use.

In real life, the ammeter and voltmeter readings in an arrangement such as that shown in Fig. 4-7 would decline with time. Ultimately, you'd be left with a dead, cold battery, a couple of zeroed-out meters, a potentiometer, and some wire.

The resistances in the diagrams like Fig. 4-7 are always put to some use in electrical and electronic circuits. Instead of resistors, you might have light bulbs, appliances (60-Hz utility ac behaves much like dc in many cases), motors, computers, and radios. Voltage division is one important way in which resistors are employed. This, along with more details about current, voltage, and resistance in dc circuits, is discussed in the next chapter.

Quiz

Refer to the text in this chapter if necessary. A good score is at least 18 correct answers. The answers are in the back of the book.

1. Suppose you double the voltage in a simple dc circuit, and cut the resistance in half. The current will become:

    A. Four times as great.

    B. Twice as great.

    C. The same as it was before.

    D. Half as great.

2. A wiring diagram would most likely be found in:

  A. An engineer's general circuit idea notebook.

  B. An advertisement for an electrical device.

  C. The service/repair manual for a radio receiver.

  D. A procedural flowchart.

For questions 3 through 11, see Fig. 4-7.

3. Given a dc voltage source delivering 24 V and a circuit resistance of 3.3K Ω, what is the current?

  A. 0.73 A.

  B. 138 A.

  C. 138 mA.

  D. 7.3 mA.

4. Suppose that a circuit has 472 Ω of resistance and the current is 875 mA. Then the source voltage is:

  A. 413 V.

  B. 0.539 V.

  C. 1.85 V.

  D. None of the above.

5. The dc voltage in a circuit is 550 mV and the current is 7.2 mA. Then the resistance is:

  A. 0.76 Ω.

  B. 76 Ω.

  C. 0.0040 Ω.

  D. None of the above.

6. Given a dc voltage source of 3.5 kV and a circuit resistance of 220 Ω, what is the current?

  A. 16 mA.

  B. 6.3 mA.

  C. 6.3 A.

  D. None of the above.

7. A circuit has a total resistance of 473,332 Ω and draws 4.4 mA. The best expression for the voltage of the source is:

  A. 2082 V.

  B. 110 kV.

  C. 2.1 kV.

  D. 2.08266 kV.

8. A source delivers 12 V and the current is 777 mA. Then the best expression for the resistance is:

    A. 15 Ω.

    B. 15.4 Ω.

    C. 9.3 Ω.

    D. 9.32 Ω.

9. The voltage is 250 V and the current is 8.0 mA. The power dissipated by the potentiometer is:

    A. 31 mW.

    B. 31 W.

    C. 2.0 W.

    D. 2.0 mW.

10. The voltage from the source is 12 V and the potentiometer is set for 470 Ω. The power is about:

    A. 310 mW.

    B. 25.5 mW.

    C. 39.2 W.

    D. 3.26 W.

11. The current through the potentiometer is 17 mA and its value is 1.22 KΩ. The power is:

    A. 0.24 uW.

    B. 20.7 W.

    C. 20.7 mW.

    D. 350 mW.

12. Suppose six resistors are hooked up in series, and each of them has a value of 540 Ω. Then the total resistance is:

    A. 90 Ω.

    B. 3.24 KΩ.

    C. 540 Ω.

    D. None of the above.

13. Four resistors are connected in series, each with a value of 4.0 KΩ. The total resistance is:

    A. 1 KΩ.

    B. 4 KΩ.

    C. 8 KΩ.

    D. 16 KΩ.

14. Suppose you have three resistors in parallel, each with a value of 68,000 Ω. Then the total resistance is:

 A. 23 Ω.

 B. 23 KΩ.

 C. 204 Ω.

 D. 0.2M Ω.

15. There are three resistors in parallel, with values of 22 Ω, 27 Ω, and 33 Ω. A 12-V battery is connected across this combination, as shown in Fig. 4-11. What is the current drawn from the battery by this resistance combination?

 A. 1.3 A.

 B. 15 mA.

 C. 150 mA.

 D. 1.5 A.

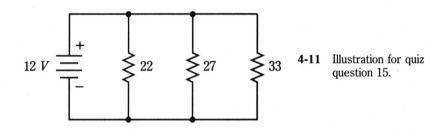

4-11 Illustration for quiz question 15.

16. Three resistors, with values of 47 Ω, 68 Ω, and 82 Ω, are connected in series with a 50-V dc generator, as shown in Fig. 4-12. The total power consumed by this network of resistors is:

 A. 250 mW.

 B. 13 mW.

 C. 13 W.

 D. Not determinable from the data given.

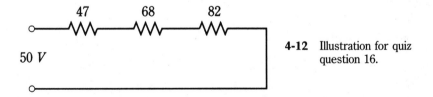

4-12 Illustration for quiz question 16.

17. You have an unlimited supply of 1-W, 100-Ω resistors. You need to get a 100-Ω, 10-W resistor. This can be done most cheaply by means of a series-parallel matrix of:

 A. 3 x 3 resistors.

B. 4 x 3 resistors.

C. 4 x 4 resistors.

D. 2 x 5 resistors.

18. You have an unlimited supply of 1-W, 1000-Ω resistors, and you need a 500-Ω resistance rated at 7 W or more. This can be done by assembling:

   A. Four sets of two 1000-Ω resistors in series, and connecting these four sets in parallel.

   B. Four sets of two 1000-Ω resistors in parallel, and connecting these four sets in series.

   C. A 3 x 3 series-parallel matrix of 1000-Ω resistors.

   D. Something other than any of the above.

19. You have an unlimited supply of 1-W, 1000-Ω resistors, and you need to get a 3000-Ω, 5-watt resistance. The best way is to:

   A. Make a 2 x 2 series-parallel matrix.

   B. Connect three of the resistors in parallel.

   C. Make a 3 x 3 series-parallel matrix.

   D. Do something other than any of the above.

20. Good engineering practice usually requires that a series-parallel resistive network be made:

   A. From resistors that are all very rugged.

   B. From resistors that are all the same.

   C. From a series combination of resistors in parallel.

   D. From a parallel combination of resistors in series.

A good score is at least 18 correct answers. The answers are in the back of the book.

5
CHAPTER

Direct-current circuit analysis

IN THIS CHAPTER, YOU'LL LEARN MORE ABOUT DC CIRCUITS AND HOW THEY behave. These principles apply to almost all circuits in utility ac applications, too.

Sometimes it's necessary to analyze networks that don't have obvious practical use. But even a passive network of resistors can serve to set up the conditions for operation of a complex electrical device such as a radio amplifier or a digital calculator, by providing specific voltages or currents.

Current through series resistances

Have you ever used those tiny holiday lights that come in strings? If one bulb burns out, the whole set of bulbs goes dark; then you have to find out which bulb is bad, and replace it to get the lights working again. Each bulb works with something like 10 V; there are about a dozen bulbs in the string. You plug in the whole bunch and the 120-V utility mains drive just the right amount of current through each bulb.

In a series circuit, such as a string of light bulbs (Fig. 5-1), the current at any given point is the same as the current at any other point. The ammeter, A, is shown in the line between two of the bulbs. If it were moved anywhere else along the current path, it would indicate the same current. This is true in any series dc circuit, no matter what the components actually are and regardless of whether or not they all have the same resistance.

If the bulbs in Fig. 5-1 were of different resistances, some of them would consume more power than others. In case one of the bulbs in Fig. 5-1 burns out, and its socket is then shorted out instead of filled with a replacement bulb, the current through the whole chain will increase, because the overall resistance of the string would go down. This would force each of the remaining bulbs to carry too much current. Another bulb would probably burn out before long. If it, too, were replaced with a short circuit, the current

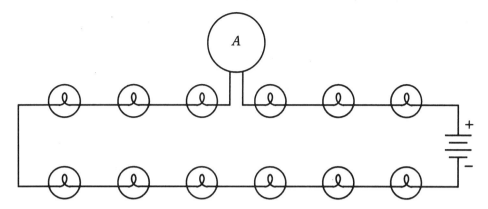

5-1 Light bulbs in series. An ammeter, A, is placed in the circuit to measure current.

would be increased still further. A third bulb would probably blow out almost right away after the string was plugged in.

Voltages across series resistances

The bulbs in the string of Fig. 5-1, being all the same, each get the same amount of voltage from the source. If there are a dozen bulbs in a 120-V circuit, each bulb will have a potential difference of 10 V across it. This will be true no matter how large or small the bulbs are, as long as they're all identical.

In a series circuit, the voltage is divided up among the components. The sum total of the potential differences across each resistance is equal to the supply voltage. This is always true no matter how large or how small the resistances and whether or not they're all the same value.

If you think about this for a moment, it's easy to see why it's true. Look at the schematic diagram of Fig. 5-2. Each resistor carries the same current. Each resistor Rn has a potential difference En across it, equal to the product of the current and the resistance of that particular resistor. These En's are in series, like cells in a battery, so they add together. What if the En's across all the resistors added up to something more or less than the supply voltage, E? Then there would have to be a "phantom EMF" some place, adding or taking away voltage. But there is no such. An EMF cannot come out of nowhere. This principle will be formalized later in this chapter.

Look at this another way. The voltmeter V in Fig. 5-2 shows the voltage E of the battery, because the meter is hooked up across the battery. The meter V also shows the sum of the En's across the set of resistors, because it's connected across the set of resistors. The meter says the same thing whether you think of it as measuring the battery voltage E, or as measuring the sum of the En's across the series combination of resistors. Therefore, E is equal to the sum of the En's.

This is a fundamental rule in series dc circuits. It also holds for 60-Hz utility ac circuits almost all the time.

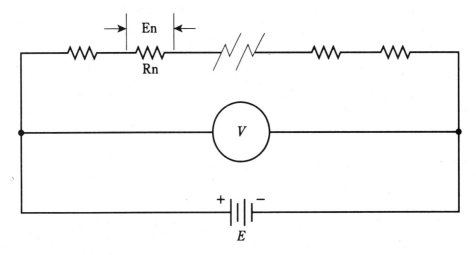

5-2   Analysis of voltage in a series circuit. See text for discussion.

How do you find the voltage across any particular resistor Rn in a circuit like the one in Fig. 5-2? Remember Ohm's Law for finding voltage: $E = IR$. The voltage is equal to the product of the current and the resistance. Remember, too, that you must use volts, ohms, and amperes when making calculations. In order to find the current in the circuit, I, you need to know the total resistance and the supply voltage. Then $I = E/R$. First find the current in the whole circuit; then find the voltage across any particular resistor.

Problem 5-1

In Fig. 5-2, there are 10 resistors. Five of them have values of 10 Ω, and the other five have values of 20 Ω. The power source is 15 Vdc. What is the voltage across one of the 10-Ω resistors? Across one of the 20-Ω resistors?

First, find the total resistance: $R = (10 \times 5) + (20 \times 5) = 50 + 100 = 150 \; \Omega$. Then find the current: $I = E/R = 15/150 = 0.10 \text{ A} = 100 \text{ mA}$. This is the current through each of the resistors in the circuit.

If Rn = 10 Ω, then En = I(Rn) = $0.1 \times 10 = 1.0$ V.
If Rn = 20 Ω, then En = I(Rn) = $0.1 \times 20 = 2.0$ V.

You can check to see whether all of these voltages add up to the supply voltage. There are five resistors with 1.0 V across each, for a total of 5.0 V; there are also five resistors with 2.0 V across each, for a total of 10 V. So the sum of the voltages across the resistors is 5.0 V + 10 V = 15 V.

Problem 5-2

In the circuit of Fig. 5-2, what will happen to the voltages across the resistors if one of the 20-Ω resistors is shorted out?

In this case the total resistance becomes $R = (10 \times 5) + (20 \times 4) = 50 + 80 = 130$ Ω. The current is therefore $I = E/R = 15/130 = 0.12 \text{ A}$. This is the current at any point in the circuit. This is rounded off to two significant figures.

The voltage En across Rn = 10 Ω is equal to En = I(Rn) = $0.12 \times 10 = 1.2$ V.

The voltage En across Rn = 20 Ω is En = I(Rn) = 0.12 × 20 = 2.4 V.

Checking the total voltage, we add (5 × 1.2) + (4 × 2.4) = 6.0 + 9.6 = 15.6 V. This rounds off to 16 V. Where did the extra volt come from?

The above is an example of what can happen when you round off to significant figures and then go through a problem a different way. The rechecking process is not part of the original problem. The answers you got the first time are perfectly alright. The figure 16 V is the result of a kind of mathematical trick, a "gremlin."

If this phenomenon bothers you, go ahead and keep all the digits your calculator will hold, while you do Problem 5-2 and recheck. The current in the circuit, as obtained by means of a calculator, is 0.115384615 A. When you find the voltages to all these extra digits and recheck, the error will be so tiny that it will cancel itself out, and you'll get a final rounded-off figure of 15 V rather than 16 V.

Some engineers wait until they get the final answer in a problem before they round off to the allowed number of significant digits. This is because the mathematical bugaboo just described can result in large errors, especially in *iterative processes*, involving calculations that are done over and over many times.

You'll probably never be faced with situations like this unless you plan to become an electrical engineer.

Voltage across parallel resistances

Imagine now a set of ornamental light bulbs connected in parallel (Fig. 5-3). This is the method used for outdoor holiday lighting, or for bright indoor lighting. It's much easier to fix a parallel-wired string of holiday lights if one bulb should burn out than it is to fix a series-wired string. And the failure of one bulb does not cause catastrophic system failure. In fact, it might be awhile before you notice that the bulb is dark, because all the other ones will stay lit, and their brightness will not change.

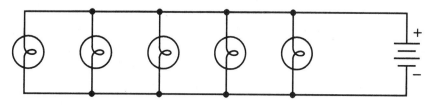

5-3   Light bulbs in parallel.

In a parallel circuit, the voltage across each component is always the same and it is always equal to the supply or battery voltage. The current drawn by each component depends only on the resistance of that particular device. In this sense, the components in a parallel-wired circuit work independently, as opposed to the series-wired circuit in which they all interact.

If any one branch of a parallel circuit is taken away, the conditions in the other branches will remain the same. If new branches are added, assuming the power supply can handle the load, conditions in previously existing branches will not be affected.

Currents through parallel resistances

Refer to the schematic diagram of Fig. 5-4. The resistors are called Rn. The total parallel resistance in the circuit is R. The battery voltage is E. The current in branch n, containing resistance Rn, is measured by ammeter A and is called In.

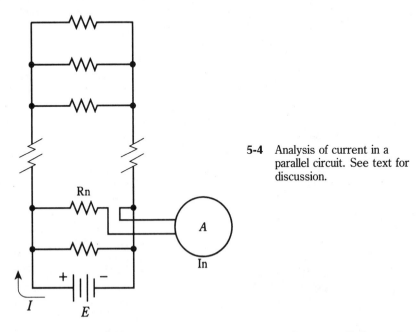

5-4 Analysis of current in a parallel circuit. See text for discussion.

The sum of all the In's in the circuit is equal to the total current, I, drawn from the source. That is, the current is divided up in the parallel circuit, similarly to the way that voltage is divided up in a series circuit.

If you're astute, you'll notice that the direction of current flow in Fig. 5-4 is out from the positive battery terminal. But don't electrons flow out of the minus terminal? Yes–but scientists consider current to flow from plus to minus. This is an example of a mathematical convention or "custom." Such things often outlast their appropriateness. Back in the early days of electrical experimentation, physicists had to choose a direction for the flow of current, and plus-to-minus seemed more logical than minus-to-plus. The exact nature of electric current flow wasn't known then. This notation has not been changed. It was feared that tampering with it would just cause confusion; some people would acknowledge the change while others would not. This might lead to motors running the wrong way, magnets repelling when they should attract, transistors being blown out, and other horrors. Just look at the mess caused by the conflict between Fahrenheit and Celsius temperatures, or between miles and kilometers.

Problem 5-3

Suppose that the battery in Fig. 5-4 delivers 12 V. Further suppose that there are 12 resistors, each with a value of 120 Ω in the parallel circuit. What is the total current, I, drawn from the battery?

First, find the total resistance. This is easy, because all the resistors have the same value. Just divide Rn = 120 by 12 to get R = 10 Ω. Then the current, *I*, is found by Ohm's Law: *I* = *E*/*R* = 12/10 = 1.2 A.

Problem 5-4

In the circuit of Fig. 5-4, what does the ammeter A say?

This involves finding the current in any given branch. The voltage is 12 V across every branch; Rn = 120. Therefore In, the ammeter reading, is found by Ohm's Law: *In* = *E*/*Rn* = 12/120 = 0.10 A = 100 mA.

All of the In's should add to get the total current, *I*. There are 12 identical branches, each carrying 0.1 A; therefore the sum is 0.1 × 12 = 1.2 A. It checks out. And there aren't any problems this time with significant figures.

Those two problems were designed to be easy. Here are two that are a little more involved.

Problem 5-5

Three resistors are in parallel across a battery that supplies *E* = 12 V. The resistances are R1 = 22 Ω, R2 = 47 Ω, and R3 = 68 Ω. These resistors carry currents *I*1, *I*2, and *I*3 respectively. What is the current, I3, through R3?

This is done by means of Ohm's Law, as if R3 were the only resistor in the circuit. There's no need to worry about the parallel combination. The other branches do not affect *I*3. Thus *I*3 = *E*/*R*3 = 12/68 = 0.18 A = 180 mA.

That problem wasn't hard at all. But it would have seemed that way, had you needlessly calculated the total parallel resistance of R1, R2, and R3.

Problem 5-6

What is the total current drawn by the circuit described in Problem 5-5?

There are two ways to go at this. One method involves finding the total resistance, *R*, of R1, R2, and R3 in parallel, and then calculating *I* based on *R*. Another, perhaps easier, way is to find the currents through R1, R2, and R3 individually, and then add them up.

Using the first method, first change the resistances *Rn* into conductances *Gn*. This gives G1 = 1/R1 = 1/22 = 0.04545 siemens, G2 = 1/R2 = 1/47 = 0.02128 siemens, and G3 = 1/R3 = 1/68 = 0.01471 siemens. Adding these gives G = 0.08144 siemens. The resistance is therefore *R* = 1/G = 1/0.08144 = 12.279 Ω. Use Ohm's Law to find *I* = *E*/*R* = 12/12.279 = 0.98 A = 980 mA. Note that extra digits are used throughout the calculation, rounding off only at the end.

Now let's try the other method. Find *I*1 = E/R1 = 12/22 = 0.5455 A, I2 = E/R2 = 12/47 = 0.2553 A, and I3 = E/R3 = 12/68 = 0.1765 A. Adding these gives *I* = *I*1 + *I*2 + *I*3 = 0.5455 + 0.2553 + 0.1765 = 0.9773 A, rounded off to 0.98 A.

Allowing extra digits during the calculation saved my having to explain away a mathematical artifact. It could save you similar chagrin some day. Doing the problem both ways helped me to be sure I didn't make any mistakes in finding the answer to this problem. It could have the same benefit for you, when the option presents itself.

Power distribution in series circuits

Let's switch back now to series circuits. This is a good exercise: getting used to thinking in different ways and to changing over quickly and often.

When calculating the power in a circuit containing resistors in series, all you need to do is find out the current, I, that the circuit is carrying. Then it's easy to calculate the power Pn, based on the formula $Pn = I^2Rn$.

Problem 5-7

Suppose we have a series circuit with a supply of 150 V and three resistors: R1 = 330 Ω, R2 = 680 Ω, and R3 = 910 Ω. What is the power dissipated by R2?

You must find the current in the circuit. To do this, calculate the total resistance first. Because the resistors are in series, the total is $R = 330 + 680 + 910 = 1920\ \Omega$. Then the current is $I = 150/1920 = 0.07813$ A = 78.1 mA. The power in R2 is $P2 = I^2R2 = 0.07813 \times 0.07813 \times 680 = 4.151$ W. Round this off to two significant digits, because that's all we have in the data, to obtain 4.2 W.

The total power dissipated in a series circuit is equal to the sum of the wattages dissipated in each resistor. In this way, the distribution of power in a series circuit is like the distribution of the voltage.

Problem 5-8

Calculate the total power in the circuit of Problem 5-7 by two different methods.

The first method is to figure out the power dissipated by each of the three resistors separately, and then add the figures up. The power $P2$ is already known. Let's bring it back to the four significant digits while we calculate: $P2 = 4.151$ W. Recall that the current in the circuit is $I = 0.07813$ A. Then P1 = $0.07813 \times 0.07813 \times 330 = 2.014$ W, and P3 = $0.07813 \times 0.07813 \times 910 = 5.555$ W. Adding these gives $P = 2.014 + 4.151 + 5.555 = 11.720$ W. Round this off to 12 W.

The second method is to find the series resistance of all three resistors. This is $R = 1920\ \Omega$, as found in Problem 5-7. Then $P = I^2R = 0.07813 \times 0.07813 \times 1920 = 11.72$ W, again rounded to 12 W.

You might recognize this as an electrical analog of the distributive law you learned in junior-high-school algebra.

Power distribution in parallel circuits

When resistances are wired in parallel, they each consume power according to the same formula, $P = I^2R$. But the current is not the same in each resistance. An easier method to find the power Pn, dissipated by resistor Rn, is by using the formula $Pn = E^2/Rn$, where E is the voltage of the supply. Recall that this voltage is the same across every resistor in a parallel circuit.

Problem 5-9

A circuit contains three resistances $R1 = 22\ \Omega$, $R2 = 47\ \Omega$, and $R3 = 68\ \Omega$ across a voltage $E = 3.0$ V. Find the power dissipated by each resistor.

First find E^2, because you'll be needing that number often: $E^2 = 3.0 \times 3.0 = 9.0$. Then $P1 = 9.0/22 = 0.4091$ W, $P2 = 9.0/47 = 0.1915$ W, $P3 = 9.0/68 = 0.1324$ W. These can be rounded off to $P1 = 0.41$ W, $P2 = 0.19$ W, and $P3 = 0.13$ W. But remember the values to four places for the next problem.

In a parallel circuit, the total power consumed is equal to the sum of the wattages dissipated by the individual resistances. In this respect, the parallel circuit acts like the series circuit. Power cannot come from nowhere, nor can it vanish. It must all be accounted for.

Problem 5-10

Find the total consumed power of the resistor circuit in Problem 5-9 using two different methods.

The first method involves adding *P1, P2,* and *P3.* Let's use the four-significant-digit values for "error reduction insurance." The sum is $P = 0.4091 + 0.1915 + 0.1324 = 0.7330$ W. This can be rounded to 0.73 W or 730 mW.

The second method involves finding the resistance R of the parallel combination. You can do this calculation yourself, keeping track for four digits for insurance reasons, getting $R = 12.28 \ \Omega$. Then $P = E^2/R = 9.0/12.28 = 0.7329$ W. This can be rounded to 0.73 W or 730 mW.

In pure mathematics and logic, the results are all deduced from a few simple, intuitively appealing principles called *axioms.* You might already know some of these, such as Euclid's geometry postulates. In electricity and electronics, complex circuit analysis can be made easier if you are acquainted with certain axioms, or *laws.* You've already seen some of these in this chapter. They are:

- The current in a series circuit is the same at every point along the way.
- The voltage across any component in a parallel circuit is the same as the voltage across any other, or across the whole set.
- The voltages across elements in a series circuit always add up to the supply voltage.
- The currents through elements in a parallel circuit always add up to the total current drawn from the supply.
- The total power consumed in a series or parallel circuit is always equal to the sum of the wattages dissipated in each of the elements.

Now you will learn two of the most famous laws in electricity and electronics. These make it possible to analyze extremely complicated series-parallel networks. That's not what you'll be doing in this course, but given the previous axioms and *Kirchhoff's Laws* that follow, you could if you had to.

Kirchhoff's first law

The physicist Gustav Robert Kirchhoff (1824-1887) was a researcher and experimentalist in electricity back in the time before radio, before electric lighting, and before much was understood about how currents flow.

Kirchhoff reasoned that current must work something like water in a network of pipes, and that the current going into any point has to be the same as the current going out. This is true for any point in a circuit, no matter how many branches lead into or out of the point. Two examples are shown in Fig. 5-5.

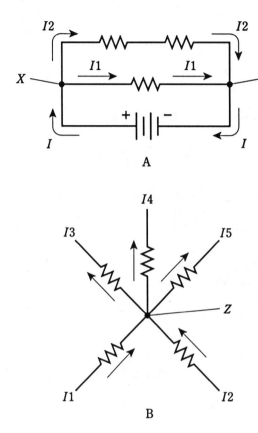

A

5-5   Kirchhoff's First Law. At A, the current into either X or Y is the same as the current out of that point: $I = I1 + I2$. At B, the current into Z equals the current out of Z: $I1 + I2 = I3 + I4 + I5$. Also see quiz questions 13 and 14.

B

In a network of water pipes that does not leak, and into which no water is added along the way, the total number of cubic feet going in has to be the same as the total volume going out. Water can't form from nothing, nor can it disappear, inside a closed system of pipes. Charge carriers, thought Kirchhoff, must act the same way in an electric circuit.

This is *Kirchhoff's First Law*. An alternative name might be the law of *conservation of current*.

Problem 5-11

Refer to Fig. 5-5A. Suppose all three resistors have values of 100 Ω, and that $I1 = 2.0$ A while $I2 = 1.0$ A. What is the battery voltage?

First, find the current I drawn from the battery. It must be 3.0 A; $I = I1 + I2 = 2.0 + 1.0 = 3.0$ A. Next, find the resistance of the whole combination. The two 100-Ω resistors in series give a value of 200 Ω, and this is in parallel with 100 Ω. You can do the calculations and find that the total resistance, R, across the battery, E, is 66.67 Ω. Then $E = IR = 66.67 \times 3.0 = 200$ volts. (Some battery.)

Problem 5-12

In Fig. 5-5B, suppose each of the two resistors below point Z has a value of 100 Ω, and all three resistors above Z are 10.0 Ω. The current through each 100-Ω resistor is 500 mA. What is the current through any of the 10.0-Ω resistors, assuming it is equally distributed? What is the voltage, then, across any of the 10.0-Ω resistors?

The total current into Z is 500 mA + 500 mA = 1.00 A. This must be divided three ways equally among the 10-Ω resistors. Therefore, the current through any one of them is 1.00/3 A = 0.333 A = 333 mA.

The voltage across any one of the 10.0-Ω resistors is found by Ohm's Law: $E = IR$ = 0.333 $\times$ 10.0 = 3.33 V.

Kirchhoff's second law

The sum of all the voltages, as you go around a circuit from some fixed point and return there from the opposite direction, and taking polarity into account, is always zero.

This might sound strange. Surely there is voltage in your electric hair dryer, or radio, or computer. Yes, there is, between different points. But no point can have an EMF with respect to itself. This is so simple that it's almost laughable. A point in a circuit is always shorted out to itself.

What Kirchhoff really was saying, when he wrote his second law, is a more general version of the second and third points previously mentioned. He reasoned that voltage cannot appear out of nowhere, nor can it vanish. All the potential differences must balance out in any circuit, no matter how complicated and no matter how many branches there are.

This is *Kirchhoff's Second Law*. An alternative name might be the law of *conservation of voltage*.

Consider the rule you've already learned about series circuits: The voltages across all the individual resistors add up to the supply voltage. Yes, they do, but the polarities of the EMFs across the resistors are opposite to that of the battery. This is shown in Fig. 5-6. It's a subtle thing. But it becomes clear when a series circuit is drawn with all the components, including the battery or other EMF source, in line with each other, as in Fig. 5-6.

Problem 5-13

Refer to the diagram of Fig. 5-6. Suppose the four resistors have values of 50, 60, 70 and 80 Ω, and that the current through them is 500 mA. What is the supply voltage, E?

Find the voltages *E1, E2, E3,* and *E4* across each of the resistors. This is done via Ohm's Law. In the case of *E1*, say with the 50-Ω resistor, calculate *E1* = 0.500 $\times$ 50 = 25 V. In the same way, you can calculate *E2* = 30 V, *E3* = 35 V, and *E4* = 40 V. The supply voltage is the sum *E1* + *E2* + *E3* + *E4* = 25 + 30 + 35 + 40 V = 130 V.

Kirchhoff's Second Law tells us that the polarities of the voltages across the resistors are in the opposite direction from that of the supply in the above example.

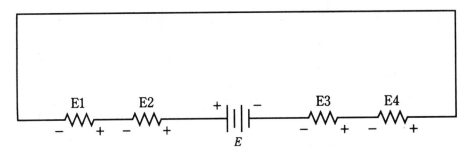

5-6   Kirchhoff's Second Law. The sum of the voltages across the resistors is equal to, but has opposite polarity from, the supply voltage E. Thus $E + E1 + E2 + E3 + E4 = 0$. Also see quiz questions 15 and 16.

Problem 5-14

In Fig. 5-6, suppose the battery provides 20 V. Let the resistors, having voltage drops *E1, E2, E3,* and *E4*, have their ohmic values in the ratio 1:2:3:4 respectively. What is *E3*?

This problem does not tell you the current in the circuit, nor the exact resistance values. But you don't need to know these things. Regardless of what the actual ohmic values are, the ratio *E1:E2:E3:E4* will be the same. This is a sort of *corollary* to Kirchhoff's Second Law. You can just invent certain ohmic values with the necessary ratio. Let's have them be R1 = 1 Ω, R2 = 2 Ω, R3 = 3 Ω, and R4 = 4 Ω. Then the total resistance is R = R1 + R2 + R3 + R4 = 1 + 2 + 3 + 4 = 10 Ω. You can calculate the current as $I = E/R = 20/10 = 2$ A. Then the voltage *E3*, across R3, is given by Ohm's Law as $E3 = I(R3) = 2 \times 3 = 6$ V.

You are encouraged to calculate the other voltages and observe that they add up to 20 V.

In this problem, there is freedom to literally pick numbers out of the air so that calculations are easy. You could have chosen ohmic values like 47, 94, 141, and 188 Ω (these too are in the ratio 1:2:3:4), and you'd still get *E3* = 6 V. (Go ahead and try it.) But that would have made needless work for yourself.

Series combinations of resistors are often used by electronic engineers to obtain various voltage ratios, to make circuits work just right. These resistance circuits are called *voltage divider networks*.

Voltage divider networks

Earlier, you were assured that this course would not drag you through ridiculously complicated circuits. You can imagine, by now nightmarish series-parallel matrixes of resistors drawn all over whole sheets of paper, captioned with wicked queries: "What is the current through R135?" But that stuff is best left to professional engineers, and even they aren't likely to come across it very often. Their job is to make things as neat and efficient as possible. If an engineer actually is faced with such a scenario, the reaction will probably be, "How can this circuit be simplified?"

Resistances in series produce ratios of voltages, and these ratios can be tailored to meet certain needs.

When designing voltage divider networks, the resistance values should be as small as possible, without causing too much current drain on the supply. In practice the optimum values depend on the nature of the circuit being designed. This is a matter for engineers, and specific details are beyond the scope of this course. The reason for choosing the smallest possible resistances is that, when the divider is used with a circuit, you do not want that circuit to upset the operation of the divider. The voltage divider "fixes" the intermediate voltages best when the resistance values are as small as the current-delivering capability of the power supply will allow.

Figure 5-7 illustrates the principle of voltage division. The individual resistances are *R1, R2, R3, ..., Rn*. The total resistance is $R = R1 + R2 + R3 + ... + Rn$. The supply voltage is *E*, and the current in the circuit is therefore $I = E/R$. At the various points P1, P2, P3, ..., Pn, voltages will be *E1, E2, E3, ..., En*. The last voltage, *En*, is the same as the supply voltage, *E*. All the other voltages are less than *E*, so $E1 < E2 < E3 < ... < En = E$. (The symbol $<$ means "is less than.")

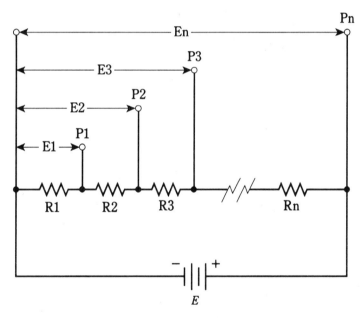

5-7 General arrangement for voltage divider circuit. Designators are discussed in the text. Also see quiz questions 19 and 20.

The voltages at the various points increase according to the sum total of the resistances up to each point, in proportion to the total resistance, multiplied by the supply voltage. The voltage *E1* is equal to $E \times R1/R$. The voltage *E2* is equal to $E \times (R1 + R2)/R$. The voltage *E3* is $E \times (R1 + R2 + R3)/R$. You can mentally continue this process to get each one of the voltages at points all the way up to $En = E \times (R1 + R2 + R3 + ... + Rn)/R = E \times R/R = E \times 1 = E$.

Usually there are only two or three intermediate voltages in a voltage-divider network. So designing such a circuit isn't as complicated as the above formulas might lead you to think.

The following problems are similar to those encountered by electronic engineers.

Problem 5-15

In a transistorized amplifier, the battery supplies 9.0 V. The minus terminal is at common (chassis) ground. At some point, you need to get +2.5 V. Give an example of a pair of resistors that can be connected in series, such that +2.5 V will be provided at some point.

See the schematic diagram of Fig. 5-8. There are infinitely many different combinations of resistances that will work here. You pick some total value, say $R = R1 + R2 = 1000$ Ω. You know that the ratio $R1:R$ will always be the same as the ratio $E1:E$. In this case $E1 = 2.5$ V, so $E1:E = 2.5/9.0 = 0.28$. Therefore $R1:R$ should be 0.28. Because $R = 1000$ Ω, this means $R1 = 280$ Ω. The value of $R2$ will be the difference $1000 - 280$ Ω $= 720$ Ω.

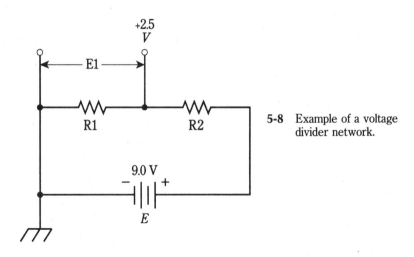

5-8   Example of a voltage divider network.

Problem 5-16

What is the current drawn by the resistances in the previous problem?

Simply use Ohm's Law to get $I = E/R = 9.0/1000 = 9.0$ mA.

Problem 5-17

Suppose that it is permissible to draw up to 100 mA of current in the problem shown by Fig. 5-8. You, the engineer, want to design the circuit to draw this maximum current, because that will offer the best *voltage regulation* for the circuit to be operated from the network. What values of resistances $R1$ and $R2$ should you use?

Calculate the total resistance first, using Ohm's Law: $R = E/I = 9.0/0.1 = 90$ Ω. The ratio desired is $R1:R2 = 2.5/9.0 = 0.28$. Then you would use $R1 = 0.28 \times 90 = 25$ Ω. The value of $R2$ is the remainder: $R2 = 90 - 25 = 65$ Ω.

Quiz

Refer to the text in this chapter if necessary. A good score is at least 18 correct answers. The answers are in the back of the book.

1. In a series-connected string of holiday ornament bulbs, if one bulb gets shorted out, which of these is most likely?

    A. All the other bulbs will go out.

    B. The current in the string will go up.

    C. The current in the string will go down.

    D. The current in the string will stay the same.

2. Four resistors are connected in series across a 6.0-V battery. The values are R1 = 10 Ω, R2 = 20 Ω, R3 = 50 Ω, and R4 = 100 Ω as shown in Fig. 5-9. The voltage across R2 is:

    A. 0.18 V.

    B. 33 mV.

    C. 5.6 mV.

    D. 670 mV.

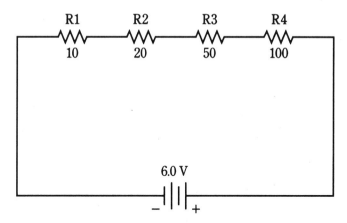

5-9 Illustration for quiz questions 2, 3, 8, and 9.

3. In question 2 (Fig. 5-9), the voltage across the combination of R3 and R4 is:

    A. 0.22 V.

    B. 0.22 mV.

    C. 5.0 V.

    D. 3.3 V.

4. Three resistors are connected in parallel across a battery that delivers 15 V. The values are R1 = 470 Ω, R2 = 2.2 KΩ, R3 = 3.3 KΩ (Fig. 5-10). The voltage across R2 is:

    A. 4.4 V.

B. 5.0 V.

C. 15 V.

D. Not determinable from the data given.

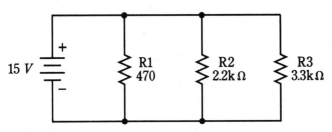

5-10   Illustration for quiz questions 4, 5, 6, 7, 10, and 11.

5. In the example of question 4 (Fig. 5-10), what is the current through R2?

A. 6.8 mA.

B. 43 mA.

C. 150 mA.

D. 6.8 A.

6. In the example of question 4 (Fig. 5-10), what is the total current drawn from the source?

A. 6.8 mA.

B. 43 mA.

C. 150 mA.

D. 6.8 A.

7. In the example of question 4 (Fig. 5-10), suppose that resistor R2 opens up. The current through the other two resistors will:

A. Increase.

B. Decrease.

C. Drop to zero.

D. No change.

8. Four resistors are connected in series with a 6.0-V supply, with values shown in Fig. 5-9 (the same as question 2). What is the power dissipated by the whole combination?

A. 200 mW.

B. 6.5 mW.

C. 200 W.

D. 6.5 W.

9. In Fig. 5-9, what is the power dissipated by R4?

A. 11 mW.

B. 0.11 W.

C. 0.2 W.

D. 6.5 mW.

10. Three resistors are in parallel in the same configuration and with the same values as in problem 4 (Fig. 5-10). What is the power dissipated by the whole set?

A. 5.4 W.

B. 5.4 uW.

C. 650 W.

D. 650 mW.

11. In Fig. 5-10, the power dissipated by R1 is:

A. 32 mW.

B. 480 mW.

C. 2.1 W.

D. 31 W.

12. Fill in the blank in the following sentence. In a either a series or a parallel circuit, the sum of the ____ s in each component is equal to the total ____ provided by the supply.

A. Current.

B. Voltage.

C. Wattage.

D. Resistance.

13. Refer to Fig. 5-5A. Suppose the resistors each have values of 33 Ω. The battery provides 24 V. The current I1 is:

A. 1.1 A.

B. 730 mA.

C. 360 mA.

D. Not determinable from the information given.

14. Refer to Fig. 5-5B. Let each resistor have a value of 820 Ω. Suppose the top three resistors all lead to light bulbs of the exact same wattage. If I1 = 50 mA and I2 = 70 mA, what is the power dissipated in the resistor carrying current I4?

A. 33 W.

B. 40 mW.

C. 1.3 W.

D. It can't be found using the information given.

15. Refer to Fig. 5-6. Suppose the resistances R1, R2, R3, and R4 are in the ratio 1:2:4:8 from left to right, and the battery supplies 30 V. Then the voltage E2 is:

A. 4 V.

B. 8 V.

C. 16 V.

D. Not determinable from the data given.

16. Refer to Fig. 5-6. Let the resistances each be 3.3 KΩ, and the battery 12 V. If the plus terminal of a dc voltmeter is placed between R1 and R2 (with voltages *E1* and *E2*), and the minus terminal of the voltmeter is placed between R3 and R4 (with voltages *E3* and *E4*), what will the meter register?

A. 0 V.

B. 3 V.

C. 6 V.

D. 12 V.

17. In a voltage divider network, the total resistance:

A. Should be large to minimize current drain.

B. Should be as small as the power supply will allow.

C. Is not important.

D. Should be such that the current is kept to 100 mA.

18. The maximum voltage output from a voltage divider:

A. Is a fraction of the power supply voltage.

B. Depends on the total resistance.

C. Is equal to the supply voltage.

D. Depends on the ratio of resistances.

19. Refer to Fig. 5-7. The battery E is 18.0 V. Suppose there are four resistors in the network: R1 = 100 Ω, R2 = 22.0 Ω, R3 = 33.0 Ω, R4 = 47.0 Ω. The voltage E3 at P3 is:

A. 4.19 V.

B. 13.8 V.

C. 1.61 V.

D. 2.94 V.

20. Refer to Fig. 5-7. The battery is 12 V; you want intermediate voltages of 3.0, 6.0 and 9.0 V. Suppose that a maximum of 200 mA is allowed through the network. What values should the resistors, R1, R2, R3, and R4 have, respectively?

A. 15 Ω, 30 Ω, 45 Ω, 60 Ω.

B. 60 Ω, 45 Ω, 30 Ω, 15 Ω.

C. 15 Ω, 15 Ω, 15 Ω, 15 Ω.

D. There isn't enough information to design the circuit.

A good score is at least 18 correct answers. The answers are in the back of the book.

6
CHAPTER

Resistors

AS YOU'VE ALREADY SEEN, ANY ELECTRICAL DEVICE HAS SOME RESISTANCE; none is a perfect conductor. You've also seen some examples of circuits containing components designed to oppose the flow of current. This chapter more closely examines resistors— devices that oppose, control, or limit electrical current.

Why, you might ask, would anyone want to put things into a circuit to reduce the current? Isn't it true that resistors always dissipate some power as heat, and that this invariably means that a circuit becomes less efficient than it would be without the resistor? Well, it's true that resistors always dissipate some power as heat. But resistors can optimize the ability of a circuit to generate or amplify a signal, making the circuit maximally efficient at whatever it is designed to do.

Purpose of the resistor

Resistors can play any of numerous different roles in electrical and electronic equipment. Here are a few of the more common ways resistors are used.

Voltage division

You've already learned a little about how voltage dividers can be designed using resistors. The resistors dissipate some power in doing this job, but the resulting voltages are needed for the proper *biasing* of electronic transistors or vacuum tubes. This ensures that an amplifier or oscillator will do its job in the most efficient, reliable possible way.

Biasing

In order to work efficiently, transistors or tubes need the right bias. This means that the control electrode—the *base*, *gate*, or *grid*—must have a certain voltage or current. Networks of resistors accomplish this. Different bias levels are needed for different types of circuits. A radio transmitting amplifier would usually be biased differently than an oscillator or a low-level receiving amplifier. Sometimes voltage division is required for biasing. Other times it isn't necessary. Figure 6-1 shows a transistor whose base is biased using a pair of resistors in a voltage-dividing configuration.

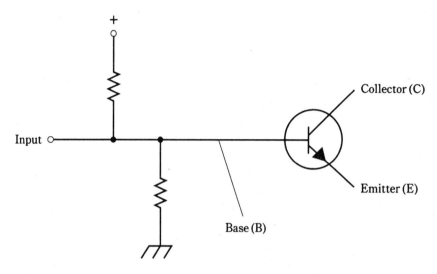

6-1   Voltage divider for biasing the base of a transistor.

Current limiting

Resistors interfere with the flow of electrons in a circuit. Sometimes this is essential to prevent damage to a component or circuit. A good example is in a receiving amplifier. A resistor can keep the transistor from using up a lot of power just getting hot. Without resistors to limit or control the current, the transistor might be overstressed carrying direct current that doesn't contribute to the signal. An improperly designed amplifier might need to have its transistor replaced often, because a resistor wasn't included in the design where it was needed, or because the resistor isn't the right size. Figure 6-2 shows a current-limiting resistor connected in series with a transistor. Usually it is in the emitter circuit, as shown in this diagram, but it can also be in the collector circuit.

Power dissipation

Dissipating power as heat is not always bad. Sometimes a resistor can be used as a "dummy" component, so that a circuit "sees" the resistor as if it were something more complicated. In radio, for example, a resistor can be used to take the place of an antenna. A transmitter can then be tested in such a way that it doesn't interfere with signals on the airwaves. The transmitter output heats the resistor, without radiating any signal. But as far as the transmitter "knows," it's hooked up to a real antenna (Fig. 6-3).

Another case in which power dissipation is useful is at the input of a power amplifier. Sometimes the circuit *driving* the amplifier (supplying its input signal) has too much power

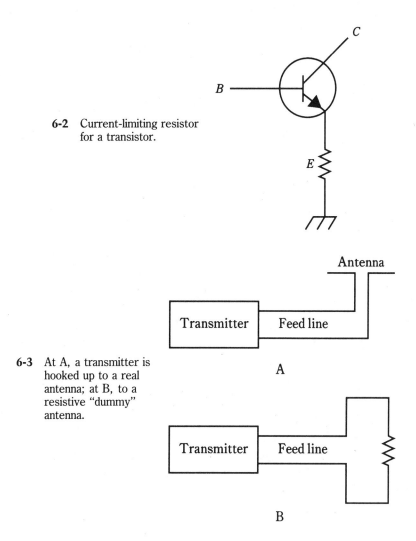

6-2 Current-limiting resistor for a transistor.

6-3 At A, a transmitter is hooked up to a real antenna; at B, to a resistive "dummy" antenna.

for the amplifier input. A resistor, or network of resistors, can dissipate this excess so that the power amplifier doesn't get too much drive.

Bleeding off charge

In a high-voltage, direct-current (dc) power supply, capacitors are used to smooth out the fluctuations in the output. These capacitors acquire an electric charge, and they store it for awhile. In some power supplies, these *filter capacitors* hold the full output voltage of the supply, say something like 750 V, even after the supply has been turned off, and even after it is unplugged from the wall outlet. If you attempt to repair such a power supply, you might get clobbered by this voltage. *Bleeder resistors*, connected across the filter capacitors, drain their stored charge so that servicing the supply is not dangerous (Fig. 6-4).

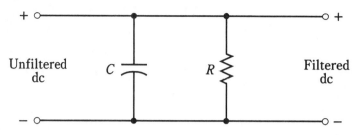

6-4   A bleeder resistor is connected across the filter capacitor in a power supply.

It's always a good idea to short out all filter capacitors, using a screwdriver with an insulated handle, before working on a high-voltage dc power supply. I recall an instance when I was repairing the supply for a radio power amplifier. The capacitors were holding about 2 kV. My supervisor, not very well acquainted with electronics, was looking over my shoulder. I said, "Gonna be a little pop, now," and took a Phillips screwdriver, making sure I had hold of the insulated handle only, and shorted the filter capacitor to the chassis. **Bang!** I gave my supervisor a brief explanation while he took some deep breaths.

Even if a supply has bleeder resistors, they take awhile to get rid of the residual charge. For safety, always do what I did, whether your supervisor is around or not.

Impedance matching

A more subtle, more sophisticated use for resistors is in the *coupling* in a chain of amplifiers, or in the input and output circuits of amplifiers. In order to produce the greatest possible amplification, the *impedances* must agree between the output of a given amplifier and the input of the next. The same is true between a source of signal and the input of an amplifier. Also, this applies between the output of the last amplifier in a chain, and the *load*, whether that load is a speaker, a headset, a FAX machine, or whatever.

Impedance is the alternating-current (ac) cousin of resistance in direct-current (dc) circuits. This is discussed in the next section of this book.

The carbon-composition resistor

Probably the cheapest method of making a resistor is to mix up finely powdered carbon (a fair electrical conductor) with some nonconductive substance, press the resulting clay-like stuff into a cylindrical shape, and insert wire leads in the ends (Fig. 6-5). The resistance of the final product will depend on the ratio of carbon to the nonconducting material, and also on the physical distance between the wire leads. The nonconductive material is usually phenolic, similar to plastic. This results in a *carbon-composition* resistor.

Carbon-composition resistors can be made to have quite low resistances, all the way up to extremely high resistances. This kind of resistor has the advantage of being pretty much *nonreactive*. That means that it introduces almost pure resistance into the circuit, and not much capacitance or inductance. This makes the carbon-composition resistor useful in radio receivers and transmitters.

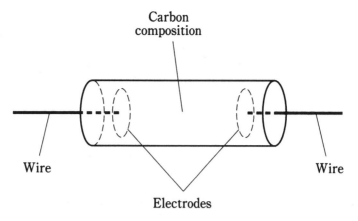

6-5   Construction of a carbon-composition resistor.

Carbon-composition resistors dissipate power according to how big, physically, they are. Most of the carbon-composition resistors you see in electronics stores can handle ¼ W or ½ W. There are ⅛-W units for miniaturized, low- power circuitry, and 1-W or 2-W components for circuits where some electrical ruggedness is needed. Occasionally you'll see a much larger unit, but these are rare.

The wirewound resistor

A more obvious way to get resistance is to use a length of wire that isn't a good conductor. Nichrome is most often used for this. The wire can be wound around a cylindrical form, like a coil (Fig. 6-6). The resistance is determined by how well the wire metal conducts, by its diameter or *gauge*, and by its length. This component is called a *wirewound resistor*.

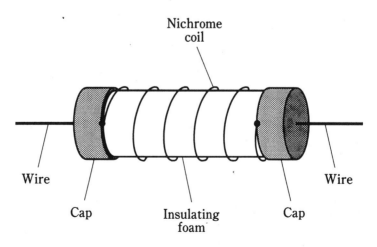

6-6   Construction of a wirewound resistor.

One of the advantages of wirewound resistors is that they can be made to have values within a very close range; that is, they are precision components. Another advantage is that wirewound resistors can be made to handle large amounts of power. Some wirewounds might actually do well as electric heaters, dissipating hundreds, or even thousands of watts.

A disadvantage of wirewound resistors, in some applications, is that they act like inductors. This makes them unsuitable for use in most radio-frequency circuits.

Wirewound resistors usually have low to moderate values of resistance.

Film type resistors

Carbon, nichrome, or some mixture of ceramic and metal (*cermet*) can be applied to a cylindrical form as a film, or thin layer, in order to obtain a desired value of resistance. This type of resistor is called a *carbon-film resistor* or *metal-film resistor*. It looks like a carbon-composition type, but the construction technique is different (Fig. 6-7).

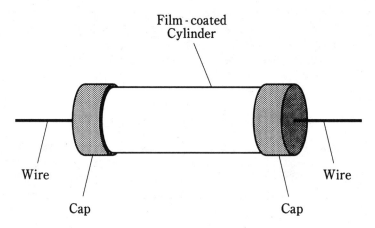

6-7   Construction of a film type resistor.

The cylindrical form is made of an insulating substance, such as porcelain. The film is deposited on this form by various methods, and the value tailored as desired. Metal-film units can be made to have nearly exact values. Film type resistors usually have low to medium-high resistance.

A major advantage of film type resistors is that they, like carbon-composition units, do not have much inductance or capacitance. A disadvantage, in some applications, is that they can't handle as much power as the more massive carbon-composition units or as wirewound types.

Integrated-circuit resistors

Increasingly, whole electronic circuits are being fabricated on semiconductor wafers known as *integrated circuits (ICs)*. It is possible nowadays to put a whole radio receiver into a couple of ICs, or *chips*, whose total volume is about the same as that of the tip of your little finger. In 1930, a similar receiver would have been as large as a television set.

Resistors can be fabricated onto the semiconductor chip that makes up an IC. The thickness, and the types and concentrations of impurities added, control the resistance of the component.

IC resistors can only handle a tiny amount of power because of their small size. But because IC circuits in general are designed to consume minimal power, this is not a problem. The small signals produced by ICs can be amplified using circuits made from discrete components if it is necessary to obtain higher signal power.

The potentiometer

All of the resistors mentioned are fixed in value. It is impossible to change or adjust their resistances. Of course, their values will change if they overheat, or if you chip pieces of them out, but they're meant to provide an unchanging opposition to the flow of electric current.

It might have occurred to you that a variable resistor can be made by hooking up a bunch of fixed resistors in series or parallel, and then switching more or fewer of them in and out. This is almost never done in electronic circuits because there's a better way to get a variable resistance: use a *potentiometer*.

The construction of a potentiometer is shown in simplified form in Fig. 6-8. A resistive strip is bent into a nearly complete circle, and terminals are connected to either end. This forms a fixed resistance. To obtain the variable resistance, a sliding contact is attached to a rotatable shaft and bearing, and is connected to a third terminal. The resistance between this middle terminal, and either of the end terminals, can vary from zero up to the resistance of the whole strip.

Some potentiometers use a straight strip of resistive material, and the control moves up and down, or from side to side. This type of variable resistor, called a *slide potentiometer*, is used in graphic equalizers, as the volume controls in some stereo amplifiers, and in some other applications when a linear scale is preferable to a circular scale.

Potentiometers are made to handle only very low levels of current, at low voltage.

Linear taper

One type of potentiometer uses a strip of resistive material whose density is constant all the way around. This results in a *linear taper*. The resistance between the center terminal and either end terminal changes at a steady rate as the control shaft is turned.

Suppose a linear taper potentiometer has a value of zero to 280 Ω. In most units the shaft rotates about 280 degrees, or a little more than three-quarters of a circle. Then the resistance between the center and one end terminal will increase right along with the number of degrees that the shaft is turned. The resistance between the center and the other end terminal will be equal to 280 minus the number of degrees the shaft is turned. Engineers say that the resistance is a *linear function* of the shaft position.

Linear taper potentiometers are commonly used in electronic test instruments and in various consumer electronic devices. A graph of resistance versus shaft displacement for a linear taper potentiometer is shown in Fig. 6-9.

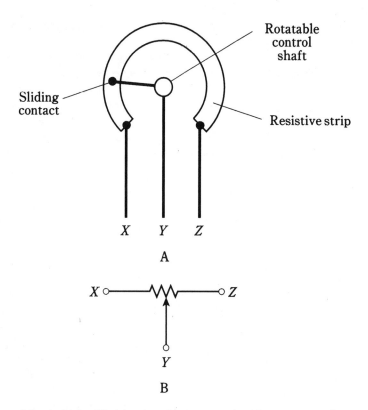

6-8 At A, simplified drawing of the construction of a rotary potentiometer. At B, schematic symbol.

Audio or logarithmic taper

There are some applications for which linear taper potentiometers don't work well. The volume control of a radio receiver is a good example. Your ear/brain perceives sound level according to the *logarithm* of its true level. If you use a linear taper potentiometer as the volume control of a transistor radio or other sound system, the level will seem to go up too slowly in some parts of the control range and too fast in other parts of the control range.

To compensate for the way in which people perceive sound level, an *audio taper* potentiometer is used. In this device, the resistance between the center and end terminal increases in a nonlinear way. This type of potentiometer is sometimes called a *logarithmic-taper* device.

If the shaft is all the way counterclockwise, the volume at the speaker is zero or near zero. If you turn the shaft 30 degrees clockwise, the volume increases to some perceived level; call it one sound unit. If you then turn the volume 30 degrees further clockwise, the volume will seem to go up to two sound units. But in fact it has increased much more than this, in terms of actual sound power.

You perceive sound not as a direct function of the true volume, but in units that are based on the logarithm of the intensity. Audio-taper potentiometers are manufactured so that as you turn the shaft, the sound seems to increase in a smooth, natural way. A graph of resistance versus shaft displacement for an audio-taper potentiometer is shown in Fig. 6-10.

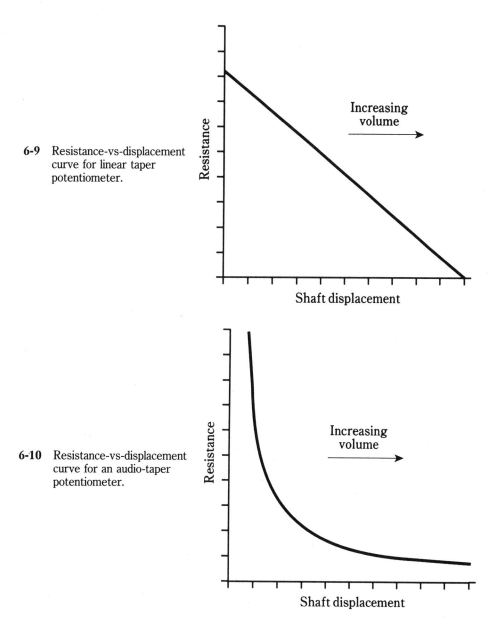

6-9   Resistance-vs-displacement curve for linear taper potentiometer.

6-10   Resistance-vs-displacement curve for an audio-taper potentiometer.

This is a good time to sidetrack for a moment and examine how sound sensation is measured.

The decibel

Perceived levels of sound, and of other phenomena such as light and radio signals, change according to the logarithm of the actual power level. Units have been invented to take this into account.

The fundamental unit of sound change is called the *decibel*, abbreviated *dB*. A change of 1 dB is the minimum increase in sound level that you can detect, if you are expecting it. A change of -1 dB is the minimum detectable decrease in sound volume, when you are anticipating the change. Increases in volume are positive decibel values; decreases in volume are negative values.

If you aren't expecting the level of sound to change, then it takes about 3 dB or -3 dB of change to make a noticeable difference.

Calculating decibel values

Decibel values are calculated according to the logarithm of the ratio of change. Suppose a sound produces a power of P watts on your eardrums, and then it changes (either getting louder or softer) to a level of Q watts. The change in decibels is obtained by dividing out the ratio Q/P, taking its base-10 logarithm, and then multiplying the result by 10:

$$dB = 10 \log (Q/P)$$

As an example, suppose a speaker emits 1 W of sound, and then you turn up the volume so that it emits 2 W of sound power. Then $P = 1$ and $Q = 2$, and $dB = 10 \log (2/1) = 10 \log 2 = 10 \times 0.3 = 3$ dB. This is the minimum detectable level of volume change if you aren't expecting it: a doubling of the actual sound power.

If you turn the volume level back down again, then $P/Q = 1/2 = 0.5$, and you can calculate $dB = 10 \log 0.5 = 10 \times -0.3 = -3$ dB.

A change of plus or minus 10 dB is an increase or decrease in sound power of 10 times. A change of plus or minus 20 dB is a hundredfold increase or decrease in sound power. It is not unusual to encounter sounds that range in loudness over plus/minus 60 dB or more—a millionfold variation.

Sound power in terms of decibels

The above formula can be worked inside-out, so that you can determine the final sound power, given the initial sound power and the decibel change.

Suppose the initial sound power is P, and the change in decibels is dB. Let Q be the final sound power. Then $Q = P$ antilog (dB/10).

As an example, suppose the initial power, P, is 10 W, and the change is -3 dB. Then the final power, Q, is $Q = 10$ antilog $(-3/10) = 10 \times 0.5 = 5$ W.

Decibels in real life

A typical volume control potentiometer might have a resistance range such that you can adjust the level over about plus/minus 80 dB. The audio taper ensures that the decibel increase or decrease is a straightforward function of the rotation of the shaft.

Sound levels are sometimes specified in decibels relative to the *threshold of hearing,* or the lowest possible volume a person can detect in a quiet room, assuming their hearing is normal. This threshold is assigned the value 0 dB. Other sound levels can then be quantified, as a number of decibels such as 30 dB or 75 dB.

If a certain noise is given a loudness of 30 dB, it means it's 30 dB above the threshold of hearing, or 1,000 times as loud as the quietest detectable noise. A noise at 60 dB is 1,000,000 times as powerful as the threshold of hearing. Sound-level meters are used to determine the dB levels of various noises and acoustic environments.

A typical conversation might be at a level of about 70 dB. This is 10,000,000 times the threshold of hearing, in terms of actual sound power. The roar of the crowd at a rock concert might be 90 dB, or 1,000,000,000 times the threshold of hearing.

A sound at 100 dB, typical of the music at a large rock concert, is 10,000,000,000 times as loud, in terms of power, as a sound at the threshold of hearing. If you are sitting in the front row, and if it's a loud band, your ears might get walloped with peaks of 110 dB. That is 100 billion times the minimum sound power you can detect in a quiet room.

The rheostat

A variable resistor can be made from a wirewound element, rather than a solid strip of material. This is called a *rheostat*. A rheostat can have either a rotary control or a sliding control. This depends on whether the nichrome wire is wound around a doughnut-shaped form (*toroid*) or a cylindrical form (*solenoid*).

Rheostats always have inductance, as well as resistance. They share the advantages and disadvantages of fixed wirewound resistors.

A rheostat is not continuously adjustable, as a potentiometer is. This is because the movable contact slides along from turn to turn of the wire coil. The smallest possible increment is the resistance in one turn of the coil. The rheostat resistance therefore adjusts in a series of little jumps.

Rheostats are used in high-voltage, high-power applications. A good example is in a variable-voltage power supply. This kind of supply uses a transformer that steps up the voltage from the 117-V utility mains, and diodes to change the ac to dc. The rheostat can be placed between the utility outlet and the transformer (Fig. 6-11). This results in a variable voltage at the power-supply output. A potentiometer would be destroyed instantly in this application.

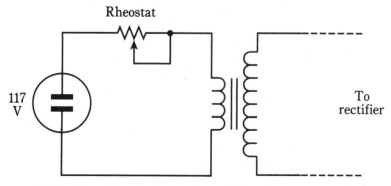

6-11   Connection of a rheostat in a variable-voltage power supply.

Resistor values

In theory, a resistor can have any value from the lowest possible (such as a shaft of solid silver) to the highest (open air). In practice, it is unusual to find resistors with values less than about 0.1 Ω, or more than about 100 MΩ.

Resistors are manufactured in standard values that might at first seem rather odd to you. The standard numbers are 1.0, 1.2, 1.5, 1.8, 2.2, 2.7, 3.3, 3.9, 4.7, 5.6, 6.8, and 8.2. Units are commonly made with values derived from these values, multiplied by some power of 10. Thus you will see units of 47 Ω, 180 Ω, 6.8 KΩ or 18 MΩ, but not 380 Ω or 650 KΩ. Maybe you've wondered at some of the resistor values that have been used in problems and quiz questions in previous chapters. Now you know that these choices weren't totally arbitrary; they were picked to represent values you might find in real circuits.

In addition to the above values, there are others that are used for resistors made with greater precision, or tighter *tolerance*. These are power-of-10 multiples of 1.1, 1.3, 1.6, 2.0, 2.4, 3.0, 3.6, 4.3, 5.1, 6.2, 7.5, and 9.1.

You don't have to memorize these numbers. They'll become familiar enough over time, as you work with electrical and electronic circuits.

Tolerance

The first set of numbers above represents standard resistance values available in tolerances of plus or minus 10 percent. This means that the resistance might be as much as 10 percent more or 10 percent less than the indicated amount. In the case of a 470-ohm resistor, for example, the value can be off by as much as 47 ohms and still be within tolerance. That's a range of 423 to 517 ohms. The tolerance is calculated according to the specified value of the resistor, not the actual value. You might measure the value of a 470-ohm resistor and find it to be 427 ohms, and it would be within 10 percent of the specified value; if it measures 420 ohms, it's outside the 10-percent range and is a "reject."

The second set, along with the first set, of numbers represents standard resistance values available in tolerances of plus or minus 5 percent. A 470-ohm, 5-percent resistor will have an actual value of 470 ohms plus or minus 24 ohms, or a range of 446 to 494 ohms.

Some resistors are available in tolerances tighter than 5 percent. These precision units are employed in circuits where a little error can make a big difference. In most audio and radio-frequency oscillators and amplifiers, 10-percent or 5- percent tolerance is good enough. In many cases, even a 20-percent error is all right.

Power rating

All resistors are given a specification that determines how much power they can safely dissipate. Typical values are ¼ W, ½ W and 1 W. Units also exist with ratings of ⅛ W or 2 W. These dissipation ratings are for continuous duty.

You can figure out how much current a given resistor can handle, by using the formula for power (P) in terms of current (I) and resistance (R): $P = I^2R$. Just work this formula backwards, plugging in the power rating for P and the resistance of the unit for R, and solve for I. Or you can find the square root of P/R. Remember to use amperes for current, ohms for resistance, and watts for power.

The power rating for a given resistor can, in effect, be increased by using a network of 2×2, 3×3, 4×4, etc., units in series-parallel. You've already learned about this. If you need a 47-ohm, 45-W resistor, but all you have is a bagful of 47-ohm, 1-W resistors, you can make a 7×7 network in series-parallel, and this will handle 49 W. It might look terrible, but it'll do the job.

Power ratings are specified with a margin for error. A good engineer never tries to take advantage of this and use, say, a ¼-W unit in a situation where it will need to draw 0.27 W. In fact, good engineers usually include their own safety margin. Allowing 10 percent, a ¼-W resistor should not be called upon to handle more than about 0.225 W. But it's silly, and needlessly expensive, to use a 2-W resistor where a ¼-W unit will do, unless, of course, the 2-W resistor is all that's available.

Temperature compensation

All resistors change value somewhat when the temperature changes dramatically. And because resistors dissipate power, they can get hot just because of the current they carry. Often, this current is so tiny that it doesn't appreciably heat the resistor. But in some cases it does, and the resistance might change. Then the circuit will behave differently than it did when the resistor was still cool.

There are various ways to approach problems of resistors changing value when they get hot.

One method is to use specially manufactured resistors that do not appreciably change value when they get hot. Such units are called *temperature-compensated*. But one of these can cost several times as much as an ordinary resistor.

Another approach is to use a power rating that is much higher than the actual dissipated power in the resistor. This will keep the resistor from getting very hot. Usually, it's a needless expense to do this, but if the small change in value cannot be tolerated, it's sometimes the most cost effective.

Still another scheme is to use a series-parallel network of resistors that are all identical, in the manner you already know about, to increase the power dissipation rating. Alternatively, you can take several resistors, say three of them, each with about three times the intended resistance, and connect them all in parallel. Or you can take several resistors, say four of them, each with about ¼ the intended resistance, and connect them in series.

It is unwise to combine several resistors with greatly different values. This can result in one of them taking most of the load while the others loaf, and the combination will be no better than the single hot resistor you started with.

You might get the idea of using two resistors with half (or twice) the value you need, but with *opposite* resistance-versus-temperature characteristics, and connecting them in series (or in parallel). Then the one whose resistance decreases with heat (*negative temperature coefficient*) will have a canceling-out effect on the one whose resistance goes up (*positive temperature coefficient*). This is an elegant theory, but in practice you probably won't be able to find two such resistors without spending at least as much money as you would need to make a 3 × 3 series-parallel network. And you can't be sure that the opposing effects will exactly balance. It would be better, in such a case, to make a 2 × 2 series-parallel array of ordinary resistors.

The color code

Some resistors have color bands that indicate their values and tolerances. You'll see three, four, or five bands around carbon-composition resistors and film resistors. Other units are large enough so that the values can be printed on them in ordinary numerals.

On resistors with *axial leads*, the bands (first, second, third, fourth, fifth) are arranged as shown in Fig. 6-12A. On resistors with *radial leads*, the bands are arranged

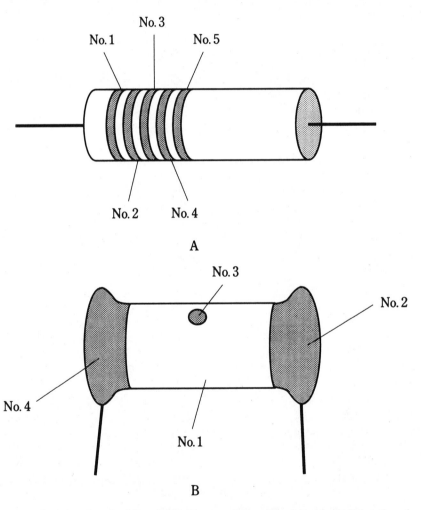

6-12   At A, location of color-code bands on a resistor with axial leads. At B, location of color codings on a resistor having radial leads.

as shown in Fig. 6-12B. The first two bands represent numbers 0 through 9; the third band represents a multiplier of 10 to some power. For the moment, don't worry about the fourth and fifth bands. Refer to Table 6-1.

Table 6-1. Resistor color code.

Color of band	Numeral (Bands no. 1 and 2)	Multiplier (Band no. 3)
Black	0	1
Brown	1	10
Red	2	100
Orange	3	1K
Yellow	4	10K
Green	5	100K
Blue	6	1M
Violet	7	10M
Gray	8	100M
White	9	1000M

See text for discussion of bands no. 4 and 5.

Suppose you find a resistor whose first three bands are yellow, violet, and red, in that order. Then the resistance is 4,700 Ω or 4.7 KΩ. Read yellow = 4, violet = 7, red = x 100.

As another example, suppose you stick your hand in a bag and pull out a unit with bands of blue, gray, orange. Refer to Table 6-1 and determine blue = 6, gray = 8, orange = x 1000. Therefore, the value is 68,000 Ω = 68 KΩ.

After a few hundred real-life experiences with this color code, you'll have it memorized. If you aren't going to be using resistors that often, you can always keep a copy of Table 6-1 handy and use it when you need it.

The fourth band, if there is one, indicates tolerance. If it's silver, it means the resistor is rated at plus or minus 10 percent. If it's gold, the resistor is rated at plus or minus 5 percent. If there is no fourth band, the resistor is rated at plus or minus 20 percent.

The fifth band, if there is one, indicates the percentage that the value might change in 1,000 hours of use. A brown band indicates a maximum change of 1 percent of the rated value. A red band indicates 0.1 percent; an orange band indicates 0.01 percent; a yellow band indicates 0.001 percent. If there is no fifth band, it means that the resistor might deviate by more than 1 percent of the rated value after 1,000 hours of use.

A good engineer always tests a resistor with an ohmmeter before installing it. If the unit happens to be labeled wrong, it's easy to catch while assembling a complex electronic circuit. But once the circuit is all together, and it won't work because some resistor is mislabeled (and this happens), it's a gigantic pain to find the problem.

Quiz

Refer to the text in this chapter if necessary. A good score is at least 18 correct. Answers are in the back of the book.

1. Biasing in an amplifier circuit:
   - A. Keeps it from oscillating.
   - B. Matches it to other amplifier stages in a chain.
   - C. Can be done using voltage dividers.
   - D. Maximizes current flow.

2. A transistor can be protected from needless overheating by:
   - A. Current-limiting resistors.
   - B. Bleeder resistors.
   - C. Maximizing the driving power.
   - D. Shorting out the power supply when the circuit is off.

3. Bleeder resistors:
   - A. Are connected across the capacitor in a power supply.
   - B. Keep a transistor from drawing too much current.
   - C. Prevent an amplifier from being overdriven.
   - D. Optimize the efficiency of an amplifier.

4. Carbon-composition resistors:
   - A. Can handle lots of power.
   - B. Have capacitance or inductance along with resistance.
   - C. Are comparatively nonreactive.
   - D. Work better for ac than for dc.

5. The best place to use a wirewound resistor is:
   - A. In a radio-frequency amplifier.
   - B. When the resistor doesn't dissipate much power.
   - C. In a high-power, radio-frequency circuit.
   - D. In a high-power, direct-current circuit.

6. A metal-film resistor:
   - A. Is made using solid carbon/phenolic paste.
   - B. Has less reactance than a wirewound type.
   - C. Can dissipate large amounts of power.
   - D. Has considerable inductance.

7. A meter-sensitivity control in a test instrument would probably be:
   - A. A set of switchable, fixed resistors.

B. A linear-taper potentiometer.

C. A logarithmic-taper potentiometer.

D. A wirewound resistor.

8. A volume control in a stereo compact-disc player would probably be:

A. A set of switchable, fixed resistors.

B. A linear-taper potentiometer.

C. A logarithmic-taper potentiometer.

D. A wirewound resistor.

9. If a sound triples in actual power level, approximately what is the decibel increase?

A. 3 dB.

B. 5 dB.

C. 6 dB.

D. 9 dB.

10. Suppose a sound changes in volume by −13 dB. If the original sound power is 1 W, what is the final sound power?

A. 13 W.

B. 77 mW.

C. 50 mW.

D. There is not enough information to tell.

11. The sound from a transistor radio is at a level of 50 dB. How many times the threshold of hearing is this, in terms of actual sound power?

A. 50.

B. 169.

C. 5,000.

D. 100,000.

12. An advantage of a rheostat over a potentiometer is that:

A. A rheostat can handle higher frequencies.

B. A rheostat is more precise.

C. A rheostat can handle more current.

D. A rheostat works better with dc.

13. A resistor is specified as having a value of 68 Ω, but is measured with an ohmmeter as 63 Ω. The value is off by:

A. 7.4 percent.

B. 7.9 percent.

C. 5 percent.

D. 10 percent.

14. Suppose a resistor is rated at 3.3 KΩ, plus or minus 5 percent. This means it can be expected to have a value between:

   A. 2,970 and 3,630 Ω.

   B. 3,295 and 3,305 Ω.

   C. 3,135 and 3,465 Ω.

   D. 2.8 KΩ and 3.8 KΩ.

15. A package of resistors is rated at 56 Ω, plus or minus 10 percent. You test them with an ohmmeter. Which of the following values indicates a reject?

   A. 50.0 Ω.

   B. 53.0 Ω.

   C. 59.7 Ω.

   D. 61.1 Ω.

16. A resistor has a value of 680 Ω, and you expect it will have to draw 1 mA maximum continuous current. What power rating is best for this application?

   A. 1/4 W.

   B. 1/2 W.

   C. 1 W.

   D. 2 W.

17. Suppose a 1-KΩ resistor will dissipate 1.05 W, and you have many 1-W resistors of all common values. If there's room for 20-percent resistance error, the cheapest solution is to use:

   A. Four 1 KΩ, 1-W resistors in series-parallel.

   B. Two 2.2 KΩ, 1-W resistors in parallel.

   C. Three 3.3 KΩ, 1-W resistors in parallel.

   D. One 1 KΩ, 1-W resistor, since manufacturers allow for a 10-percent margin of safety.

18. Red, red, red, gold indicates a resistance of:

   A. 22 Ω.

   B. 220 Ω.

   C. 2.2 KΩ.

   D. 22 KΩ.

19. The actual resistance of the above unit can be expected to vary by how much above or below the specified value?

   A. 11 Ω.

   B. 110 Ω.

   C. 22 Ω.

   D. 220 Ω.

20. A resistor has three bands: gray, red, yellow. This unit can be expected to have a value within approximately what range?

    A. 660 KΩ to 980 KΩ.

    B. 740 KΩ to 900 KΩ.

    C. 7.4 KΩ to 9.0 KΩ.

    D. The manufacturer does not make any claim.

7
CHAPTER

Cells and batteries

ONE OF THE MOST COMMON AND MOST VERSATILE SOURCES OF DC IS THE *CELL*. The term *cell* means self-contained compartment, and it can refer to any of various different things in (and out of) science. In electricity and electronics, a cell is a unit source of dc energy. There are dozens of different types of electrical cells.

When two or more cells are connected in series, the result is known as a *battery*.

Kinetic and potential energy

Energy can exist in either of two main forms. *Kinetic energy* is the kind you probably think of right away when you imagine energy. A person running, a car moving down a freeway, a speeding aircraft, a chamber of superheated gas—all these things are visible manifestations of kinetic energy, or energy in action. The dissipation of electrical power, over time, is a form of kinetic energy too.

Potential energy is not as vividly apparent. When you raise a block of concrete into the air, you are creating potential energy. You remember the units called *foot pounds*, the best way to measure such energy, from school physics classes. If you raise a one-pound weight a foot, it gains one foot pound of potential energy. If you raise it 100 feet, it gains 100 foot pounds. If you raise a 100-pound weight 100 feet, it will gain 100×100, or 10,000, foot pounds of potential energy. This energy becomes spectacularly evident if you happen to drop a 100-pound weight from a tenth-story window. (But don't!)

Electrochemical energy

In electricity, one important form of potential energy exists in the atoms and molecules of some chemicals under special conditions.

Early in the history of electrical science, laboratory physicists found that when metals came into contact with certain chemical solutions, voltages appeared between the pieces of metal. These were the first *electrochemical cells*.

A piece of lead and a piece of lead dioxide immersed in an acid solution (Fig. 7-1) will show a persistent voltage. This can be detected by connecting a galvanometer between the pieces of metal. A resistor of about 1,000 ohms should *always* be used in series with the galvanometer in experiments of this kind; connecting the galvanometer directly will cause too much current to flow, possibly damaging the galvanometer and causing the acid to "boil."

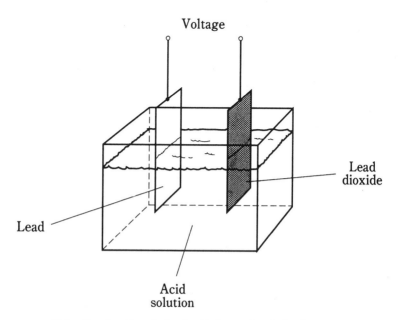

7-1   Construction of a lead-acid electrochemical cell.

The chemicals and the metal have an inherent ability to produce a constant exchange of charge carriers. If the galvanometer and resistor are left hooked up between the two pieces of metal for a long time, the current will gradually decrease, and the electrodes will become coated. The acid will change, also. The *chemical energy*, a form of potential energy in the acid, will run out. All of the potential energy in the acid will have been turned into kinetic electrical energy as current in the wire and galvanometer. In turn, this current will have heated the resistor (another form of kinetic energy), and escaped into the air and into space.

Primary and secondary cells

Some electrical cells, once their potential (chemical) energy has all been changed to electricity and used up, must be thrown away. They are no good anymore. These are called *primary cells*.

Other kinds of cells, like the lead-and-acid unit depicted above, can get their chemical energy back again. Such a cell is a *secondary cell.*

Primary cells include the ones you usually put in a flashlight, in a transistor radio, and in various other consumer devices. They use dry electrolyte pastes along with metal electrodes. They go by names such as *dry cell, zinc-carbon cell, alkaline cell,* and others. Go into a department store and find the panel of batteries, and you'll see various sizes and types of primary cells, such as AAA batteries, D batteries, camera batteries, and watch batteries. You should know by now that these things are cells, not true batteries. This is a good example of a misnomer that has gotten so widespread that store clerks might look at you funny if you ask for a couple of cells. You'll also see real batteries, such as the little 9-V transistor batteries and the large 6-V lantern batteries.

Secondary cells can also be found increasingly in consumer stores. *Nickel-cadmium (Ni-Cd or NICAD)* cells are probably the most common. They're available in the some of the same sizes as nonrechargeable dry cells. The most common sizes are AA, C, and D. These cost several times as much as ordinary dry cells, and a charging unit also costs a few dollars. But if you take care of them, these rechargeable cells can be used hundreds of times and will pay for themselves several times over if you use a lot of "batteries" in your everyday life.

The battery in your car is made from secondary cells connected in series. These cells recharge from the alternator or from an outside charging unit. This battery has cells like the one in Fig. 7-1. It is extremely dangerous to short-circuit the terminals of such a battery, because the acid (sulfuric acid) can "boil" out and burn your skin and eyes.

An important note is worth making here: *Never* short-circuit any cell or battery, because it might burst or explode.

The Weston standard cell

Most electrochemical cells produce about 1.2 V to 1.8 V of electric potential. Different types vary slightly. A mercury cell has a voltage that is a little bit less than that of a zinc-carbon or alkaline cell. The voltage of a cell can also be affected by variables in the manufacturing process. Normally, this is not significant. Most consumer type dry cells can be assumed to produce 1.5 Vdc.

There are certain types of cells whose voltages are predictable and exact. These are called *standard cells.* One example of a standard cell is the *Weston cell.* It produces 1.018 V at room temperature. This cell uses a solution of cadmium sulfate. The positive electrode is made from mercury sulfate, and the negative electrode is made using mercury and cadmium. The whole device is set up in a container as shown in Fig. 7-2.

When properly constructed and used at room temperature, the voltage of the Weston standard cell is always the same, and this allows it to be used as a dc voltage standard. There are other kinds of standard cells, but the Weston cell is the most common.

Storage capacity

Recall that the unit of energy is the watt hour (Wh) or the kilowatt hour (kWh). Any electrochemical cell or battery has a certain amount of electrical energy that can be gotten

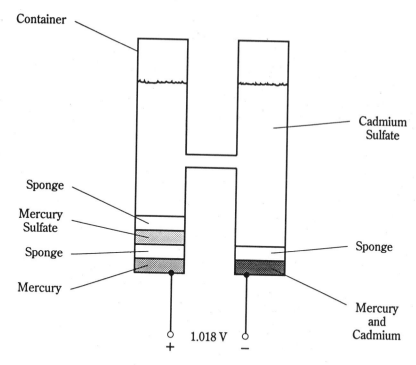

7-2 A Weston standard cell.

from it, and this can be specified in watt hours or kilowatt hours. More often though it's given in *ampere hours (Ah)*.

A battery with a rating of 2 Ah can provide 2 A for an hour, or 1 A for 2 hours. Or it can provide 100 mA for 20 hours. Within reason, the product of the current in amperes, and the use time in hours, can be as much as, but not more than 2. The limitations are the *shelf life* at one extreme, and the *maximum safe current* at the other. Shelf life is the length of time the battery will last if it is sitting on a shelf without being used; this might be years. The maximum safe current is represented by the lowest load resistance (heaviest load) that the battery can work into before its voltage drops because of its own *internal resistance.* A battery is never used with loads that are too heavy, because it can't supply the necessary current anyway, and it might "boil", burst, or blow up.

Small cells have storage capacity of a few milliampere hours (mAh) up to 100 or 200 mAh. Medium-sized cells might supply 500 mAh or 1 Ah. Large automotive or truck batteries can provide upwards of 50 Ah. The energy capacity in watt hours is the ampere-hour capacity times the battery voltage.

When an ideal cell or battery is used, it delivers a fairly constant current for awhile, and then the current starts to fall off (Fig. 7-3). Some types of cells and batteries approach this ideal behavior, called a *flat discharge curve*, and others have current that declines gradually, almost right from the start. When the current that a battery can provide has tailed off to about half of its initial value, the cell or battery is said to be "weak." At this time, it should be replaced. If it's allowed to run all the way out, until the current actually goes to zero, the cell or battery is "dead." Some rechargeable cells and batteries, especially the nickel-cadmium type, should never be used until the current goes down to zero, because this can ruin them.

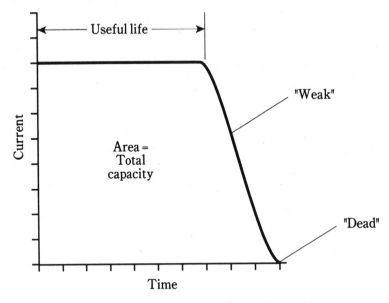

7-3 A flat discharge curve. This is considered ideal.

The area under the curve in Fig. 7-3 is the total capacity of the cell or battery in ampere hours. This area is always pretty much the same for any particular type and size of cell or battery, regardless of the amount of current drawn while it's in use.

Common dime-store cells and batteries

The cells you see in grocery, department, drug, and hardware stores that are popular for use in household convenience items like flashlights and transistor radios are usually of the *zinc-carbon* or *alkaline* variety. These provide 1.5 V and are available in sizes known as AAA (very small), AA (small), C (medium large), and D (large). You have probably seen all of these sizes hanging in packages on a pegboard. Batteries made from these cells are usually 6 V or 9 V.

One type of cell and battery that has become available recently, the nickel-cadmium rechargeable type, is discussed in some detail a bit later in this chapter.

Zinc-carbon cells

These cells have a fairly long shelf life. A cylindrical zinc-carbon cutaway diagram is shown at Fig. 7-4. The zinc forms the case and is the negative electrode. A carbon rod serves as the positive electrode. The electrolyte is a paste of manganese dioxide and carbon. Zinc-carbon cells are inexpensive and are good at moderate temperatures, and in applications where the current drain is moderate to high. They are not very good in extreme cold.

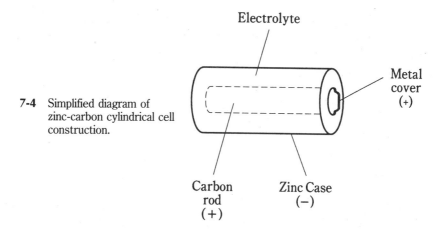

7-4 Simplified diagram of zinc-carbon cylindrical cell construction.

Alkaline cells

The alkaline cell uses granular zinc for the negative electrode, potassium hydroxide as the electrolyte, and a device called a *polarizer* as the positive electrode. The geometry of construction is similar to that of the zinc-carbon cell. An alkaline cell can work at lower temperatures than a zinc-carbon cell. It also lasts longer in most electronic devices, and is therefore preferred for use in transistor radios, calculators, and portable cassette players. Its shelf life is much longer than that of a zinc-carbon cell. As you might expect, it costs more.

Transistor batteries

Those little 9-V things with the funny connectors on top consist of six tiny zinc-carbon or alkaline cells in series. Each of the six cells supplies 1.5 V.

Even though these batteries have more voltage than individual cells, the total energy available from them is less than that from a C cell or D cell. This is because the electrical energy that can be gotten from a cell or battery is directly proportional to the amount of chemical energy stored in it, and this, in turn, is a direct function of the *volume* (size) of the cell. C or D size cells have more volume than a transistor battery, and therefore contain more stored energy, assuming the same chemical type.

The ampere-hour capacity of a transistor battery is very small. But transistor radios don't need much current. These batteries are also used in other low-current electronic devices, such as remote-control garage-door openers, TV channel changers, remote video-cassette recorder (VCR) controls, and electronic calculators.

Lantern batteries

These get their name from the fact that they find much of their use in lanterns. These are the batteries with a good, solid mass so they last a long time. One type has spring contacts on the top. The other type has thumbscrew terminals. Besides keeping a lantern lit for awhile, these big batteries, usually rated at 6 V and consisting of four good-size zinc-carbon or alkaline cells, can provide enough energy to operate a low-power radio transceiver. Two of them in series make a 12-V battery that can power a 5-W Citizen

Band (CB) or ham radio. They're also good for scanner radio receivers in portable locations, for camping lamps, and for other medium-power needs.

Miniature cells and batteries

In recent years, cells and batteries—especially cells—have become available in many different sizes and shapes besides the old cylindrical cells, transistor batteries and lantern batteries. These are used in watches, cameras, and other microminiature electronic gizmos.

Silver-oxide types

Silver-oxide cells are usually made into a button-like shapes, and can fit inside even a small wristwatch. They come in various sizes and thicknesses, all with similar appearances. They supply 1.5 V, and offer excellent energy storage for the weight. They also have a flat discharge curve, like the one shown in the graph of Fig. 7-3. The previously described zinc-carbon and alkaline cells and batteries have a current output that declines with time in a steady fashion, as shown in Fig. 7- 5. This is known as a *declining discharge curve.*

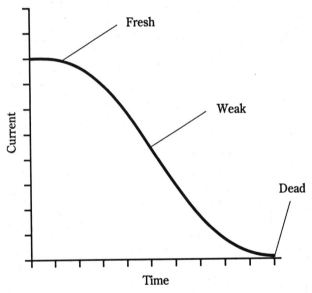

7-5   A declining discharge curve.

Silver-oxide cells can be stacked to make batteries. Several of these miniature cells, one on top of the other, might provide 6 V or 9 V for a transistor radio or other light-duty electronic device. The resulting battery is about the size of an AAA cylindrical cell.

Mercury types

Mercury cells, also called *mercuric oxide* cells, have advantages similar to silver-oxide cells. They are manufactured in the same general form. The main difference, often not of significance, is a somewhat lower voltage per cell: 1.35 V. If six of these cells are stacked to make a battery, the resulting voltage will be about 8.1 V rather than 9 V. One additional cell can be added to the stack, yielding about 9.45 V.

There has been some decrease in the popularity of mercury cells and batteries in recent years. This is because of the fact that mercury is highly toxic. When mercury cells and batteries are dead, they must be discarded. Eventually the mercury, a chemical element, leaks into the soil and ground water. Mercury pollution has become a significant concern in places that might really surprise you.

Lithium types

Lithium cells have become popular since the early eighties. There are several variations in the chemical makeup of these cells; they all contain lithium, a light, highly reactive metal. Lithium cells can be made to supply 1.5 V to 3.5 V, depending on the particular chemistry used. These cells, like their silver-oxide cousins, can be stacked to make batteries.

The first applications of lithium batteries was in memory backup for electronic microcomputers. Lithium cells and batteries have superior shelf life, and they can last for years in very-low-current applications such as memory backup or the powering of a digital liquid-crystal-display (LCD) watch or clock. These cells also provide energy capacity per unit volume that is vastly greater than other types of electrochemical cells.

Lithium cells and batteries are used in low-power devices that must operate for a long time without power-source replacement. Heart pacemakers and security systems are two examples of such applications.

Lead-acid cells and batteries

You've already seen the basic configuration for a lead-acid cell. This has a solution of sulfuric acid, along with a lead electrode (negative) and a lead-dioxide electrode (positive). These batteries are rechargeable.

Automotive batteries are made from sets of lead-acid cells having a free-flowing liquid acid. You cannot tip such a battery on its side, or turn it upside-down, without running the risk of having some of the acid electrolyte get out.

Lead-acid batteries are also available in a construction that uses a semisolid electrolyte. These batteries are popular in consumer electronic devices that require a moderate amount of current. Notebook or laptop computers, and portable video-cassette recorders (VCRs), are the best examples.

A large lead-acid battery, such as the kind in your car, can store several tens of ampere-hours. The smaller ones, like those in notebook computers, have less capacity but more versatility. Their overwhelming advantage is their ability to be used many times at reasonable cost.

Nickel-cadmium cells and batteries

You've probably seen, or at least heard of, *NICAD* cells and batteries. They have become quite common in consumer devices such as those little radios and cassette players you can wear while doing aerobics or just sitting around. (These entertainment units are not too safe for walking or jogging in traffic. And never wear them while riding a bicycle.) You can buy two sets of cells and switch them every couple of hours of use, charging one set while using the other. Plug-in charger units cost only a few dollars.

Types of NICAD cells

Nickel-cadmium cells are made in several types. *Cylindrical cells* are the standard cells; they look like dry cells. *Button cells* are those little things that are used in cameras, watches, memory backup applications, and other places where miniaturization is important. *Flooded cells* are used in heavy-duty applications and can have a charge capacity of as much as 1,000 Ah. *Spacecraft cells* are made in packages that can withstand the vacuum and temperature changes of a spaceborne environment.

Uses of NICADs

There are other uses for NICADs besides in portable entertainment equipment. Most orbiting satellites are in darkness half the time, and in sunlight half the time. Solar panels can be used while the satellite in sunlight, but during the times that the earth eclipses the sun, batteries are needed to power the electronic equipment on board the satellite. The solar panels can charge a set of NICADs, in addition to powering the satellite, for half of each orbit. The NICADs can provide the power during the dark half of each orbit.

Nickel-cadmium batteries are available in packs of cells. These packs can be plugged into the equipment, and might even form part of the case for a device. An example of this is the battery pack for a handheld ham radio tranceiver. Two of these packs can be bought, and they can be used alternately, with one installed in the "handie-talkie" (HT) while the other is being charged.

NICAD neuroses

There are some things you need to know about NICAD cells and batteries, in order to get the most out of them.

One rule, already mentioned, is that you should never discharge them all the way until they "die." This can cause the polarity of a cell, or of one or more cells in a battery, to reverse. Once this happens, the cell or battery is ruined.

Another phenomenon, peculiar to this type of cell and battery, is called *memory*. If a NICAD is used over and over, and is discharged to exactly the same extent every time (say, two-thirds of the way), it might start to "go to sleep" at that point in its discharge cycle. This is uncommon; lab scientists have trouble forcing it to occur so they can study it. But when it does happen, it can give the illusion that the cell or battery has lost some of its storage capacity. Memory problems can be solved. Use the cell or battery almost all the way up, and then fully charge it. Repeat the process, and the memory will be "erased."

NICADS do best using wall chargers that take several hours to fully replenish the cells or batteries. There are *high-rate* or *quick* chargers available, but these can sometimes force too much current through a NICAD. It's best if the charger is made especially for the cell or battery type being charged. An electronics dealer, such as the manager at a Radio Shack store, should be able to tell you which chargers are best for which cells and batteries.

Photovoltaic cells and batteries

The *photovoltaic* cell is completely different from any of the electrochemical cells. It's also known as a *solar* cell. This device converts visible light, infrared, and/or ultraviolet directly into electric current.

Solar panels

Several, or many, photovoltaic cells can be combined in series-parallel to make a *solar panel*. An example is shown in Fig. 7-6. Although this shows a 3×3 series-parallel array, the matrix does not have to be symmetrical. And it's often very large. It might consist of, say, 50 parallel sets of 20 series-connected cells. The series scheme boosts the voltage to the desired level, and the parallel scheme increases the current-delivering ability of the panel. It's not unusual to see hundreds of solar cells combined in this way to make a large panel.

Construction and performance

The construction of a photovoltaic cell is shown in Fig. 7-7. The device is a flat semiconductor *P-N junction*, and the assembly is made transparent so that light can fall directly on the *P-type* silicon. The metal ribbing, forming the positive electrode, is interconnected by means of tiny wires. The negative electrode is a metal backing, placed in contact with the *N-type* silicon. Most solar cells provide about 0.5 V. If there is very low current demand, dim light will result in the full output voltage from a solar cell. As the current demand increases, brighter light is needed to produce the full output voltage. There is a maximum limit to the current that can be provided from a solar cell, no matter how bright the light. This limit is increased by connecting solar cells in parallel.

Practical applications

Solar cells have become cheaper and more efficient in recent years, as researchers have looked to them as a long-term alternative energy source. Solar panels are used in satellites. They can be used in conjunction with rechargeable batteries, such as the lead-acid or nickel-cadmium types, to provide power independent of the commercial utilities.

A completely independent solar/battery power system is called a *stand-alone* system. It generally uses large solar panels, large-capacity lead-acid or NICAD batteries, power converters to convert the dc into ac, and a rather sophisticated charging circuit. These systems are best suited to environments where there is sunshine a high percentage of the time.

Solar cells, either alone or supplemented with rechargeable batteries, can be connected into a home electric system in an *interactive* arrangement with the electric utilities. When the solar power system can't provide for the needs of the household all by itself, the utility company can take up the slack. Conversely, when the solar power system supplies more than enough for the needs of the home, the utility company can buy the excess.

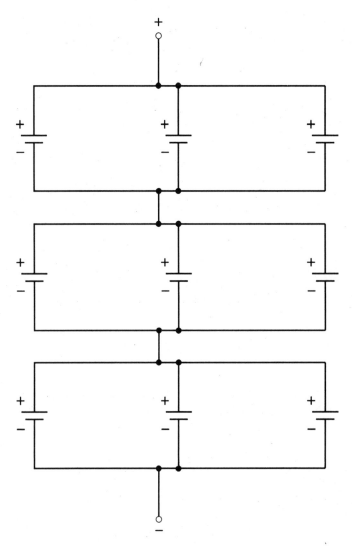

7-6   Connection of cells in series-parallel.

How large of a battery?

You might get the idea that you can connect hundreds, or even thousands, of cells in series and obtain batteries with fantastically high EMFs. Why not put 1,000 zinc-carbon cells in series, for example, and get 1.5 kV? Or put 2,500 solar cells in series and get 1.25 kV? Maybe it's possible to put a billion solar cells in series, out in some vast sun-scorched desert wasteland, and use the resulting 500 MV (megavolts) to feed the greatest high-tension power line the world has ever seen.

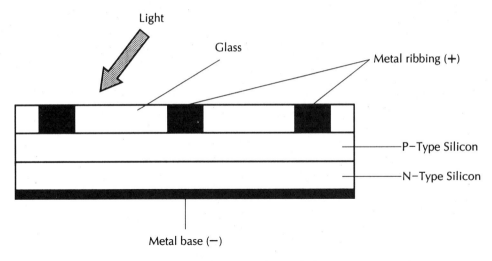

7-7 Cross-sectional view of silicon photovoltaic (solar) cell construction.

There are several reasons why these schemes aren't good ideas. First, high voltages for practical purposes can be generated cheaply and efficiently by power converters that work from 117-V or 234-V utility mains. Second, it would be difficult to maintain a battery of thousands, millions or billions of cells in series. Imagine a cell holder with 1,000 sets of contacts. And not one of them can open up, lest the whole battery become useless, because all the cells must be in series. (Solar panels, at least, can be permanently wired together. Not so with batteries that must often be replaced.) And finally, the *internal resistances* of the cells would add up and limit the current, as well as the output voltage, that could be derived by connecting so many cells in series. This is not so much of a problem with series-parallel combinations, as in solar panels, as long as the voltages are reasonable. But it is a big factor if all the cells are in series, with the intent of getting a huge voltage. This effect will occur with any kind of cell, whether electrochemical or photovoltaic.

In the days of the Second World War, portable two-way radios were built using vacuum tubes. These were powered by batteries supplying 103.5 V. The batteries were several inches long and about an inch in diameter. They were made by stacking many little zinc-carbon cells on top of each other, and enclosing the whole assembly in a single case. You could get a nasty jolt from one of those things. They were downright dangerous! A fresh 103.5-V battery would light up a 15-W household incandescent bulb to almost full brilliance. But the 117-V outlet would work better, and for a lot longer.

Nowadays, handheld radio transceivers will work from NICAD battery packs or batteries of ordinary dry cells, providing 6 V, 9 V, or 12 V. Even the biggest power transistors rarely use higher voltages. Automotive or truck batteries can produce more than enough power for almost any mobile or portable communications system. And if a really substantial setup is desired, gasoline-powered generators are available, and they will supply the needed energy at far less cost than batteries. There's just no use for a megabattery of a thousand, a million or a zillion volts.

Quiz

Refer to the text in this chapter if necessary. A good score is 18 correct. Answers are in the back of the book.

1. The chemical energy in a battery or cell:
    A. Is a form of kinetic energy.
    B. Cannot be replenished once it is gone.
    C. Changes to kinetic energy when the cell is used.
    D. Is caused by electric current.

2. A cell that cannot be recharged is:
    A. A dry cell.
    B. A wet cell.
    C. A primary cell.
    D. A secondary cell.

3. A Weston cell is generally used:
    A. As a current reference source.
    B. As a voltage reference source.
    C. As a power reference source.
    D. As an energy reference source.

4. The voltage in a battery is:
    A. Less than the voltage in a cell of the same kind.
    B. The same as the voltage in a cell of the same kind.
    C. More than the voltage in a cell of the same kind.
    D. Always a multiple of 1.018 V.

5. A direct short-circuit of a battery can cause:
    A. An increase in its voltage.
    B. No harm other than a rapid discharge of its energy.
    C. The current to drop to zero.
    D. An explosion.

6. A cell of 1.5 V supplies 100 mA for seven hours and twenty minutes, and then it is replaced. It has supplied:
    A. 7.33 Ah.
    B. 733 mAh.
    C. 7.33 Wh.
    D. 733 mWh.

7. A 12-V auto battery is rated at 36 Ah. If a 100-W, 12-Vdc bulb is connected across this battery, about how long will the bulb stay lit, if the battery has been fully charged?

    A. 4 hours and 20 minutes.

    B. 432 hours.

    C. 3.6 hours.

    D. 21.6 minutes.

8. Alkaline cells:

    A. Are cheaper than zinc-carbon cells.

    B. Are generally better in radios than zinc-carbon cells.

    C. Have higher voltages than zinc-carbon cells.

    D. Have shorter shelf lives than zinc-carbon cells.

9. The energy in a cell or battery depends mainly on:

    A. Its physical size.

    B. The current drawn from it.

    C. Its voltage.

    D. All of the above.

10. In which of the following places would a "lantern" battery most likely be found?

    A. A heart pacemaker.

    B. An electronic calculator.

    C. An LCD wall clock.

    D. A two-way portable radio.

11. In which of the following places would a transistor battery be the best power-source choice?

    A. A heart pacemaker.

    B. An electronic calculator.

    C. An LCD wristwatch.

    D. A two-way portable radio.

12. In which of the following places would you most likely choose a lithium battery?

    A. A microcomputer memory backup.

    B. A two-way portable radio.

    C. A portable audio cassette player.

    D. A rechargeable flashlight.

13. Where would you most likely find a lead-acid battery?

    A. In a portable audio cassette player.

    B. In a portable video camera/recorder.

    C. In an LCD wall clock.

    D. In a flashlight.

14. A cell or battery that keeps up a constant current-delivering capability almost until it dies is said to have:

    A. A large ampere-hour rating.

    B. Excellent energy capacity.

    C. A flat discharge curve.

    D. Good energy storage per unit volume.

15. Where might you find a NICAD battery?

    A. In a satellite.

    B. In a portable cassette player.

    C. In a handheld radio transceiver.

    D. In more than one of the above.

16. A disadvantage of mercury cells and batteries is that:

    A. They don't last as long as other types.

    B. They have a flat discharge curve.

    C. They pollute the environment.

    D. They need to be recharged often.

17. Which kind of battery should never be used until it "dies"?

    A. Silver-oxide.

    B. Lead-acid.

    C. Nickel-cadmium.

    D. Mercury.

18. The current from a solar panel is increased by:

    A. Connecting solar cells in series.

    B. Using NICAD cells in series with the solar cells.

    C. Connecting solar cells in parallel.

    D. Using lead-acid cells in series with the solar cells.

19. An interactive solar power system:

    A. Allows a homeowner to sell power to the utility.

    B. Lets the batteries recharge at night.

    C. Powers lights but not electronic devices.

    D. Is totally independent from the utility.

20. One reason why it is impractical to make an extremely high-voltage battery of cells is that:

    A. There's a danger of electric shock.

    B. It is impossible to get more than 103.5 V with electrochemical cells.

    C. The battery would weigh too much.

    D. There isn't any real need for such a thing.

8
CHAPTER

Magnetism

THE STUDY OF MAGNETISM IS A SCIENCE IN ITSELF. ELECTRIC AND MAGNETIC phenomena interact; a detailed study of magnetism and electromagnetism could easily fill a book. Magnetism was mentioned briefly near the end of chapter 2. Here, the subject is examined more closely. The intent is to get you familiar with the general concepts of magnetism, insofar as it is important for a basic understanding of electricity and electronics.

The geomagnetic field

The earth has a core made up largely of iron, heated to the extent that some of it is liquid. As the earth rotates, the iron flows in complex ways. It is thought that this flow is responsible for the huge magnetic field that surrounds the earth. The sun has a magnetic field, as does the whole Milky Way galaxy. These fields might have originally magnetized the earth.

Geomagnetic poles and axis

The *geomagnetic field*, as it is called, has poles, just as a bar magnet does. These poles are near, but not at, the geographic poles. The north geomagnetic pole is located in the frozen island region of northern Canada. The south geomagnetic pole is near Antarctica. The *geomagnetic axis* is somewhat tilted relative to the axis on which the earth rotates. Not only this, but it does not exactly run through the center of the earth. It's like an apple core that's off center.

The solar wind

The geomagnetic field would be symmetrical around the earth, but charged particles from the sun, constantly streaming outward through the solar system, distort the lines of flux.

134

This *solar wind* literally "blows" the geomagnetic field out of shape, as shown in Fig. 8-1. At and near the earth's surface, the lines of flux are not affected very much, and the geomagnetic field is nearly symmetrical.

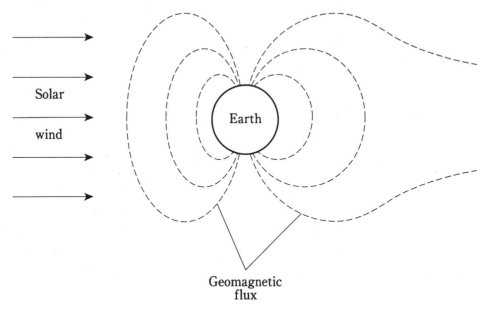

8-1   The geomagnetic field is distorted by the solar wind.

The magnetic compass

The presence of the geomagnetic field was first noticed in ancient times. Some rocks, called *lodestones*, when hung by strings would always orient themselves a certain way. This was correctly attributed to the presence of a "force" in the air. Even though it was some time before the details were fully understood, this effect was put to use by early seafarers and land explorers. Today, a *magnetic compass* can still be a valuable navigation aid, used by mariners, backpackers, and others who travel far from familiar landmarks.

The geomagnetic field and the magnetic field around a compass needle interact, so that a force is exerted on the little magnet inside the compass. This force works not only in a horizontal plane (parallel to the earth's surface), but vertically at most latitudes. The vertical component is zero only at the *geomagnetic equator*, a line running around the globe equidistant from both *geomagnetic poles*. As the *geomagnetic latitude* increases, either towards the north or the south geomagnetic pole, the magnetic force pulls up and down on the compass needle more and more. You have probably noticed this when you hold a compass. One end of the needle seems to insist on touching the compass face, while the other end tilts up toward the glass. The needle tries to align itself parallel to the magnetic *lines of flux*.

Magnetic force

Magnets "stick" to some metals. Iron, nickel, and alloys containing either or both of these elements, are known as *ferromagnetic* materials. When a magnet is brought near a piece of ferromagnetic material, the atoms in the material become lined up, so that the metal is temporarily magnetized. This produces a *magnetic force* between the atoms of the ferromagnetic substance and those in the magnet.

If a magnet is brought near another magnet, the force is even stronger. Not only is it more powerful, but it can be repulsive or attractive, depending on the way the magnets are turned. The force gets stronger as the magnets are brought near each other.

Some magnets are so strong that no human being can ever pull them apart if they get "stuck" together, and no person can bring them all the way together against their mutual repulsive force. This is especially true of *electromagnets*, discussed later in this chapter. The tremendous forces available are of use in industry. A huge electromagnet can be used to carry heavy pieces of scrap iron from place to place. Other electromagnets can provide sufficient repulsion to suspend one object above another. This is called *magnetic levitation* and is the basis for some low-friction, high-speed trains now being developed.

Electric charge in motion

Whenever the atoms in a ferromagnetic material are aligned, a *magnetic field* exists. A magnetic field can also be caused by the motion of electric charge carriers, either in a wire or in free space.

The magnetic field around a permanent magnet arises from the same cause as the field around a wire that carries an electric current. The responsible factor in either case is the motion of electrically charged particles. In a wire, the electrons move along the conductor, being passed from atom to atom. In a permanent magnet, the movement of orbiting electrons occurs in such a manner that a sort of current is produced just by the way they move within individual atoms.

Magnetic fields can be produced by the motion of charged particles through space. The sun is constantly ejecting protons and helium nuclei. These particles carry a positive electric charge. Because of this, they have magnetic fields. When these fields interact with the geomagnetic field, the particles are forced to change direction. Charged particles from the sun are *accelerated* toward the geomagnetic poles. If there is a *solar flare*, the sun ejects far more charged particles than normal. When these arrive at the geomagnetic poles, the result can actually disrupt the geomagnetic field. Then there is a *geomagnetic storm*. This causes changes in the earth's ionosphere, affecting long-distance radio communications at certain frequencies. If the fluctuations are intense enough, even wire communications and electric power transmission can be interfered with. Microwave transmissions are generally immune to the effects of a geomagnetic storm, although the wire links can be affected. Aurora (northern or southern lights) are frequently observed at night during these events.

Flux lines

Perhaps you have seen the experiment in which iron filings are placed on a sheet of paper, and then a magnet is placed underneath the paper. The filings arrange themselves in a pattern that shows, roughly, the "shape" of the magnetic field in the vicinity of the magnet. A bar magnet has a field with a characteristic form (Fig. 8-2).

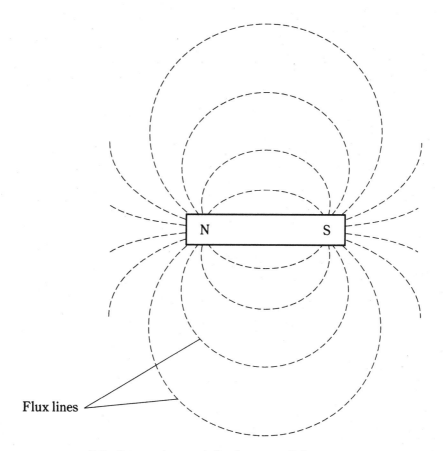

8-2 Pattern of magnetic flux lines around a bar magnet.

Another experiment involves passing a current-carrying wire through the paper at a right angle, as shown in Fig. 8-3. The iron filings will be grouped along circles centered at the point where the wire passes through the paper.

Physicists consider magnetic fields to be comprised of *flux lines*. The intensity of the field is determined according to the number of flux lines passing through a certain cross section, such as a square centimeter or a square meter. The lines don't really exist as geometric threads in space, or as anything solid, but it is intuitively appealing to imagine them, and the iron filings on the paper really do bunch themselves into lines when there is a magnetic field of sufficient strength to make them move.

Magnetic polarity

A magnetic field has a direction at any given point in space near a current-carrying wire or a permanent magnet. The flux lines run parallel with the direction of the field. A magnetic field is considered to begin at the north magnetic pole, and to terminate at the south magnetic pole. In the case of a permanent magnet, it is obvious where these poles are.

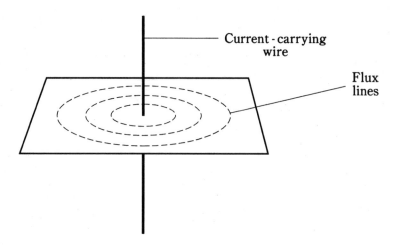

8-3 Pattern of magnetic flux lines around a current-carrying wire.

With a current-carrying wire, the magnetic field just goes around and around endlessly, like a dog chasing its own tail.

A charged electric particle, such as a proton, hovering in space, is a *monopole*, and the electric flux lines around it aren't closed (Fig. 8-4). A positive charge does not have to be mated with a negative charge. The electric flux lines around any stationary, charged particle run outward in all directions for a theoretically infinite distance.

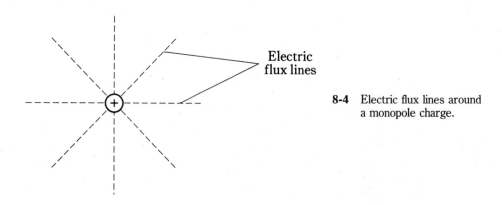

8-4 Electric flux lines around a monopole charge.

But a magnetic field is different. All magnetic flux lines are *closed loops*. With permanent magnets, there is always a starting point (the north pole) and an ending point (the south pole). Around the current-carrying wire, the loops are circles. This can be plainly seen in the experiments with iron filings on paper. Never do magnetic flux lines run off into infinity. Never is a magnetic pole found without an accompanying, opposite pole.

Dipoles and monopoles

A pair of magnetic poles is called a *dipole*. A lone pole, like the positive pole in a proton, is called a *monopole*.

Magnetic monopoles do not ordinarily exist in nature. If they could somehow be conjured up, all sorts of fascinating things might happen. Scientists are researching this to see if they can create artificial magnetic monopoles.

At first you might think that the magnetic field around a current-carrying wire is caused by a monopole, or that there aren't any poles at all, because the concentric circles don't actually converge anywhere. But in fact, you can think of any half-plane, with the edge along the line of the wire, as a magnetic dipole, and the lines of flux as going around once from the "north face" of the half-plane to the "south face."

The lines of flux in a magnetic field always connect the two poles. Some flux lines are straight; some are curved. The greatest flux density, or field strength, around a bar magnet is near the poles, where the lines converge. Around a current- carrying wire, the greatest field strength is near the wire.

Magnetic field strength

The overall magnitude of a magnetic field is measured in units called *webers*, abbreviated Wb. A smaller unit, the *maxwell (Mx)*, is sometimes used if a magnetic field is very weak. One weber is equivalent to 100,000,000 maxwells. Scientists would use exponential notation and say that one 1 Wb = 10^8 Mx. Conversely, 1 Mx = 0.00000001 Wb = 10^{-8} Wb.

The tesla and the gauss

If you have a certain permanent magnet or electromagnet, you might see its "strength" expressed in terms of webers or maxwells. But usually you'll hear units called *teslas* or *gauss*. These units are expressions of the concentration, or intensity, of the magnetic field within a certain cross section. The *flux density*, or number of lines per square meter or per square centimeter, are more useful expressions for magnetic effects than the overall quantity of magnetism. A flux density of one tesla is equal to one weber per square meter. A flux density of one gauss is equal to one maxwell per square centimeter. It turns out that the gauss is 0.0001, or 10^{-4}, tesla. Conversely, the tesla is 10,000, or 10^4, gauss.

If you are confused by the distinctions between webers and teslas, or between maxwells and gauss, think of a light bulb. A 100-watt lamp might emit a total of 20 watts of visible-light power. If you enclose the bulb completely, then 20 W will fall on the interior walls of the chamber, no matter how large or small the chamber might be. But this is not a very useful notion of the brightness of the light. You know that a single 100-watt light bulb gives plenty of light for a small walk-in closet, but it is nowhere near adequate to illuminate a gymnasium. The important consideration is the number of watts per unit area. When we say the bulb gives off x watts or milliwatts of light, it's like saying a magnet has y webers or maxwells of magnetism. When we say that the bulb provides x watts or milliwatts per square meter, it's analogous to saying that a magnetic field has a flux density of y teslas or gauss.

The ampere-turn and the gilbert

When working with electromagnets, another unit is employed. This is the *ampere-turn (At)*. It is a unit of *magnetomotive force*. A wire, bent into a circle and carrying 1 A of current, will produce 1 At of magnetomotive force. If the wire is bent into a loop having 50 turns, and the current stays the same, the resulting magnetomotive force will be 50 At. If the current is then reduced to 1/50 A or 20 mA, the magnetomotive force will go back down to 1 At.

The *gilbert* is also sometimes used to express magnetomotive force. This unit is equal to 0.796 At. Thus, to get ampere-turns from gilberts, multiply by 0.796; to get gilberts from ampere-turns, multiply by 1.26.

Electromagnets

Any electric current, or movement of charge carriers, produces a magnetic field. This field can become extremely intense in a tightly coiled wire having many turns, and that carries a large electric current. When a ferromagnetic core is placed inside the coil, the magnetic lines of flux are concentrated in the core, and the field strength in and near the core becomes tremendous. This is the principle of an *electromagnet* (Fig. 8-5).

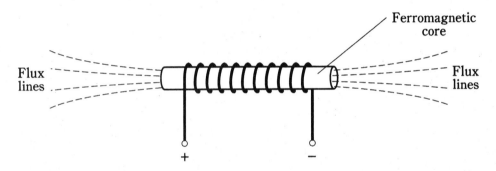

8-5   A ferromagnetic core concentrates the lines of magnetic flux.

Electromagnets are almost always cylindrical in shape. Sometimes the cylinder is long and thin; in other cases it is short and fat. But whatever the ratio of diameter to length for the core, the principle is always the same: the magnetic field produced by the current results in magnetization of the core.

Direct-current types

Electromagnets can use either direct or alternating current. The type with which you are probably familiar is the dc electromagnet.

You can build a dc electromagnet by taking a large bolt, like a stove bolt, and wrapping a few dozen or a few hundred turns of wire around it. These items are available in almost any hardware store. Be sure the bolt is made of ferromagnetic material. (If a permanent magnet "sticks" to the bolt, the bolt is ferromagnetic.) Ideally, the bolt should be at least 3/8 inch in diameter and several inches long. You need to use insulated wire, preferably made of solid, soft copper. "Bell wire" works very well.

Just be sure all the wire turns go in the same direction. A large 6-V lantern battery can provide plenty of dc to work the electromagnet. Never leave the coil connected to the battery for more than a few seconds at a time. And never use a car battery for this experiment! (The acid might boil out.)

Direct-current electromagnets have defined north and south poles, just like permanent magnets. The main difference is that an electromagnet can get much stronger than any permanent magnet. You should see evidence of this if you do the above experiment with a large enough bolt and enough turns of wire.

Aternating-current types

You might get the idea that the electromagnet can be made far stronger if, rather than using a lantern battery for the current source, you plug the wires into a wall outlet. In theory, this is true. In practice, you'll probably blow the fuse or circuit breaker. Do not try this. The electrical circuits in some buildings are not adequately protected and it can create a fire hazard. Also, you can get a lethal shock from the 117-V utility mains.

Some electromagnets use ac, and these magnets will "stick" to ferromagnetic objects. But the polarity of the magnetic field reverses every time the direction of the current reverses. That means there are 120 fluctuations per second, or 60 complete north-to-south-to-north polarity changes (Fig. 8-6) every second. If a permanent magnet, or a dc electromagnet, is brought near either "pole" of an ac electromagnet, there will be no net force. This is because the poles will be alike half the time, and opposite half the time, producing an equal amount of attractive and repulsive force.

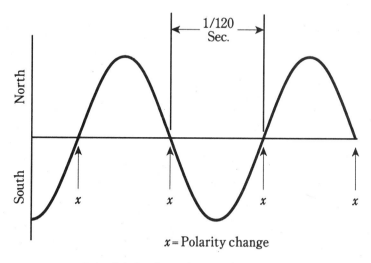

8-6 Polarity change in an ac electromagnet.

For an ac electromagnet to work, the core material must have high *permeability*, but low *retentivity*. These terms will now be discussed.

Permeability

Some substances cause the magnetic lines of flux to get closer together than they are in the air. Some materials can cause the lines of flux to become farther apart than they are in the air.

The first kind of material is *ferromagnetic*, and is of primary importance in magnetism. Ferromagnetic substances are the ones that can be "magnetized." Iron and nickel are examples. Various alloys are even more ferromagnetic than pure iron or pure nickel.

The other kind of material is called *diamagnetic*. Wax, dry wood, bismuth, and silver are substances that actually decrease the magnetic flux density. No diamagnetic material reduces the strength of a magnetic field by anywhere near the factor that ferromagnetic substances can increase it.

Permeability is measured on a scale relative to a vacuum, or free space. Free space is assigned permeability 1. If you have a coil of wire with an air core, and a current is forced through the wire, then the flux in the coil core is at a certain density, just about the same as it would be in a vacuum. Therefore, the permeability of pure air is about equal to 1. If you place an iron core in the coil, the flux density increases by a factor of about 60 to several thousand times. Therefore, the permeability of iron can range from 60 (impure) to as much as 8,000 (highly refined).

If you use certain *permalloys* as the core material in electromagnets, you can increase the flux density, and therefore the local strength of the field, by as much as 1,000,000 times. Such substances thus have permeability as great as 1,000,000.

If for some reason you feel compelled to make an electromagnet that is as weak as possible, you could use dry wood or wax for the core material. But usually, diamagnetic substances are used to keep magnetic objects apart, while minimizing the interaction between them.

Diamagnetic metals have the useful property that they conduct electric current very well, but magnetic current very poorly. They can be used for *electrostatic shielding*, a means of allowing magnetic fields to pass through while blocking electric fields.

Table 8-1 gives the permeability ratings for some common materials.

Retentivity

Certain ferromagnetic materials stay magnetized better than others. When a substance, such as iron, is subjected to a magnetic field as intense as it can handle, say by enclosing it in a wire coil carrying a massive current, there will be some *residual magnetism* left when the current stops flowing in the coil. *Retentivity*, also sometimes called *remanence*, is a measure of how well the substance will "memorize" the magnetism, and thereby become a permanent magnet.

Retentivity is expressed as a percentage. If the flux density in the material is x tesla or gauss when it is subjected to the greatest possible magnetomotive force, and then goes down to y tesla or gauss when the current is removed, the retentivity is equal to $100(y/x)$.

As an example, suppose that a metal rod can be magnetized to 135 gauss when it is enclosed by a coil carrying an electric current. Imagine that this is the maximum possible flux density that the rod can be forced to have. For any substance, there is always such a

Table 8-1. Permeability of some common materials.

Substance	Permeability (approx.)
Aluminum	Slightly more than 1
Bismuth	Slightly less than 1
Cobalt	60–70
Ferrite	100–3000
Free space	1
Iron	60–100
Iron, refined	3000–8000
Nickel	50–60
Permalloy	3000–30,000
Silver	Slightly less than 1
Steel	300–600
Super permalloys	100,000–1,000,000
Wax	Slightly less than 1
Wood, dry	Slightly less than 1

maximum; further increasing the current in the wire will not make the rod any more magnetic. Now suppose that the current is shut off, and 19 gauss remain in the rod. Then the retentivity, B_r, is:

$$B_r = 100(19/135) = 100 \times 0.14 = 14 \text{ percent}$$

Various different substances have good retentivity; these are excellent for making permanent magnets. Other materials have poor retentivity. They might work well as electromagnets, but not as permanent magnets.

Sometimes it is desirable to have a substance with good ferromagnetic properties, but poor retentivity. This is the case when you want to have an electromagnet that will operate from dc, so that it maintains a constant polarity, but that will lose its magnetism when the current is shut off.

If a ferromagnetic substance has poor retentivity, it's easy to make it work as the core for an ac electromagnet, because the polarity is easy to switch. If the retentivity is very high, the material is "sluggish" and will not work well for ac electromagnets.

Permanent magnets

Any ferromagnetic material, or substance whose atoms can be permanently aligned, can be made into a *permanent magnet*. These are the magnets you probably played with as a child. Some alloys can be made into stronger magnets than others.

One alloy that is especially suited to making strong magnets is *alnico*. This word derives from the metals that comprise it: *al*uminum, *ni*ckel and *co*balt. Other elements are often added, including copper and titanium. But any piece of iron or steel can be *magnetized*, at least to some extent. You might have used a screwdriver, for example, that was magnetized, so that it could hold on to screws when installing or removing them from hard-to-reach places.

Permanent magnets are best made from materials with high retentivity. Magnets are made by using a high-retentivity ferromagnetic material as the core of an electromagnet for an extended period of time. This experiment is not a good one to do at home with a battery, because there is a risk of battery explosion.

If you want to magnetize a screwdriver a little bit so that it will hold onto screws just stroke the shaft of the screwdriver with the end of a bar magnet several dozen times. But remember that once you have magnetized a tool, it is difficult to completely demagnetize it.

The solenoid

A cylindrical coil, having a movable ferromagnetic core, can be useful for various things. This is a *solenoid*. Electrical relays, bell ringers, electric "hammers," and other mechanical devices make use of the principle of the solenoid.

A ringer device

Figure 8-7 is a simplified diagram of a bell ringer. Its solenoid is an electromagnet, except that the core is not completely solid, but has a hole going along its axis. The coil has several layers, but the wire is always wound in the same direction, so that the electromagnet is quite powerful. A movable steel rod runs through the hole in the electromagnet core.

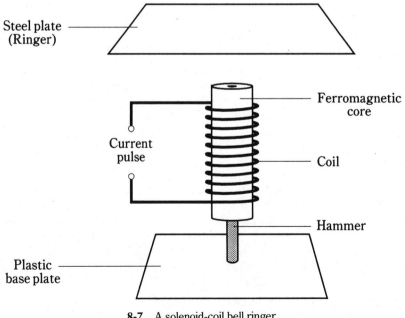

8-7   A solenoid-coil bell ringer.

When there is no current flowing in the coil, the steel rod is held down by the force of gravity. But when a pulse of current passes through the coil, the rod is pulled forcibly upward so that it strikes the ringer plate. This plate is like one of the plates in a xylophone. The current pulse is short, so that the steel rod falls back down again to its resting position, allowing the plate to reverberate: *Gonggg!* Some office telephones are equipped with ringers that produce this noise, rather than the conventional ringing or electronic bleeping emitted by most phone sets.

A relay

In some electronic devices, it is inconvenient to place a switch exactly where it should be. For example, you might want to switch a communications line from one branch to another from a long distance away. In many radio transmitters, the wiring carries high-frequency alternating currents that must be kept within certain parts of the circuit, and not routed out to the front panel for switching. A *relay* makes use of a solenoid to allow remote-control switching.

A diagram of a relay is shown in Fig. 8-8. The movable lever, called the *armature*, is held to one side by a spring when there is no current flowing through the electromagnet. Under these conditions, terminal X is connected to Y, but not to Z. When a sufficient current is applied, the armature is pulled over to the other side. This disconnects terminal X from terminal Y, and connects X to Z.

There are numerous types of relays used for different purposes. Some are meant for use with dc, and others are for ac; a few will work with either type of current. A *normally closed* relay completes the circuit when there is no current flowing in its electromagnet, and breaks the circuit when current flows. A *normally open* relay is just the opposite. ("Normal" in this sense means no current in the coil.) The relay in the illustration (Fig. 8-8) can be used either as a normally open or normally closed relay, depending on which contacts are selected. It can also be used to switch a line between two different circuits.

Some relays have several sets of contacts. Some relays are meant to remain in one state (either with current or without) for a long time, while others are meant to switch several times per second. The fastest relays work dozens of times per second. These are used for such purposes as keying radio transmitters in Morse code or radioteletype.

The dc motor

Magnetic fields can produce considerable mechanical forces. These forces can be harnessed to do work. The device that converts direct-current energy into rotating mechanical energy is a *dc motor*.

Motors can be microscopic in size, or as big as a house. Some tiny motors are being considered for use in medical devices that can actually circulate in the bloodstream or be installed in body organs. Others can pull a train at freeway speeds.

In a dc motor, the source of electricity is connected to a set of coils, producing magnetic fields. The attraction of opposite poles, and the repulsion of like poles, is switched in such a way that a constant *torque*, or rotational force, results. The greater the current that flows in the coils, the stronger the torque, and the more electrical energy is needed.

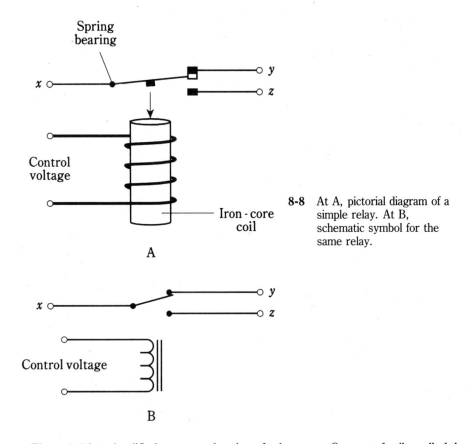

8-8   At A, pictorial diagram of a simple relay. At B, schematic symbol for the same relay.

Figure 8-9 is a simplified, cutaway drawing of a dc motor. One set of coils, called the *armature coil*, goes around with the motor shaft. The other set of coils, called the *field coil*, is stationary. The current direction is periodically reversed during each rotation by means of the *commutator*. This keeps the force going in the same angular direction, so the motor continues to rotate rather than oscillating back and forth. The shaft is carried along by its own inertia, so that it doesn't come to a stop during those instants when the current is being switched in polarity.

Some dc motors can also be used to generate direct current. These motors contain permanent magnets in place of one of the sets of coils. When the shaft is rotated, a pulsating direct current appears across the coil.

Magnetic data storage

Magnetic fields can be used to store data in different forms. Common media for data storage include the *magnetic tape*, the *magnetic disk*, and *magnetic bubble memory*.

Magnetic tape

Recording tape is the stuff you find in cassette players. It is also sometimes seen on reel-to-reel devices. These days, magnetic tape is used for home entertainment, especially hi-fi music and home video.

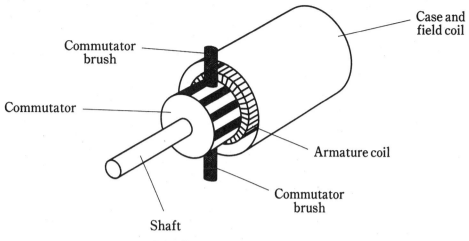

8-9 Cutaway view of a dc motor.

The tape itself consists of millions of particles of iron oxide, attached to a plastic or mylar strip. A fluctuating magnetic field, produced by the *recording head*, polarizes these particles. As the field changes in strength next to the recording head, the tape passes by at a constant, controlled speed. This produces regions in which the iron-oxide particles are polarized in either direction (Fig. 8-10).

When the tape is run at the same speed through the recorder in the playback mode, the magnetic fields around the individual particles cause a fluctuating field that is detected by the *pickup head*. This field has the same pattern of variations as the original field from the recording head.

Magnetic tape is available in various widths and thicknesses, for different applications. The thicker tapes result in cassettes that don't play as long, but the tape is more resistant to stretching. The speed of the tape determines the fidelity of the recording. Higher speeds are preferred for music and video, and lower speeds for voice.

The data on a magnetic tape can be distorted or erased by external magnetic fields. Therefore, tapes should be protected from such fields. Keep magnetic tape away from magnets. Extreme heat can also result in loss of data, and possibly even physical damage to the tape.

Magnetic disk

The age of the personal computer has seen the development of ever-more-compact data-storage systems. One of the most versatile is the *magnetic disk*.

A magnetic disk can be either rigid or flexible. Disks are available in various sizes. *Hard disks* store the most data, and are generally found inside of computer units. *Floppy disks* or *diskettes* are usually either 3½ or 5¼ inch in diameter, and can be inserted and removed from recording/playback machines called *disk drives*.

The principle of the magnetic disk, on the micro scale, is the same as that of the magnetic tape. The information is stored in *digital* form; that is, there are only two different ways that the particles are magnetized. This results in almost perfect, error-free storage.

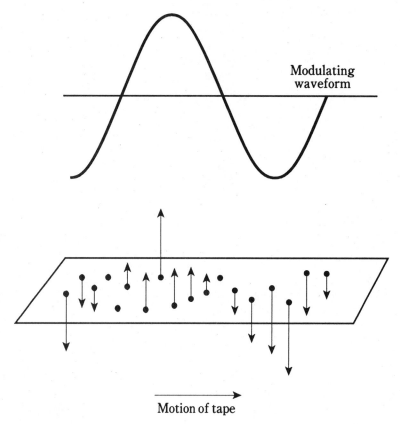

8-10   On recording tape, particles are magnetized in a pattern that follows the modulating waveform.

On a larger scale, the disk works differently than the tape, simply because of the difference in geometry. On a tape, the information is spread out over a long span, and some bits of data are far away from others. But on a disk, no two bits are ever farther apart than the diameter of the disk. This means that data can be stored and retrieved much more quickly onto, or from, a disk than is possible with a tape.

A typical diskette can store an amount of digital information equivalent to a short novel.

The same precautions should be observed when handling and storing magnetic disks, as are necessary with magnetic tape.

Magnetic bubble memory

Bubble memory is a sophisticated method of storing data that gets rid of the need for moving parts such as are required in tape machines and disk drives. This type of memory is used in large computer systems, because it allows the storage, retrieval, and transfer of great quantities of data. The bits of data are stored as tiny magnetic fields, in a medium that is made from magnetic film and semiconductor materials.

A full description of the way bubble memory systems are made, and the way they work, is too advanced for this book. Bubble memory makes use of all the advantages of magnetic data storage, as well as the favorable aspects of electronic data storage. Advantages of electronic memory include rapid storage and recovery, and high density (a lot of data can be put in a tiny volume of space). Advantages of magnetic memory include *nonvolatility* (it can be stored for a long time without needing a constant current source), high density and comparatively low cost.

Quiz

Refer to the text in this chapter if necessary. A good score is at least 18 correct. Answers are in the back of the book.

1. The geomagnetic field:

    A. Makes the earth like a huge horseshoe magnet.

    B. Runs exactly through the geographic poles.

    C. Is what makes a compass work.

    D. Is what makes an electromagnet work.

2. Geomagnetic lines of flux:

    A. Are horizontal at the geomagnetic equator.

    B. Are vertical at the geomagnetic equator.

    C. Are always slanted, no matter where you go.

    D. Are exactly symmetrical around the earth, even far out into space.

3. A material that can be permanently magnetized is generally said to be:

    A. Magnetic.

    B. Electromagnetic.

    C. Permanently magnetic.

    D. Ferromagnetic.

4. The force between a magnet and a piece of ferromagnetic metal that has not been magnetized:

    A. Can be either repulsive or attractive.

    B. Is never repulsive.

    C. Gets smaller as the magnet gets closer to the metal.

    D. Depends on the geomagnetic field.

5. Magnetic flux can always be attributed to:

    A. Ferromagnetic materials.

    B. Aligned atoms.

    C. Motion of charged particles.

    D. The geomagnetic field.

6. Lines of magnetic flux are said to originate:
   A. In atoms of ferromagnetic materials.
   B. At a north magnetic pole.
   C. Where the lines converge to a point.
   D. In charge carriers.

7. The magnetic flux around a straight, current-carrying wire:
   A. Gets stronger with increasing distance from the wire.
   B. Is strongest near the wire.
   C. Does not vary in strength with distance from the wire.
   D. Consists of straight lines parallel to the wire.

8. The gauss is a unit of:
   A. Overall magnetic field strength.
   B. Ampere-turns.
   C. Magnetic flux density.
   D. Magnetic power.

9. A unit of overall magnetic field quantity is the:
   A. Maxwell.
   B. Gauss.
   C. Tesla.
   D. Ampere-turn.

10. If a wire coil has 10 turns and carries 500 mA of current, what is the magnetomotive force in ampere-turns?
    A. 5000.
    B. 50.
    C. 5.0.
    D. 0.02.

11. If a wire coil has 100 turns and carries 1.30 A of current, what is the magnetomotive force in gilberts?
    A. 130.
    B. 76.9.
    C. 164.
    D. 61.0.

12. Which of the following is *not* generally possible in a geomagnetic storm?
    A. Charged particles streaming out from the sun.
    B. Fluctuations in the earth's magnetic field.
    C. Disruption of electrical power transmission.

D. Disruption of microwave radio links.

13. An ac electromagnet:

   A. Will attract only other magnetized objects.

   B. Will attract pure, unmagnetized iron.

   C. Will repel other magnetized objects.

   D. Will either attract or repel permanent magnets, depending on the polarity.

14. An advantage of an electromagnet over a permanent magnet is that:

   A. An electromagnet can be switched on and off.

   B. An electromagnet does not have specific polarity.

   C. An electromagnet requires no power source.

   D. Permanent magnets must always be cylindrical.

15. A substance with high retentivity is best suited for making:

   A. An ac electromagnet.

   B. A dc electromagnet.

   C. An electrostatic shield.

   D. A permanent magnet.

16. A relay is connected into a circuit so that a device gets a signal only when the relay coil carries current. The relay is probably:

   A. An ac relay.

   B. A dc relay.

   C. Normally closed.

   D. Normally open.

17. A device that reverses magnetic field polarity to keep a dc motor rotating is:

   A. A solenoid.

   B. An armature coil.

   C. A commutator.

   D. A field coil.

18. A high tape-recorder motor speed is generally used for:

   A. Voices.

   B. Video.

   C. Digital data.

   C. All of the above.

19. An advantage of a magnetic disk, as compared with magnetic tape, for data storage and retrieval is that:

   A. A disk lasts longer.

   B. Data can be stored and retrieved more quickly with disks than with tapes.

C. Disks look better.

D. Disks are less susceptible to magnetic fields.

20. A bubble memory is best suited for:

A. A large computer.

B. A home video entertainment system.

C. A portable cassette player.

D. A magnetic disk.

Test: Part one

1. An application in which an analog meter would almost always be preferred over a digital meter is:

    A. A signal-strength indicator in a radio receiver.

    B. A meter that shows power-supply voltage.

    C. A utility watt-hour meter.

    D. A clock.

    E. A device in which a direct numeric display is wanted.

2. Which of the following statements is false?

    A. The current in a series dc circuit is divided up among the resistances.

    B. In a parallel dc circuit, the voltage is the same across each component.

    C. In a series dc circuit, the sum of the voltages across all the components, going once around a complete circle, is zero.

    D. The net resistance of a parallel set of resistors is less than the value of the smallest resistor.

    E. The total power consumed in a series circuit is the sum of the wattages consumed by each of the components.

3. The ohm is a unit of:

    A. Electrical charge quantity.

    B. The rate at which charge carriers flow.

    C. Opposition to electrical current.

    D. Electrical conductance.

    E. Potential difference.

4. A wiring diagram differs from a schematic diagram in that:

    A. A wiring diagram is less detailed.

    B. A wiring diagram shows component values.

    C. A schematic does not show all the interconnections between the components.

    D. A schematic shows pictures of components, while a wiring diagram shows the electronic symbols.

    E. A schematic shows the electronic symbols, while a wiring diagram shows pictures of the components.

5. Which of the following is a good use, or place, for a wirewound resistor?

    A. To dissipate a large amount of dc power.

    B. In the input of a radio-frequency amplifier.

    C. In the output of a radio-frequency amplifier.

    D. In an antenna, to limit the transmitter power.

    E. Between ground and the chassis of a power supply.

6. The number of protons in the nucleus of an element is the:

    A. Electron number.

    B. Atomic number.

    C. Valence number.

    D. Charge number.

    E. Proton number.

7. A hot-wire ammeter:

    A. Can measure ac as well as dc.

    B. Registers current changes very fast.

    C. Can indicate very low voltages.

    D. Measures electrical energy.

    E. Works only when current flows in one direction.

8. Which of the following units indicates the rate at which energy is expended?

    A. The volt.

    B. The ampere.

    C. The coulomb.

    D. The ampere hour.

    E. The watt.

9. Which of the following correctly states Ohm's Law?

    A. Volts equal amperes divided by ohms.

    B. Ohms equal amperes divided by volts.

    C. Amperes equal ohms divided by volts.

    D. Amperes equal ohms times volts.

    E. Ohms equal volts divided by amperes.

10. The current going into a point in a dc circuit is always equal to the current:

    A. Delivered by the power supply.

    B. Through any one of the resistances.

    C. Flowing out of that point.

    D. At any other point.

    E. In any single branch of the circuit.

11. A loudness meter in a hi-fi system is generally calibrated in:

    A. Volts.

    B. Amperes.

    C. Decibels.

    D. Watt hours.

    E. Ohms.

12. A charged atom is known as:

    A. A molecule.

    B. An isotope.

    C. An ion.

    D. An electron.

    E. A fundamental particle.

13. A battery delivers 12 V to a bulb. The current in the bulb is 3 A. What is the resistance of the bulb?

    A. 36 Ω.

    B. 4 Ω.

    C. 0.25 Ω.

    D. 108 Ω.

    E. 0.75 Ω.

14. Peak values are always:

    A. Greater than average values.

    B. Less than average values.

    C. Greater than or equal to average values.

    D. Less than or equal to average values.

E. Fluctuating.

15. A resistor has a value of 680 ohms, and a tolerance of plus or minus 5 percent. Which of the following values indicates a reject?

A. 648 Ω.

B. 712 Ω.

C. 699 Ω.

D. 636 Ω.

E. 707 Ω.

16. A primitive device for indicating the presence of an electric current is:

A. An electrometer.

B. A galvanometer.

C. A voltmeter.

D. A coulometer.

E. A wattmeter.

17. A disadvantage of mercury cells is that they:

A. Pollute the environment when discarded.

B. Supply less voltage than other cells.

C. Can reverse polarity unexpectedly.

D. Must be physically large.

E. Must be kept right-side-up.

18. A battery supplies 6.0 V to a bulb rated at 12 W. How much current does the bulb draw?

A. 2.0 A.

B. 0.5 A.

C. 72 A.

D. 40 mA.

E. 72 mA.

19. Of the following, which is not a common use of a resistor?

A. Biasing for a transistor.

B. Voltage division.

C. Current limiting.

D. Use as a "dummy" antenna.

E. Increasing the charge in a capacitor.

20. When a charge builds up without a flow of current, the charge is said to be:

A. Ionizing.

B. Atomic.

C. Molecular.

D. Electronic.

E. Static.

21. The sum of the voltages, going around a dc circuit, but not including the power supply, has:

    A. Equal value, and the same polarity, as the supply.

    B. A value that depends on the ratio of the resistances.

    C. Different value from, but the same polarity as, the supply.

    D. Equal value as, but opposite polarity from, the supply.

    E. Different value, and opposite polarity, from the supply.

22. A watt hour meter measures:

    A. Voltage.

    B. Current.

    C. Power.

    D. Energy.

    E. Charge.

23. Every chemical element has its own unique type of particle, called its:

    A. Molecule.

    B. Electron.

    C. Proton.

    D. Atom.

    E. Isotope.

24. An advantage of a magnetic disk over magnetic tape for data storage is that:

    A. Data is too closely packed on the tape.

    B. The disk is immune to the effects of magnetic fields.

    C. Data storage and retrieval is faster on disk.

    D. Disks store computer data in analog form.

    E. Tapes cannot be used to store digital data.

25. A 6-V battery is connected across a series combination of resistors. The resistance values are 1, 2, and 3 Ω. What is the current through the 2-Ω resistor?

    A. 1 A.

    B. 3 A.

    C. 12 A.

    D. 24 A.

    E. 72 A.

26. A material that has extremely high electrical resistance is known as:

    A. A semiconductor.

    B. A paraconductor.

    C. An insulator.

    D. A resistor.

    E. A diamagnetic substance.

27. Primary cells:

    A. Can be used over and over.

    B. Have higher voltage than other types of cells.

    C. All have exactly 1.500 V.

    D. Cannot be recharged.

    E. Are made of zinc and carbon.

28. A rheostat:

    A. Is used in high-voltage and/or high-power dc circuits.

    B. Is ideal for tuning a radio receiver.

    C. Is often used as a bleeder resistor.

    D. Is better than a potentiometer for low-power audio.

    E. Offers the advantage of having no inductance.

29. A voltage typical of a dry cell is:

    A. 12 V.

    B. 6 V.

    C. 1.5 V.

    D. 117 V.

    E. 0.15 V.

30. A geomagnetic storm:

    A. Causes solar wind.

    B. Causes charged particles to bombard the earth.

    C. Can disrupt the earth's magnetic field.

    D. Ruins microwave communications.

    E. Has no effect near the earth's poles.

31. An advantage of an alkaline cell over a zinc-carbon cell is that:

    A. The alkaline cell provides more voltage.

    B. The alkaline cell can be recharged.

    C. An alkaline cell works at lower temperatures.

    D. The alkaline cell is far less bulky for the same amount of energy capacity.

    E. There is no advantage of alkaline over zinc-carbon cells.

32. A battery delivers 12 V across a set of six 4-Ω resistors in a series voltage-dividing combination. This provides six different voltages, differing by an increment of:

    A. 1/4 V.

    B. 1/3 V.

C. 1 V.

D. 2 V.

E. 3 V.

33. A unit of electrical charge quantity is the:

A. Volt.

B. Ampere.

C. Watt.

D. Tesla.

E. Coulomb.

34. A unit of sound volume is:

A. The volt per square meter.

B. The volt.

C. The watt hour.

D. The decibel.

E. The ampere per square meter.

35. A 24-V battery is connected across a set of four resistors in parallel. Each resistor has a value of 32 ohms. What is the total power dissipated by the resistors?

A. 0.19 W.

B. 3 W.

C. 192 W.

D. 0.33 W.

E. 72 W.

36. The main difference between a "lantern" battery and a "transistor" battery is:

A. The lantern battery has higher voltage.

B. The lantern battery has more energy capacity.

C. Lantern batteries cannot be used with electronic devices such as transistor radios.

D. Lantern batteries can be recharged, but transistor batteries cannot.

E. The lantern battery is more compact.

37. NICAD batteries are most extensively used:

A. In disposable flashlights.

B. In large lanterns.

C. As car batteries.

D. In handheld radio transcievers.

E. In remote garage-door-opener control boxes.

38. A voltmeter should have:

A. Very low internal resistance.

    B. Electrostatic plates.

    C. A sensitive amplifier.

    D. High internal resistance.

    E. The highest possible full-scale value.

39. The purpose of a bleeder resistor is to:

    A. Provide bias for a transistor.

    B. Serve as a voltage divider.

    C. Protect people against the danger of electric shock.

    D. Reduce the current in a power supply.

    E. Smooth out the ac ripple in a power supply.

40. A dc electromagnet:

    A. Has constant polarity.

    B. Requires a core with high retentivity.

    C. Will not attract or repel a permanent magnet.

    D. Has polarity that periodically reverses.

    E. Cannot be used to permanently magnetize anything.

41. The rate at which charge carriers flow is measured in:

    A. Amperes.

    B. Coulombs.

    C. Volts.

    D. Watts.

    E. Watt hours.

42. A 12-V battery is connected to a set of three resistors in series. The resistance values are 1, 2, and 3 ohms. What is the voltage across the 3-Ω resistor?

    A. 1 V.

    B. 2 V.

    C. 4 V.

    D. 6 V.

    E. 12 V.

43. Nine 90-ohm resistors are connected in a 3 x 3 series-parallel network. The total resistance is:

    A. 10 Ω.

    B. 30 Ω.

    C. 90 Ω.

    D. 270 Ω.

    E. 810 Ω.

44. A device commonly used for remote switching of wire communications signals is:

    A. A solenoid.

    B. An electromagnet.

    C. A potentiometer.

    D. A photovoltaic cell.

    E. A relay.

45. NICAD memory:

    A. Occurs often when NICADs are misused.

    B. Indicates that the cell or battery is dead.

    C. Does not occur very often.

    D. Can cause a NICAD to explode.

    E. Causes NICADs to reverse polarity.

46. A 100-W bulb burns for 100 hours. It has consumed:

    A. 0.10 kWh.

    B. 1.00 kWh.

    C. 10.0 kWh.

    D. 100 kWh.

    E. 1000 kWh.

47. A material with high permeability:

    A. Increases magnetic field quantity.

    B. Is necessary if a coil is to produce a magnetic field.

    C. Always has high retentivity.

    D. Concentrates magnetic lines of flux.

    E. Reduces flux density.

48. A chemical compound:

    A. Consists of two or more atoms.

    B. Contains an unusual number of neutrons.

    C. Is technically the same as an ion.

    D. Has a shortage of electrons.

    E. Has an excess of electrons.

49. A 6.00-V battery is connected to a parallel combination of two resistors, whose values are 8.00 Ω and 12.0 Ω. What is the power dissipated in the 8-Ω resistor?

    A. 0.300 W.

    B. 0.750 W.

    C. 1.25 W.

    D. 1.80 W.

    E. 4.50 W.

50. The main problem with a bar-graph meter is that:

    A. Is isn't very sensitive.

    B. It isn't stable.

    C. It can't give a very precise reading.

    D. You need special training to read it.

    E. It shows only peak values.

2
PART

Alternating current

Alternating current basics

DIRECT CURRENT (DC) IS SIMPLE. IT CAN BE EXPRESSED IN TERMS OF JUST TWO variables: polarity (or direction), and magnitude. Alternating current (ac) is somewhat more complicated, because there are three things that can vary. Because of the greater number of parameters, alternating-current circuits behave in more complex ways than direct-current circuits. This chapter will acquaint you with the most common forms of alternating current. A few of the less often-seen types are also mentioned.

Definition of alternating current

Recall that direct current has a polarity, or direction, that stays the same over a long period of time. Although the magnitude might vary—the number of amperes, volts, or watts can fluctuate—the charge carriers always flow in the same direction through the circuit.

In alternating current, the polarity reverses again and again at regular intervals. The magnitude usually changes because of this constant reversal of polarity, although there are certain cases where the magnitude doesn't change even though the polarity does.

The rate of change of polarity is the third variable that makes ac so much different from dc. The behavior of an ac wave depends largely on this rate: the *frequency*.

Period and frequency

In a *periodic* ac wave, the kind that is discussed in this chapter (and throughout the rest of this book), the function of *magnitude versus time* repeats itself over and over, so that the same pattern recurs countless times. The length of time between one repetition of the pattern, or one *cycle*, and the next is called the *period* of the wave. This is illustrated in Fig. 9-1 for a simple ac wave. The period of a wave can, in theory, be anywhere from a

165

minuscule fraction of a second to many centuries. Radio-frequency currents reverse polarity millions or billions of times a second. The charged particles held captive by the magnetic field of the sun, and perhaps also by the much larger magnetic fields around galaxies, might reverse their direction over periods measured in thousands or millions of years. Period, when measured in seconds, is denoted by T.

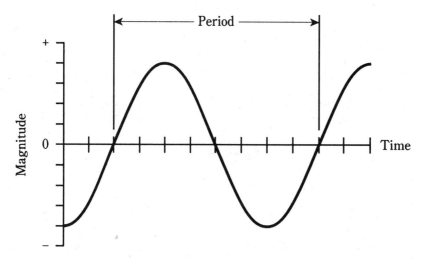

9-1 A sine wave. The period is the length of time for one complete cycle.

The *frequency*, denoted f, of a wave is the reciprocal of the period. That is, $f = 1/T$ and $T = 1/f$. Originally, frequency was specified in *cycles per second*, abbreviated *cps*. High frequencies were sometimes given in *kilocycles*, *megacycles* or *gigacycles*, representing thousands, millions or billions of cycles per second. But nowadays, the unit is known as the *hertz*, abbreviated *Hz*. Thus, 1 Hz = 1 cps, 10 Hz = 10 cps, and so on.

Higher frequencies are given in *kilohertz (kHz)*, *megahertz (MHz)* or *gigahertz (GHz)*. The relationships are:

$$1 \text{ kHz} = 1000 \text{ Hz}$$
$$1 \text{ MHz} = 1000 \text{ kHz} = 1,000,000 \text{ Hz}$$
$$1 \text{ GHz} = 1000 \text{ MHz} = 1,000,000,000 \text{ Hz}$$

Sometimes an even bigger unit, the *terahertz (THz)*, is needed. This is a trillion (1,000,000,000,000) hertz. Electrical currents generally do not attain such frequencies, although *electromagnetic radiation* can.

Some ac waves have only one frequency. These waves are called *pure*. But often, there are components at multiples of the main, or *fundamental*, frequency. There might also be components at "odd" frequencies. Some ac waves are extremely complex, consisting of hundreds, thousands, or even infinitely many different component frequencies. In this book, most of the attention will be given to ac waves that have just one frequency.

The sine wave

Sometimes, alternating current has a *sine-wave*, or *sinusoidal*, nature. This means that the direction of the current reverses at regular intervals, and that the current-versus-time curve is shaped like the trigonometric sine function. The waveform in Fig. 9-1 is a sine wave.

Any ac wave that consists of a single frequency will have a perfect sine waveshape. And any perfect sine-wave current contains only one component frequency. In practice, a wave might be so close to a sine wave that it looks exactly like the sine function on an oscilloscope, when in reality there are traces of other frequencies present. Imperfections are often too small to see. But pure, single-frequency ac not only looks perfect, but actually is a perfect replication of the trigonometric sine function.

The current at the wall outlets in your house has an almost perfect sine waveshape, with a frequency of 60 Hz.

The square wave

Earlier in this chapter, it was said that there can be an alternating current whose magnitude never changes. You might at first think this is impossible. How can polarity reverse without some change in the level? The *square wave* is an example of this.

On an oscilloscope, a perfect square wave looks like a pair of parallel, dotted lines, one with positive polarity and the other with negative polarity (Fig. 9-2A). The oscilloscope shows a graph of voltage on the vertical scale, versus time on the horizontal scale. The transitions between negative and positive for a true square wave don't show up on the oscilloscope, because they're instantaneous. But perfection is rare. Usually, the transitions can be seen as vertical lines (Fig. 9-2B).

A square wave might have equal negative and positive peaks. Then the absolute magnitude of the wave is constant, at a certain voltage, current or power level. Half of the time it's +x, and the other half it's −x volts, amperes or watts. Some square waves are lopsided, with the positive and negative magnitudes differing.

Sawtooth waves

Some ac waves rise and fall in straight lines as seen on an oscilloscope screen. The slope of the line indicates how fast the magnitude is changing. Such waves are called *sawtooth waves* because of their appearance.

Sawtooth waves are generated by certain electronic test devices. These waves provide ideal signals for control purposes. Integrated circuits can be wired so that they produce sawtooth waves having an exact desired shape.

Fast-rise, slow-decay

In Fig. 9-3, one form of sawtooth wave is shown. The positive-going slope (rise) is extremely steep, as with a square wave, but the negative-going slope (fall or decay) is gradual. The period of the wave is the time between points at identical positions on two successive pulses.

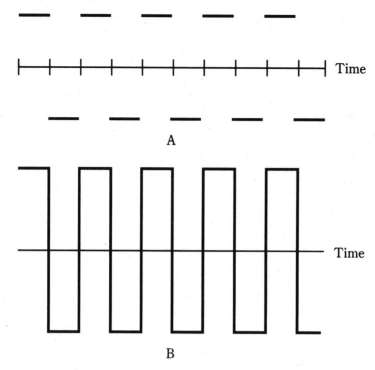

9-2   At A, a perfect square wave. At B, the more common rendition of a square wave.

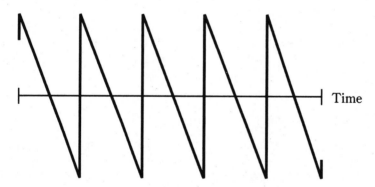

9-3   Fast-rise, slow-decay sawtooth.

Slow-rise, fast-decay

Another form of sawtooth wave is just the opposite, with a gradual positive-going slope and a vertical negative-going transition. This type of wave is sometimes called a *ramp*, because it looks like an incline going upwards (Fig. 9-4). This waveshape is useful for scanning in television sets and oscilloscopes. It tells the electron beam to move, or trace, at a constant speed from left to right during the upwards sloping part of the wave. Then it retraces, or brings the electron beam back, at a high speed for the next trace.

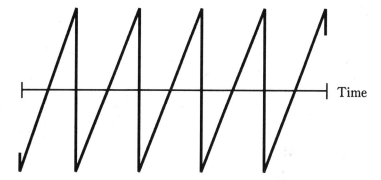

9-4 Slow-rise, fast-decay sawtooth, also called a ramp.

Variable rise and decay

You can probably guess that sawtooth waves can have rise and decay slopes in an infinite number of different combinations. One example is shown in Fig. 9-5. In this case, the positive-going slope is the same as the negative-going slope. This is a *triangular wave*.

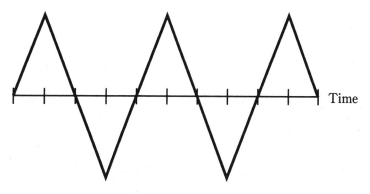

9-5 Triangular wave.

Complex and irregular waveforms

The shape of an ac wave can get exceedingly complicated, but as long as it has a definite period, and as long as the polarity keeps switching back and forth between positive and negative, it is ac.

Fig. 9-6 shows an example of a complex ac wave. You can see that there is a period, and therefore a definable frequency. The period is the time between two points on succeeding wave repetitions.

With some waves, it can be difficult or almost impossible to tell the period. This is because the wave has two or more components that are nearly the same magnitude. When this happens, the *frequency spectrum* of the wave will be multifaceted. The energy is split up among two or more frequencies.

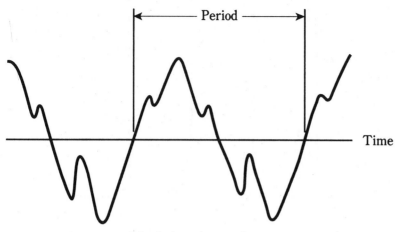

9-6 An irregular waveform.

Frequency spectrum

An oscilloscope shows a graph of magnitude versus time. Because time is on the horizontal axis, the oscilloscope is said to be a *time-domain* instrument.

Sometimes you want to see magnitude as a function of frequency, rather than as a function of time. This can be done with a *spectrum analyzer*. It is a *frequency-domain* instrument, with a cathode-ray display similar to an oscilloscope. Its horizontal axis shows frequency, from some adjustable minimum (extreme left) to some adjustable maximum (extreme right).

An ac sine wave, as displayed on a spectrum analyzer, appears as a single "pip," or vertical line (Fig. 9-7A). This means that all of the energy in the wave is concentrated at one single frequency.

Many ac waves contain *harmonic* energy along with the fundamental, or main, frequency. A harmonic frequency is a whole-number multiple of the fundamental frequency. For example, if 60 Hz is the fundamental, then harmonics can exist at 120 Hz, 180 Hz, 240 Hz, and so on. The 120-Hz wave is the *second harmonic*; the 180-Hz wave is the *third harmonic*.

In general, if a wave has a frequency equal to n times the fundamental, then that wave is the *nth harmonic*. In the illustration of Fig. 9-7B, a wave is shown along with several harmonics, as it would look on the display screen of a spectrum analyzer.

The frequency spectra of square waves and sawtooth waves contain harmonic energy in addition to the fundamental. The wave shape depends on the amount of energy in the harmonics, and the way in which this energy is distributed among the harmonic frequencies. A detailed discussion of these relationships is far too sophisticated for this book.

Irregular waves can have practically any imaginable frequency distribution. An example is shown at Fig. 9-8. This is a display of a voice-modulated radio signal. Much of the energy is concentrated at the center of the pattern, at the frequency shown by the vertical line. But there is also plenty of energy "splattered around" this *carrier* frequency. On an oscilloscope, this signal would look like a fuzzy sine wave, indicating that it is ac, although it contains a potpourri of minor components.

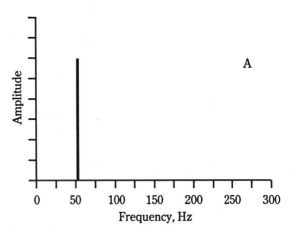

9-7 At A, pure 60-Hz sine
wave on spectrum
analyzer. At B, 60-Hz
wave containing harmonics.

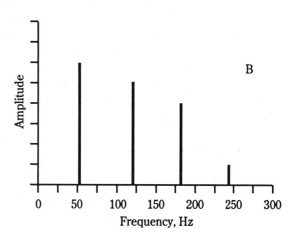

9-8 Modulated radio signal on
spectrum analyzer.

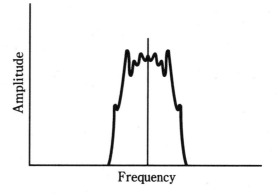

Little bits of a cycle

Engineers break the ac cycle down into small parts for analysis and reference. One complete cycle can be likened to a single revolution around a circle.

Degrees

One method of specifying the *phase* of an ac cycle is to divide it into 360 equal parts, called *degrees* or *degrees of phase*. The value 0 degrees is assigned to the point in the cycle where the magnitude is 0 and positive-going. The same point on the next cycle is given the value 360 degrees. Then halfway through the cycle is 180 degrees; a quarter cycle is 90 degrees, and so on. This is illustrated in Fig. 9-9.

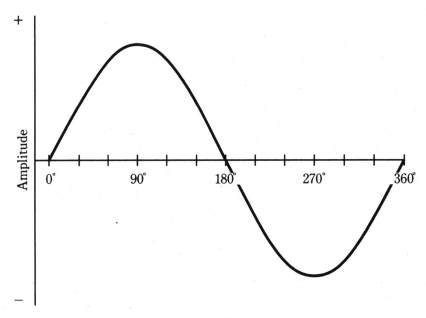

9-9 A cycle is divided into 360 degrees.

Radians

The other method of specifying phase is to divide the cycle into 6.28 equal parts. This is approximately the number of radii of a circle that can be laid end-to-end around the circumference. A *radian* of phase is equal to about 57.3 degrees. This unit of phase is something you won't often be needing to use, because it's more common among physicists than among engineers.

Sometimes, the frequency of an ac wave is measured in radians per second, rather than in hertz (cycles per second). Because there are about 6.28 radians in a complete cycle of 360 degrees, the *angular frequency* of a wave, in radians per second, is equal to about 6.28 times the frequency in hertz.

Phase difference

Two ac waves might have exactly the same frequency, but they can still have different effects because they are "out of sync" with each other. This is especially true when ac waves are added together to produce a third, or *composite*, signal.

If two ac waves have the same frequency and the same magnitude, but differ in phase by 180 degrees (a half cycle), they will cancel each other out, and the net signal will be zero. If the two waves are in phase, the resulting signal will have the same frequency, but twice the amplitude of either signal alone.

If two ac waves have the same frequency but different magnitudes, and differ in phase by 180 degrees, the resulting composite signal will have the same frequency as the originals, and a magnitude equal to the difference between the two. If two such waves are exactly in phase, the composite will have the same frequency as the originals, and a magnitude equal to the sum of the two.

If the waves have the same frequency but differ in phase by some odd amount such as 75 degrees or 310 degrees, the resulting signal will have the same frequency, but will not have the same waveshape as either of the original signals. The variety of such cases is infinite.

Household utility current, as you get it from wall outlets, consists of a 60-Hz sine wave with just one phase component. But the energy is transmitted over long distances in three phases, each differing by 120 degrees or ⅓ cycle. This is what is meant by *three-phase ac*. Each of the three ac waves carries ⅓ of the total power in a utility transmission line.

Amplitude of alternating current

Amplitude is sometimes called magnitude, level or intensity. Depending on the quantity being measured, the magnitude of an ac wave might be given in amperes (for current), volts (for voltage), or watts (for power).

Instantaneous amplitude

The *instantaneous amplitude* of an ac wave is the amplitude at some precise moment in time. This constantly changes. The manner in which it varies depends on the waveform. You have already seen renditions of common ac waveforms in this chapter. Instantaneous amplitudes are represented by individual points on the wave curves.

Peak amplitude

The *peak amplitude* of an ac wave is the maximum extent, either positive or negative, that the instantaneous amplitude attains.

In many waves, the positive and negative peak amplitudes are the same. But sometimes they differ. Figure 9-9 is an example of a wave in which the positive peak amplitude is the same as the negative peak amplitude. Figure 9-10 is an illustration of a wave that has different positive and negative peak amplitudes.

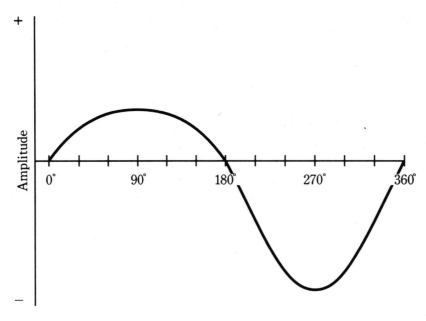

9-10 A wave with unequal positive and negative peaks.

Peak-to-peak amplitude

The *peak-to-peak (pk-pk) amplitude* of a wave is the net difference between the positive peak amplitude and the negative peak amplitude (Fig. 9-11). Another way of saying this is that the pk-pk amplitude is equal to the positive peak amplitude plus the negative peak amplitude. Peak-to-peak is a way of expressing how much the wave level "swings" during the cycle.

In many waves, the pk-pk amplitude is just twice the peak amplitude. This is the case when the positive and negative peak amplitudes are the same.

Root-mean-square amplitude

Often, it is necessary to express the effective level of an ac wave. This is the voltage, current or power that a dc source would have to produce, in order to have the same general effect. When you say a wall outlet has 117 V, you mean 117 effective volts. The most common figure for effective ac levels is called the *root-mean-square*, or *rms*, value.

For a perfect sine wave, the rms value is equal to 0.707 times the peak value, or 0.354 times the pk-pk value. Conversely, the peak value is 1.414 times the rms value, and the pk-pk value is 2.828 times the rms value. The rms figures are most often used with perfect sine waves, such as the utility voltage, or the effective voltage of a radio signal.

For a perfect square wave, the rms value is the same as the peak value. The pk-pk value is twice the rms value or the peak value.

For sawtooth and irregular waves, the relationship between the rms value and the peak value depends on the shape of the wave. But the rms value is never more than the peak value for any waveshape.

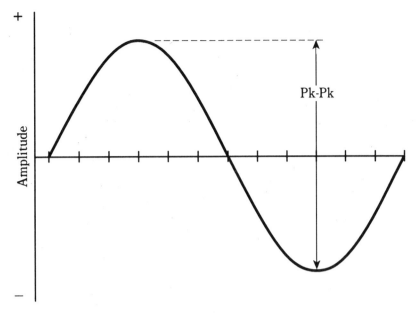

9-11 Peak-to-peak amplitude.

The name "root mean square" was not chosen just because it sounds interesting. It literally means that the value of a wave is mathematically operated on, by taking the square root of the mean of the square of all its values. You don't really have to be concerned with this process, but it's a good idea to remember the above numbers for the relationships between peak, pk-pk, and rms values for sine waves and square waves.

For 117 V rms at a utility outlet, the peak voltage is considerably greater. The pk-pk voltage is far greater.

Superimposed direct current

Sometimes a wave can have components of both ac and dc. The simplest example of an ac/dc combination is illustrated by the connection of a dc source, such as a battery, in series with an ac source, like the utility mains. An example is shown in the schematic diagram of Fig. 9-12. Imagine connecting a 12-V automotive battery in series with the wall outlet. *(Do not try this experiment in real life!)* Then the ac wave will be displaced either positively or negatively by 12 V, depending on the polarity of the battery. This will result in a sine wave at the output, but one peak will be 24 V (twice the battery voltage) more than the other.

Any ac wave can have dc components along with it. If the dc component exceeds the peak value of the ac wave, then fluctuating, or *pulsating,* dc will result. This would happen, for example, if a 200-Vdc source were connected in series with the utility output. Pulsating dc would appear, with an average value of 200 V but with instantaneous values much higher and lower. The waveshape in this case is illustrated by Fig. 9-13.

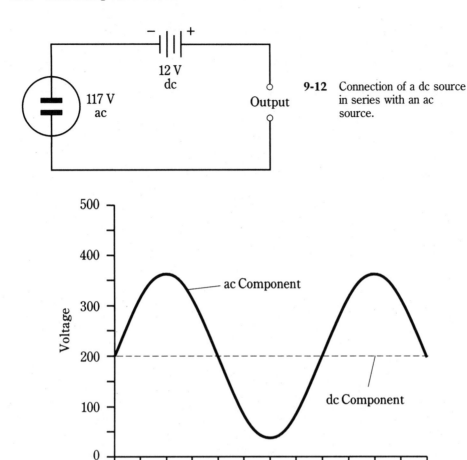

9-12 Connection of a dc source in series with an ac source.

9-13 Waveform resulting from 117 Vac in series with + 200 Vdc.

"Hybrid" ac/dc combinations are not often generated deliberately. But such waveforms are sometimes seen at certain points in electronic circuitry.

The ac generator

Alternating current is easily generated by means of a rotating magnet in a coil of wire (Fig. 9-14A), or by a rotating coil of wire inside a powerful magnet (Fig. 9-14B). In either case, the ac appears between the ends of the length of wire.

The ac voltage that a generator can develop depends on the strength of the magnet, the number of turns in the wire coil, and the speed at which the magnet or coil rotates. The ac frequency depends only on the speed of rotation. Normally, for utility ac, this speed is 3,600 revolutions per minute (rpm), or 60 complete revolutions per second (rps), so that the frequency is 60 Hz.

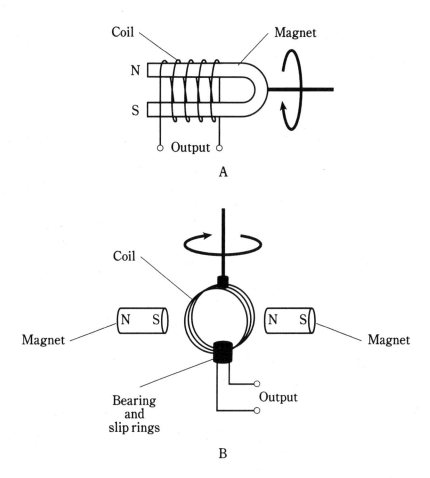

9-14   Two forms of ac generator. At A, the magnet rotates; at B, the coil rotates.

When a load, such as a light bulb or heater, is connected to an ac generator, it becomes more difficult to turn the generator. The more power needed from a generator, the greater the amount of power required to drive it. This is why it is not possible to connect a generator to, for instance, your stationary bicycle, and pedal an entire city into electrification. There's no way to get something for nothing. The electrical power that comes out of a generator can never be more than the mechanical power driving it. In fact, there is always some energy lost, mainly as heat in the generator. Your legs might generate 50 W of power to run a small radio, but nowhere near enough to provide electricity for a household.

The *efficiency* of a generator is the ratio of the power output to the driving power, both measured in the same units (such as watts or kilowatts), multiplied by 100 to get a percentage. No generator is 100 percent efficient. But a good one can come fairly close to this ideal.

At power plants, the generators are huge. Each one is as big as a house. The generators are driven by massive turbines. The turbines are turned by various natural sources of energy. Often, steam drives the turbines, and the steam is obtained via heat derived from the natural energy source.

Why ac?

You might wonder why ac is even used. Isn't it a lot more complicated than dc?

Well, ac is easy to generate from turbines, as you've just seen. Rotating coil-and-magnet devices always produce ac, and in order to get dc from this, *rectification* and *filtering* are necessary. These processes can be difficult to achieve with high voltages.

Alternating current lends itself well to being transformed to lower or higher voltages, according to the needs of electrical apparatus. It is not so easy to change dc voltages.

Electrochemical cells produce dc directly, but they are impractical for the needs of large populations. To serve millions of consumers, the immense power of falling or flowing water, the ocean tides, wind, burning fossil fuels, safe nuclear fusion, or of geothermal heat are needed. (Nuclear fission will work, but it is under scrutiny nowadays because it produces dangerous radioactive by-products.) All of these energy sources can be used to drive turbines that turn ac generators.

Technology is advancing in the realm of solar-electric energy; someday a significant part of our electricity might come from photovoltaic power plants. These would generate dc.

Thomas Edison is said to have favored dc over ac for electrical power transmission in the early days, as utilities were first being planned. His colleagues argued that ac would work better. It took awhile to convince Mr. Edison to change his mind. He eventually did. But perhaps he knew something that his contemporaries did not foresee.

There is one advantage to direct current in utility applications. This is for the transmission of energy over great distances using wires. Direct currents, at extremely high voltages, are transported more efficiently than alternating currents. The wire has less effective resistance with dc than with ac, and there is less energy lost in the magnetic fields around the wires.

Direct-current *high-tension* transmission lines are being considered for future use. Right now, the main problem is expense. Sophisticated power-conversion equipment is needed. If the cost can be brought within reason, Edison's original sentiments will be at least partly vindicated. His was a long view.

Quiz

Refer to the text in this chapter if necessary. A good score is at least 18 correct. Answers are in the back of the book.

1. Which of the following can vary with ac, but not with dc?

   A. Power.

   B. Voltage.

   C. Frequency.

   D. Magnitude.

2. The length of time between a point in one cycle and the same point in the next cycle of an ac wave is the:

    A. Frequency.

    B. Magnitude.

    C. Period.

    D. Polarity.

3. On a spectrum analyzer, a pure ac signal, having just one frequency component, would look like:

    A. A single pip.

    B. A perfect sine wave.

    C. A square wave.

    D. A sawtooth wave.

4. The period of an ac wave is:

    A. The same as the frequency.

    B. Not related to the frequency.

    C. Equal to 1 divided by the frequency.

    D. Equal to the amplitude divided by the frequency.

5. The sixth harmonic of an ac wave whose period is 0.001 second has a frequency of:

    A. 0.006 Hz.

    B. 167 Hz.

    C. 7 kHz.

    D. 6 kHz.

6. A degree of phase represents:

    A. 6.28 cycles.

    B. 57.3 cycles.

    C. 1/6.28 cycle.

    D. 1/360 cycle.

7. Two waves have the same frequency but differ in phase by 1/20 cycle. The phase difference in degrees is:

    A. 18.

    B. 20.

    C. 36.

    D. 5.73.

8. A signal has a frequency of 1770 Hz. The angular frequency is:

    A. 1770 radians per second.

    B. 11,120 radians per second.

    C. 282 radians per second.

    D. Impossible to determine from the data given.

9. A triangular wave:

    A. Has a fast rise time and a slow decay time.

    B. Has a slow rise time and a fast decay time.

    C. Has equal rise and decay rates.

    D. Rises and falls abruptly.

10. Three-phase ac:

    A. Has waves that add up to three times the originals.

    B. Has three waves, all of the same magnitude.

    C. Is what you get at a common wall outlet.

    D. Is of interest only to physicists.

11. If two waves have the same frequency and the same amplitude, but opposite phase, the composite wave is:

    A. Twice the amplitude of either wave alone.

    B. Half the amplitude of either wave alone.

    C. A complex waveform, but with the same frequency as the originals.

    D. Zero.

12. If two waves have the same frequency and the same phase, the composite wave:

    A. Has a magnitude equal to the difference between the two originals.

    B. Has a magnitude equal to the sum of the two originals.

    C. Is complex, with the same frequency as the originals.

    D. Is zero.

13. In a 117-V utility circuit, the peak voltage is:

    A. 82.7 V.

    B. 165 V.

    C. 234 V.

    D. 331 V.

14. In a 117-V utility circuit, the pk-pk voltage is:

    A. 82.7 V.

    B. 165 V.

    C. 234 V.

    D. 331 V.

15. In a perfect sine wave, the pk-pk value is:

    A. Half the peak value.

    B. The same as the peak value.

    C. 1.414 times the peak value.

    D. Twice the peak value.

16. If a 45-Vdc battery is connected in series with the 117-V utility mains as shown in Fig. 9-15, the peak voltages will be:

    A. +210 V and −120 V.

    B. +162 V and −72 V.

    C. +396 V and −286 V.

    D. Both equal to 117 V.

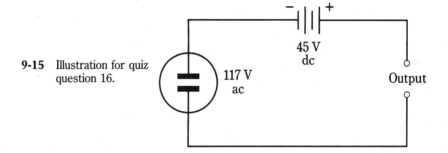

9-15 Illustration for quiz question 16.

17. In the situation of question 16, the pk-pk voltage will be:

    A. 117 V.

    B. 210 V.

    C. 331 V.

    D. 396 V.

18. Which one of the following does *not* affect the power output available from a particular ac generator?

    A. The strength of the magnet.

    B. The number of turns in the coil.

    C. The type of natural energy source used.

    D. The speed of rotation of the coil or magnet.

19. If a 175-V dc source were connected in series with the utility mains from a standard wall outlet, the result would be:

    A. Smooth dc.

    B. Smooth ac.

   C. Ac with one peak greater than the other.

   D. Pulsating dc.

20. An advantage of ac over dc in utility applications is:

   A. Ac is easier to transform from one voltage to another.

   B. Ac is transmitted with lower loss in wires.

   C. Ac can be easily gotten from dc generators.

   D. Ac can be generated with less dangerous by-products.

10
CHAPTER

Inductance

THIS CHAPTER DELVES INTO DEVICES THAT OPPOSE THE FLOW OF AC BY TEMPO-
rarily storing some of the electrical energy as a magnetic field. Such devices are called
inductors. The action of these components is known as *inductance*.

Inductors often, but not always, consist of wire coils. Sometimes a length of wire, or
a pair of wires, is used as an inductor. Some active electronic devices display inductance,
even when you don't think of the circuit in those terms.

Inductance can appear where it isn't wanted. Noncoil inductance becomes increasingly common as the frequency of an alternating current increases. At very-high,
ultra-high, and microwave radio frequencies, this phenomenon becomes a major consideration in the design of communications equipment.

The property of inductance

Suppose you have a wire a million miles long. What will happen if you make this wire into
a huge loop, and connect its ends to the terminals of a battery (Fig. 10-1)?

You can surmise that a current will flow through the loop of wire. But this is only part
of the picture.

If the wire was short, the current would begin to flow immediately and it would attain
a level limited by the resistance in the wire and in the battery. But because the wire is
extremely long, it will take awhile for the electrons from the negative terminal to work
their way around the loop to the positive terminal.

The effect of the current moves along the wire at a little less than the speed of light.
In this case, it's about 180,000 miles per second, perhaps 97 percent of the speed of light
in free space. It will take a little time for the current to build up to its maximum level. The
first electrons won't start to enter the positive terminal until more than five seconds have
passed.

The magnetic field produced by the loop will be small at first, because current is

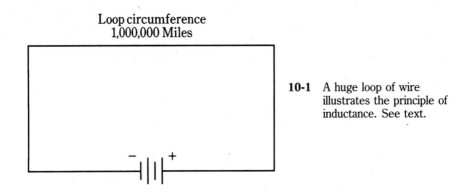

10-1   A huge loop of wire illustrates the principle of inductance. See text.

flowing in only part of the loop. The flux will increase over a period of a few seconds, as the electrons get around the loop. Figure 10-2 is an approximate graph of the overall magnetic field versus time. After about 5.5 seconds, current is flowing around the whole loop, and the magnetic field has reached its maximum.

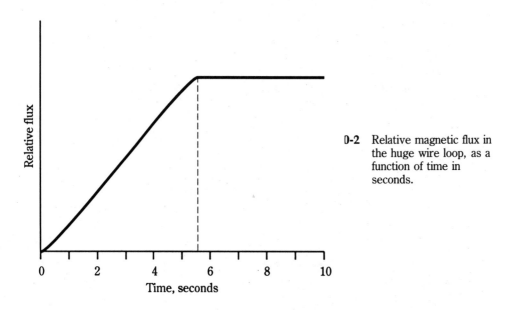

0-2   Relative magnetic flux in the huge wire loop, as a function of time in seconds.

A certain amount of energy is stored in this magnetic field. The ability of the loop to store energy in this way is the property of inductance. It is abbreviated by the letter L.

Practical inductors

Of course, it's not easy to make wire loops even approaching a million miles in circumference. But lengths of wire can be coiled up. When this is done, the magnetic flux is increased many times for a given length of wire compared with the flux produced by a single-turn loop. This is how inductors are made in practical electrical and electronic devices.

For any coil, the magnetic flux density is multiplied when a ferromagnetic core is placed within the coil of wire. Remember this from the study of magnetism. The increase in flux density has the effect of multiplying the inductance of a coil, so that it is many times greater with a ferromagnetic core than with an air core.

The current that an inductor can handle depends on the size of the wire. The inductance does not; it is a function of the number of turns in the coil, the diameter of the coil, and the overall shape of the coil.

In general, inductance of a coil is directly proportional to the number of turns of wire. Inductance is also directly proportional to the diameter of the coil. The length of a coil, given a certain number of turns and a certain diameter, has an effect also: the longer the coil, the less the inductance.

The unit of inductance

When a battery is connected across a wire-coil inductor (or any kind of inductor), it takes awhile for the current flow to establish itself throughout the inductor. The current changes at a rate that depends on the inductance: the greater the inductance, the slower the rate of change of current for a given battery voltage.

The unit of inductance is an expression of the ratio between the rate of current change and the voltage across an inductor. An inductance of one *henry*, abbreviated *H*, represents a potential difference of one volt across an inductor within which the current is increasing or decreasing at one ampere per second.

The henry is an extremely large unit of inductance. Rarely will you see an inductor anywhere near this large, although some power-supply filter chokes have inductances up to several henrys. Usually, inductances are expressed in *millihenrys (mH), microhenrys (uH)*, or even in *nanohenrys (nH)*. You should know your prefix multipliers fairly well by now, but in case you've forgotten, 1 mH = 0.001 H = 10^{-3} H, 1 uH = 0.001 mH = 0.000001 H = 10^{-6} H, and 1 nH = 0.001 uH = 10^{-9} H.

Very small coils, with few turns of wire, produce small inductances, in which the current changes quickly and the voltages are small. Huge coils with ferromagnetic cores, and having many turns of wire, have large inductances, in which the current changes slowly and the voltages are large.

Inductors in series

As long as the magnetic fields around inductors do not interact, inductances in series add like resistances in series. The total value is the sum of the individual values. It's important to be sure that you are using the same size units for all the inductors when you add their values.

Problem 10-1

Three 40-uH inductors are connected in series, and there is no interaction, or *mutual inductance*, among them (Fig. 10-3). What is the total inductance?

You can just add up the values. Call the inductances of the individual components L_1, L_2, and L_3, and the total inductance L. Then $L = L_1 + L_2 + L_3 = 40 + 40 + 40 = 120$ uH.

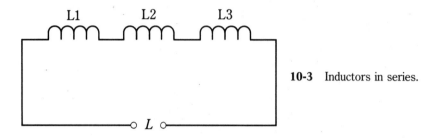

10-3   Inductors in series.

Problem 10-2

Suppose there are three inductors, with no mutual inductance, and their values are 20.0 mH, 55.0 uH, and 400 nH. What is the total inductance of these components if they are connected in series as shown in Fig. 10-3?

First, convert the inductances to the same units. You might use microhenrys, because that's the "middle-sized" unit here. Call $L_1 = 20.0$ mH $= 20,000$ uH; $L_2 = 55.0$ uH; $L_3 = 400$ nH $= 0.400$ uH. Then the total inductance is $L = 20,000 + 55.0 + 0.400$ uH $= 20,055.4$ uH. The values of the original separate components were each given to three significant figures, so you should round the final figure off to 20,100 uH.

Note that subscripts are now used in designators. An example is L_2 (rather than L2). Some engineers like subscripts, while others don't want to bother with them. You should get used to seeing them both ways. They're both alright. If there are several inductors in series, and one of them has a value much larger than the values of the others, then the total inductance will be only a little bit more than the value of the largest inductor.

Inductors in parallel

If there is no mutual inductance among two or more parallel-connected inductors, their values add up like the values of resistors in parallel. Suppose you have inductances L_1, L_2, L_3, ..., L_n all connected in parallel. Then you can find the reciprocal of the total inductance, 1/L, using the following formula:

$$1/L = 1/L_1 + 1/L_2 + 1/L_3 + ... + 1/L_n$$

The total inductance, L, is found by taking the reciprocal of the number you get for 1/L.

Again, as with inductances in series, it's important to remember that all the units have to agree. Don't mix microhenrys with millihenrys, or henrys with nanohenrys. The units you use for the individual component values will be the units you get for the final answer.

Problem 10-3

Suppose there are three inductors, each with a value of 40 uH, connected in parallel with no mutual inductance, as shown in Fig. 10-4. What is the net inductance of the set?

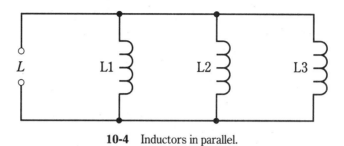

10-4   Inductors in parallel.

Call the inductances L_1 = 40 uH, L_2 = 40 uH and L_3 = 40 uH. Use the formula above to obtain 1/L = 1/40 + 1/40 + 1/40 = 3/40 = 0.075. Then L = 1/0.075 = 13.333 uH. This should be rounded off to 13 uH, because the original inductances are specified to only two significant digits.

Problem 10-4

Imagine that there are four inductors in parallel, with no mutual inductance. Their values are L_1 = 75 mH, L_2 = 40 mH, L_3 = 333 uH, and L_4 = 7.0 H. What is the total net inductance?

You can use henrys, millihenrys, or microhenrys as the standard units in this problem. Suppose you decide to use henrys. Then L_1 = 0.075 H, L_2 = 0.040 H, L_3 = 0.000333 H, and L_4 = 7.0 H. Use the above formula to obtain 1/L = 13.33 + 25 + 3003 + 0.143 = 3041.473. The reciprocal of this is the inductance L = 0.00032879 H = 328.79 uH. This can be rounded off to 330 uH because of significant-digits considerations.

This is just about the same as the 333-uH inductor alone. In real life, you could have only the single 330-uH inductor in this circuit, and the inductance would be essentially the same as with all four inductors.

If there are several inductors in parallel, and one of them has a value that is far smaller than the values of all the others, then the total inductance is just a little smaller than the value of the smallest inductor.

Interaction among inductors

In practical electrical circuits, there is almost always some mutual inductance between or among coils when they are wound in a cylindrical shape. The magnetic fields extend significantly outside solenoidal coils, and mutual effects are almost inevitable. The same is true between and among lengths of wire, especially at very-high, ultra-high, and microwave radio frequencies. Sometimes, mutual inductance is all right, and doesn't have a detrimental effect on the behavior of a circuit. But it can be a bad thing.

Mutual inductance can be minimized by using *shielded* wires and *toroidal* inductors. The most common shielded wire is *coaxial cable*. Toroidal inductors are discussed a little later in this chapter.

Coefficient of coupling

The *coefficient of coupling*, specified by the letter k, is a number ranging from 0 (no coupling) to 1 (maximum possible coupling). Two coils that are separated by a sheet of solid iron would have essentially k = 0; two coils wound on the same form, one right over the other, would have practically k = 1.

Mutual inductance

The mutual inductance is specified by the letter M, and is expressed in the same units as inductance: henrys, millihenrys, microhenrys, or nanohenrys. The value of M is a function of the values of the inductors, and also of the coefficient of coupling.

For two inductors, having values of L_1 and L_2 (both expressed in the same size units), and with a coefficient of coupling k, the mutual inductance M is found by multiplying the inductance values, taking the square root of the result, and then multiplying by k. Mathematically,

$$M = k \, (L_1 L_2)^{1/2}$$

Effects of mutual inductance

Mutual inductance can operate either to increase the inductance of a pair of series-connected inductors, or to decrease it. This is because the magnetic fields might reinforce each other, or they might act against each other.

When two inductors are connected in series, and there is *reinforcing* mutual inductance between them, the total inductance L is given in the formula:

$$L = L_1 + L_2 + 2M$$

where L_1 and L_2 are the values of the individual inductors, and M is the mutual inductance. All inductances must be expressed in the same size units.

Problem 10-5

Suppose two coils, having values of 30 uH and 50 uH, are connected in series so that their fields reinforce, as shown in Fig. 10-5, and that the coefficient of coupling is 0.5. What is the total inductance of the combination?

First, calculate M from k. According to the formula for this, given above, M = 0.5(50 × 30)$^{1/2}$ = 19.4 uH. Then the total inductance is equal to L = L_1 + L_2 + 2M = 30 + 50 + 38.8 = 118.8 uH, rounded to 120 uH because only two significant digits are justified.

When two inductors are connected in series and the mutual inductance is *in opposition*, the total inductance L is given by the formula:

$$L = L_1 + L_2 - 2M$$

where, again, L_1 and L_2 are the values of the individual inductors.

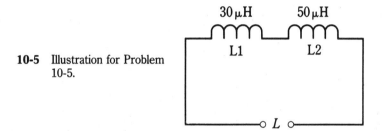

10-5 Illustration for Problem 10-5.

Problem 10-6

There are two coils with values L_1 = 835 uH and L_2 = 2.44 mH. They are connected in series so that their coefficient of coupling is 0.922, acting so that the coils oppose each other, as shown in Fig. 10-6. What is the net inductance of the pair?

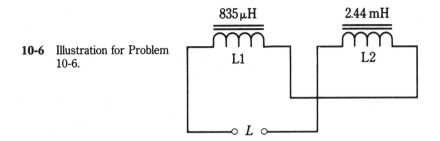

10-6 Illustration for Problem 10-6.

First, calculate M. Notice that the coil inductances are specified in different units. Convert them both to microhenrys, so that L_2 becomes 2440 uH. Then M = 0.922(835 × 2440)$^{1/2}$ = 1316 uH. The total inductance is therefore L = L_1 + L_2 − 2M = 835 + 2440 − 2632 = 643 uH.

It is possible for mutual inductance to increase the total series inductance of a pair of coils by as much as a factor of 2, if the coupling is total and if the flux reinforces. Conversely, it is possible for the inductances of two coils to cancel each other. If two equal-valued inductors are connected in series so that their fluxes oppose, the result will be theoretically zero inductance.

Air-core coils

The simplest inductors (besides plain, straight lengths of wire) are coils. A coil can be wound on a plastic, wooden or other nonferromagnetic material, and it will work very well, although no *air-core* inductor can have very much inductance. In practice, the maximum attainable inductance for such coils is about 1 mH.

Air-core coils are used mostly at radio frequencies, in transmitters, receivers, and antenna networks. In general, the higher the frequency of an alternating current, the less inductance is needed to produce significant effects. Air-core coils can be made to have almost unlimited current-carrying capacity, just by using heavy-gauge wire and making the radius of the coil large. Air does not dissipate much energy in the form of heat; it is almost *lossless*. For these reasons, air-core coils can be made highly efficient.

Powdered-iron and ferrite cores

Ferromagnetic substances can be crushed into dust and then bound into various shapes, providing core materials that greatly increase the inductance of a coil having a given number of turns. Depending on the mixture used, the increase in flux density can range from a factor of a few times, up through hundreds, thousands, and even millions of times. A small coil can thus be made to have a large inductance.

Powdered-iron cores are common at radio frequencies. *Ferrite* has a higher permeability than powdered iron, causing a greater concentration of magnetic flux lines within the coil. Ferrite is used at lower radio frequencies and at audio frequencies, as well as at medium and high radio frequencies.

The main trouble with ferromagnetic cores is that, if the coil carries more than a certain amount of current, the core will *saturate*. This means that the ferromagnetic material is holding as much flux as it possibly can. Any further increase in coil current will not produce a corresponding increase in the magnetic flux in the core. The result is that the inductance changes, decreasing with coil currents that are more than the critical value.

In extreme cases, ferromagnetic cores can waste considerable power as heat. If a core gets hot enough, it might fracture. This will permanently change the inductance of the coil, and will also reduce its current-handling ability.

Permeability tuning

Solenoidal, or cylindrical, coils can be made to have variable inductance by sliding ferromagnetic cores in and out of them. This is a common practice in radio communications. The frequency of a radio circuit can be adjusted in this way, as you'll learn later in this book.

Because moving the core in and out changes the effective permeability within a coil of wire, this method of tuning is called *permeability tuning*. The in/out motion can be precisely controlled by attaching the core to a screw shaft, and anchoring a nut at one end of the coil (Fig. 10-7). As the screw shaft is rotated clockwise, the core enters the coil, so that the inductance increases. As the screw shaft is rotated counterclockwise, the core moves out of the coil and the inductance decreases.

Toroids

Inductor coils do not have to be wound on cylindrical forms, or on cylindrical ferromagnetic cores. In recent years, a new form of coil has become increasingly common. This is the *toroid*. It gets its name from the donut shape of the ferromagnetic core. The coil is wound over a core having this shape (Fig. 10-8).

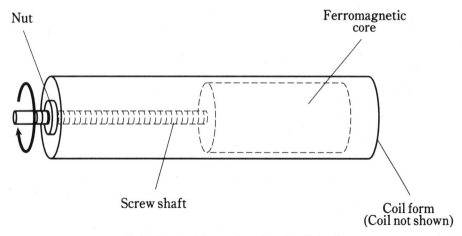

10-7 Permeability tuning of a solenoidal coil.

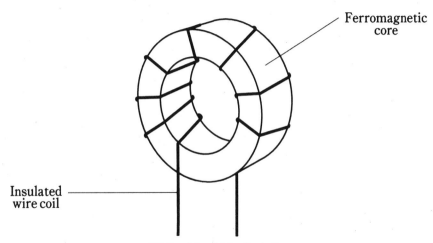

10-8 A toroidal coil winding.

There are several advantages to toroidal coils over solenoidal, or cylindrical, ones. First, fewer turns of wire are needed to get a certain inductance with a toroid, as compared with a solenoid. Second, a toroid can be physically smaller for a given inductance and current-carrying capacity. Third, and perhaps most important, essentially all of the flux in a toroidal inductor is contained within the core material. This reduces unwanted mutual inductances with components near the toroid.

There are some disadvantages, or limitations in the flexibility, of toroidal coils. It is more difficult to permeability-tune a toroidal coil than it is to tune a solenoidal one. It's been done, but the hardware is cumbersome. Toroidal coils are harder to wind than solenoidal ones.

Sometimes, mutual inductance between or among physically separate coils is wanted; with a toroid, the coils have to be wound on the same form for this to be possible.

Pot cores

There is another way to confine the magnetic flux in a coil so that unwanted mutual inductance does not occur. This is to extend a solenoidal core completely around the outside of the coil, making the core into a shell (Fig. 10-9). This is known as a *pot core*. Whereas in most inductors the coil is wound around the form, in a pot core the form is wrapped around the coil.

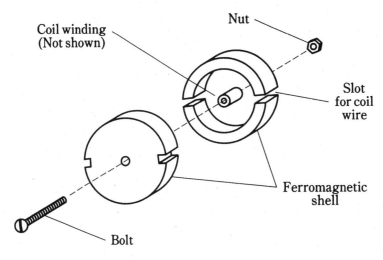

10-9 A pot core shell. Coil winding is not shown.

The core comes in two halves, inside one of which the coil is wound. Then the parts are assembled and held together by a bolt and nut. The entire assembly looks like a miniature oil tank. The wires come out of the core through small holes.

Pot cores have the same advantages as toroids. The core tends to prevent the magnetic flux from extending outside the physical assembly. Inductance is greatly increased compared to solenoidal windings having a comparable number of turns. In fact, pot cores are even better than toroids if the main objective is to get an extremely large inductance within a small volume of space.

The main disadvantage of a pot core is that tuning, or adjustment of the inductance, is all but impossible. The only way to do it is by switching in different numbers of turns, using *taps* at various points on the coil.

Filter chokes

The largest values of inductance that can be obtained in practice are on the order of several henrys. The primary use of a coil this large is to smooth out the pulsations in direct current that result when ac is *rectified* in a power supply. This type of coil is known as a *filter choke*. You'll learn more about power supplies later in this book.

Inductors at audio frequency

Inductors at audio frequencies range in value from a few millihenrys up to about 1 H. They are almost always toroidally wound, or are wound in a pot core, or comprise part of an audio transformer.

Inductors can be used in conjunction with moderately large values of capacitance in order to obtain audio *tuned circuits*. However, in recent years, audio tuning has been taken over by active components, particularly *integrated circuits*.

Inductors at radio frequency

The radio frequencies range from 9 kHz to well above 100 GHz. At the low end of this range, inductors are similar to those at audio frequencies. As the frequency increases, cores having lower permeability are used. Toroids are quite common up through about 30 MHz. Above that frequency, air-core coils are more often used.

In radio-frequency (rf) circuits, coils are routinely connected in series or in parallel with capacitors to obtain tuned circuits. Other arrangements yield various characterstics of *attenuation versus frequency*, serving to let signals at some frequencies pass, while rejecting signals at other frequencies. You'll learn more about this in the chapter on *resonance*.

Transmission-line inductors

At radio frequencies of more than about 100 MHz, another type of inductor becomes practical. This is the type formed by a length of *transmission line*.

A transmission line is generally used to get energy from one place to another. In radio communications, transmission lines get energy from a transmitter to an antenna, and from an antenna to a receiver.

Types of transmission line

Transmission lines usually take either of two forms, the *parallel-wire* type or the *coaxial* type.

A parallel-wire transmission line consists of two wires running alongside each other with a constant spacing (Fig. 10-10). The spacing is maintained by polyethylene rods molded at regular intervals to the wires, or by a solid web of polyethylene. You have seen this type of line used with television receiving antennas. The substance separating the wires is called the *dielectric* of the transmission line.

A coaxial transmission line has a wire conductor surrounded by a tubular braid or pipe (Fig. 10-11). The wire is kept at the center of this tubular *shield* by means of polythylene beads, or more often, by solid or foamed polyethylene dielectric, all along the length of the line.

Line inductance

Short lengths of any type of transmission line behave as inductors, as long as the line is less than 90 electrical degrees in length. At 100 MHz, 90 electrical degrees, or ¼

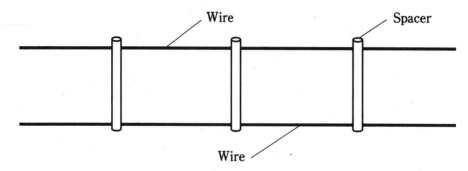

10-10   Parallel-wire transmission line.

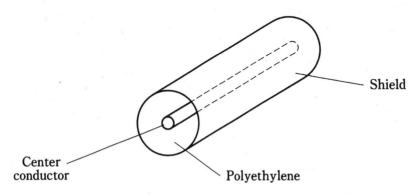

10-11   Coaxial transmission line.

wavelength, in free space is just 75 cm, or a little more than 2 ft. In general, if f is the frequency in megahertz, then ¼ wavelength (s) in free space, in centimeters, is given by:

$$s = 7500/f$$

The length of a quarter-wavelength section of transmission line is shortened from the free-space quarter wavelength by the effects of the dielectric. In practice, ¼ wavelenth along the line can be anywhere from about 0.66 of the free-space length (for coaxial lines with solid polyethylene dielectric) to about 0.95 of the free-space length (for parallel-wire line with spacers molded at intervals of several inches).

The factor by which the wavelength is shortened is called the *velocity factor* of the line. This is because the shortening of the wavelength is a result of a slowing-down of the speed with which the radio signals move in the line, as compared with their speed in space (the speed of light). If the velocity factor of a line is given by v, then the above formula for the length of a quarter-wave line, in centimeters, becomes:

$$s = 7500v/f$$

Very short lengths of line—a few electrical degrees—produce small values of inductance. As the length approaches ¼ wavelength, the inductance increases.

Transmission line inductors behave differently than coils in one important way: the inductance of a transmission-line section changes as the frequency changes. At first, the inductance will become larger as the frequency increases. At a certain limiting frequency, the inductance becomes infinite. Above that frequency, the line becomes *capacitive* instead. You'll learn about capacitance shortly.

A detailed discussion of frequency, transmission line type and length, and inductance is beyond the level of this book. Texts on radio engineering are recommended for further information on this subject.

Unwanted inductances

Any length of wire has some inductance. As with a transmission line, the inductance of a wire increases as the frequency increases. Wire inductance is therefore more significant at radio frequencies than at audio frequencies.

In some cases, especially in radio communications equipment, the inductance of, and among, wires can become a major bugaboo. Circuits can oscillate when they should not. A receiver might respond to signals that it's not designed to intercept. A transmitter can send out signals on unauthorized and unintended frequencies. The frequency response of any circuit can be altered, degrading the performance of the equipment.

Sometimes the effects of *stray inductance* are so small that they are not important; this might be the case in a stereo hi-fi set located at a distance from other electronic equipment. In some cases, stray inductance can cause life-threatening malfunctions. This might happen with certain medical devices.

The most common way to minimize stray inductance is to use coaxial cables between and among sensitive circuits or components. The shield of the cable is connected to the *common ground* of the apparatus.

Quiz

Refer to the text in this chapter if necessary. A good score is 18 correct. Answers are in the back of the book.

1. An inductor works by:
    A. Charging a piece of wire.
    B. Storing energy as a magnetic field.
    C. Choking off high-frequency ac.
    D. Introducing resistance into a circuit.

2. Which of the following does *not* affect the inductance of a coil?
    A. The diameter of the wire.
    B. The number of turns.
    C. The type of core material.
    D. The length of the coil.

3. In a small inductance:
    A. Energy is stored and released slowly.

B. The current flow is always large.

C. The current flow is always small.

D. Energy is stored and released quickly.

4. A ferromagnetic core is placed in an inductor mainly to:

A. Increase the current carrying capacity.

B. Increase the inductance.

C. Limit the current.

D. Reduce the inductance.

5. Inductors in series, assuming there is no mutual inductance, combine:

A. Like resistors in parallel.

B. Like resistors in series.

C. Like batteries in series with opposite polarities.

D. In a way unlike any other type of component.

6. Two inductors are connected in series, without mutual inductance. Their values are 33 mH and 55 mH. The net inductance of the combination is:

A. 1.8 H.

B. 22 mH.

C. 88 mH.

D. 21 mH.

7. If the same two inductors (33 mH and 55 mH) are connected in parallel without mutual inductance, the combination will have a value of:

A. 1.8 H.

B. 22 mH.

C. 88 mH.

D. 21 mH.

8. Three inductors are connected in series without mutual inductance. Their values are 4 nH, 140 uH, and 5 H. For practical purposes, the net inductance will be very close to:

A. 4 nH.

B. 140 uH.

C. 5 H.

D. None of these.

9. Suppose the three inductors mentioned above are connected in parallel without mutual inductance. The net inductance will be close to:

A. 4 nH.

B. 140 uH.

C. 5 H.

D. None of these.

10. Two inductors, each of 100 uH, are in series. The coefficient of coupling is 0.40. The net inductance, if the coil fields reinforce each other, is:

    A. 50 uH.

    B. 120 uH.

    C. 200 uH.

    D. 280 uH.

11. If the coil fields oppose in the foregoing series-connected arrangement, the net inductance is:

    A. 50 uH.

    B. 120 uH.

    C. 200 uH.

    D. 280 uH.

12. Two inductors, having values of 44 mH and 88 mH, are connected in series with a coefficient of coupling equal to 1.0 (maximum possible mutual inductance). If their fields reinforce, the net inductance (to two significant digits) is:

    A. 7.5 mH.

    B. 132 mH.

    C. 190 mH.

    D. 260 mH.

13. If the fields in the previous situation oppose, the net inductance will be:

    A. 7.5 mH.

    B. 132 mH.

    C. 190 mH.

    D. 260 mH.

14. With permeability tuning, moving the core further into a solenoidal coil:

    A. Increases the inductance.

    B. Reduces the inductance.

    C. Has no effect on the inductance, but increases the current-carrying capacity of the coil.

    D. Raises the frequency.

15. A significant advantage, in some situations, of a toroidal coil over a solenoid is:

    A. The toroid is easier to wind.

    B. The solenoid cannot carry as much current.

    C. The toroid is easier to tune.

D. The magnetic flux in a toroid is practically all within the core.

16. A major feature of a pot-core winding is:
    A. High current capacity.
    B. Large inductance in small volume.
    C. Efficiency at very high frequencies.
    D. Ease of inductance adjustment.

17. As an inductor core material, air:
    A. Has excellent efficiency.
    B. Has high permeability.
    C. Allows large inductance in a small volume.
    D. Has permeability that can vary over a wide range.

18. At a frequency of 400 Hz, the most likely form for an inductor would be:
    A. Air-core.
    B. Solenoidal.
    C. Toroidal.
    D. Transmission-line.

19. At a frequency of 95 MHz, the best form for an inductor would be:
    A. Air-core.
    B. Pot core.
    C. Either of the above.
    D. Neither of the above.

20. A transmission-line inductor made from coaxial cable, having velocity factor of 0.66, and working at 450 MHz, would be shorter than:
    A. 16.7 m.
    B. 11 m.
    C. 16.7 cm.
    D. 11 cm.

11
CHAPTER

Capacitance

ELECTRICAL COMPONENTS CAN OPPOSE AC IN THREE DIFFERENT WAYS, TWO OF which you've learned about already.

Resistance slows down the rate of transfer of charge carriers (usually electrons) by "brute force." In this process, some of the energy is invariably converted from electrical form to heat. Resistance is said to *consume power* for this reason. Resistance is present in dc as well as in ac circuits, and works the same way for either direct or alternating current.

Inductance impedes the flow of ac charge carriers by temporarily storing the energy as a magnetic field. But this energy is given back later.

Capacitance, about which you'll learn in this chapter, impedes the flow of ac charge carriers by temporarily storing the energy as an *electric* field. This energy is given back later, just as it is in an inductance. Capacitance is not generally important in pure-dc circuits. It can have significance in circuits where dc is pulsating, and not steady.

Capacitance, like inductance, can appear when it is not wanted or intended. As with inductance, this effect tends to become more evident as the ac frequency increases.

The property of capacitance

Imagine two very large, flat sheets of metal such as copper or aluminum, that are excellent electrical conductors. Suppose they are each the size of the state of Nebraska, and are placed one over the other, separated by just a foot of space. What will happen if these two sheets of metal are connected to the terminals of a battery, as shown in Fig. 11-1?

The two plates will become charged electrically, one positively and the other negatively. You might think that this would take a little while, because the sheets are so big. This is an accurate supposition.

If the plates were small, they would both become charged almost instantly, attaining a relative voltage equal to the voltage of the battery. But because the plates are gigantic,

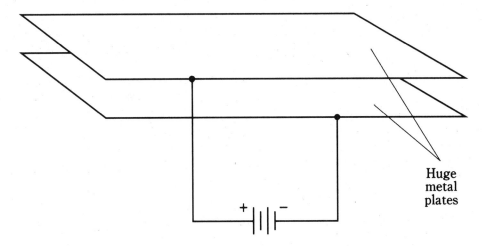

11-1   A huge pair of parallel plates illustrates the principle of capacitance. See text.

it will take awhile for the negative one to "fill up" with electrons, and it will take an equal amount of time for the other one to get electrons "sucked out." Finally, however, the voltage between the two plates will be equal to the battery voltage, and an electric field will exist in the space between the plates.

This electric field will be small at first; the plates don't charge right away. But the charge will increase over a period of time, depending on how large the plates are, and also depending on how far apart they are. Figure 11-2 is a relative graph showing the intensity of the electric field between the plates as a function of time, elapsed from the instant the plates are connected to the battery terminals.

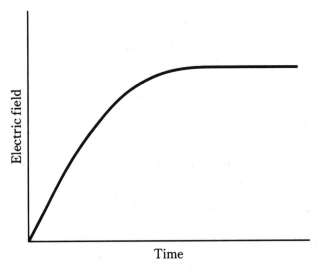

11-2   Relative electric field intensity between the huge metal plates, as a function of time.

Energy will be stored in this electric field. The ability of the plates, and of the space between them, to store this energy is the property of *capacitance*. It is denoted by the letter *C*.

Practical capacitors

It's out of the question to make a capacitor of the above dimensions. But two sheets, or strips, of foil can be placed one on top of the other, separated by a thin, nonconducting sheet such as paper, and then the whole assembly can be rolled up to get a large effective surface area. When this is done, the electric flux becomes great enough so that the device exhibits significant capacitance. In fact, two sets of several plates each can be meshed together, with air in between them, and the resulting capacitance will be significant at high ac frequencies.

In a capacitor, the electric flux concentration is multiplied when a *dielectric* of a certain type is placed between the plates. Plastics work very well for this purpose. This increases the effective surface area of the plates, so that a physically small component can be made to have a large capacitance.

The voltage that a capacitor can handle depends on the thickness of the metal sheets or strips, on the spacing between them, and on the type of dielectric used.

In general, capacitance is directly proportional to the surface area of the conducting plates or sheets. Capacitance is *inversely proportional* to the separation between conducting sheets; in other words, the closer the sheets are to each other, the greater the capacitance. The capacitance also depends on the *dielectric constant* of the material between the plates. A vacuum has a dielectric constant of 1; some substances have dielectric constants that multiply the effective capacitance many times.

The unit of capacitance

When a battery is connected between the plates of a capacitor, it takes some time before the electric field reaches its full intensity. The voltage builds up at a rate that depends on the capacitance: the greater the capacitance, the slower the rate of change of voltage in the plates.

The unit of capacitance is an expression of the ratio between the amount of current flowing and the rate of voltage change across the plates of a capacitor. A capacitance of one *farad*, abbreviated F, represents a current flow of one ampere while there is a potential-difference increase or decrease of one volt per second. A capacitance of one farad also results in one volt of potential difference for an electric charge of one coulomb.

The farad is a huge unit of capacitance. You'll almost never see a capacitor with a value of 1 F. Commonly employed units of capacitance are the *microfarad (uF)* and the *picofarad (pF)*. A capacitance of 1 uF represents a millionth (10^{-6}) of a farad, and 1 pF is a millionth of a microfarad, or a trillionth of a farad (10^{-12} F).

Some quite large capacitances can be stuffed into physically small components. Conversely, some capacitors with small values take up large volumes. The bulkiness of a capacitor is proportional to the voltage that it can handle, more than it is related to the capacitance. The higher the rated voltage, the bigger, physically, the component will be.

Capacitors in series

With capacitors, there is almost never any mutual interaction. This makes capacitors somewhat easier to work with than inductors.

Capacitors in series add together like resistors in parallel. If you connect two capacitors of the same value in series, the result will be half the capacitance of either component alone. In general, if there are several capacitors in series, the composite value will be less than any of the single components. It is important that you always use the same size units when determining the capacitance of any combination. Don't mix microfarads with picofarads. The answer that you get will be in whichever size units you use for the individual components.

Suppose you have several capacitors with values C_1, C_2, C_3, ..., C_n all connected in series. Then you can find the reciprocal of the total capacitance, $1/C$, using the following formula:

$$1/C = 1/C_1 + 1/C_2 + 1/C_3 + ... + 1/C_n$$

The total capacitance, C, is found by taking the reciprocal of the number you get for $1/C$.

If two or more capacitors are connected in series, and one of them has a value that is extremely tiny compared with the values of all the others, the composite capacitance can be taken as the value of the smallest component.

Problem 11-1

Two capacitors, with values of $C_1 = 0.10$ uF and $C_2 = 0.050$ uF, are connected in series (Fig. 11-3). What is the total capacitance?

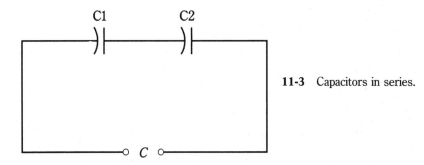

11-3   Capacitors in series.

Using the above formula, first find the reciprocals of the values. They are $1/C_1 = 10$ and $1/C_2 = 20$. Then $1/C = 10 + 20 = 30$, and $C = 1/30 = 0.033$ uF.

Problem 11-2

Two capacitors with values of 0.0010 uF and 100 pF are connected in series. What is the total capacitance?

Convert to the same size units. A value of 100 pF represents 0.000100 uF. Then you can say that $C_1 = 0.0010$ uF and $C_2 = 0.00010$ uF. The reciprocals are $1/C_1 = 1000$ and

$1/C_2 = 10,000$. Therefore, $1/C = 1,000 + 10,000 = 11,000$, and $C = 0.000091$ uF. This number is a little awkward, and you might rather say it's 91 pF.

In the above problem, you could have chosen pF to work with, rather than uF. In either case, there is some tricky decimal placement involved. It's important to double-check calculations when numbers get like this. Calculators will take care of the decimal placement problem, sometimes using exponent notation and sometimes not, but a calculator can only work with what you put into it! If you put a wrong number in, you will get a wrong answer, perhaps off by a factor of 10, 100, or even 1,000.

Problem 11-3

Five capacitors, each of 100 pF, are in series. What is the total capacitance?

If there are n capacitors in series, all of the same value so that $C_1 = C_2 = C_3 = \ldots = C_n$, the total value C is just 1/n of the capacitance of any of the components alone. Because there are five 100-pF capacitors here, the total is $C = 100/5 = 20$ pF.

Capacitors in parallel

Capacitances in parallel add like resistances in series. That is, the total capacitance is the sum of the individual component values. Again, you need to be sure that you use the same size units all the way through.

If two or more capacitors are connected in parallel, and one of the components is much, much larger than any of the others, the total capacitance can be taken as simply the value of the biggest one.

Problem 11-4

Three capacitors are in parallel, having values of $C_1 = 0.100$ uF, $C_2 = 0.0100$ uF, and $C_3 = 0.00100$ uF, as shown in Fig. 11-4. What is the total capacitance?

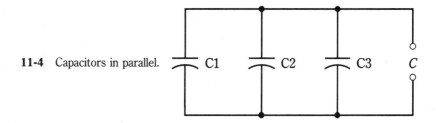

11-4   Capacitors in parallel.

Just add them up: $C = 0.100 + 0.0100 = 0.00100 = 0.11100$ uF. Because the values are given to three significant figures, the final answer should be stated as $C = 0.111$ uF.

Problem 11-5

Two capacitors are in parallel, one with a value of 100 uF and one with a value of 100 pF. What is the effective total capacitance?

In this case, without even doing any calculations, you can say that the total is 100 uF for practical purposes. The 100-pF unit is only a millionth of the capacitance of the 100-uF component; therefore, the smaller capacitor contributes essentially nothing to the composite total.

Dielectric materials

Just as certain solids can be placed within a coil to increase the inductance, materials exist that can be sandwiched in between the plates of a capacitor to increase the capacitance. The substance between the plates is called the *dielectric* of the capacitor.

Air works quite well as a dielectric. It has almost no loss. But it is difficult to get very much capacitance using air as the dielectric. Some solid material is usually employed as the dielectric for most *fixed* capacitors, that is, for types manufactured to have a constant, unchangeable value of capacitance.

Dielectric materials conduct electric fields well, but they are not good conductors of electric currents. In fact, the materials are known as good insulators.

Solid dielectrics increase the capacitance for a given surface area and spacing of the plates. Solid dielectrics also allow the plates to be rolled up, squashed, and placed very close together (Fig. 11-5). Both of these act to increase the capacitance per unit volume, allowing reasonable capacitances to exist in a small volume.

Foil

Foil

Dielectric

11-5   Foil sheets can be rolled up with dielectric material sandwiched in between.

Paper capacitors

In the early days of radio, capacitors were commonly made by placing paper, soaked with mineral oil, between two strips of foil, rolling the assembly up, attaching wire leads to the two pieces of foil, and enclosing the rolled-up foil and paper in a cylindrical case.

These capacitors can still sometimes be found in electronic equipment. They have values ranging from about 0.001 uF to 0.1 uF, and can handle low to moderate voltages, usually up to about 1000 V.

Mica capacitors

When you were a child, you might have seen mica, a naturally occurring, transparent substance that flakes off in thin sheets. This material makes an excellent dielectric for capacitors.

Mica capacitors can be made by alternately stacking metal sheets and layers of mica, or by applying silver ink to the sheets of mica. The metal sheets are wired together into two meshed sets, forming the two terminals of the capacitor. This scheme is shown in Fig. 11-6.

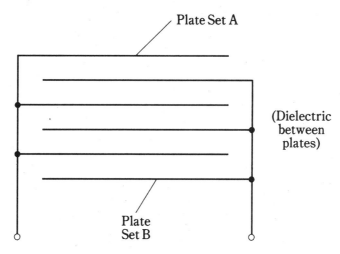

11-6   Meshing of plates to increase capacitance.

Mica capacitors have low loss; that is, they waste very little power as heat, provided their voltage rating is not exceeded. Voltage ratings can be up to several thousand volts if thick sheets of mica are used. But mica capacitors tend to be large physically in proportion to their capacitance. The main application for mica capacitors is in radio receivers and transmitters. Their capacitances are a little lower than those of paper capacitors, ranging from a few tens of picofarads up to about 0.05 uF.

Ceramic capacitors

Porcelain is another material that works well as a dielectric. Sheets of metal are stacked alternately with wafers of ceramic to make these capacitors. The meshing/layering geometry of Fig. 11-6 is used. Ceramic, like mica, has quite low loss, and therefore allows for high efficiency.

For low values of capacitance, just one layer of ceramic is needed, and two metal plates can be glued to the disk-shaped porcelain, one on each side. This type of component is known as a *disk-ceramic* capacitor. Alternatively, a tube or cylinder of ceramic can be employed, and metal ink applied to the inside and outside of the tube. Such units are called *tubular* capacitors.

Ceramic capacitors have values ranging from a few picofrads to about 0.5 uF. Their voltage ratings are comparable to those of paper capacitors.

Plastic-film capacitors

Various different plastics make good dielectrics for the manufacture of capacitors. Polyester, polyethylene, and polystyrene are commonly used. The substance called *mylar* that you might have seen used to tint windows makes a good dielectric for capacitors.

The method of manufacture is similar to that for paper capacitors when the plastic is flexible. Stacking methods can be used if the plastic is more rigid. The geometries can vary, and these capacitors are therefore found in several different shapes.

Capacitance values for plastic-film units range from about 50 pF to several tens of microfarads. Most often they are in the range of 0.001 uF to 10 uF. Plastic capacitors are employed at audio and radio frequencies, and at low to moderate voltages. The efficiency is good, although not as high as that for mica-dielectric or air-dielectric units.

Electrolytic capacitors

All of the above-mentioned types of capacitors provide relatively small values of capacitance. They are also *nonpolarized*, meaning that they can be hooked up in a circuit in either direction. An *electrolytic* capacitor provides considerably greater capacitance than any of the above types, but it must be connected in the proper direction in a circuit to work right. Therefore, an electrolytic capacitor is a *polarized* component.

Electrolytic capacitors are made by rolling up aluminum foil strips, separated by paper saturated with an *electrolyte* liquid. The electrolyte is a conducting solution. When dc flows through the component, the aluminum *oxidizes* because of the electrolyte. The oxide layer is nonconducting, and forms the dielectric for the capacitor. The layer is extremely thin, and this results in a high capacitance per unit volume.

Electrolytic capacitors can have values up to thousands of microfarads, and some units can handle thousands of volts. These capacitors are most often seen in audio-frequency circuits and in dc power supplies.

Tantalum capacitors

Another type of electrolytic capacitor uses tantalum rather than aluminum. The tantalum can be foil, as is the aluminum in a conventional electrolytic capacitor. It might also take the form of a porous pellet, the irregular surface of which provides a large area in a small volume. An extremely thin oxide layer forms on the tantalum.

Tantalum capacitors have high reliability and excellent efficiency. They are often used in military applications because they do not fail often. They can be used in audio-frequency and digital circuits in place of aluminum electrolytics.

Semiconductor capacitors

A little later in this book, you'll learn about *semiconductors*. These materials, in their many different forms, have revolutionized electrical and electronic circuit design in the past several decades.

These materials can be employed to make capacitors. A semiconductor *diode* conducts current in one direction, and refuses to conduct in the other direction. When a voltage source is connected across a diode so that it does not conduct, the diode acts as a capacitor. The capacitance varies depending on how much of this *reverse voltage* is applied to the diode. The greater the reverse voltage, the smaller the capacitance. This makes the diode a variable capacitor. Some diodes are especially manufactured to serve this function. Their capacitances fluctuate rapidly along with pulsating dc. These components are called *varactor diodes* or simply *varactors*.

Capacitors can be formed in the semiconductor materials of an *integrated circuit (IC)* in much the same way. Sometimes, IC diodes are fabricated to serve as varactors. Another way to make a capacitor in an IC is to sandwich an oxide layer into the semiconductor material, between two layers that conduct well.

You have probably seen ICs in electronic equipment; almost any personal computer has dozens of them. They look like little boxes with many prongs (Fig. 11-7).

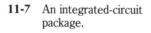 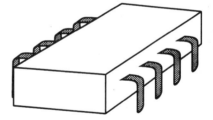

11-7   An integrated-circuit package.

Semiconductor capacitors usually have small values of capacitance. They are physically tiny, and can handle only low voltages. The advantages are miniaturization, and an ability, in the case of the varactor, to change in value at a rapid rate.

Variable capacitors

Capacitors can be varied in value by adjusting the mutual surface area between the plates, or by changing the spacing between the plates. The two most common types of variable capacitors (besides varactors) are the *air variable* and the *trimmer*. You might also encounter *coaxial capacitors*.

Air variables

By connecting two sets of metal plates so that they mesh, and by affixing one set to a rotatable shaft, a variable capacitor is made. The rotatable set of plates is called the *rotor*, and the fixed set is called the *stator*. This is the type of component you might have seen in older radio receivers, used to tune the frequency. Such capacitors are still used in transmitter output tuning networks. Figure 11-8 is a functional rendition of an air-variable capacitor.

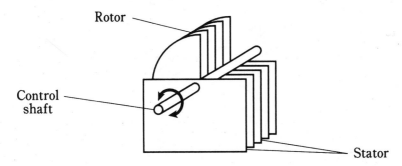

11-8 Simplified drawing of an air-variable capacitor.

Air variables have maximum capacitance that depends on the number of plates in each set, and also on the spacing between the plates. Common maximum values are 50 pF to about 1,000 pF; minimum values are a few picofarads. The voltage-handling capability depends on the spacing between the plates; some air variables can handle many kilovolts.

Air variables are used primarily at radio frequencies. They are highly efficient, and are nonpolarized, although the rotor is usually connected to common ground (the chassis or circuit board).

Trimmers

When it is not necessary to change the value of a capacitor very often, a trimmer might be used. It consists of two plates, mounted on a ceramic base and separated by a sheet of mylar, mica, or some other dielectric. The plates are "springy" and can be squashed together more or less by means of a screw (Fig. 11-9). Sometimes two sets of several plates are interleaved to increase the capacitance.

Trimmers can be connected in parallel with an air variable, so that the range of the air variable can be adjusted. Some air-variable capacitors have trimmers built in.

Typical maximum values for trimmers range from a few picofarads up to about 200 pF. They handle low to moderate voltages, and are highly efficient. They are nonpolarized.

Coaxial capacitors

Recall from the previous chapter that sections of transmission lines can work as inductors. They can act as capacitors, too.

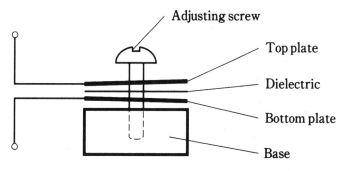

11-9 A trimmer capacitor.

If a section of transmission line is less than ¼ wavelength long, and is left open at the far end (rather than shorted out), it will act as a capacitor. The capacitance will increase with length.

The most common transmission-line capacitor uses two telescoping sections of tubing. This is called a *coaxial capacitor* and works because there is a certain effective surface area between the inner and the outer tubing sections. A sleeve of plastic dielectric is placed between the sections of tubing, as shown in Fig. 11-10. This allows the capacitance to be adjusted by sliding the inner section in or out of the outer section.

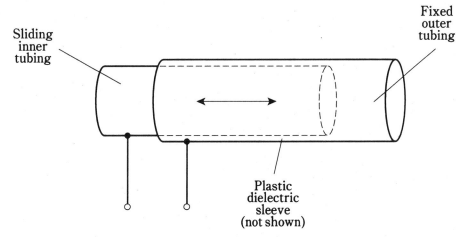

11-10 A coaxial variable capacitor.

Coaxial capacitors are used in radio-frequency applications, particularly in antenna systems. Their values are generally from a few picofarads up to about 100 pF.

Tolerance

Capacitors are rated according to how nearly their values can be expected to match the rated capacitance. The most common tolerance is 10 percent; some capacitors are rated at 5 percent or even at 1 percent. In all cases, the tolerance ratings are *plus-or-minus*.

Therefore, a 10-percent capacitor can range from 10 percent less than its assigned value to 10 percent more.

Problem 11-6

A capacitor is rated at 0.001 uF, plus-or-minus 10 percent. What is the actual range of capacitances it can have?

First, multiply 0.001 by 10 percent to get the plus-or-minus variation. This is 0.001 × 0.1 = 0.0001 uF. Then add and subtract this from the rated value to get the maximum and minimum possible capacitances. The result is 0.0011 uF to 0.0009 uF.

You might prefer to work with picofarads instead of microfarads, if the small numbers make you feel uneasy. Just change 0.001 uF to 1000 pF. Then the variation is plus-or-minus 1000 × 0.1 = 100 pF, and the range becomes 1100 pF to 900 pF.

Temperature coefficient

Some capacitors increase in value as the temperature increases. These components have a *positive temperature coefficient*. Some capacitors' values get less as the temperature rises; these have *negative temperature coefficient*. Some capacitors are manufactured so that their values remain constant over a certain temperature range. Within this span of temperatures, such capacitors have *zero temperature coefficient*.

The temperature coefficient is specified in percent per degree Celsius.

Sometimes, a capacitor with a negative temperature coefficient can be connected in series or parallel with a capacitor having a positive temperature coefficient, and the two opposite effects will cancel each other out over a range of temperatures. In other instances, a capacitor with a positive or negative temperature coefficient can be used to cancel out the effect of temperature on other components in a circuit, such as inductors and resistors.

You won't have to do calculations involving temperature coefficients if you're in a management position; you can delegate these things to the engineers. If you plan to become an engineer, you'll most likely have computer software that will perform the calculations for you.

Interelectrode capacitance

Any two pieces of conducting material, when they are brought near each other, will act as a capacitor. Often, this *interelectrode capacitance* is so small that it can be neglected. It rarely amounts to more than a few picofarads.

In ac circuits and at audio frequencies, interelectrode capacitance is not usually significant. But it can cause problems at radio frequencies. The chances for trouble increase as the frequency increases. The most common phenomena are *feedback*, and a change in the frequency characteristics of a circuit.

Interelectrode capacitance is minimized by keeping wire leads as short as possible. It can also be reduced by using shielded cables, and by enclosing circuits in metal housings if interaction might produce trouble. This is why, if you've ever opened up a sophisticated communications radio, you might have seen numerous metal enclosures inside the main box.

Quiz

Refer to the text in this chapter if necessary. A good score is 18 correct. Answers are in the back of the book.

1. Capacitance acts to store electrical energy as:
   A. Current.
   B. Voltage.
   C. A magnetic field.
   D. An electric field.

2. As capacitor plate area increases, all other things being equal:
   A. The capacitance increases.
   B. The capacitance decreases.
   C. The capacitance does not change.
   D. The voltage-handling ability increases.

3. As the spacing between plates in a capacitor is made smaller, all other things being equal:
   A. The capacitance increases.
   B. The capacitance decreases.
   C. The capacitance does not change.
   D. The voltage-handling ability increases.

4. A material with a high dielectric constant:
   A. Acts to increase capacitance per unit volume.
   B. Acts to decrease capacitance per unit volume.
   C. Has no effect on capacitance.
   D. Causes a capacitor to become polarized.

5. A capacitance of 100 pF is the same as:
   A. 0.01 uF.
   B. 0.001 uF.
   C. 0.0001 uF.
   D. 0.00001 uF.

6. A capacitance of 0.033 uF is the same as:
   A. 33 pF.

B. 330 pF.

C. 3300 pF.

D. 33,000 pF.

7. Five 0.050-uF capacitors are connected in parallel. The total capacitance is:

A. 0.010 uF.

B. 0.25 uF.

C. 0.50 uF.

D. 0.025 uF.

8. If the same five capacitors are connected in series, the total capacitance will be:

A. 0.010 uF.

B. 0.25 uF.

C. 0.50 uF.

D. 0.025 uF.

9. Two capacitors are in series. Their values are 47 pF and 33 pF. The composite value is:

A. 80 pF.

B. 47 pF.

C. 33 pF.

D. 19 pF.

10. Two capacitors are in parallel. Their values are 47 pF and 470 uF. The combination capacitance is:

A. 47 pF.

B. 517 pF.

C. 517 uF.

D. 470 uF.

11. Three capacitors are in parallel. Their values are 0.0200 uF, 0.0500 uF and 0.10000 uF. The total capacitance is:

A. 0.0125 uF.

B. 0.170 uF.

C. 0.1 uF.

D. 0.125 uF.

12. Air works well as a dielectric mainly because it:

A. Has a high dielectric constant.

B. Is not physically dense.

C. Has low loss.

D. Allows for large capacitance in a small volume.

13. Which of the following is *not* a characteristic of mica capacitors?

    A. High efficiency.

    B. Small size.

    C. Capability to handle high voltages.

    D. Low loss.

14. A disk ceramic capacitor might have a value of:

    A. 100 pF.

    B. 33 uF.

    C. 470 uF.

    D. 10,000 uF.

15. A paper capacitor might have a value of:

    A. 0.001 pF.

    B. 0.01 uF.

    C. 100 uF.

    D. 3300 uF.

16. An air-variable capacitor might have a range of:

    A. 0.01 uF to 1 uF.

    B. 1 uF to 100 uF.

    C. 1 pF to 100 pF.

    D. 0.001 pF to 0.1 pF.

17. Which of the following types of capacitors is polarized?

    A. Paper.

    B. Mica.

    C. Interelectrode.

    D. Electrolytic.

18. If a capacitor has a negative temperature coefficient:

    A. Its value decreases as the temperature rises.

    B. Its value increases as the temperature rises.

    C. Its value does not change with temperature.

    D. It must be connected with the correct polarity.

19. A capacitor is rated at 33 pF, plus or minus 10 percent. Which of the following capacitances is outside the acceptable range?

    A. 30 pF.

    B. 37 pF.

    C. 35 pF.

    D. 31 pF.

20. A capacitor, rated at 330 pF, shows an actual value of 317 pF. How many percent off is its value?

    A. 0.039.

    B. 3.9.

    C. 0.041.

    D. 4.1.

12
CHAPTER

Phase

AN ALTERNATING CURRENT REPEATS THE SAME WAVE TRACE OVER AND OVER. Each 360-degree cycle is identical to every other. The wave can have any imaginable shape, but as long as the polarity reverses periodically, and as long as every cycle is the same, the wave can be called true ac.

In this chapter, you'll learn more about the most common type of ac, the sine wave. You'll get an in-depth look at the way engineers and technicians think of ac sine waves. There will be a discussion of the circular-motion model of the sine wave. You'll see how these waves add together, and how they can cancel out.

Instantaneous voltage and current

You've seen "stop motion" if you've done much work with a video-cassette recorder (VCR). In fact, you've probably seen it if you've watched any television sportscasts. Suppose that it were possible for you to stop time in real life, any time you wanted. Then you could examine any instant of time in any amount of detail that would satisfy your imagination.

Recall that an ac sine wave has a unique, characteristic shape, as shown in Fig. 12-1. This is the way the graph of the function $y = \sin x$ appears on the coordinate plane. (The abbreviation *sin* stands for *sine* in trigonometry.) Suppose that the peak voltage is plus-or-minus 1 V, as shown. Further imagine that the period is 1 second, so that the frequency is 1 Hz. Let the wave begin at time t = 0. Then each cycle begins every time the value of t is a whole number; at every such instant, the voltage is zero and positive-going.

If you freeze time at t = 446.00 seconds, say, the voltage will be zero. Looking at the diagram, you can see that the voltage will also be zero every so-many-and-a-*half* seconds; that is, it will be zero at t = 446.5 seconds. But instead of getting more positive at these instants, the voltage will be swinging towards the negative.

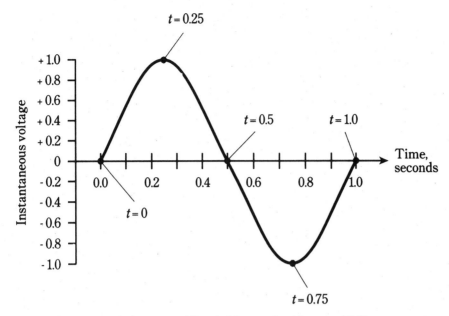

12-1   A sine wave with period 1 second and frequency 1 Hz.

If you freeze time at so-many-and-a-*quarter* seconds, say t = 446.25 seconds, the voltage will be +1 V. The wave will be exactly at its positive peak. If you stop time at so-many-and-*three-quarter* seconds, say t = 446.75 seconds, the voltage will be exactly at its negative peak, −1 V.

At intermediate times, say, so-many-and-*three-tenths* seconds, the voltage will have intermediate values.

Rate of change

By examining the diagram of Fig. 12-1, you can see that there are times the voltage is increasing, and times it is decreasing. *Increasing*, in this context, means "getting more positive," and *decreasing* means "getting more negative." The most rapid increase in voltage occurs when t = 0.0 and t = 1.0 in Fig. 12-1. The most rapid decrease takes place when t = 0.5.

Notice that when t = 0.25, and also when t = 0.75, the instantaneous voltage is not changing. This condition exists for a vanishingly small moment. You might liken the value of the voltage at t = 0.25 to the altitude of a ball you've tossed straight up into the air, when it reaches its highest point. Similarly, the value of voltage at t = 0.75 is akin to the position of a swing at its lowest altitude.

If n is any whole number, then the situation at t = n.25 is the same as it is for t = 0.25; also, for t = n.75, things are just the same as they are when t = 0.75. The single cycle shown in Fig. 12-1 represents every possible condition of the ac sine wave having a frequency of 1 Hz and a peak value of plus-or-minus 1 V.

Suppose that you graph the *rate of change in the voltage* of the wave in Fig. 12-1 against time. What will this graph look like? It turns out that it will have a shape that is a sine wave, but it will be displaced to the left of the original wave by one-quarter of a cycle. If you plot the relative rate of change against time as shown in Fig. 12-2, you get the *derivative*, or rate of change, of the sine wave. This is a *cosine* wave, having the same general, characteristic shape as the sine wave. But the *phase* is different.

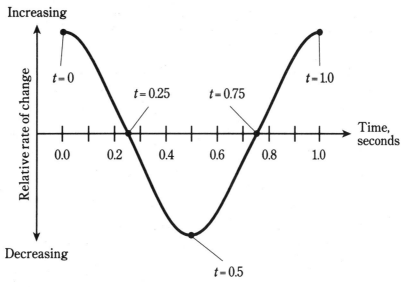

12-2   A sine wave representing the rate of change in instantaneous amplitude of the wave in Fig. 12-1.

Sine waves as circular motion

A sine wave represents the most efficient possible way that a quantity can alternate back and forth. The reasons for this are rather complex, and a thorough discussion of it would get into that fuzzy thought territory where science begins to overlap with esthetics, mathematics, and philosophy. You need not worry about it. You might recall, however, that a sine wave has just one frequency component, and represents a pure wave for this reason.

Suppose that you were to swing a glowing ball around and around at the end of a string, at a rate of one revolution per second. The ball would describe a circle in space (Fig. 12-3A). Imagine that you swing the ball around so that it is always at the same level; that is, so that it takes a path that lies in a horizontal plane. Imagine that you do this in a perfectly dark gymnasium. Now if a friend stands some distance away, with his or her eyes right in the plane of the ball's path, what will your friend see? All that will be visible is the glowing ball, oscillating back and forth (Fig. 12-3B). The ball will seem to move toward the right, slow down, then stop and reverse its direction, going back towards the left. It will move faster and faster, then slower again, reaching its left-most point, at which it will stop and turn around again. This will go on and on, with a frequency of 1 Hz, or a complete cycle per second, because you are swinging the ball around at one revolution per second.

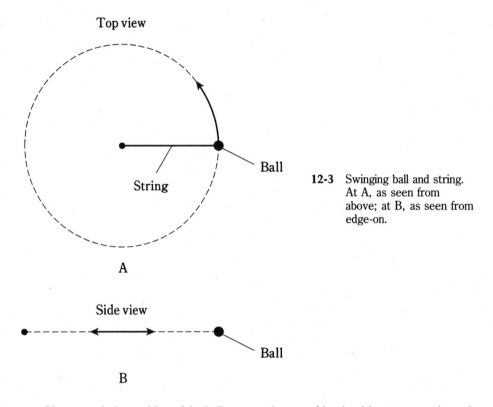

Top view

String

Ball

A

Side view

Ball

B

12-3 Swinging ball and string. At A, as seen from above; at B, as seen from edge-on.

If you graph the position of the ball, as seen by your friend, with respect to time, the result will be a sine wave (Fig. 12-4). This wave has the same characteristic shape as all sine waves.

It is true that some sine waves are taller than others, and some are stretched out horizontally more than others. But the general waveform is the same in every case. By multiplying or dividing the amplitude and the wavelength of any sine wave, it can be made to fit exactly along the curve of any other sine wave. The standard sine wave is the function $y = \sin x$ in the (x,y) coordinate plane.

You might whirl the ball around faster or slower than one revolution per second. The string might be made longer or shorter. This would alter the height and/or the frequency of the sine wave graphed in Fig. 12-4. But the fundamental rule would always apply: the sine wave can be reduced to circular motion.

Degrees of phase

Back in Chapter 9, *degrees of phase* were discussed. If you wondered then why phase is spoken of in terms of angular measure, the reason should be clearer now. A circle has 360

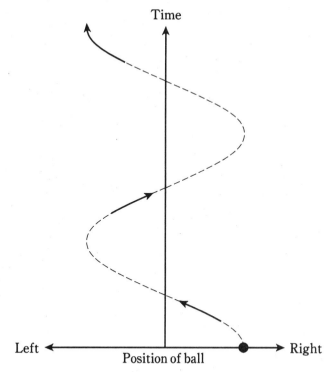

12-4 Position of ball as seen edge-on, as a function of time.

degrees. A sine wave can be represented as circular motion. Exact moments along the sine curve correspond to specific angles, or positions, around a circle.

Figure 12-5 shows the way a *rotating vector* is used to represent a sine wave. At A, the vector points "east," and this is assigned the value of 0 degrees, where the wave amplitude is zero and is increasing positively. At B, the vector points "north"; this is the 90-degree instant, where the wave has attained its maximum positive amplitude. At C, the vector points "west." This is 180 degrees, the instant where the wave has gone back to zero amplitude, and is getting more negative. At D, the wave points "south." This is 270 degrees, and represents the maximum negative amplitude. When a full circle has been completed, or 360 degrees, the vector once again points "east." Thus, 360 degrees is the same as 0 degrees. In fact, a value of x degrees represents the same condition as x plus or minus any multiple of 360 degrees.

The four points for the model of Fig. 12-5 are shown on a sine wave graph in Fig. 12-6.

You can think of the vector as going around and around, at a rate that corresponds to one revolution per cycle of the wave. If the wave has a frequency of 1 Hz, the vector goes around at a rate of a revolution per second (1 rps). If the wave has a frequency of 100 Hz, the speed of the vector is 100 rps, or a revolution every 0.01 second. If the wave is 1 MHz, then the speed of the vector is 1,000,000 or 10^6 rps, and it goes once around every 10^{-6}, or 0.000001, second.

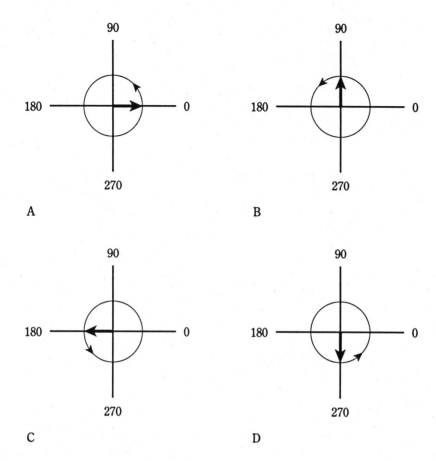

12-5   Vector representation of sine wave at the start of a cycle (A), at ¼ cycle (B), at ½ cycle (C), and at ¾ (D).

The peak amplitude of the wave can be thought of in terms of the length of the vector. In Fig. 12-5, time is represented by the angle counterclockwise from "due east," and amplitude is independent of time. This differs from the more common rendition of the sine wave, such as the one in Fig. 12-6.

In a sense, whatever force "causes" the wave is always there, whether there's any instantaneous voltage or not. The wave is created by angular motion (revolution) of this force. This is visually apparent in the rotating-vector model. The reasons for thinking of ac as a *vector quantity*, having *magnitude* and *direction*, will become more clear in the following chapters, as you learn about reactance and impedance.

If a wave has a frequency of f Hz, then the vector makes a complete 360-degree revolution every 1/f seconds. The vector goes through 1 degree of phase every 1/(360f) seconds.

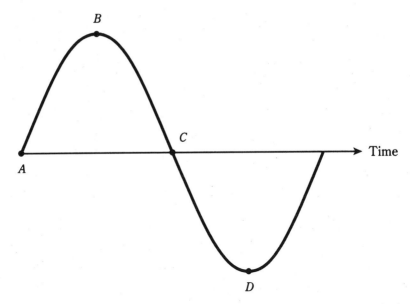

12-6   The four points for the model of Fig. 12-5, shown on a standard amplitude-versus-time graph of a sine wave.

Radians of phase

An angle of 1 *radian* is about 57.3 degrees. A complete circle is 6.28 radians around. If a wave has a frequency of f Hz, then the vector goes through 1 radian of phase every 1/(57.3f) seconds. The number of radians per second for an ac wave is called the *angular frequency*.

Radians are used mainly by physicists. Engineers and technicians generally use degrees when talking about phase, and Hertz when talking about frequency.

Phase coincidence

When two sine waves have the same frequency, they can behave much differently if their cycles begin at different times. Whether or not the *phase difference*, often called the *phase angle* and specified in degrees, matters depends on the nature of the circuit.

Phase angle can have meaning only when two waves have identical frequencies. If the frequencies differ, even by just a little bit, the relative phase constantly changes, and you can't specify a single number. In the following discussions of phase angle, assume that the two waves always have identical frequencies.

Phase coincidence means that two waves begin at exactly the same moment. They are "lined up." This is shown in Fig. 12-7 for two waves having different amplitudes. (If the amplitudes were the same, you would see only one wave.) The phase difference in this case is 0 degrees. You might say it's any multiple of 360 degrees, too, but engineers and technicians almost never speak of any phase angle of less than 0 or more than 360 degrees.

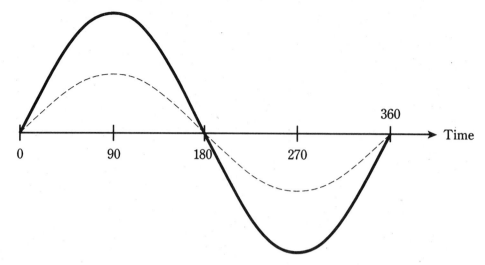

12-7   Two sine waves in phase coincidence.

If two sine waves are in phase coincidence, the peak amplitude of the *resultant* wave, which will also be a sine wave, is equal to the sum of the peak amplitudes of the two *composite* waves. The phase of the resultant is the same as that of the composite waves.

Phase opposition

When two waves begin exactly ½ cycle, or 180 degrees, apart, they are said to be in *phase opposition*. This is illustrated by the drawing of Fig. 12-8. In this situation, engineers sometimes also say that the waves are *out of phase,* although this expression is a little nebulous because it could be taken to mean some phase difference other than 180 degrees.

If two sine waves have the same amplitude and are in phase opposition, they will exactly cancel each other out. This is because the instantaneous amplitudes of the two waves are equal and opposite at every moment in time.

If two sine waves have different amplitudes and are in phase opposition, the peak value of the resultant, which will be a sine wave, is equal to the difference between the peak values of the two composite waves. The phase of the resultant will be the same as the phase of the stronger of the two composite waves.

The sine wave has the unique property that, if its phase is shifted by 180 degrees, the resultant wave is the same as turning the original wave "upside-down." Not all waveforms have this property. Perfect square waves do, but some rectangular and sawtooth waves don't, and irregular waveforms almost never do.

Leading phase

Two waves can differ in phase by any amount from 0 degrees (in phase), through 180 degrees (phase opposition), to 360 degrees (back in phase again).

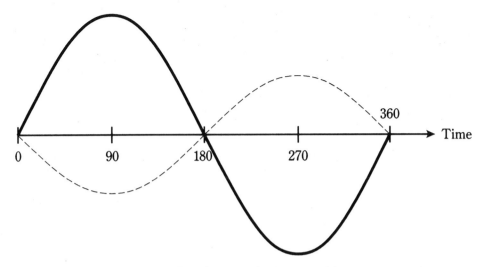

12-8 Two sine waves in phase opposition.

Suppose there are two sine waves, wave X and wave Y, with identical frequency. If wave X begins a fraction of a cycle *earlier* than wave Y, then wave X is said to be *leading* wave Y in phase. For this to be true, X must begin its cycle less than 180 degrees before Y. Figure 12-9 shows wave X leading wave Y by 90 degrees of phase. The difference could be anything greater than 0 degrees, up to 180 degrees.

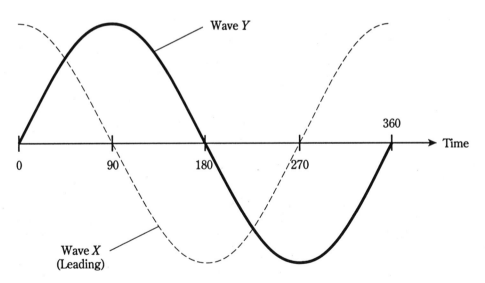

12-9 Wave X leads wave Y by 90 degrees.

Note that if wave X (the dotted line in Fig. 12-9) is leading wave Y (the solid line), then wave X is somewhat to the *left* of wave Y. In a time line, the left is earlier and the right is later.

Lagging phase

Suppose that wave X begins its cycle more than 180 degrees, but less than 360 degrees, ahead of wave Y. In this situation, it is easier to imagine that wave X starts its cycle *later* than wave Y, by some value between 0 and 180 degrees. Then wave X is not leading, but instead is *lagging*, wave Y. Figure 12-10 shows wave X lagging wave Y by 90 degrees. The difference could be anything between 0 and 180 degrees.

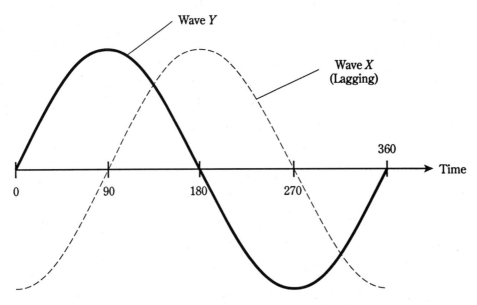

12-10   Wave X lags wave Y by 90 degrees.

You can surmise by now that leading phase and lagging phase are different ways of looking at similar "animals." In practice, ac sine waves are oscillating rapidly, sometimes thousands, millions or even billions of times per second. If two waves have the same frequency and different phase, how do you know that one wave is really leading the other by some small part of a cycle, instead of lagging by a cycle and a fraction, or by a few hundred, thousand, million, or billion cycles and a fraction? The answer lies in the real-life effects of the waves. Engineers and technicians think of phase differences, for sine waves having the same frequency, as always being between 0 and 180 degrees, either leading or lagging. It rarely matters, in practice, whether one wave started a few seconds earlier or later than the other.

So, while you might think that the diagram of Fig. 12-9 shows wave X lagging wave Y by 270 degrees, or that the diagram of Fig. 12-10 shows wave X leading wave Y by 270 degrees, you would get an odd look from an engineer if you said so aloud. And if you said something like "This wave is leading that one by 630 degrees, you might actually be laughed at.

Note that if wave X (the dotted line in Fig. 12-10) is lagging wave Y (the solid line), then wave X is somewhat to the *right* of wave Y.

Vector diagrams of phase relationships

The circular renditions of sine waves, such as are shown in the four drawings of Fig. 12-5, are well-suited to showing phase relationships.

If a sine wave X is *leading* a sine wave Y by some number of degrees, then the two waves can be drawn as vectors, with vector **X** being that number of degrees *counterclockwise* from vector **Y**. If wave X *lags* Y by some number of degrees, then **X** will be *clockwise* from **Y** by that amount.

If two waves are in phase, their vectors overlap (line up). If they are in phase opposition, they point in exactly opposite directions.

The drawings of Fig. 12-11 show four phase relationships between waves X and Y. At A, X is in phase with Y. At B, X leads Y by 90 degrees. At C, X and Y are 180 degrees opposite in phase; at D, X lags Y by 90 degrees. In all cases, you can think of the vectors rotating *counterclockwise* at the rate of f revolutions per second, if their frequency is f Hz.

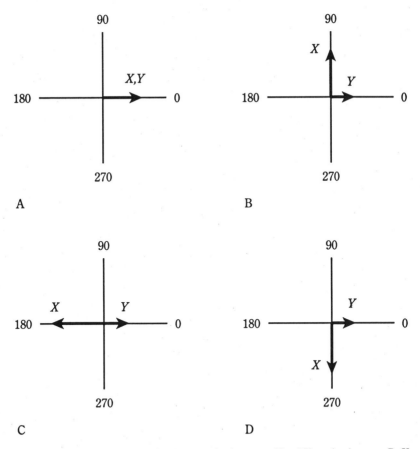

12-11   Vector representation of phase. At A, waves X and Y are in phase; at B, X leads Y by 90 degrees; at C, X, and Y are 180 degrees out of phase; at D, X lags Y by 90 degrees.

Quiz

Refer to the text in this chapter if necessary. A good score is 18 correct. Answers are in the back of the book.

1. Which of the following is *not* a general characteristic of an ac wave?

    A. The wave shape is identical for each cycle.

    B. The polarity reverses periodically.

    C. The electrons always flow in the same direction.

    D. There is a definite frequency.

2. A sine wave:

    A. Always has the same general appearance.

    B. Has instantaneous rise and fall times.

    C. Is in the same phase as a cosine wave.

    D. Rises very fast, but decays slowly.

3. The derivative of a sine wave:

    A. Is shifted in phase by 1/2 cycle from the sine wave.

    B. Is a representation of the rate of change.

    C. Has instantaneous rise and fall times.

    D. Rises very fast, but decays slowly.

4. A phase difference of 180 degrees in the circular model represents:

    A. ¼ revolution.

    B. ½ revolution.

    C. A full revolution.

    D. Two full revolutions.

5. You can add or subtract a certain number of degrees of phase to or from a wave, and end up with exactly the same wave again. This number is:

    A. 90.

    B. 180.

    C. 270.

    D. 360.

6. You can add or subtract a certain number of degrees of phase to or from a sine wave, and end up with an inverted (upside-down) representation of the original. This number is:

    A. 90.

    B. 180.

    C. 270.

    D. 360.

7. A wave has a frequency of 300 kHz. One complete cycle takes:

    A. 1/300 second.

    B. 0.00333 second.

    C. 1/3,000 second.

    D. 0.00000333 second.

8. If a wave has a frequency of 440 Hz, how long does it take for 10 degrees of phase?

    A. 0.00273 second.

    B. 0.000273 second.

    C. 0.0000631 second.

    D. 0.00000631 second.

9. Two waves are in phase coincidence. One has a peak value of 3 V and the other a peak value of 5 V. The resultant will be:

    A. 8 V peak, in phase with the composites.

    B. 2 V peak, in phase with the composites.

    C. 8 V peak, in phase opposition with respect to the composites.

    D. 2 V peak, in phase opposition with respect to the composites.

10. Shifting the phase of an ac sine wave by 90 degrees is the same thing as:

    A. Moving it to the right or left by a full cycle.

    B. Moving it to the right or left by 1/4 cycle.

    C. Turning it upside-down.

    D. Leaving it alone.

11. A phase difference of 540 degrees would more often be spoken of as:

    A. An offset of more than one cycle.

    B. Phase opposition.

    C. A cycle and a half.

    D. 1.5 Hz.

12. Two sine waves are in phase opposition. Wave X has a peak amplitude of 4 V and wave Y has a peak amplitude of 8 V. The resultant has a peak amplitude of:

    A. 4 V, in phase with the composites.

    B. 4 V, out of phase with the composites.

    C. 4 V, in phase with wave X.

    D. 4 V, in phase with wave Y.

13. If wave X leads wave Y by 45 degrees of phase, then:

    A. Wave Y is 1/4 cycle ahead of wave X.

B. Wave *Y* is ¼ cycle behind wave *X*.

C. Wave *Y* is ⅛ cycle behind wave *X*.

D. Wave *Y* is ¹⁄₁₆ cycle ahead of wave *X*.

14. If wave *X* lags wave *Y* by ⅓ cycle, then:

A. *Y* is 120 degrees earlier than *X*.

B. *Y* is 90 degrees earlier than *X*.

C. *Y* is 60 degrees earlier than *X*.

D. *Y* is 30 degrees earlier than *X*.

15. In the drawing of Fig. 12-12:

A. *X* lags *Y* by 45 degrees.

B. *X* leads *Y* by 45 degrees.

C. *X* lags *Y* by 135 degrees.

D. *X* leads *Y* by 135 degrees.

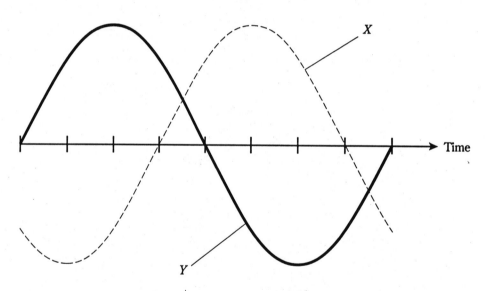

12-12   Illustration for quiz question 15.

16. Which of the drawings in Fig. 12-13 represents the situation of Fig. 12-12?

A. A.

B. B.

C. C.

D. D.

17. In vector diagrams such as those of Fig. 12-13, length of the vector represents:
    A. Average amplitude.
    B. Frequency.
    C. Phase difference.
    D. Peak amplitude.

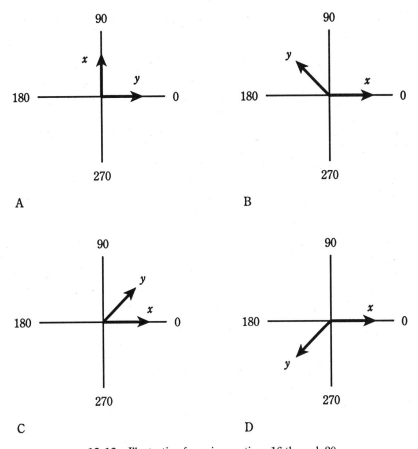

12-13   Illustration for quiz questions 16 through 20.

18. In vector diagrams such as those of Fig. 12-13, the angle between two vectors represents:
    A. Average amplitude.
    B. Frequency.
    C. Phase difference.
    D. Peak amplitude.

19. In vector diagrams such as those of Fig. 12-13, the distance from the center of the graph represents:

    A. Average amplitude.

    B. Frequency.

    C. Phase difference.

    D. Peak amplitude.

20. In diagrams like those of Fig. 12-13, the progression of time is sometimes depicted as:

    A. Movement to the right.

    B. Movement to the left.

    C. Rotation counterclockwise.

    D. Rotation clockwise.

13
CHAPTER

Inductive reactance

IN DC CIRCUITS, RESISTANCE IS A SIMPLE THING. IT CAN BE EXPRESSED AS A number, from zero (a perfect conductor) to extremely large values, increasing without limit through the millions, billions, and even trillions of ohms. Physicists call resistance a *scalar* quantity, because it can be expressed on a one-dimensional scale. In fact, dc resistance can be represented along a half line, or ray, as shown in Fig. 13-1.

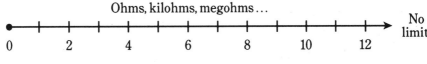

13-1   Resistance can be represented on a ray.

Given a certain dc voltage, the current decreases as the resistance increases, in accordance with Ohm's Law, as you already know. The same law holds for ac through a resistance, if the ac voltage and current are both specified as peak, pk-pk, or rms values.

Coils and direct current

Suppose that you have some wire that conducts electricity very well. What will happen if you wind a length of the wire into a coil and connect it to a source of dc, as shown in Fig. 13-2? The wire will draw a large amount of current, possibly blowing a fuse or overstressing a battery. It won't matter whether the wire is a single-turn loop, or whether it's lying haphazardly on the floor, or whether it's wrapped around a stick. The current will be large. In amperes, it will be equal to $I = E/R$, where I is the current, E is the dc voltage, and R is the resistance of the wire (a low resistance).

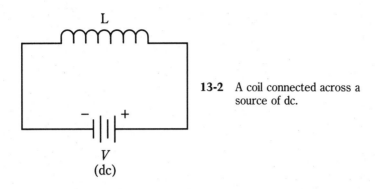

13-2 A coil connected across a source of dc.

You can make an electromagnet, as you've already seen, by passing dc through a coil wound around an iron rod. But there will still be a large, constant current in the coil. The coil will probably get more or less hot, as energy is dissipated in the resistance of the wire. The battery, too, or power supply components, will become warm or hot.

If the voltage of the battery or power supply is increased, the wire in the coil, iron core or not, will get hotter. Ultimately, if the supply can deliver the necessary current, the wire will melt.

Coils and alternating current

Suppose you change the voltage source, connected across the coil, from dc to ac (Fig. 13-3). Imagine that you can vary the frequency of the ac, from a few hertz to hundreds of hertz, then kilohertz, then megahertz.

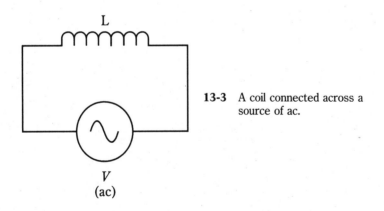

13-3 A coil connected across a source of ac.

At first, the ac current will be high, just as it is with dc. But the coil has a certain amount of inductance, and it takes some time for current to establish itself in the coil. Depending on how many turns there are, and on whether the core is air or a ferromagnetic material, you'll reach a point, as the ac frequency increases, when the coil starts to get "sluggish." That is, the current won't have time to get established in the coil before the polarity of the voltage reverses.

At high ac frequencies, the *current* through the coil will have difficulty following the *voltage* placed across the coil. Just as the coil starts to "think" that it's making a good short circuit, the ac voltage wave will pass its peak, go back to zero, and then try to pull the electrons the other way.

This sluggishness in a coil for ac is, in effect, similar to dc resistance. As the frequency is raised higher and higher, the effect gets more and more pronounced. Eventually, if you keep on increasing the frequency of the ac source, the coil will not even begin to come near establishing a current with each cycle. It will act like a large resistance. Hardly any ac current will flow through it.

The opposition that the coil offers to ac is called *inductive reactance*. It, like resistance, is measured in ohms. It can vary just as resistance does, from near zero (a short piece of wire) to a few ohms (a small coil) to kilohms or megohms (bigger and bigger coils).

Like resistance, inductive reactance affects the current in an ac circuit. But, unlike simple resistance, reactance changes with frequency. And the effect is not just a decrease in the current, although in practice this will occur. It is a change in the way the current flows with respect to the voltage.

Reactance and frequency

Inductive reactance is one of two kinds of reactance; the other, called *capacitive reactance*, will be discussed in the next chapter. Reactance in general is symbolized by the capital letter X. Inductive reactance is indicated by the letter X with a subscript L: X_L.

If the frequency of an ac source is given, in hertz, as f, and the inductance of a coil in henrys is L, then the inductive reactance is:

$$X_L = 6.28fL$$

This same formula applies if the frequency, f, is in kilohertz and the inductance, L, is in millihenrys. And it also applies if f is in megahertz and L is in microhenrys. Just remember that if frequency is in thousands, inductance must be in thousand*ths*, and if frequency is in millions, inductance must be in million*ths*.

Inductive reactance increases *linearly* with increasing ac frequency. This means that the function of X_L vs f is a straight line when graphed.

Inductive reactance also increases linearly with inductance. Therefore, the function of X_L vs L also appears as a straight line on a graph.

The value of X_L is *directly proportional* to f; X_L is also directly proportional to L. These relationships are graphed, in relative form, in Fig. 13-4.

Problem 13-1

Suppose a coil has an inductance of 0.50 H, and the frequency of the ac passing through it is 60 Hz. What is the inductive reactance?

Using the above formula, $X_L = 6.28 \times 60 \times 0.50 = 188.4$ ohms. You should round this off to two significant digits, or 190 ohms, because the inductance and frequency are both expressed to only two significant digits.

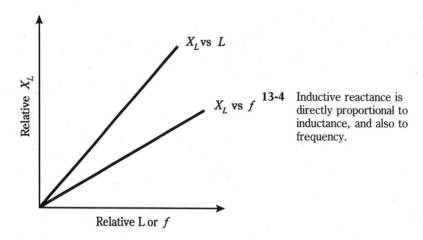

13-4 Inductive reactance is directly proportional to inductance, and also to frequency.

Problem 13-2

What will be the inductive reactance of the above coil if the supply is a battery that supplies pure dc?

Because dc has a frequency of zero, $X_L = 6.28 \times 0 \times 0.50 = 0$. That is, there will be no inductive reactance. Inductance doesn't generally have any practical effect with pure dc.

Problem 13-3

If a coil has an inductive reactance of 100 Ω at a frequency of 5.00 MHz, what is its inductance?

In this case, you need to plug numbers into the formula and solve for the unknown L. Start out with the equation $100 = 6.28 \times 5.00 \times L = 31.4 \times L$. Then, recall that because the frequency is in megahertz, or millions of hertz, the inductance will come out in microhenrys, or millionths of a henry. You can divide both sides of the equation by 31.4, getting $L = 100/31.4 = 3.18$ uH.

Points in the RL plane

Inductive reactance can be plotted along a half line, just as can resistance. In a circuit containing both resistance and inductance, the characteristics become two-dimensional. You can orient the resistance and reactance half lines perpendicular to each other to make a quarter-plane coordinate system, as shown in Fig. 13-5. Resistance is usually plotted horizontally, and inductive reactance is plotted vertically, going upwards.

In this scheme, RL combinations form *impedances*. You'll learn all about this in chapter 15. Each point on the *RL plane* corresponds to one unique impedance value. Conversely, each RL impedance value corresponds to one unique point on the plane.

For reasons made clear in chapter 15, impedances on the RL plane are written in the form $R + jX_L$, where R is the resistance and X_L is the inductive reactance.

If you have a pure resistance, say $R = 5$ Ω, then the *complex* impedance is $5 + j0$, and is at the point $(5,0)$ on the RL plane. If you have a pure inductive reactance, such as

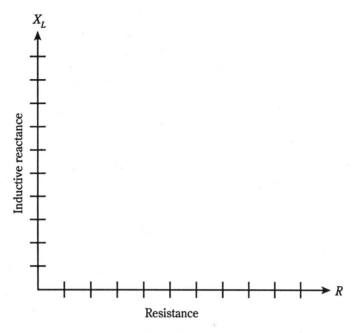

13-5   The RL quarter-plane.

$X_L = 3 \, \Omega$, then the complex impedance is $0 + j3$, and is at the point $(0,3)$ on the RL plane. These points, and others, are shown in Fig. 13-6.

In real life, all coils have some resistance, because no wire is a perfect conductor. All resistors have at least a tiny bit of inductive reactance, because they take up some physical space. So there is really no such thing as a mathematically perfect pure resistance like $5 + j0$, or a mathematically perfect pure reactance like $0 + j3$. Sometimes you can get pretty close, but absolutely pure resistances or reactances never exist, if you want to get really theoretical.

Often, resistance and inductive reactance are both deliberately placed in a circuit. Then you get impedances values such as $2 + j3$ or $4 + j1.5$. These are shown in Fig. 13-6 as points on the RL plane.

Remember that the values for X_L are *reactances*, not the actual inductances. Therefore, they vary with the frequency in the RL circuit. Changing the frequency has the effect of making the points move in the RL plane. They move vertically, going upwards as the ac frequency increases, and downwards as the ac frequency decreases. If the ac frequency goes down to zero, the inductive reactance vanishes. Then $X_L = 0$, and the point is along the resistance axis of the RL plane.

Vectors in the RL Plane

Engineers sometimes like to represent points in the RL plane (and in other types of coordinate planes, too) as vectors. This gives each point a definite magnitude and a precise direction.

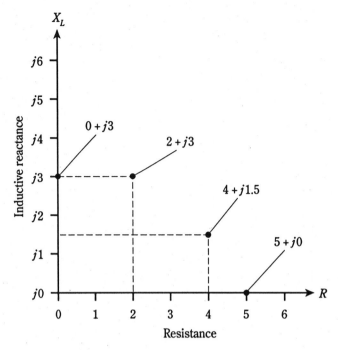

13-6 Four points in the RL impedance plane. See text for discussion.

In Fig. 13-6, there are four different points shown. Each one is represented by a certain distance to the right of the origin (0,0), and a specific distance upwards from the origin. The first of these is the resistance, R, and the second is the inductive reactance, X_L. Thus, the RL combination is a two-dimensional quantity. There is no way to uniquely define RL combinations as single numbers, or *scalars*, because there are two different quantities that can vary independently.

Another way to think of these points is to draw lines from the origin out to them. Then you can think of the points as rays, each having a certain length, or magnitude, and a definite direction, or angle counterclockwise from the resistance axis. These rays, going out to the points, are vectors (Fig. 13-7). You've already been introduced to these things.

Vectors seem to engender apprehension in some people, as if they were invented by scientists for the perverse pleasure of befuddling ordinary folks. "What are you taking this semester?" asks Jane. "*Vector* analysis!" Joe shudders (if he's one of the timid types), or beams (if he wants to impress Jane).

This attitude is completely groundless. Just think of vectors as arrows that have a certain length, and that point in some direction.

In Fig. 13-7, the points of Fig. 13-6 are shown as vectors. The only difference is that there is some more ink on the paper.

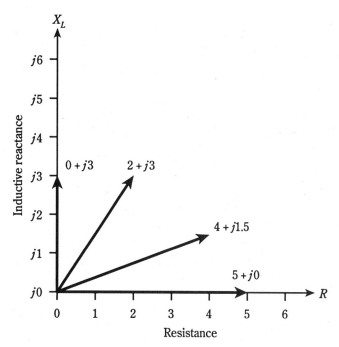

13-7   Four vectors in the RL impedance plane.

Current lags voltage

Inductance, as you recall, stores electrical energy as a magnetic field. When a voltage is placed across a coil, it takes awhile for the current to build up to full value.

When ac is placed across a coil, the current *lags* the voltage in phase.

Pure inductance

Suppose that you place an ac voltage across a low-loss coil, with a frequency high enough so that the inductive reactance, X_L, is much larger than the resistance, R. In this situation, the current is one-quarter of a cycle behind the voltage. That is, the current lags the voltage by 90 degrees (Fig. 13-8).

At very low frequencies, large inductances are normally needed in order for this current lag to be a full ¼ cycle. This is because any coil has some resistance; no wire is a perfect conductor. If some wire were found that had a mathematically zero resistance, and if a coil of any size were wound from this wire, then the current would lag the voltage by 90 degrees in this inductor, no matter what the ac frequency.

When the value of X_L is very large compared with the value of R in a circuit—that is, when there is an essentially pure inductance—the vector in the RL plane points straight up along the X_L axis. Its angle is 90 degrees from the R axis, which is considered the *zero line* in the RL plane.

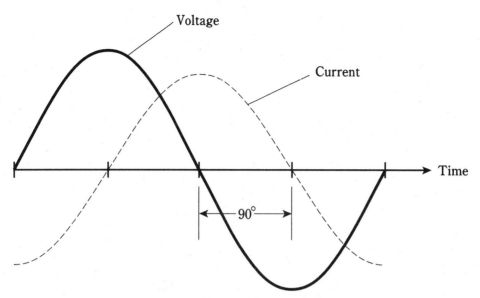

13-8 In a pure inductance, the current lags the voltage by 90 degrees.

Inductance and resistance

When the resistance in a resistance-inductance circuit is significant compared with the inductive reactance, the current lags the voltage by something less than 90 degrees (Fig. 13-9). If R is small compared with X_L, the current lag is almost 90 degrees; as R gets larger, the lag decreases. A circuit with resistance and inductance is called an *RL circuit*.

The value of R in an RL circuit might increase relative to X_L because resistance is deliberately placed in series with the inductance. Or, it might happen because the ac frequency gets so low that X_L decreases until it is in the same ball park with the loss resistance R in the coil winding. In either case, the situation can be schematically represented by a coil in series with a resistor (Fig. 13-10).

If you know the values of X_L and R, you can find the *angle of lag*, also called the *RL phase angle*, by plotting the point $R + jX_L$ on the RL plane, drawing the vector from the origin $0 + j0$ out to that point, and then measuring the angle of the vector, counterclockwise from the resistance axis. You can use a protractor to measure this angle, or you can compute its value using trigonometry.

In fact, you don't need to know the actual values of X_L and R in order to find the angle of lag. All you need to know is their *ratio*. For example, if $X_L = 5\ \Omega$ and $R = 3\ \Omega$, you will get the same angle as you would get if $X_L = 50\ \Omega$ and $R = 30\ \Omega$, or if $X_L = 20\ \Omega$ and $R = 12\ \Omega$. The angle of lag will be the same for any values of X_L and R in the ratio of 5:3.

It's easy to find the angle of lag whenever you know the ratio of R to X_L. You'll see some examples shortly.

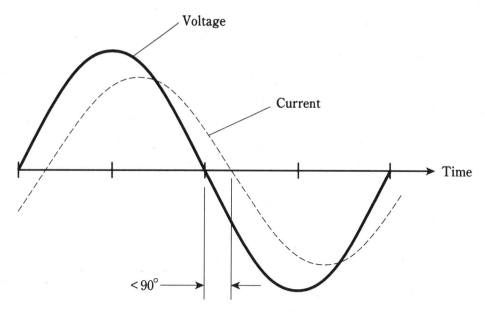

13-9   In a circuit with inductance and resistance, the current lags the voltage by less than 90 degrees.

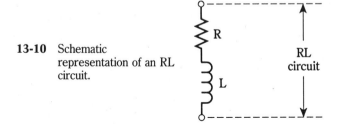

13-10   Schematic representation of an RL circuit.

Pure resistance

As the resistance in an RL circuit becomes large with respect to the inductive reactance, the angle of lag gets smaller and smaller. The same thing happens if the inductive reactance gets small compared with the resistance. When R is many times greater than X_L, whatever their actual magnitudes might be, the vector in the RL plane lies almost on the R axis, going "east" or to the right. The RL phase angle is nearly zero. The current comes into phase with the voltage.

In a pure resistance, with no inductance at all, the current is precisely in phase with the voltage (Fig. 13-11). A pure resistance doesn't store any energy, as an inductive circuit does, but sends the energy out as heat, light, electromagnetic waves, sound, or some other form that never comes back into the circuit.

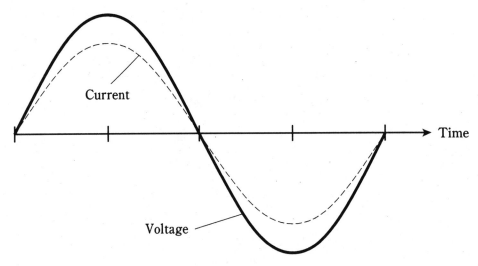

13-11 In a circuit with only resistance, the current is in phase with the voltage.

How much lag?

If you know the ratio of the inductive reactance to the resistance, X_L/R, in an RL circuit, then you can find the phase angle. Of course, you can find the angle of lag if you know the actual values of X_L and R.

Pictorial method

It isn't necessary to construct an entire RL plane to find phase angles. You can use a ruler that has centimeter (cm) and millimeter (mm) markings, and a protractor.

First, draw a line a little more than 10 cm long, going from left to right on a sheet of paper. Use the ruler and a sharp pencil. Then, with the protractor, construct a line off the left end of this first line, going vertically upwards. Make this line at least 10 cm long.

The horizontal line, or the one going to the right, is the R axis of a crude coordinate system. The vertical line, or the one going upwards, is the X_L axis.

If you know the values of X_L and R, divide them down, or multiply them up (in your head) so that they're both between 0 and 100. For example, if $X_L = 680\ \Omega$ and $R = 840$ Ω, you can divide them both by 10 to get $X_L = 68$ and $R = 84$. Plot these points lightly by making hash marks on the vertical and horizontal lines you've drawn. The R mark will be 84 mm to the right of the origin, or intersection of the original two perpendicular lines. The X_L mark will be 68 mm up from the origin.

Draw a line connecting the two hash marks, as shown in Fig. 13-12. This line will run at a slant, and will form a triangle along with the two axes. Your hash marks, and the origin of the coordinate system, form *vertices* of a *right triangle*. The triangle is called "right" because one of its angles is a right angle (90 degrees.)

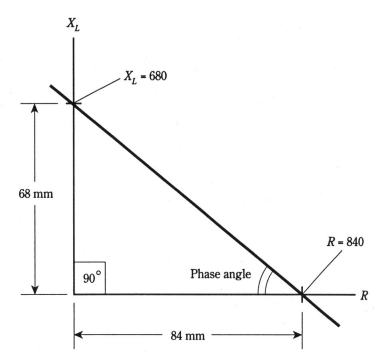

13-12 Pictorial method of finding phase angle.

Measure the angle between the slanted line and the horizontal, or R, axis. Extend one or both of the lines if necessary in order to get a good reading on the protractor. This angle will be between 0 and 90 degrees, and represents the phase angle in the RL circuit.

The actual vector, $R + jX_L$, is found by constructing a rectangle using the origin and your two hash marks as three of the four vertices, and drawing new horizontal and vertical lines to complete the figure. The vector is the diagonal of this rectangle, as shown in Fig. 13-13. The phase angle is the angle between this vector and the R axis. It will be the same as the angle of the slanted line in Fig. 13-12.

Trigonometric method

The pictorial method is inexact. It can be a bother, sometimes, to divide down and get numbers that are easy to work with. Drawing the pictures accurately requires care and patience. If the phase angle is very close to 0 degrees or 90 degrees—that is, if the ratio X_L/R is very small or large—the accuracy is especially poor in the drawing.

There's a better way. If you have a calculator that can find the *arctangent* of a number, you've got it made easy. Nowadays, if you intend to work with engineers, there's no excuse not to have a good calculator. If you don't have one, I suggest you go out and buy one right now. It should have "trig" functions and their inverses, "log" functions, exponential functions, and others that engineers use often.

If you know the values X_L and R, then the phase angle is simply the arctangent of their ratio, or arctan (X_L/R). This might also be written $\tan^{-1} (X_L/R)$. Punch a few buttons on the calculator, and you have it.

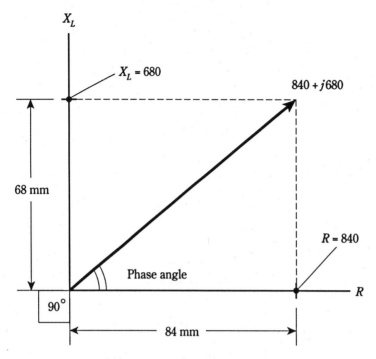

13-13 Another pictorial way of finding phase angle. This method shows the actual impedance vector.

Problem 13-4

The inductive reactance in an RL circuit is 680 Ω, and the resistance is 840 Ω. What is the phase angle?

Find the ratio $X_L/R = 680/840$. The calculator will display something like 0.809523809. Find the arctangent, or $\tan^{-1}$, getting a phase angle of 38.99099404 degrees (as shown on the calculator). Round this off to 39.0 degrees.

Problem 13-5

An RL circuit works at a frequency of 1.0 MHz. It has a resistance of 10 Ω, and an inductance of 90 uH. What is the phase angle?

This is a rather complicated problem, because it requires several steps. But each step is straightforward. You need to do them carefully, one at a time, and then recheck the whole problem once or twice when you're done calculating.

First, find the inductive reactance. This is found by the formula $X_L = 6.28fL = 6.28 \times 1.0 \times 90 = 565$ Ω. Then find the ratio $X_L/R = 565/10 = 56.5$. The phase angle is arctan $56.5 = 89$ degrees. This indicates that the circuit is almost purely reactive. The resistance contributes little to the behavior of this RL circuit at this frequency.

Problem 13-6

What is the phase angle of the above circuit at 10 kHz?

This requires that X_L be found over again, for the new frequency. Suppose you decide to use megahertz, so it will go nicely in the formula with microhenrys. A frequency of 10 kHz is the same as 0.010 MHz. Calculating, you get $X_L = 6.28fL = 6.28 \times 0.010 \times 90 = 6.28 \times 0.90 = 5.65\ \Omega$. The ratio X_L/R is then $5.65/10 = 0.565$. The phase angle is arctan $0.565 = 29$ degrees. At this frequency, the resistance and inductance both play significant roles in the RL circuit.

Quiz

Refer to the text in this chapter if necessary. A good score is 18 correct. Answers are in the back of the book.

1. As the number of turns in a coil increases, the current in the coil will eventually:
   A. Become very large.
   B. Stay the same.
   C. Decrease to near zero.
   D. Be stored in the core material.

2. As the number of turns in a coil increases, the reactance:
   A. Increases.
   B. Decreases.
   C. Stays the same.
   D. Is stored in the core material.

3. As the frequency of an ac wave gets lower, the value of X_L for a particular coil:
   A. Increases.
   B. Decreases.
   C. Stays the same.
   D. Depends on the voltage.

4. A coil has an inductance of 100 mH. What is the reactance at a frequency of 1000 Hz?
   A. $0.628\ \Omega$.
   B. $6.28\ \Omega$.
   C. $62.8\ \Omega$.
   D. $628\ \Omega$.

5. A coil shows an inductive reactance of 200 Ω at 500 Hz. What is its inductance?
   A. 0.637 H.
   B. 628 H.
   C. 63.7 mH.
   D. 628 mH.

6. A coil has an inductance of 400 uH. Its reactance is 33 Ω. What is the frequency?

    A. 13 kHz.

    B. 0.013 kHz.

    C. 83 kHz.

    D. 83 MHz.

7. An inductor has X_L = 555 Ω at f = 132 kHz. What is L?

    A. 670 mH.

    B. 670 uH.

    C. 460 mH.

    D. 460 uH.

8. A coil has L = 689 uH at f = 990 kHz. What is X_L?

    A. 682 Ω.

    B. 4.28 Ω.

    C. 4.28 KΩ.

    D. 4.28 MΩ.

9. An inductor has L = 88 mH with X_L = 100 Ω. What is f?

    A. 55.3 kHz.

    B. 55.3 Hz.

    C. 181 kHz.

    D. 181 Hz.

10. Each point in the RL plane:

    A. Corresponds to a unique resistance.

    B. Corresponds to a unique inductance.

    C. Corresponds to a unique combination of resistance and inductive reactance.

    D. Corresponds to a unique combination of resistance and inductance.

11. If the resistance R and the inductive reactance X_L both vary from zero to unlimited values, but are always in the ratio 3:1, the points in the RL plane for all the resulting impedances will fall along:

    A. A vector pointing straight up.

    B. A vector pointing "east."

    C. A circle.

    D. A ray of unlimited length.

12. Each impedance $R + jX_L$:

    A. Corresponds to a unique point in the RL plane.

    B. Corresponds to a unique inductive reactance.

C. Corresponds to a unique resistance.

D. All of the above.

13. A vector is a quantity that has:

A. Magnitude and direction.

B. Resistance and inductance.

C. Resistance and reactance.

D. Inductance and reactance.

14. In an RL circuit, as the ratio of inductive reactance to resistance, X_L/R, decreases, the phase angle:

A. Increases.

B. Decreases.

C. Stays the same.

D. Cannot be found.

15. In a purely reactive circuit, the phase angle is:

A. Increasing.

B. Decreasing.

C. 0 degrees.

D. 90 degrees.

16. If the inductive reactance is the same as the resistance in an RL circuit, the phase angle is:

A. 0 degrees.

B. 45 degrees.

C. 90 degrees.

D. Impossible to find; there's not enough data given.

17. In Fig. 13-14, the impedance shown is:

A. 8.0.

B. 90.

C. $90 + j8.0$.

D. $8.0 + j90$.

18. In Fig. 13-14, note that the R and X_L scale divisions are of different sizes. The phase angle is:

A. About 50 degrees, from the looks of it.

B. 48 degrees, as measured with a protractor.

C. 85 degrees, as calculated trigonometrically.

D. 6.5 degrees, as calculated trigonometrically.

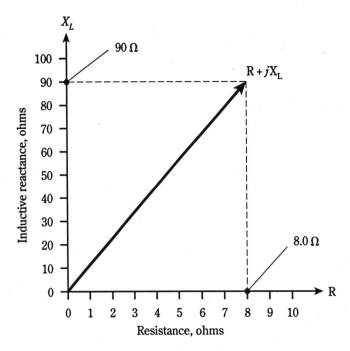

13-14 Illustration for quiz questions 17 and 18.

19. An RL circuit consists of a 100-uH inductor and a 100-Ω resistor. What is the phase angle at a frequency of 200 kHz?

A. 45.0 degrees.

B. 51.5 degrees.

C. 38.5 degrees.

D. There isn't enough data to know.

20. An RL circuit has an inductance of 88 mH. The resistance is 95 Ω. What is the phase angle at 800 Hz?

A. 78 degrees.

B. 12 degrees.

C. 43 degrees.

D. 47 degrees.

14
CHAPTER

Capacitive reactance

INDUCTIVE REACTANCE IS SOMETHING LIKE RESISTANCE, IN THE SENSE THAT IT IS a one-dimensional, or scalar, quantity that can vary from zero upwards without limit. Inductive reactance, like resistance, can be represented by a ray, and is measured in ohms.

Inductive reactance has its counterpart in the form of *capacitive reactance*. This too can be represented as a ray, starting at the same zero point as inductive reactance, but running off in the opposite direction, having *negative* ohmic values (Fig. 14-1). When the ray for capacitive reactance is combined with the ray for inductive reactance, a number line is the result, with ohmic values that range from the huge negative numbers, through zero, to huge positive numbers.

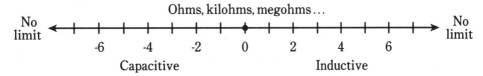

14-1   Inductive and capacitive reactance can be represented on a complete ohmic number line

Capacitors and direct current

Suppose that you have two big, flat metal plates, both of which are excellent electrical conductors. Imagine that you stack them one on top of the other, with only air in between. What will take place if you connect a source of dc across the plates (Fig. 14-2)? The plates will become electrically charged, and will reach a potential difference equal to the dc source voltage. It won't matter how big or small the plates are; their mutual voltage will always be the same as that of the source, although, if the plates are monstrously large, it might take awhile for them to become fully charged. The current, once the plates are charged, will be zero.

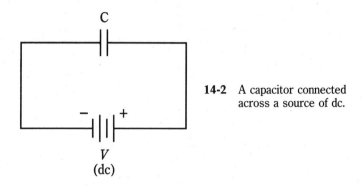

14-2 A capacitor connected across a source of dc.

If you put some insulating material, such as glass, between the plates, their mutual voltage will not change, although the charging time might increase.

If you increase the source voltage, the potential difference between the plates will follow along, more or less rapidly, depending on how large the plates are and on what is between them. If the voltage is increased without limit, arcing will eventually take place. That is, sparks will begin to jump between the plates.

Capacitors and alternating current

Suppose that the source is changed from direct to alternating current (Fig. 14-3). Imagine that you can adjust the frequency of this ac from a low value of a few hertz, to hundreds of hertz, to many kilohertz, megahertz and gigahertz.

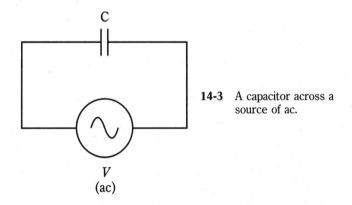

14-3 A capacitor across a source of ac.

At first, the voltage between the plates will follow just about exactly along as the ac source polarity reverses over and over. But the set of plates has a certain amount of capacitance, as you have learned. Perhaps they can charge up fast, if they are small and if the space between them is large, but they can't charge instantaneously.

As you increase the frequency of the ac voltage source, there will come a point at which the plates do not get charged up very much before the source polarity reverses. The set of plates will be "sluggish." The charge won't have time to get established with each ac cycle.

At high ac frequencies, the voltage between the plates will have trouble following the current that is charging and discharging them. Just as the plates begin to get a good charge, the ac current will pass its peak and start to discharge them, pulling electrons out of the negative plate and pumping electrons into the positive plate.

As the frequency is raised, the set of plates starts to act more and more like a short circuit. When the frequency is low, there is a small charging current, but this quickly tails off and drops to zero as the plates become fully charged. As the frequency becomes high, the current flows for more and more of every cycle before dropping off; the charging time remains constant while the *period* of the charging/discharging wave is getting shorter. Eventually, if you keep on increasing the frequency, the period of the wave will be much shorter than the charging/discharging time, and current will flow in and out of the plates in just about the same way as it would flow if the plates were shorted out.

The opposition that the set of plates offers to ac is the capacitive reactance. It is measured in ohms, just like inductive reactance, and just like resistance. But it is, by convention, assigned negative values rather than positive ones. Capacitive reactance, denoted X_C, can vary, just as resistance and inductive reactance do, from near zero (when the plates are huge and close together, and/or the frequency is very high) to a few negative ohms, to many negative kilohms or megohms.

Capacitive reactance varies with frequency. But X_C gets larger (negatively) as the frequency goes *down*. This is the opposite of what happens with inductive reactance, which gets larger (positively) as the frequency goes *up*.

Sometimes, capacitive reactance is talked about in terms of its *absolute value*, with the minus sign removed. Then you might say that X_C is increasing as the frequency decreases, or that X_C is decreasing as the frequency is raised. It's best, however, if you learn to work with negative X_C values right from the start. This will be important later, when you need to work with inductive and capacitive reactances together in the same circuits.

Reactance and frequency

In many ways capacitive reactance behaves like a mirror image of inductive reactance. But in another sense, X_C is an extension of X_L into negative values—below zero.

If the frequency of an ac source is given in hertz as f, and the capacitance of a capacitor in farads as C, then the capacitive reactance is:

$$X_C = -1/(6.28fC)$$

This same formula applies if the frequency, f, is in megahertz and the capacitance, C, is in microfarads (uF). Just remember that if the frequency is in millions, the capacitance must be in million*ths*.

Capacitive reactance varies *inversely* with the frequency. This means that the function X_C vs f appears as a curve when graphed, and this curve "blows up" as the frequency nears zero.

Capacitive reactance also varies inversely with the actual value of capacitance, given a fixed frequency. Therefore, the function of X_C vs C also appears as a curve that "blows up" as the capacitance approaches zero.

The negative of X_C is *inversely proportional* to frequency, and also to capacitance. Relative graphs of these functions are shown in Fig. 14-4.

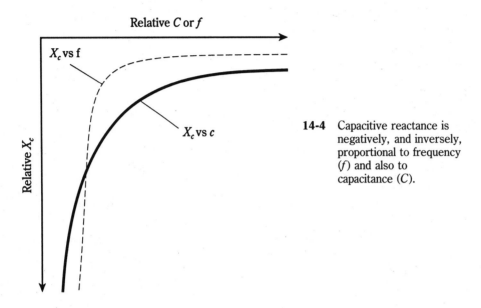

14-4 Capacitive reactance is negatively, and inversely, proportional to frequency (f) and also to capacitance (C).

Problem 14-1

A capacitor has a value of 0.00100 uF at a frequency of 1.00 MHz. What is the capacitive reactance?

Use the formula and plug in the numbers. You can do this directly, since the data is specified in *micro*farads (millionths) and in *mega*hertz (millions):

$$X_C = -1/(6.28 \times 1.00 \times 0.00100) = -1/(0.00628) = -159 \ \Omega$$

This is rounded to three significant figures, since all the data is given to this many digits.

Problem 14-2

What will be the capacitive reactance of the above capacitor, if the frequency decreases to zero? That is, if the source is dc?

In this case, if you plug the numbers into the formula, you'll get zero in the denominator. Mathematicians will tell you that this is a no-no. But in reality, you can say that the reactance will be "extremely large negatively, and for practical purposes, negative infinity."

Problem 14-3

Suppose a capacitor has a reactance of $-100 \ \Omega$ at a frequency of 10.0 MHz. What is its capacitance?

In this problem, you need to put the numbers in the formula and solve for the unknown C. Begin with the equation:

$$-100 = -1/(6.28 \times 10.0 \times C)$$

Dividing through by -100, you get:

$$1 = 1/(628 \times 10.0 \times C)$$

Multiply each side of this by C, and you obtain:

$$C = 1/(628 \times 10.0)$$

This can be solved easily enough. Divide out $C = 1/6280$ on your calculator, and you'll get $C = 0.000159$. Because the frequency is given in megahertz, this capacitance comes out in microfarads, so that $C = 0.000159$ uF. You might rather say that this is 159 pf (remember that 1 pF $= 0.000001$ uF).

Admittedly, the arithmetic for dealing with capacitive reactance is a little messier than that for inductive reactance. This is the case for two reasons. First, you have to work with reciprocals, and therefore the numbers sometimes get awkward. Second, you have to watch those negative signs. It's easy to leave them out. But they're important when looking at reactances in the coordinate plane, because the minus sign tells you that the reactance is capacitive, rather than inductive.

Points in the RC plane

Capacitive reactance can be plotted along a half line, or ray, just as can inductive reactance. In fact, capacitive and inductive reactance, considered as one, form a whole line that is made of two half lines stuck together and pointing in opposite directions. The point where they join is the zero-reactance point. This was shown back in Fig. 14-1.

In a circuit containing resistance and capacitive reactance, the characteristics are two-dimensional, in a way that is analogous to the situation with the RL plane from the previous chapter. The resistance ray and the capacitive-reactance ray can be placed end-to-end at right angles to make a quarter plane called the *RC plane* (Fig. 14-5). Resistance is plotted horizontally, with increasing values toward the right. Capacitive reactance is plotted *downwards*, with increasingly negative values as you go down.

The combinations of R and X_C in this RC plane form impedances. You'll learn about impedance in greater detail in the next chapter. Each point on the RC plane corresponds to one and only one impedance. Conversely, each specific impedance coincides with one and only one point on the plane.

Impedances that contain resistance and capacitance are written in the form $R + jX_C$. Remember that X_C is *never positive*, that is, it is always negative or zero. Because of this, engineers will often write $R - jX_C$, dropping the minus sign from X_C and replacing addition with subtraction in the complex rendition of impedance.

If the resistance is pure, say $R = 3$ Ω, then the complex impedance is $3 - j0$, and this corresponds to the point $(3, 0)$ on the RC plane. You might at this point suspect that $3 - j0$ is the same as $3 + j0$, and that you really need not even write the "$j0$" part at all. In theory, both of these notions are indeed correct. But writing the "$j0$" part indicates that you are open to

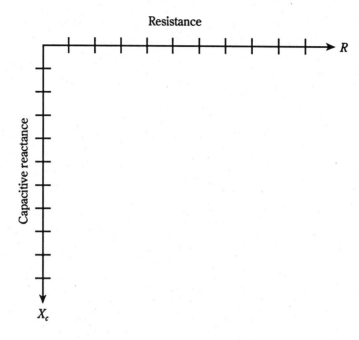

14-5 The RC quarter-plane.

the possibility that there might be reactance in the circuit, and that you're working in two dimensions. It also underscores the fact that the impedance is a pure resistance.

If you have a pure capacitive reactance, say $X_C = -4$ Ω, then the complex impedance is $0 - j4$, and this is at the point $(0, -4)$ on the RC plane. Again, it's important, for completeness, to write the "0" and not just the "$-j4$."

The points for $3 - j0$ and $0 - j4$, and two others, are plotted on the RC plane in Fig. 14-6.

In practical circuits, all capacitors have some *leakage resistance*. If the frequency goes to zero, that is, if the source is dc, a tiny current will flow because no insulator is perfect. Some capacitors have almost no leakage resistance, and come close to being perfect. But none are mathematically flawless. All resistors have a little bit of capacitive reactance, just because they occupy physical space. So there is no such thing as a mathematically pure resistance, either. The points $3 - j0$ and $0 - j4$ are idealized.

Often, resistance and capacitive reactance are both placed in a circuit deliberately. Then you get impedances such as $2 - j3$ and $5 - j5$, both shown in Fig. 14-6.

Remember that the values for X_C are *reactances*, and not the actual capacitances. They vary with the frequency in an RC circuit. If you raise or lower the frequency, the value of X_C will change. A higher frequency causes X_C to get smaller and smaller negatively (closer to zero). A lower frequency causes X_C to get larger and larger negatively (farther from zero, or lower down on the RC plane). If the frequency goes to zero, then the capacitive reactance drops off the bottom of the plane, out of sight. In that case you have two oppositely charged plates or sets of plates, and no "action."

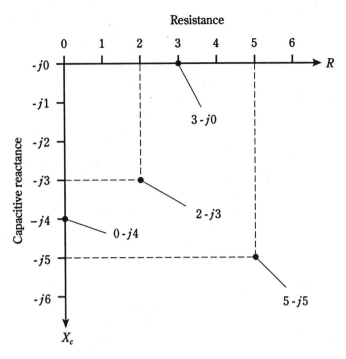

14-6 Four points in the RC plane. See text for discussion.

Vectors in the RC plane

If you work much with engineers, or if you plan to become one, you'll get familiar with the RC plane, just as you will with the RL plane. Recall from the last chapter that RL impedances can be represented as vectors. The same is true for RC impedances.

In Fig. 14-6, there are four different impedance points. Each one is represented by a certain distance to the right of the origin $(0,0)$, and a certain displacement downwards. The first of these is the resistance, R, and the second is the capacitive reactance, X_C. Therefore, the RC impedance is a two-dimensional quantity.

Doesn't this look like a mirror-image reflection of RL impedances? You could almost imagine that we're looking at an RL plane reflected in a pool of still water. This is, in fact, an excellent way to envision this situation.

The impedance points in the RC plane can be rendered as vectors, just as this can be done in the RL plane. Then the points become rays, each with a certain length and direction. The magnitude and direction for a vector, and the coordinates for the point, both uniquely define the same impedance value. The length of the vector is the distance of the point from the origin, and the direction is the angle measured *clockwise* from the resistance (R) line, and specified in *negative* degrees. The equivalent vectors, for the points in Fig. 14-6, are illustrated in Fig. 14-7.

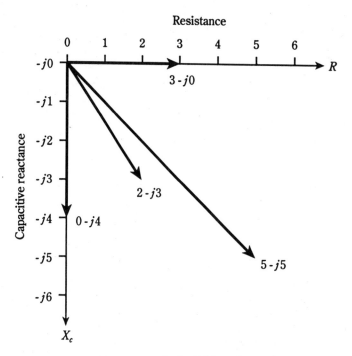

14-7 Four vectors in the RC impedance plane.

Current leads voltage

Capacitance stores energy in the form of an electric field. When a current is driven through a capacitor, it takes a little time before the plates can fully charge to the full potential difference of the source voltage.

When an ac voltage source is placed across a capacitor, the voltage in the capacitor lags the current in phase. Another way of saying this is that the current *leads* the voltage. The phase difference can range from zero, to a small part of a cycle, to a quarter of a cycle (90 degrees).

Pure capacitance

Imagine placing an ac voltage source across a capacitor. Suppose that the frequency is high enough, and/or the capacitance large enough, so that the capacitive reactance, X_C, is extremely small compared with the resistance, R. Then the current leads the voltage by a full 90 degrees (Fig. 14-8).

At very high frequencies, it doesn't take very much capacitance for this to happen. Small capacitors usually have less leakage resistance than large ones. At lower frequencies, the capacitance must be larger, although high-quality, low-loss capacitors are not too difficult to manufacture except at audio frequencies and at the 60-Hz utility frequency.

The situation depicted in Fig. 14-8 represents an essentially pure capacitive reactance. The vector in the RC plane points just about straight down. Its angle is -90 degrees from the R axis or zero line.

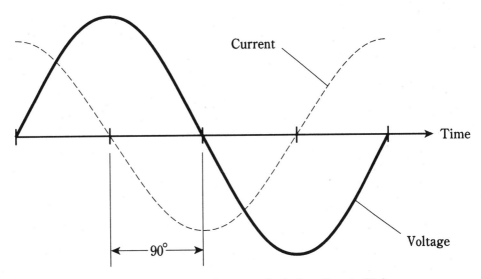

14-8 In a pure capacitance, the current leads the voltage by 90 degrees.

Capacitance and resistance

When the resistance in a resistance-capacitance circuit is significant compared with the capacitive reactance, the current leads the voltage by something less than 90 degrees (Fig. 14-9). If R is small compared with X_C, the difference is almost a quarter of a cycle. As R gets larger, or as X_C becomes smaller, the phase difference gets less. A circuit containing resistance and capacitance is called an *RC circuit*.

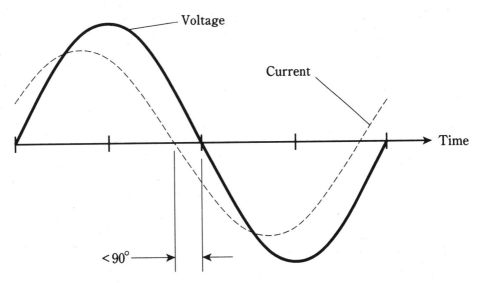

14-9 In a circuit with capacitance and resistance, the current leads the voltage by less than 90 degrees.

The value of R in an RC circuit might increase relative to X_C because resistance is deliberately put into a circuit. Or, it might happen because the frequency becomes so low that X_C rises to a value comparable with the leakage resistance of the capacitor. In either case, the situation can be represented by a resistor, R, in series with a capacitor, C (Fig. 14-10).

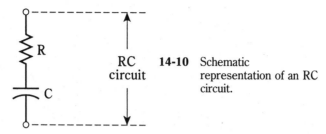

RC circuit

14-10   Schematic representation of an RC circuit.

If you know the values of X_C and R, you can find the *angle of lead*, also called the *RC phase angle*, by plotting the point $R - jX_C$ on the RC plane, drawing the vector from the origin $0 - j0$ out to that point, and then measuring the angle of the vector, clockwise from the resistance axis. You can use a protractor to measure this angle, as you did in the previous chapter for RL phase angles. Or you can use trigonometry to calculate the angle.

As with RL circuits, you only need to know the ratio of X_C to R to determine the phase angle. For example, if $X_C = -4\ \Omega$ and $R = 7\ \Omega$, you'll get the same angle as with $X_C = -400\ \Omega$ and $R = 700\ \Omega$, or with $X_C = -16\ \Omega$ and $R = 28\ \Omega$. The phase angle will be the same for any ratio of $X_C{:}R = -4{:}7$.

Pure resistance

As the resistance in an RC circuit gets large compared with the capacitive reactance, the angle of lead becomes smaller. The same thing happens if the value of X_C gets small compared with the value of R. When you call X_C "large," you mean large *negatively*. When you say that X_C is "small," you mean that it is close to zero, or small *negatively*.

When R is many times larger than X_C, whatever their actual values, the vector in the RC plane will be almost right along the R axis. Then the RC phase angle will be nearly zero, that is, just a little bit negative. The voltage will come nearly into phase with the current. The plates of the capacitor will not come anywhere near getting fully charged with each cycle. The capacitor will be said to "pass the ac" with very little loss, as if it were shorted out. But it will still have an extremely high X_C for any ac signals at much lower frequencies that might exist across it at the same time. (This property of capacitors can be put to use in electronic circuits, for example when an engineer wants to let radio signals get through while blocking audio frequencies.)

Ultimately, if the capacitive reactance gets small enough, the circuit will act as a pure resistance, and the current will be in phase with the voltage.

How much lead?

If you know the ratio of capacitive reactance to resistance, or X_C/R, in an RC circuit, then you can find the phase angle. Of course, you can find this angle of lead if you know the precise values.

Pictorial method

You can use a protractor and a ruler to find phase angles for RC circuits, just as you did with RL circuits, as long as the angles aren't too close to 0 or 90 degrees.

First, draw a line somewhat longer than 10 cm, going from left to right on the paper. Then, use the protractor to construct a line going somewhat more than 10 cm vertically downwards, starting at the left end of the horizontal line.

The horizontal line is the R axis of a crude RC plane. The line going down is the X_C axis.

If you know the values of X_C and R, divide or multiply them by a constant, chosen to make both values fall between -100 and 100. For example, if $X_C = -3800\ \Omega$ and R = $7400\ \Omega$, divide them both by 100, getting -38 and 74. Plot these points on the lines as hash marks. The X_C mark goes 38 mm down from the intersection point between your two axes (*negative* 38 mm *up* from the intersection point). The R mark goes 74 mm to the right of the intersection point.

Now, draw a line connecting the two hash marks, as shown in Fig. 14-11. This line will be at a slant, and will form a triangle along with the two axes. This is a right triangle, with the right angle at the origin of the RC plane.

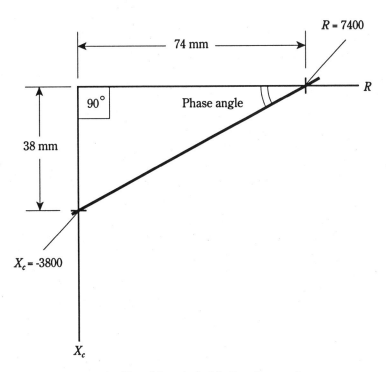

14-11   Pictorial method of finding phase angle.

Measure the angle between the slanted line and the R, or horizontal, axis. Use the protractor for this. Extend the lines, if necessary, using the ruler, to get a good reading on the protractor. This angle will be between 0 and 90 degrees. Multiply this reading by

−1 to get the RC phase angle. That is, if the protractor shows 27 degrees, the RC phase angle is −27 degrees.

The actual vector is found by constructing a rectangle using the origin and your two hash marks, making new perpendicular lines to complete the figure. The vector is the diagonal of this rectangle, running out from the origin (Fig. 14-12). The phase angle is the angle between the R axis and this vector, multiplied by −1. It will have the same measure as the angle of the slanted line you constructed in Fig. 14-11.

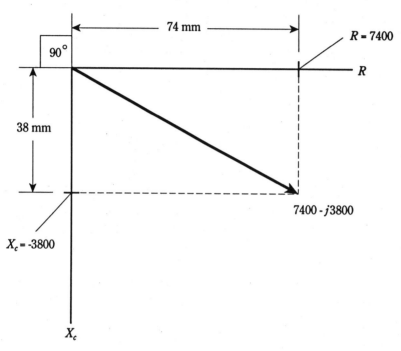

14-12   Another pictorial way of finding phase angle. This method shows the actual impedance vector.

Trigonometric method

The more accurate way to find RC phase angles is to use trigonometry. Then you don't have to draw a figure and use a protractor. You only need to punch a few buttons on your calculator. (Just make sure they're the right ones, in the right order.)

If you know the values X_C and R, find the ratio X_C/R. If you know the ratio only, call it X_C/R and enter it into the calculator. Be sure not to make the mistake of getting the ratio upside down (R/X_C). This ratio will be a negative number or zero, because X_C is always negative or zero, and R is always positive. Find the arctangent (arctan or $\tan^{-1}$) of this number. This is the RC phase angle.

Always remember, when doing problems of this kind, to use the *capacitive reactance* for X_C, and not the capacitance. This means that, if you are given the capacitance, you must use the formula for X_C and then calculate the RC phase angle.

Problem 14-4

The capacitive reactance in an RC circuit is $-3800\,\Omega$, and the resistance is $7400\,\Omega$. What is the phase angle?

Find the ratio $X_C/R = -3800/7400$. The calculator will display something like 0.513513513. Find the arctangent, or $\tan^{-1}$, getting a phase angle of 27.18111109 degrees on the calculator display. Round this off to 27.18 degrees.

Problem 14-5

An RC circuit works at a frequency of 3.50 MHz. It has a resistance of $130\,\Omega$ and a capacitance of 150 pF. What is the phase angle?

This problem is a little more involved. First, you must find the capacitive reactance for a capacitor of 150 pF. Convert this to microfarads, getting C = 0.000150 uF. Remember that *micro*farads go with *mega*hertz (millionths go with millions to cancel each other out). Then:

$$X_C = -1/(6.28 \times 3.50 \times 0.000150)$$
$$= -1/0.003297 = -303\,\Omega$$

Now you can find the ratio $X_C/R = -303/130 = -2.33$; the phase angle is arctan (-2.33) $= -66.8$ degrees.

Problem 14-6

What is the phase angle in the above circuit if the frequency is raised to 7.10 MHz?

You need to find the new value for X_C, because it will change as a result of the frequency change. Calculating,

$$X_C = -1/(6.28 \times 7.10 \times 0.000150)$$
$$= -1/0.006688 = -150\,\Omega$$

The ratio $X_C/R = -150/130 = -1.15$; the phase angle is therefore arctan $(-1.15) = -49.0$ degrees.

Quiz

Refer to the text in this chapter if necessary. A good score is at least 18 correct. Answers are in the back of the book.

1. As the size of the plates in a capacitor increases, all other things being equal:

   A. The value of X_C increases negatively.

   B. The value of X_C decreases negatively.

   C. The value of X_C does not change.

   D. You can't say what happens to X_C without more data.

2. If the dielectric material between the plates of a capacitor is changed, all other things being equal:

   A. The value of X_C increases negatively.

B. The value of X_C decreases negatively.

C. The value of X_C does not change.

D. You can't say what happens to X_C without more data.

3. As the frequency of a wave gets lower, all other things being equal, the value of X_C for a capacitor:

A. Increases negatively.

B. Decreases negatively.

C. Does not change.

D. Depends on the current.

4. A capacitor has a value of 330 pF. What is its capacitive reactance at a frequency of 800 kHz?

A. -1.66 Ω.

B. -0.00166 Ω.

C. -603 Ω.

D. -603K Ω.

5. A capacitor has a reactance of -4.50 Ω at 377 Hz. What is its capacitance?

A. 9.39 uF.

B. 93.9 uF.

C. 7.42 uF.

D. 74.2 uF.

6. A capacitor has a value of 47 uF. Its reactance is -47 Ω. What is the frequency?

A. 72 Hz.

B. 7.2 MHz.

C. 0.000072 Hz.

D. 7.2 Hz.

7. A capacitor has $X_C = -8800$ Ω at f = 830 kHz. What is C?

A. 2.18 uF.

B. 21.8 pF.

C. 0.00218 uF.

D. 2.18 pF.

8. A capacitor has $C = 166$ pF at f = 400 kHz. What is X_C?

A. -2.4K Ω.

B. -2.4 Ω.

C. -2.4×10^{-6} Ω.

D. -2.4M Ω.

9. A capacitor has $C = 4700$ uF and $X_C = -33$ Ω. What is f?

   A. 1.0 Hz.

   B. 10 Hz.

   C. 1.0 kHz.

   D. 10 kHz.

10. Each point in the RC plane:

    A. Corresponds to a unique inductance.

    B. Corresponds to a unique capacitance.

    C. Corresponds to a unique combination of resistance and capacitance.

    D. Corresponds to a unique combination of resistance and reactance.

11. If R increases in an RC circuit, but X_C is always zero, then the vector in the RC plane will:

    A. Rotate clockwise.

    B. Rotate counterclockwise.

    C. Always point straight towards the right.

    D. Always point straight down.

12. If the resistance R increases in an RC circuit, but the capacitance and the frequency are nonzero and constant, then the vector in the RC plane will:

    A. Get longer and rotate clockwise.

    B. Get longer and rotate counterclockwise.

    C. Get shorter and rotate clockwise.

    D. Get shorter and rotate counterclockwise.

13. Each impedance $R - jX_C$:

    A. Represents a unique combination of resistance and capacitance.

    B. Represents a unique combination of resistance and reactance.

    C. Represents a unique combination of resistance and frequency.

    D. All of the above.

14. In an RC circuit, as the ratio of capacitive reactance to resistance, $-X_C/R$, gets closer to zero, the phase angle:

    A. Gets closer to -90 degrees.

    B. Gets closer to 0 degrees.

    C. Stays the same.

    D. Cannot be found.

15. In a purely resistive circuit, the phase angle is:

    A. Increasing.

    B. Decreasing.

C. 0 degrees.

D. −90 degrees.

16. If the ratio of X_C/R is 1, the phase angle is:

   A. 0 degrees.

   B. −45 degrees.

   C. −90 degrees.

   D. Impossible to find; there's not enough data given.

17. In Fig. 14-13, the impedance shown is:

   A. 8.02 + j323.

   B. 323 + j8.02.

   C. 8.02 − j323.

   D. 323 − j8.02.

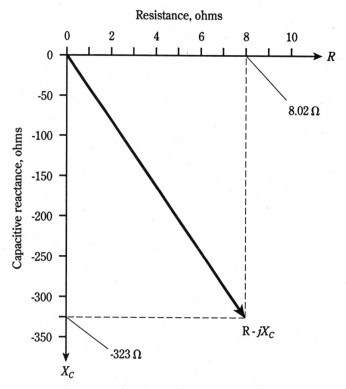

14-13   Illustration for quiz questions 17 and 18.

18. In Fig. 14-13, note that the R and X_C scale divisions are not the same size. The phase angle is:

   A. 1.42 degrees.

B. About −60 degrees, from the looks of it.

C. −58.3 degrees.

D. −88.6 degrees.

19. An RC circuit consists of a 150-pF capacitor and a 330-Ω resistor in series. What is the phase angle at a frequency of 1.34 MHz?

A. −67.4 degrees.

B. −22.6 degrees.

C. −24.4 degrees.

D. −65.6 degrees.

20. An RC circuit has a capacitance of 0.015 uF. The resistance is 52 Ω. What is the phase angle at 90 kHz?

A. −24 degrees.

B. −0.017 degrees.

C. −66 degrees.

D. None of the above.

15
CHAPTER

Impedance and admittance

YOU'VE SEEN HOW INDUCTIVE AND CAPACITIVE REACTANCE CAN BE REPRE-
sented along a line perpendicular to resistance. In this chapter, you'll put all three of these
quantities—R, X_L, and X_C—together, forming a complete, working definition of *imped-*
ance. You'll also get acquainted with *admittance*, impedance's evil twin.

To express the behavior of alternating-current (ac) circuits, you need two dimensions, because ac has variable frequency along with variable current. One dimension
(resistance) will suffice for dc, but not for ac.

In this chapter and the two that follow, the presentation is rather mathematical. You
can get a grasp of the general nature of the subject matter without learning how to do all
of the calculations presented. The mathematics is given for those of you who wish to gain
a firm understanding of how ac circuits behave.

Imaginary numbers

What does the lower-case j actually mean in expressions of impedance such as $4 + j7$ and
$45 - j83$? This was briefly discussed earlier in this book, but what is this thing j, really?

Mathematicians use the lower-case letter i to represent j. (Mathematicians and
physicists/engineers often differ in notation as well as in philosophy.) This *imaginary*
number is the square root of -1. It is the number that, when multiplied by itself, gives
-1. So $i = j$, and $j \times j = -1$.

The entire set of imaginary numbers derives from this single unit. The square of an
imaginary number is negative—always. No *real number* has this property. Whether a real
number is positive, zero or negative, its square can never be negative— never.

The notion of j (or i, if you're a mathematician) came about simply because some
mathematicians wondered what the square root of -1 would behave like, if there were
such a thing. So the mathematicians "imagined" the existence of this animal, and found
that it had certain properties. Eventually, the number i was granted a place among the

realm of numbers. Mathematically, it's as real as the real numbers. But the original term "imaginary" stuck, so that this number carries with it a mysterious aura.

It's not important, in this context, to debate the reality of the abstract, but to reassure you that imaginary numbers are not particularly special, and are not intended or reserved for just a few eccentric geniuses. "Imaginary" numbers are as real as the "real" ones. And just as unreal, in that neither kind are concrete; you can hold neither type of number in your hand, nor eat them, nor throw them in a wastebasket.

The *unit imaginary number j* can be multiplied by any real number, getting an infinitude of imaginary numbers forming an *imaginary number line* (Fig. 15-1). This is a duplicate of the *real number line* you learned about in school. It must be at a right angle to the real number line when you think of real and imaginary numbers at the same time.

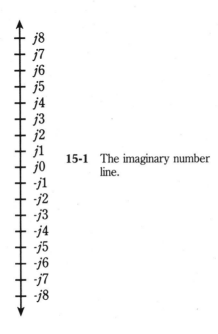

15-1 The imaginary number line.

Complex numbers

When you add a real number and an imaginary number, such as $4 + j7$ or $45 - j83$, you have a *complex number*. This term doesn't mean complicated; it would better be called composite. But, again, the original name stuck, even if it wasn't the best possible thing to call it.

Real numbers are one-dimensional. They can be depicted on a line. Imaginary numbers are also one-dimensional for the same reason. But complex numbers need two dimensions to be completely defined.

Adding and subtracting complex numbers

Adding complex numbers is just a matter of adding the real parts and the complex parts separately. The sum of $4 + j7$ and $45 - j83$ is therefore $(4 + 45) + j(7 - 83) = 49 + j(-76) = 49 - j76$.

Subtracting complex numbers works similarly. The difference $(4 + j7) - (45 - j83)$ is found by multiplying the second complex number by -1 and then adding the result, getting $(4 + j7) + (-1(45 - j83)) = (4 + j7) + (-45 + j83) = -41 + j90$.

The general formula for the sum of two complex numbers $(a + jb)$ and $(c + jd)$ is:

$$(a + jb) + (c + jd) = (a + c) + j(b + d)$$

The plus and minus number signs get tricky when working with sums and differences of complex numbers. Just remember that any difference can be treated as a sum: multiply the second number by -1 and then add. You might want to do some exercises to get yourself acquainted with the way these numbers behave, but in working with engineers, you will not often be called upon to wrestle with complex numbers at the level of "nitty-gritty."

If you plan to become an engineer, you'll need to practice adding and subtracting complex numbers. But it's not difficult once you get used to it by doing a few sample problems.

Multiplying complex numbers

You should know how complex numbers are multiplied, to have a full understanding of their behavior. When you multiply these numbers, you only need to treat them as sums of number pairs, that is, as *binomials*.

It's easier to give the general formula than to work with specifics here. The product of $(a + jb)$ and $(c + jd)$ is equal to $ac + jad + jbc + jjbd$. Simplifying, remember that $jj = -1$, so you get the final formula:

$$(a + jb)(c + jd) = (ac - bd) + j(ad + bc)$$

As with the addition and subtraction of complex numbers, you must be careful with signs (plus and minus). And also, as with addition and subtraction, you can get used to doing these problems with a little practice. Engineers sometimes (but not too often) have to multiply complex numbers.

The complex number plane

Real and imaginary numbers can be thought of as points on a line. Complex numbers lend themselves to the notion of points on a plane. This plane is made by taking the real and imaginary number lines and placing them together, at right angles, so that they intersect at the zero points, 0 and $j0$. This is shown in Fig. 15-2. The result is a *Cartesian* coordinate plane, just like the ones you use to make graphs of everyday things like bank-account balance versus time.

Notational neuroses

On this plane, a complex number might be represented as $a + jb$ (in engineering or physicists' notation), or as $a + bi$ (in mathematicians' notation), or as an *ordered pair* (a, b). "Wait," you ask. "Is there a misprint here? Why does b go after the j, but in front of the i?" The answer: Mathematicians and engineers/physicists just don't think alike, and this is but one of myriad ways in which this is apparent. In other words, it's a matter of

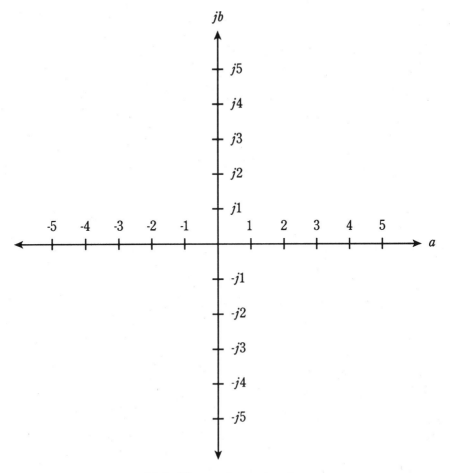

15-2   The complex number plane.

notational convention, and that is all. (It's also a somewhat humorous illustration of the different angle that an engineer takes in approaching a problem, as opposed to a mathematician.)

Complex number vectors

Complex numbers can also be represented as vectors in the complex plane. This gives each complex number a unique magnitude and direction. The magnitude is the distance of the point $a + jb$ from the origin $0 + j0$. The direction is the angle of the vector, measured counterclockwise from the $+a$ axis. This is shown in Fig. 15-3.

Absolute value

The *absolute value* of a complex number $a + jb$ is the length, or magnitude, of its vector in the complex plane, measured from the origin $(0,0)$ to the point (a, b).

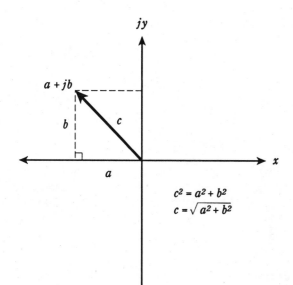

15-3 Magnitude and direction of a vector in the complex number plane.

In the case of a *pure real* number $a + j0$, the absolute value is simply the number itself, a, if it is positive, and $-a$ if a is negative.

In the case of a *pure imaginary* number $0 + jb$, the absolute value is equal to b if b (which is a real number) is positive, and $-b$ if b is negative.

If the number is neither pure real or pure imaginary, the absolute value must be found by using a formula. First, square both a and b. Then add them. Finally, take the square root. This is the length of the vector $a + jb$. The situation is illustrated in Fig. 15-4.

15-4 Calculation of absolute value, or vector length.

$$c^2 = a^2 + b^2$$
$$c = \sqrt{a^2 + b^2}$$

Problem 15-1

Find the absolute value of the complex number $-22 - j0$.

Note that this is a pure real. Actually, it is the same as $-22 + j0$, because $j0 = 0$. Therefore, the absolute value of this complex number is $-(-22) = 22$.

Problem 15-2

Find the absolute value of $0 - j34$.

This is a pure imaginary number. The value of b in this case is -34, because $0 - j34 = 0 + j(-34)$. Therefore, the absolute value is $-(-34) = 34$.

Problem 15-3

Find the absolute value of $3 - j4$.

In this number, $a = 3$ and $b = -4$, because $3 - j4$ can be rewritten as $3 + j(-4)$. Squaring both of these, and adding the results, gives $3^2 + (-4)^2 = 9 + 16 = 25$. The square root of 25 is 5; therefore, the absolute value of this complex number is 5.

You might notice this "3, 4, 5" relationship and recall the Pythagorean theorem for finding the length of the *hypotenuse* of a right triangle. The formula for finding the length of a vector in the complex-number plane comes directly from this theorem.

If you don't remember the Pythagorean theorem, don't worry; just remember the formula for the length of a vector.

The RX plane

Recall the planes for resistance (R) and inductive reactance (X_L) from chapter 13. This is the same as the upper-right quadrant of the complex-number plane shown in Fig. 15-2.

Similarly, the plane for resistance and capacitive reactance (X_C) is the same as the lower-right quadrant of the complex number plane.

Resistances are represented by nonnegative real numbers. Reactances, whether they are inductive (positive) or capacitive (negative), correspond to imaginary numbers.

No negative resistance

There is no such thing, strictly speaking, as negative resistance. That is to say, one cannot have anything better than a perfect conductor. In some cases, a supply of direct current, such as a battery, can be treated as a negative resistance; in other cases, you can have a device that acts as if its resistance were negative under certain changing conditions. But generally, in the *RX (resistance-reactance) plane*, the resistance value is always positive. This means that you can remove the negative axis, along with the upper-left and lower-left quadrants, of the complex-number plane, obtaining a half plane as shown in Fig. 15-5.

Reactance in general

Now you should get a better idea of why capacitive reactance, X_C, is considered negative. In a sense, it is an extension of inductive reactance, X_L, into the realm of negatives, in a

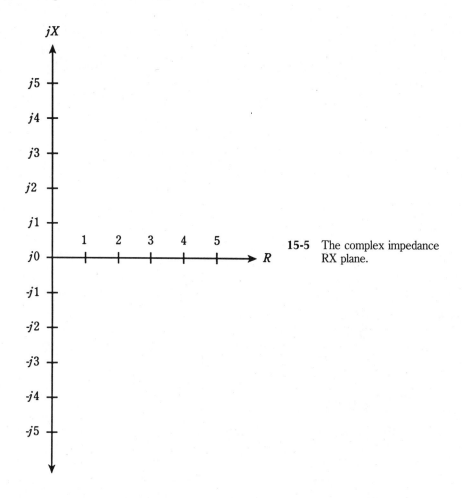

15-5   The complex impedance RX plane.

way that cannot generally occur with resistance. Capacitors act like "negative inductors." Interesting things happen when capacitors and inductors are combined, which is discussed in the next couple of chapters.

Reactance can vary from extremely large negative values, through zero, to extremely large positive values. Engineers and physicists always consider reactance to be imaginary. In the mathematical model of impedance, capacitances and inductances manifest themselves "perpendicularly" to resistance.

The general symbol for reactance is X; this encompasses both inductive reactance (X_L) and capacitive reactance (X_C).

Vector representation of impedance

Any impedance $R + jX$ can be represented by a complex number of the form $a + jb$. Just let $R = a$ and $X = b$.

It should be easy visualize, now, how the impedance vector changes as either R or X, or both, are varied. If X remains constant, an increase in R will cause the vector to get longer. If R remains constant and X_L gets larger, the vector will also grow longer. If R stays the same as X_C gets larger (negatively), the vector will grow longer yet again.

Think of the point $a + jb$, or $R + jX$, moving around in the plane, and imagine where the corresponding points on the axes lie. These points are found by drawing dotted lines from the point $R + jX$ to the R and X axes, so that the lines intersect the axes at right angles (Fig. 15-6).

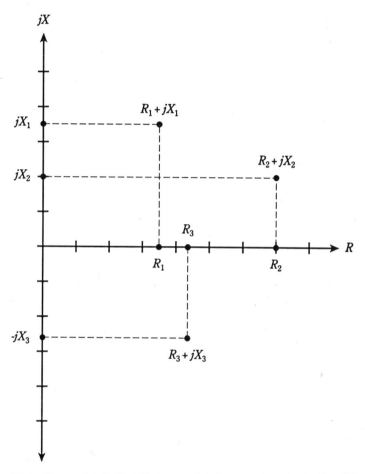

15-6 Some points in the RX plane, and their components on the R and X axes.

Now think of the points for R and X moving toward the right and left, or up and down, on their axes. Imagine what happens to the point $R + jX$ in various scenarios. This is how impedance changes as the resistance and reactance in a circuit are varied.

Resistance is one-dimensional. Reactance is also one-dimensional. But impedance is two-dimensional. To fully define impedance, you must render it on a half plane, specifying the resistance and the reactance, which are independent.

Absolute-value impedance

There will be times when you'll hear that the "impedance" of some device or component is a certain number of ohms. For example, in audio electronics, there are "8-Ω" speakers and "600-Ω" amplifier inputs. How can manufacturers quote a single number for a quantity that is two-dimensional, and needs two numbers to be completely expressed?

There are two answers to this. First, figures like this generally refer to devices that have purely resistive impedances. Thus, the "8-Ω" speaker really has a complex impedance of $8 + j0$, and the "600-Ω" input circuit is designed to operate with a complex impedance at, or near, $600 + j0$.

Second, you can sometimes talk about the length of the impedance vector, calling this a certain number of ohms. If you talk about "impedance" this way, you are being ambiguous, because you can have an infinite number of different vectors of a given length in the RX plane.

Sometimes, the capital letter Z is used in place of the word "impedance" in general discussions. This is what engineers mean when they say things like "$Z = 50\ \Omega$" or "$Z = 300\ \Omega$ nonreactive."

"$Z = 8\ \Omega$" in this context, if no specific complex impedance is given, can refer to the complex value $8 + j0$, or $0 + j8$, or $0 - j8$, or any value on a half circle of points in the RX plane that are at distance 8 units away from $0 + j0$. This is shown in Fig. 15-7. There exist an infinite number of different complex impedances with $Z = 8\ \Omega$.

Problems 15-1, 15-2 and 15-3 can be considered as problems in finding absolute-value impedance from complex impedance numbers.

Problem 15-4

Name seven different complex impedances having an absolute value of $Z = 10$.

It's easy name three: $0 + j10$, $10 + j0$ and $0 - j10$. These are pure inductance, pure capacitance and pure resistance, respectively.

A right triangle can exist having sides in a ratio of $6:8:10$ units. This is true because $6^2 + 8^2 = 10^2$. (Check it and see!) Therefore, you might have $6 + j8$, $6 - j8$, $8 + j6$ and $8 - j6$, all complex impedances whose absolute value is 10 ohms. Obviously, the value $Z = 10$ was chosen for this problem because such a whole-number right-triangle exists. It becomes quite a lot messier to do this problem (but by no means impossible) if $Z = 11$ instead.

If you're not specifically told what complex impedance is meant when a single-number ohmic figure is quoted, it's best to assume that the engineers are talking about *nonreactive* impedances. That means they are pure resistances, and that the imaginary, or reactive, factor is zero. Engineers will often speak of nonreactive impedances, or of complex impedance vectors, as "low-Z or high-Z." For instance, a speaker might be called "low-Z" and a microphone "high-Z."

Characteristic impedance

There is one property of electronic components that you'll sometimes hear called *impedance*, that really isn't "impedance" at all. This is *characteristic impedance* or *surge*

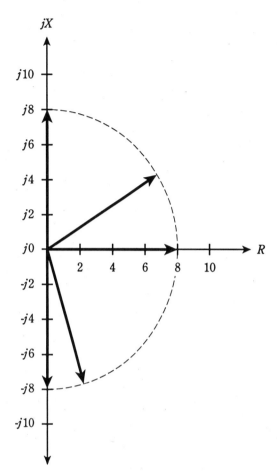

15-7 Vectors representing an absolute value impedance of 8 Ω.

impedance. It is abbreviated Z_o and is a specification of *transmission lines*. It can always be expressed as a positive real number.

Transmission lines

Any time that it is necessary to get energy or signals from one place to another, a transmission line is required. These almost always take either of two forms, *coaxial* or *two-wire*. These are illustrated qualitatively in Fig. 15-8.

Examples of transmission lines include the ribbon that goes from a television antenna to the receiver, the cable running from a hi-fi amplifier to the speakers, and the set of wires that carries electricity over the countryside.

Factors affecting Z_o

The Z_o of a parallel-wire transmission line depends on the diameter of the wires, on the spacing between the wires, and on the nature of the insulating material separating the wires.

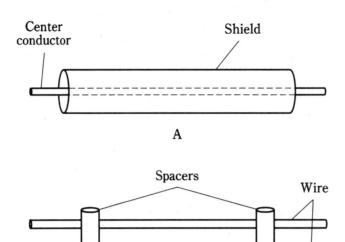

15-8 At A, coaxial transmission line. At B, parallel-wire transmission line.

In general, Z_o increases as the wire diameter gets smaller, and decreases as the wire diameter gets larger, all other things being equal.

In a coaxial line, the thicker the center conductor, the lower the Z_o if the shield stays the same size. If the center conductor stays the same size and the shield tubing increases in diameter, the Z_o will increase.

Also in general, Z_o increases as the spacing between wires, or between the center conductor and the *shield* or *braid*, gets greater, and decreases as the spacing is made less.

Solid dielectrics such as polyethylene reduce the Z_o of a transmission line, compared with air or a vacuum between the conductors.

Z_o in practice

In rigorous terms, the characteristic impedance of a line is determined according to the nature of the *load* with which the line works at highest efficiency.

Suppose that you have an 8-Ω hi-fi speaker, and you want to get audio energy to that speaker with the greatest possible efficiency, so that the least possible power is dissipated in the line. You would use large-diameter wires, of course, but for true optimization, you would want the spacing between the wires to be just right. Adjusting this spacing for optimum power transfer would result in a line Z_o of 8 Ω. Then, the greatest possible efficiency would be had with a speaker of impedance $8 + j0$.

If you can't get the wires to have the right size and spacing for a good match to $Z = 8$ Ω, you might need to use an *impedance transformer*. This makes the speaker's impedance look like something different, such as 50 Ω or 600 Ω.

Imagine that you have a 300-Ω television antenna, and you want the best possible reception. You purchase "300-Ω" ribbon line, with a value of Z_o that has been optimized by the manufacturer for use with antennas whose impedances are close to $300 + j0$.

For a system having an "impedance" of "$R\ \Omega$," the best line Z_o is also $R\ \Omega$. If R is much different from Z_o, an unnecessary amount of power will be wasted in heating up the transmission line. This might not be a significant amount of power, but it often is.

Impedance matching will be discussed in more detail in the next chapter.

Conductance

In an ac circuit, electrical *conductance* works the same way as it does in a dc circuit. Conductance is symbolized by the capital letter G. It was introduced back in chapter 2.

The relationship between conductance and resistance is simple: $G = 1/R$. The unit is the siemens. The larger the value of conductance, the smaller the resistance, and the more current will flow. Conversely, the smaller the value of G, the greater the value of R, and the less current will flow.

Susceptance

Sometimes, you'll come across the term *susceptance* in reference to an ac circuit containing a capacitive reactance or an inductive reactance. Susceptance is symbolized by the capital letter B. It is the reciprocal of reactance. That is, $B = 1/X$. Susceptance can be either capacitive or inductive. These are symbolized as B_C and B_L respectively. Therefore, $B_C = 1/X_C$, and $B_L = 1/X_L$.

There is a trick to determining susceptances in terms of reactances. Or, perhaps better stated, a trickiness. Susceptance is imaginary, just as is reactance. That is, all values of B require the use of the j operator, just as do all values of X. But $1/j = -j$. This reverses the sign when you find susceptance in terms of reactance.

If you have an inductive reactance of, say, 2 ohms, then this is expressed as $j2$ in the imaginary sense. What is $1/(j2)$? You can break this apart and say that $1/(j2) = (1/j)(1/2) = (1/j)0.5$. But what is $1/j$? Without making this into a mathematical treatise, suffice it to say that $1/j = -j$. Therefore, the reciprocal of $j2$ is $-j0.5$. *Inductive susceptance* is *negative-imaginary*.

If you have a capacitive reactance $X_C = 10$ ohms, then this is expressed as $X_C = -j10$. The reciprocal of this is $B_C = 1/(-j10) = (1/-j)(1/10) = (1/-j)0.1$. What is $1/-j$? Again, without going into deep theoretical math, it is equal to j. Therefore, the reciprocal of $-j10$ is $j0.1$. *Capacitive susceptance* is *positive-imaginary*.

This is exactly reversed from the situation with reactances.

Problem 15-5

Suppose you have a capacitor of 100 pF at a frequency of 3.00 MHz. What is B_C?

First, find the reactance X_C by the formula:

$$X_C = -1/(6.28fC)$$

Remembering that 100 pF = 0.000100 uF, you can substitute in this formula for $f = 3.00$ and $C = 0.000100$, getting:

$$X_C = -1/(6.28 \times 3.00 \times 0.000100)$$
$$= -1/0.001884 = -531 \ \Omega = -j531$$

The susceptance, B_C, is equal to $1/X_C$. Thus, $B_C = 1/(-j531) = j0.00188$. Remember that capacitive susceptance is positive. This can "short-circuit" any frustration you might have in manipulating the minus signs in these calculations.

Note that above, you found a reciprocal of a reciprocal. You did something and then immediately turned around and undid it, slipping a minus sign in because of the idiosyncrasies of that little j operator. In the future, you can save work by remembering that the formula for capacitive susceptance, simplified, is:

$$B_C = 6.28fC \text{ siemens} = j(6.28fC)$$

This resembles the formula for inductive reactance.

Problem 15-6

An inductor has $L = 163$ uH at a frequency of 887 kHz. What is B_L?
  First, calculate X_L, the inductive reactance:

$$X_L = 6.28fL = 6.28 \times 0.887 \times 163$$
$$= 908 \ \Omega = j908$$

The susceptance, B_L, is equal to $1/X_L$. Therefore, $B_L = 1/j908 = -j0.00110$. Remember that inductive susceptance is negative.

  The formula for inductive susceptance is similar to that for capacitive reactance:
$$B_L = -1/(6.28fL) \text{ siemens} = -j(1/(6.28fL))$$

Admittance

Conductance and susceptance combine to form *admittance*, symbolized by the capital letter Y.
  Admittance, in an ac circuit, is analogous to conductance in a dc circuit.

Complex admittance

Admittance is a complex quantity and represents the ease with which current can flow in an ac circuit. As the absolute value of impedance gets larger, the absolute value of admittance becomes smaller, in general. Huge impedances correspond to tiny admittances, and vice-versa.
  Admittances are written in complex form just like impedances. But you need to keep track of which quantity you're talking about! This will be obvious if you use the symbol, such as $Y = 3 - j0.5$ or $Y = 7 + j3$. When you see Y instead of Z, you know that negative j factors (such as $-j0.5$) mean that there is a net inductance in the circuit, and positive j factors (such as $+j3$) mean there is net capacitance.

Admittance is the complex composite of conductance and susceptance. Thus, admittance takes the form:

$$Y = G + jB$$

The *j* factor might be negative, of course, so there are times you'll write $Y = G - jB$.

Parallel circuits

Recall how resistances combine with reactances in series to form complex impedances? In chapters 13 and 14, you saw series RL and RC circuits. Perhaps you wondered why parallel circuits were ignored in those discussions. The reason is that admittance, rather than impedance, is best for working with parallel ac circuits. Therefore, the subject of parallel circuits was deferred.

Resistance and reactance combine in rather messy fashion in parallel circuits, and it can be hard to envision what's happening. But conductance (*G*) and susceptance (*B*) just add together in parallel circuits, yielding admittance (*Y*). This greatly simplifies the analysis of parallel ac circuits.

The situation is similar to the behavior of resistances in parallel when you work with dc. While the formula is a bit cumbersome if you need to find the value of a bunch of resistances in parallel, it's simple to just add the conductances.

Now, with ac, you're working in two dimensions instead of one. That's the only difference.

Parallel circuit analysis is covered in detail in the next chapter.

The GB plane

Admittance can be depicted on a plane that looks just like the complex impedance (RX) plane. Actually, it's a half plane, because there is ordinarily no such thing as negative conductance. (You can't have something that conducts worse than not at all.) Conductance is plotted along the horizontal, or G, axis on this coordinate half plane, and susceptance is plotted along the *B* axis. The plane is shown in Fig. 15-9 with several points plotted.

Although the *GB plane* looks superficially identical to the RX plane, the difference is great indeed! The GB plane is literally blown inside-out from the RX plane, as if you had jumped into a black hole and undergone a spatial transmutation, inwards-out and outwards-in, turning zero into infinity and vice-versa. Mathematicians love this kind of stuff.

The center, or origin, of the GB plane represents that point at which there is no conduction of any kind whatsoever, either for direct current or for alternating current. In the RX plane, the origin represents a perfect *short circuit*; in the GB plane it corresponds to a perfect *open circuit*.

The open circuit in the RX plane is way out beyond sight, infinitely far away from the origin. In the GB plane, it is the short circuit that is out of view.

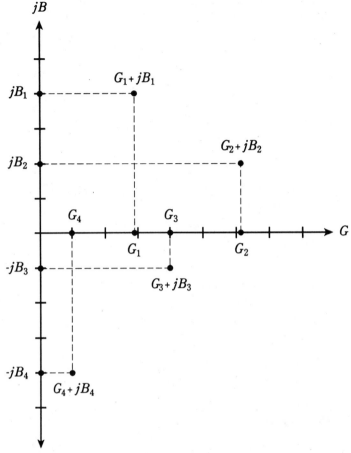

15-9 Some points in the GB plane, and their components on the G and B
axes.

Formula for conductance

As you move out towards the right ("east") along the G, or conductance, axis of the GB
plane, the conductance improves, and the current gets greater, but only for dc. The
formula for G is simply:

$$G = 1/R$$

where R is the resistance in ohms and G is the conductance in siemens, also sometimes
called mhos.

Formula for capacitive susceptance

It won't hurt to review the formulas for susceptance again. They can get a little bit
confusing, especially after having worked with reactance.

When you move upwards ("north") along the jB axis from the origin, you have ever-increasing capacitive susceptance. The formula for this quantity, B_C, is:

$$B_C = 6.28fC \text{ siemens}$$

where f is in Hertz and C is in farads. The value of B is in siemens. Alternatively, you can use frequency values in megahertz and capacitances in microfarads. The complex value is $jB = j(6.28fC)$.

Moving upwards along the jB axis indicates increasing capacitance values.

Formula for inductive susceptance

When you go down ("south") along the jB axis from the origin, you encounter increasingly negative susceptance. This is inductive susceptance; the formula for it is:

$$B_L = -1/(6.28fL) \text{ siemens}$$

where f is in hertz and L is in henrys. Alternatively, f can be expressed in megahertz, and L can be given in microhenrys. The complex value is $jB = -j(1/(6.28fL))$.

Moving downwards along the jB axis indicates decreasing values of inductance.

Vector representation of admittance

Complex admittances can be shown as vectors, just as can complex impedances. In Fig. 15-10, the points from Fig. 15-9 are rendered as vectors.

Generally, longer vectors indicate greater flow of current, and shorter ones indicate less current.

Imagine a point moving around on the GB plane, and think of the vector getting longer and shorter, and changing direction. Vectors pointing generally "northeast," or upwards and to the right, correspond to conductances and capacitances in parallel. Vectors pointing in a more or less "southeasterly" direction, or downwards and to the right, are conductances and inductances in parallel.

Why all these different expressions?

Do you think that the foregoing discussions are an elaborate mental gymnastics routine? Why do you need all these different quantities: resistance, capacitance, capacitive reactance, inductance, inductive reactance, impedance, conductance, capacitive susceptance, inductive susceptance, admittance?

Well, gymnastics are sometimes necessary to develop skill. Sometimes you need to "break a mental sweat." Each of these expressions is important.

The quantities that were dealt with before this chapter, and also early in this chapter, are of use mainly with series RLC (resistance-inductance-capacitance) circuits. The ones introduced in the second half of this chapter are important when you need to analyze parallel RLC circuits. Practice them and play with them, especially if they intimidate you. After awhile they'll become familiar.

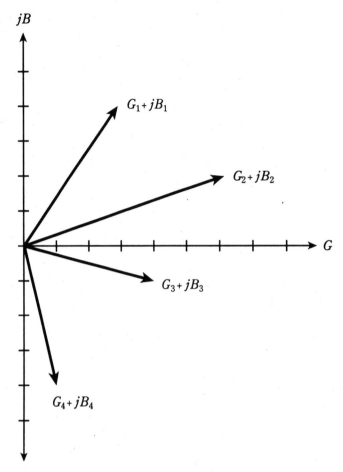

15-10   Vectors representing the points of Fig. 15-9.

Think in two dimensions. Draw your own RX and GB planes. (Be thankful there are only two dimensions, and not three! Some scientists need to deal in dozens of dimensions.)

If you want to be an engineer, you'll need to know how to handle these expressions. If you plan to manage engineers, you'll want to know what these quantities are, at least, when the engineers talk about them.

If the math seems a bit thick right now, hang in there. Impedance and admittance are the most mathematical subjects you'll have to deal with.

Quiz

Refer to the text in this chapter if necessary. A good score is 18 or more correct. Answers are in the back of the book.

1. The square of an imaginary number:

   A. Can never be negative.

B. Can never be positive.

C. Might be either positive or negative.

D. Is equal to j.

2. A complex number:

A. Is the same thing as an imaginary number.

B. Has a real part and an imaginary part.

C. Is one-dimensional.

D. Is a concept reserved for elite imaginations.

3. What is the sum of $3 + j7$ and $-3 - j7$?

A. $0 + j0$.

B. $6 + j14$.

C. $-6 - j14$.

D. $0 - j14$.

4. What is $(-5 + j7) - (4 - j5)$?

A. $-1 + j2$.

B. $-9 - j2$.

C. $-1 - j2$.

D. $-9 + j12$.

5. What is the product $(-4 - j7)(6 - j2)$?

A. $24 - j14$.

B. $-38 - j34$.

C. $-24 - j14$.

D. $-24 + j14$.

6. What is the magnitude of the vector $18 - j24$?

A. 42.

B. -42.

C. 30.

D. -30.

7. The impedance vector $5 + j0$ represents:

A. A pure resistance.

B. A pure inductance.

C. A pure capacitance.

D. An inductance combined with a capacitance.

8. The impedance vector $0 - j22$ represents:

A. A pure resistance.

B. A pure inductance.

C. A pure capacitance.

D. An inductance combined with a resistance.

9. What is the absolute-value impedance of $3.0 - j6.0$?

    A. $Z = 9.0\ \Omega$.

    B. $Z = 3.0\ \Omega$.

    C. $Z = 45\ \Omega$.

    D. $Z = 6.7\ \Omega$.

10. What is the absolute-value impedance of $50 - j235$?

    A. $Z = 240\ \Omega$.

    B. $Z = 58{,}000\ \Omega$.

    C. $Z = 285\ \Omega$.

    D. $Z = -185\ \Omega$.

11. If the center conductor of a coaxial cable is made to have smaller diameter, all other things being equal, what will happen to the Z_o of the transmission line?

    A. It will increase.

    B. It will decrease.

    C. It will stay the same.

    D. There is no way to know.

12. If a device is said to have an impedance of $Z = 100\ \Omega$, this would most often mean that:

    A. $R + jX = 100 + j0$.

    B. $R + jX = 0 + j100$.

    C. $R + jX = 0 - j100$.

    D. You need to know more specific information.

13. A capacitor has a value of 0.050 uF at 665 kHz. What is the capacitive susceptance?

    A. $j4.79$.

    B. $-j4.79$.

    C. $j0.209$.

    D. $-j0.209$.

14. An inductor has a value of 44 mH at 60 Hz. What is the inductive susceptance?

    A. $-j0.060$.

    B. $j0.060$.

    C. $-j17$.

    D. $j17$.

15. Susceptance and conductance add to form:
    A. Impedance.
    B. Inductance.
    C. Reactance.
    D. Admittance.

16. Absolute-value impedance is equal to the square root of:
    A. $G^2 + B^2$.
    B. $R^2 + X^2$.
    C. Z_o.
    D. Y.

17. Inductive susceptance is measured in:
    A. Ohms.
    B. Henrys.
    C. Farads.
    D. Siemens.

18. Capacitive susceptance is:
    A. Positive and real-valued.
    B. Negative and real-valued.
    C. Positive and imaginary.
    D. Negative and imaginary.

19. Which of the following is false?
    A. $B_C = 1/X_C$.
    B. Complex impedance can be depicted as a vector.
    C. Characteristic impedance is complex.
    D. $G = 1/R$.

20. In general, the greater the absolute value of the impedance in a circuit:
    A. The greater the flow of alternating current.
    B. The less the flow of alternating current.
    C. The larger the reactance.
    D. The larger the resistance.

16
CHAPTER

RLC circuit analysis

WHENEVER YOU SEE AC CIRCUITS WITH INDUCTANCE AND/OR CAPACITANCE AS well as resistance, you should switch your mind into "2D" mode. You must be ready to deal with two-dimensional quantities.

While you can sometimes talk and think about impedances as simple ohmic values, there are times you can't. If you're sure that there is no reactance in an ac circuit, then it's all right to say "$Z = 600$ ohms," or "This speaker is 8 ohms," or "The input impedance to this amplifier is 1,000 ohms."

As soon as you see coils and/or capacitors, you should envision the complex-number plane, either RX (resistance-reactance) or GB (conductance-admittance). The RX plane applies to series-circuit analysis. The GB plane applies to parallel-circuit analysis.

Complex impedances in series

When you see resistors, coils, and capacitors in series, you should envision the RX plane.

Each component, whether it is a resistor, an inductor, or a capacitor, has an impedance that can be represented as a vector in the RX plane. The vectors for resistors are constant regardless of the frequency. But the vectors for coils and capacitors vary with frequency, as you have learned.

Pure reactances

Pure inductive reactances (X_L) and capacitive reactances (X_C) simply add together when coils and capacitors are in series. Thus, $X = X_L + X_C$. In the RX plane, their vectors add, but because these vectors point in exactly opposite directions— inductive reactance upwards and capacitive reactance downwards—the resultant sum vector will also inevitably point either straight up or down (Fig. 16-1).

284

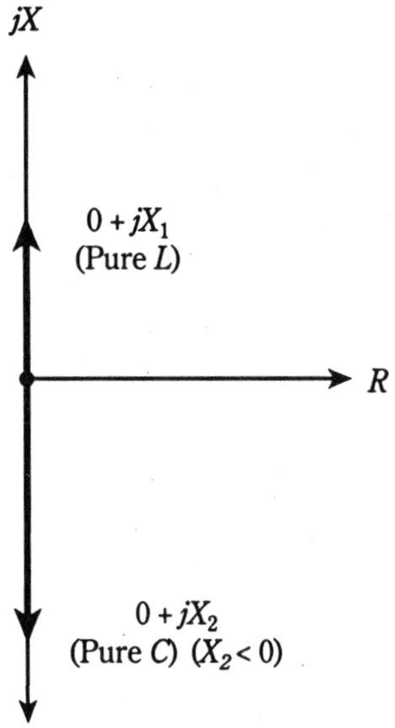

$$jX$$

$0 + jX_1$
(Pure L)

R

16-1 Pure inductance and pure capacitance are represented by reactance vectors that point straight up and down.

$0 + jX_2$
(Pure C) ($X_2 < 0$)

Problem 16-1

A coil and capacitor are connected in series, with $jX_L = j200$ and $jX_C = -j150$. What is the net reactance vector jX?

Just add the values: $jX = jX_L + jX_C = j200 + (-j150) = j(200 - 150) = j50$. This is an *inductive* reactance, because it is *positive imaginary*.

Problem 16-2

A coil and capacitor are connected in series, with $jX_L = j30$ and $jX_C = -j110$. What is the net reactance vector jX?

Again, add: $jX = j30 + (-j110) = j(30 - 110) = -j80$. This is a *capacitive* reactance, because it is *negative imaginary*.

Problem 16-3

A coil of $L = 5.00$ uH and a capacitor of $C = 200$ pF are in series. The frequency is $f = 4.00$ MHz. What is the net reactance vector jX?

First calculate:

$$jX_L = j6.28fL$$
$$= j(6.28 \times 4.00 \times 5.00) = j126$$

Then calculate:

$$jX_C = -j(1/(6.28fC))$$
$$= -j(1/(6.28 \times 4.00 \times 0.000200)) = -j199$$

Finally, add:

$$jX = jX_L + jX_C$$
$$= j126 + (-j199) = -j73$$

This is a net capacitive reactance. There is no resistance in this circuit, so the impedance vector is $0 - j73$.

Problem 16-4

What is the net reactance vector jX for the above combination at a frequency of $f = 10.0$ MHz?

First calculate:

$$jX_L = j6.28fL$$
$$= j(6.28 \times 10.0 \times 5.00) = j314$$

Then calculate:

$$jX_C = -j(1/(6.28fC))$$
$$= -j(1/(6.28 \times 10.0 \times 0.000200) = -j79.6$$

Finally, add:

$$jX = jX_L + jX_C$$
$$= j314 + (-j79.6) = j234$$

This is a net inductive reactance. Again, there is no resistance, and therefore the impedance vector is pure imaginary, $0 + j234$.

Notice that the change in frequency, between Problems 16-3 and 16-4, caused the circuit to change over from a net capacitance to a net inductance. You might think that there must be some frequency, between 4.00 MHz and 10.0 MHz, at which jX_L and jX_C add up to $j0$—that is, at which they exactly cancel each other out, yielding $0 + j0$ as the complex impedance. Then the circuit, at that frequency, would appear as a short circuit. If you suspect this, you're right.

Any series combination of coil and capacitor offers theoretically zero opposition to ac at one special frequency. This is called *series resonance*, and is dealt with in the next chapter.

Adding impedance vectors

Often, there is resistance, as well as reactance, in an ac series circuit containing a coil and capacitor. This occurs when the coil wire has significant resistance (it's never a perfect conductor). It might also be the case because a resistor is deliberately connected into the circuit.

Whenever the resistance in a series circuit is significant, the impedance vectors no longer point straight up and straight down. Instead, they run off towards the "northeast" (for the inductive part of the circuit) and "southeast" (for the capacitive part). This is illustrated in Fig. 16-2.

When vectors don't lie along a single line, you need to use *vector addition* to be sure that you get the correct resultant. Fortunately, this isn't hard.

In Fig. 16-3, the geometry of vector addition is shown. Construct a *parallelogram*, using the two vectors $Z_1 = R_1 + jX_1$ and $Z_2 = R_2 + jX_2$ as two of the sides. The diagonal is the resultant. In a parallelogram, opposite angles have equal measure. These equalities are indicated by single and double arcs in the figure.

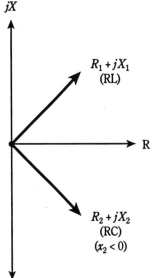

16-2 When resistance is present along with reactance, impedance vectors point "northeast" or "southeast."

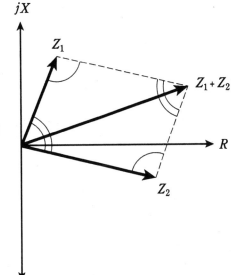

16-3 Parallelogram method of vector addition.

Formula for complex impedances in series

Given two impedances, $Z_1 = R_1 + jX_1$ and $Z_2 = R_2 + jX_2$, the net impedance Z of these in series is their vector sum, given by:

$$Z = (R_1 + R_2) + j(X_1 + X_2)$$

The reactances X_1 and X_2 might both be inductive; they might both be capacitive; or one might be inductive and the other capacitive.

Calculating a vector sum using the formula is easier than doing it geometrically with a parallelogram. The arithmetic method is also more nearly exact. The resistance and reactance components add separately. That's all there is to it.

Series RLC circuits

When a coil, capacitor and resistor are connected in series (Fig. 16-4), the resistance R can be thought of as all belonging to the coil, when you use the above formulas. (Thinking of it all as belonging to the capacitor will also work.) Then you have two vectors to add, when finding the impedance of a series RLC circuit:

$$Z = (R + jX_L) + (0 + jX_C)$$
$$= R + j(X_L + X_C)$$

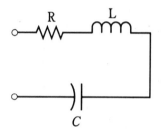

16-4   A series RLC circuit.

Problem 16-5

A resistor, coil and capacitor are connected in series with $R = 50\ \Omega$, $X_L = 22\ \Omega$, and $X_C = -33\ \Omega$. What is the net impedance, Z?

Consider the resistor to be part of the coil, obtaining two complex vectors, $50 + j22$ and $0 - j33$. Adding these gives the resistance component of $50 + 0 = 50$, and the reactive component of $j22 - j33 = -j11$. Therefore, $Z = 50 - j11$.

Problem 16-6

A resistor, coil and capacitor are connected in series with $R = 600\ \Omega$, $X_L = 444\ \Omega$ and $X_C = -444\ \Omega$. What is the net impedance, Z?

Again, consider the resistor to be part of the inductor. Then the vectors are $600 + j444$ and $0 - j444$. Adding these, the resistance component is $600 + 0 = 600$, and the reactive component is $j444 - j444 = j0$. Thus, $Z = 600 + j0$. This is a purely resistive impedance, and you can rightly call it "600 Ω."

Problem 16-7

A resistor, coil and capacitor are connected in series. The resistor has a value of 330 Ω; the capacitance is 220 pF and the inductance is 100 uH. The frequency is 7.15 MHz. What is the net complex impedance?

First, you need to calculate the inductive and capacitive reactances. Remembering the formula $X_L = 6.28fL$, multiply to obtain:

$$jX_L = j(6.28 \times 7.15 \times 100) = j4490$$

*Mega*hertz and *micro*henrys go together in the formula. As for X_C, recall the formula $X_C = -1/(6.28fC)$. Convert 220 pF to microfarads to go with megahertz in the formula; $C = 0.000220$ uF. Then:

$$jX_C = -j(1/(6.28 \times 7.15 \times 0.000220)) = -j101$$

Now, you can consider the resistance and the inductive reactance to go together, so one of the impedance vectors is $330 + j4490$. The other is $0 - j101$. Adding these gives $330 + j4389$; this rounds off to $Z = 330 + j4390$.

Problem 16-8

A resistor, coil, and capacitor are in series. The resistance is $50.0\ \Omega$; the inductance is 10.0 uH, and the capacitance is 1000 pF. The frequency is 1592 kHz. What is the complex impedance of this series RLC circuit at this frequency?

First, calculate $X_L = 6.28fL$. Convert the frequency to megahertz; 1592 kHz = 1.592 MHz. Then:

$$jX_L = j(6.28 \times 1.592 \times 10.0) = j100$$

Then calculate $X_C = 1/(6.28fC)$. Convert picofarads to microfarads, and use megahertz for the frequency. Therefore:

$$jX_C = -j(1/(6.28 \times 1.592 \times 0.001000)) = -j100$$

Let the resistance and inductive reactance go together as one vector, $50.0 + j100$. Let the capacitance alone be the other vector, $0 - j100$. The sum is $50.0 + j100 - j100 = 50.0 + j0$. This is a pure resistance of $50.0\ \Omega$. You can correctly say that the impedance is "$50.0\ \Omega$" in this case.

This concludes the analysis of series RLC circuit impedances. What about parallel circuits? To deal with these, you must calculate using conductance, susceptance and admittance, converting to impedance only at the very end.

Complex admittances in parallel

When you see resistors, coils, and capacitors in parallel, you should envision the GB (conductance-susceptance) plane.

Each component, whether it is a resistor, an inductor, or a capacitor, has an admittance that can be represented as a vector in the GB plane. The vectors for pure conductances are constant, even as the frequency changes. But the vectors for the coils and capacitors vary with frequency, in a manner similar to the way they vary in the RX plane.

Pure susceptances

Pure inductive susceptances (B_L) and capacitive susceptances (B_C) add together when coils and capacitors are in parallel. Thus, $B = B_L + B_C$. Remember that B_L is *negative* and B_C is *positive*, just the opposite from reactances.

In the GB plane, the jB_L and jB_C vectors add, but because these vectors point in exactly opposite directions—inductive susceptance down and capacitive susceptance up—the sum, jB, will also inevitably point straight down or up (Fig. 16-5).

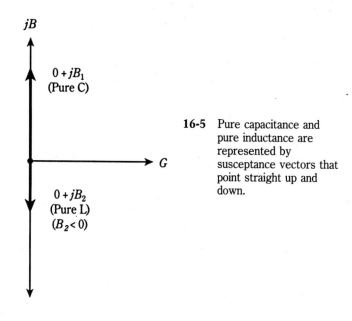

16-5 Pure capacitance and pure inductance are represented by susceptance vectors that point straight up and down.

Problem 16-9

A coil and capacitor are connected in parallel, with $jB_L = -j0.05$ and $jB_C = j0.08$. What is the net admittance vector?

Just add the values: $jB = jB_L + jB_C = -j0.05 + j0.08 = j0.03$. This is a *capacitive* susceptance, because it is *positive imaginary*. The admittance vector is $0 + j0.03$.

Problem 16-10

A coil and capacitor are connected in parallel, with $jB_L = -j0.60$ and $jB_C = j0.25$. What is the net admittance vector?

Again, add: $jB = -j0.60 + j0.25 = -j0.35$. This is an *inductive*susceptance, because it is *negative imaginary*. The admittance vector is $0 - j0.35$.

Problem 16-11

A coil of $L = 6.00$ uH and a capacitor of $C = 150$ pF are in parallel. The frequency is $f = 4.00$ MHz. What is the net admittance vector?

First calculate:

$$jB_L = -j(1/(6.28fL))$$
$$= -j(1/(6.28 \times 4.00 \times 6.00)) = -j0.00663$$

Then calculate:

$$jB_C = j(6.28fC)$$
$$= j(6.28 \times 4.00 \times 0.000150) = j0.00377$$

Finally, add:

$$jB = jB_L + jB_C$$
$$= -j0.00663 + j0.00377 = -j0.00286$$

This is a net inductive susceptance. There is no conductance in this circuit, so the admittance vector is $0 - j0.00286$.

Problem 16-12

What is the net admittance vector for the above combination at a frequency of $f = 5.31$ MHz?

First calculate:

$$jB_L = -j(1/(6.28fL))$$
$$= -j(1/(6.28 \times 5.31 \times 6.00)) = -j0.00500$$

Then calculate:

$$jB_C = j(6.28fC)$$
$$= j(6.28 \times 5.31 \times 0.000150) = j0.00500$$

Finally, add:

$$jB = jB_L + jB_C$$
$$= -j0.00500 + j0.00500 = j0$$

There is no susceptance. Because the conductance is also zero (there is nothing else in parallel with the coil and capacitor that might conduct), the admittance vector is $0 + j0$.

This situation, in which there is no conductance and no susceptance, seems to imply that this combination of coil and capacitor in parallel is an open circuit at 5.31 MHz. In theory this is true; zero admittance means no current can get through the circuit. In practice it's not quite the case. There is always a small leakage. This condition is known as *parallel resonance*. It's discussed in the next chapter.

Adding admittance vectors

In real life, there is a small amount of conductance, as well as susceptance, in an ac parallel circuit containing a coil and capacitor. This occurs when the capacitor lets a little bit of current leak through. More often, though, it is the case because a *load* is connected in parallel with the coil and capacitor. This load might be an antenna, or the input to an amplifier circuit, or some test instrument, or a transducer.

Whenever the conductance in a parallel circuit is significant, the admittance vectors no longer point straight up and down. Instead, they run off towards the "northeast" (for the capacitive part of the circuit) and "southeast" (for the inductive part). This is illustrated in Fig. 16-6.

In the problems above, you added numbers, but in fact you were adding vectors that just happened to fall along a single line, the imaginary (j) axis of the GB plane. In practical circuits, the vectors often do not lie along a single line. You've already seen how to deal with these in the RX plane. In the GB plane, the principle is the same.

Formula for complex admittances in parallel

Given two admittances, $Y_1 = G_1 + jB_1$ and $Y_2 = G_2 + jB_2$, the net admittance Y of these in parallel is their vector sum, given by:

$$Y = (G_1 + G_2) + j(B_1 + B_2)$$

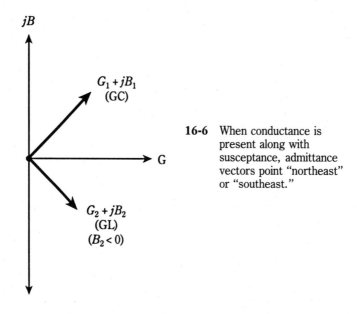

16-6 When conductance is present along with susceptance, admittance vectors point "northeast" or "southeast."

The susceptances B_1 and B_2 might both be inductive; they might both be capacitive; or one might be inductive and the other capacitive.

Parallel GLC circuits

When a coil, capacitor and resistor are connected in parallel (Fig. 16-7), the resistor should be thought of as a *conductor*, whose value in siemens is equal to the reciprocal of the value in ohms. Think of the conductance as all belonging to the inductor. (Thinking of it all as belonging to the capacitor will also work.) Then you have two vectors to add, when finding the admittance of a parallel *GLC* (conductance-inductance-capacitance) circuit:

$$Y = (G + jB_L) + (0 + jB_C)$$
$$= G + j(B_L + B_C)$$

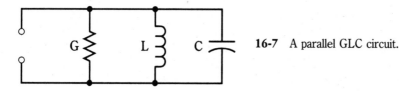

16-7 A parallel GLC circuit.

Problem 16-13

A resistor, coil and capacitor are connected in parallel with $G = 0.1$ siemens, $jB_L = -0.010$ and $jB_C = 0.020$. What is the net admittance vector?

Consider the resistor to be part of the coil, obtaining two complex vectors, $0.1 - j0.010$ and $0 + j0.020$. Adding these gives the conductance component of $0.1 + 0$

= 0.1, and the susceptance component of $-j0.010 + j0.020 = j0.010$. Therefore, the admittance vector is $0.1 + j0.010$.

Problem 16-14

A resistor, coil and capacitor are connected in parallel with $G = 0.0010$ siemens, $jB_L = -0.0022$ and $jB_C = 0.0022$. What is the net admittance vector?

Again, consider the resistor to be part of the coil. Then the complex vectors are $0.0010 - j0.0022$ and $0 + j0.0022$. Adding these, the conductance component is $0.0010 + 0 = 0.0010$, and the susceptance component is $-j0.0022 + j0.0022 = j0$. Thus, the admittance vector is $0.0010 + j0$. This is a purely conductive admittance. There is no susceptance.

Problem 16-15

A resistor, coil and capacitor are connected in parallel. The resistor has a value of 100 Ω; the capacitance is 200 pF and the inductance is 100 uH. The frequency is 1.00 MHz. What is the net complex admittance?

First, you need to calculate the inductive and capacitive susceptances. Recall $jB_L = -j(1/(6.28fL))$, and "plug in" the values, getting:

$$jB_L = -j(1/(6.28 \times 1.00 \times 100)) = -j0.00159$$

*Mega*hertz and *micro*henrys go together in the formula. As for jB_C, recall the formula $jB_C = j(6.28fC)$. Convert 200 pF to microfarads to go with megahertz in the formula; $C = 0.000200$ uF. Then:

$$jB_C = j(6.28 \times 1.00 \times 0.000200) = j0.00126$$

Now, you can consider the conductance, which is $1/100 = 0.0100$ siemens, and the inductive susceptance to go together. So one of the vectors is $0.0100 - j0.00159$. The other is $0 + j0.00126$. Adding these gives $0.0100 - j0.00033$.

Problem 16-16

A resistor, coil, and capacitor are in parallel. The resistance is 10.0 Ω; the inductance is 10.0 uH and the capacitance is 1000 pF. The frequency is 1592 kHz. What is the complex admittance of this circuit at this frequency?

First, calculate $jB_L = -j(1/(6.28fL))$. Convert the frequency to megahertz; 1592 kHz = 1.592 MHz. Then:

$$jB_L = -j/(1/(6.28 \times 1.592 \times 10.0) = -j0.0100$$

Then calculate $jB_C = j(6.28fC)$. Convert picofarads to microfarads, and use megahertz for the frequency. Therefore:

$$jB_C = j(6.28 \times 1.592 \times 0.001000) = j0.0100$$

Let the conductance and inductive susceptance go together as one vector, $0.100 - j0.0100$. (Remember that conductance is the reciprocal of resistance; here $G =$

$1/R = 1/10.0 = 0.100$.) Let the capacitance alone be the other vector, $0 + j0.0100$. Then the sum is $0.100 - j0.0100 + j0.0100 = 0.100 + j0$. This is a pure conductance of 0.100 siemens.

Converting from admittance to impedance

The GB plane is, as you have seen, similar in appearance to the RX plane, although mathematically the two are worlds apart. Once you've found a complex admittance for a parallel RLC circuit, how do you transform this back to a complex impedance? Generally, it is the impedance, not the admittance, that technicians and engineers work with.

The transformation from complex admittance, or a vector $G + jB$, to a complex impedance, or a vector $R + jX$, requires the use of the following formulas:

$$R = G/(G^2 + B^2)$$
$$X = -B/(G^2 + B^2)$$

If you know the complex admittance, first find the resistance and reactance components individually. Then assemble them into the impedance vector, $R + jX$.

Problem 16-17

The admittance vector for a certain parallel circuit is $0.010 - j0.0050$. What is the impedance vector?

In this case, $G = 0.010$ and $B = -0.0050$. Find $G^2 + B^2$ first, because you'll need to use it twice as a denominator; it is $0.010^2 + (-0.0050)^2 = 0.000100 + 0.000025 = 0.000125$. Then:

$$R = G/0.000125 = 0.010/0.000125 = 80$$
$$X = -B/0.000125 = 0.0050/0.000125 = 40$$

The impedance vector is therefore $R + jX = 80 + j40$.

Putting it all together

When you're confronted with a parallel RLC circuit, and you want to know the complex impedance $R + jX$, take these steps:

1. Find the conductance $G = 1/R$ for the resistor (it will be positive or zero).
2. Find the susceptance B_L of the inductor using the appropriate formula (it will be negative or zero).
3. Find the susceptance B_C of the capacitor using the appropriate formula (it will be positive or zero).
4. Find the net susceptance $B = B_L + B_C$ (it might be positive, negative or zero).
5. Compute R and X in terms of G and B using the appropriate formulas.
6. Assemble the vector $R + jX$.

Problem 16-18

A resistor of $10.0 \, \Omega$, a capacitor of 820 pF and a coil of 10.0 uH are in parallel. The frequency is 1.00 MHz. What is the impedance $R + jX$?

Proceed by the steps as numbered above.

1. $G = 1/R = 1/10.0 = 0.100$.
2. $B_L = -1/(6.28fL) = -1/(6.28 \times 1.00 \times 10.0) = -0.0159$.
3. $B_C = 6.28fC = 6.28 \times 1.00 \times 0.000820 = 0.00515$ (remember to convert the capacitance to microfarads, to go with megahertz).
4. $B = B_L + B_C = -0.0159 + 0.00515 = -0.0108$.
5. First find $G^2 + B^2 = 0.100^2 + (-0.0108)^2 = 0.010117$ (go to a couple of extra places to be on the safe side). Then $R = G/0.010117 = 0.100/0.010117 = 9.88$, and $X = -B/0.010117 = 0.0108/0.010117 = 1.07$.
6. The vector $R + jX$ is therefore $9.88 + j1.07$. This is the complex impedance of this parallel RLC circuit.

Problem 16-19

A resistor of 47.0 Ω, a capacitor of 500 pF, and a coil of 10.0 uH are in parallel. What is their complex impedance at a frequency of 2.252 MHz?

Proceed by the steps as numbered above.

1. $G = 1/R = 1/47.0 = 0.0213$.
2. $B_L = -1/(6.28fL) = -1/(6.28 \times 2.252 \times 10.0) = -0.00707$.
3. $B_C = 6.28fC = 6.28 \times 2.252 \times 0.000500 = 0.00707$.
4. $B = B_L + B_C = -0.00707 + 0.00707 = 0$.
5. Find $G^2 + B^2 = 0.0213^2 + 0.000^2 = 0.00045369$ (again, go to a couple of extra places). Then $R = G/0.00045369 = 0.0213/0.00045369 = 46.9$, and $X = -B/0.00045369 = 0$.
6. The vector $R + jX$ is therefore $46.9 + j0$. This is a pure resistance, almost exactly the value of the resistor in the circuit.

Reducing complicated RLC circuits

Sometimes you'll see circuits in which there are several resistors, capacitors and/or coils in series and parallel combinations. It is not the intent here to analyze all kinds of bizarre circuit situations. That would fill up hundreds of pages with formulas, diagrams, and calculations, and no one would ever read it (assuming any author could stand to write it).

A general rule applies to "complicated" RLC circuits: Such a circuit can usually be reduced to an equivalent circuit that contains one resistor, one capacitor and one inductor.

Series combinations

Resistances in series simply add. Inductances in series also add. Capacitances in series combine in a somewhat more complicated way. If you don't remember the formula, it is:

$$1/C = 1/C_1 + 1/C_2 + \ldots + 1/C_n$$

where $C_1, C_2, \ldots, C_n$ are the individual capacitances and C is the total capacitance. Once you've found $1/C$, take its reciprocal to obtain C.

An example of a "complicated" series RLC circuit is shown in Fig. 16-8A. The equivalent circuit, with just one resistor, one capacitor and one coil, is shown in Fig. 16-8B.

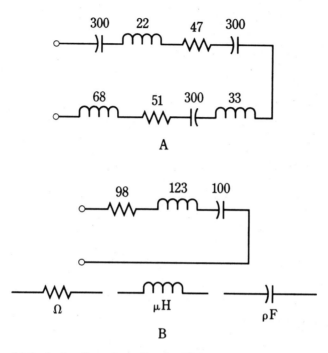

A

B

16-8 At A, a "complicated" series RLC circuit; at B, the same circuit simplified.

Parallel combinations

In parallel, resistances and inductances combine the way capacitances do in series. Capacitances just add up.

An example of a "complicated" parallel RLC circuit is shown in Fig. 16-9A. The equivalent circuit, with just one resistor, one capacitor and one coil, is shown in Fig. 16-9B.

Complicated, messy nightmares

Some RLC circuits don't fall neatly into either of the above categories. An example of such a circuit is shown in Fig. 16-10. "Complicated" really isn't the word to use here! How would you find the complex impedance at some frequency, such as 8.54 MHz?

You needn't waste much time worrying about circuits like this. But be assured: given a frequency, a complex impedance does exist.

In real life, an engineer would use a computer to solve this problem. If a program didn't already exist, the engineer would either write one, or else hire it done by a professional programmer.

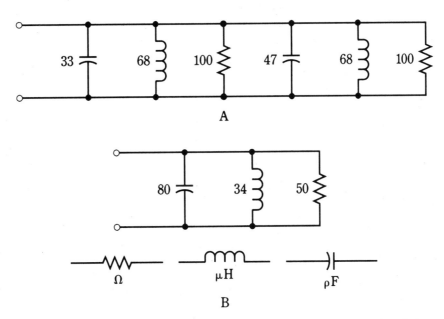

16-9 At A, a "complicated" parallel RLC circuit; at B, the same circuit simplified.

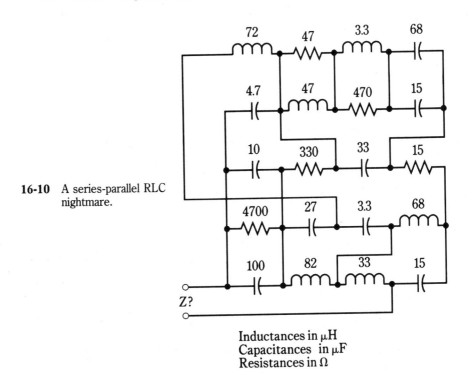

16-10 A series-parallel RLC nightmare.

Inductances in μH
Capacitances in μF
Resistances in Ω

Another way to find the complex impedance here would be to actually build the circuit, connect a signal generator to it, and measure R and X directly with an *impedance bridge*. Because "the proof of the pudding is in the eating," a performance test must eventually be done anyway, no matter how sophisticated the design theory. Engineers have to build things that work!

Ohm's law for ac circuits

Ohm's Law for a dc circuit is a simple relationship among three variables: current (I), voltage (E) and resistance (R). The formulas, again, are:

$$I = E/R$$
$$E = IR$$
$$R = E/I$$

In ac circuits containing negligible or zero reactance, these same formulas apply, as long as you are sure that you use the *effective* current and voltage.

Effective amplitudes

The effective value for an ac sine wave is the root-mean-square, or rms, value. You learned about this in chapter 9. The rms current or voltage is 0.707 times the peak amplitude. Conversely, the peak value is 1.414 times the rms value.

If you're told that an ac voltage is 35 V, or that an ac current is 570 mA, it is generally understood that this refers to a sine-wave rms level, unless otherwise specified.

Purely resistive impedances

When the impedance in an ac circuit is such that the reactance X has a negligible effect, and that practically all of the current and voltage exists through and across a resistance R, Ohm's Law for an ac circuit is expressed as:

$$I = E/Z$$
$$E = IZ$$
$$Z = E/I$$

where Z is essentially equal to R, and the values I and E are rms current and voltage.

Complex impedances

When determining the relationship among current, voltage and resistance in an ac circuit with resistance and reactance that are both significant, things get interesting.

Recall the formula for absolute-value impedance in a series RLC circuit:

$$Z^2 = R^2 + X^2$$

so Z is equal to the square root of $R^2 + X^2$. This is the length of the vector $R + jX$ in the complex impedance plane. You learned this in chapter 15. This formula applies only for series RLC circuits.

The absolute-value impedance for a parallel RLC circuit, in which the resistance is R and the reactance is X, is defined by the formula:

$$Z^2 = (RX)^2/(R^2 + X^2)$$

Thus Z is equal to RX divided by the square root of $R^2 + X^2$.

Problem 16-20

A series RX circuit (Fig. 16-11) has $R = 50.0\ \Omega$ of resistance and $X = -50.0\ \Omega$ of reactance, and 100 Vac is applied. What is the current?

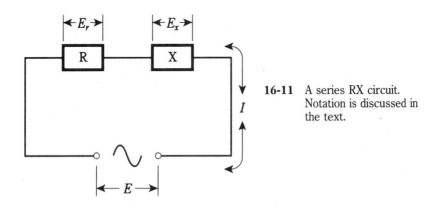

16-11   A series RX circuit. Notation is discussed in the text.

First, calculate $Z^2 = R^2 + X^2 = 50.0^2 + (-50.0)^2 = 2500 + 2500 = 5000$; Z is the square root of 5000, or 70.7 Ω. Then $I = E/Z = 100/70.7 = 1.41$ A.

Problem 16-21

What are the voltage drops across the resistance and the reactance, respectively, in the above problem?

The Ohm's Law formulas for dc will work here. Because the current is $I = 1.41$ A, the voltage drop across the resistance is equal to $E_R = IR = 1.41 \times 50.0 = 70.5$ V. The voltage drop across the reactance is the product of the current and the reactance: $E_X = IX = 1.41 \times (-50.0) = -70.5$. This is an ac voltage of equal magnitude to that across the resistance. But the phase is different.

The voltages across the resistance and the reactance—a capacitive reactance in this case, because it's negative—don't add up to 100. The meaning of the minus sign for the voltage across the capacitor is unclear, but there is no way, whether you consider this sign or not, that the voltages across the resistor and capacitor will arithmetically add up to 100. Shouldn't they? In a dc circuit, yes; in an ac circuit, generally, no.

In a resistance-reactance ac circuit, there is always a difference in phase between the voltage across the resistive part and the voltage across the reactive part. They always add up to the applied voltage *vectorially*, but not always *arithmetically*. You don't need to be concerned with the geometry of the vectors in this situation. It's enough to understand that the vectors don't fall along a single line, and this is why the voltages don't add arithmetically.

Problem 16-22

A series RX circuit (Fig. 16-11) has $R = 10.0\ \Omega$ and a net reactance $X = 40.0\ \Omega$. The applied voltage is 100. What is the current?

Calculate $Z^2 = R^2 + X^2 = 100 + 1600 = 1700$; thus $Z = 41.2$. Therefore $I = E/Z = 100/41.2 = 2.43$ A. Note that the reactance in this circuit is inductive, because it is positive.

Problem 16-23

What is the voltage across R in the preceding problem? Across X?

Knowing the current, calculate $E_R = IR = 2.43 \times 10.0 = 24.3$ V. Also, $E_X = IX = 2.43 \times 40.0 = 97.2$ V.

If you add $E_R + E_X$ arithmetically, you get 24.3 V $+ 97.2$ V $= 121.5$ V as the total across R and X. Again, the simple dc rule does not work here. The reason is the same as before.

Problem 16-24

A parallel RX circuit (Fig. 16-12) has resistance $R = 30$ ohms and a net reactance $X = -20\ \Omega$. The supply voltage is 50 V. What is the total current drawn from the supply?

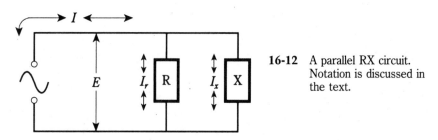

16-12   A parallel RX circuit. Notation is discussed in the text.

Find the absolute-value impedance, remembering the formula for parallel circuits: $Z^2 = (RX)^2/(R^2 + X^2) = 360,000/1300 = 277$. The impedance Z is the square root of 277, or $16.6\ \Omega$. The total current is therefore $I = E/Z = 50/16.6 = 3.01$ A.

Problem 16-25

What is the current through R above? Through X?

The Ohm's Law formulas for dc will work here. For the resistance, $I_R = E/R = 50/30 = 1.67$ A. For the reactance, $I_X = E/X = 50/(-20) = -2.5$ A.

These currents don't add up to 3.01 A, the total current, whether the minus sign is taken into account, or not. It's not really clear what the minus sign means, anyhow. The reason that the constituent currents, I_R and I_X, don't add up to the total current, I, is the same as the reason the voltages don't add up in a series RX circuit. These currents are actually 2D vectors; you're seeing them through 1D glasses.

If you want to study the geometrical details of the voltage and current vectors in series and parallel RX circuits, a good circuit theory text is recommended.

One of the most important practical aspects of ac circuit theory involves the ways that reactances, and complex impedances, behave when you try to feed power to them. That subject will start off the next chapter.

Quiz

Refer to the text in this chapter if necessary. A good score is 18 correct. Answers are in the back of the book.

1. A coil and capacitor are connected in series. The inductive reactance is 250 Ω, and the capacitive reactance is −300 Ω. What is the net impedance vector, $R + jX$?

   A. $0 + j550$.

   B. $0 - j50$.

   C. $250 - j300$.

   D. $-300 + j250$.

2. A coil of 25.0 uH and capacitor of 100 pF are connected in series. The frequency is 5.00 MHz. What is the impedance vector, $R + jX$?

   A. $0 + j467$.

   B. $25 + j100$.

   C. $0 - j467$.

   D. $25 - j100$.

3. When $R = 0$ in a series RLC circuit, but the net reactance is not zero, the impedance vector:

   A. Always points straight up.

   B. Always points straight down.

   C. Always points straight towards the right.

   D. None of the above.

4. A resistor of 150 Ω, a coil with reactance 100 Ω and a capacitor with reactance −200 Ω are connected in series. What is the complex impedance $R + jX$?

   A. $150 + j100$.

   B. $150 - j200$.

   C. $100 - j200$.

   D. $150 - j100$.

5. A resistor of 330 Ω, a coil of 1.00 uH and a capacitor of 200 pF are in series. What is $R + jX$ at 10.0 MHz?

   A. $330 - j199$.

   B. $300 + j201$.

   C. 300 + *j*142.

   D. 330 − *j*16.8.

6. A coil has an inductance of 3.00 uH and a resistance of 10.0 Ω in its winding. A capacitor of 100 pF is in series with this coil. What is $R + jX$ at 10.0 MHz?

   A. 10 + *j*3.00.

   B. 10 + *j*29.2.

   C. 10 − *j*97.

   D. 10 + *j*348.

7. A coil has a reactance of 4.00 Ω. What is the admittance vector, $G + jB$, assuming nothing else is in the circuit?

   A. 0 + *j*0.25.

   B. 0 + *j*4.00.

   C. 0 − *j*0.25.

   D. 0 − *j*4.00.

8. What will happen to the susceptance of a capacitor if the frequency is doubled, all other things being equal?

   A. It will decrease to half its former value.

   B. It will not change.

   C. It will double.

   D. It will quadruple.

9. A coil and capacitor are in parallel, with $jB_L = -j0.05$ and $jB_C = j0.03$. What is the admittance vector, assuming that nothing is in series or parallel with these components?

   A. 0 − *j*0.02.

   B. 0 − *j*0.07.

   C. 0 + *j*0.02.

   D. −0.05 + *j*0.03.

10. A coil, resistor and capacitor are in parallel. The resistance is 1 Ω; the capacitive susceptance is 1.0 siemens; the inductive susceptance is −1.0 siemens. Then the frequency is cut to half its former value. What will be the admittance vector, $G + jB$, at the new frequency?

   A. 1 + *j*0.

   B. 1 + *j*1.5.

   C. 1 − *j*1.5.

   D. 1 − *j*2.

11. A coil of 3.50 uH and a capacitor of 47.0 pF are in parallel. The frequency is 9.55 MHz. There is nothing else in series or parallel with these components. What is the admittance vector?

    A. $0 + j0.00282$.

    B. $0 - j0.00194$.

    C. $0 + j0.00194$.

    D. $0 - j0.00758$.

12. A vector pointing "southeast" in the GB plane would indicate:

    A. Pure conductance, zero susceptance.

    B. Conductance and inductive susceptance.

    C. Conductance and capacitive susceptance.

    D. Pure susceptance, zero conductance.

13. A resistor of 0.0044 siemens, a capacitor whose susceptance is 0.035 siemens, and a coil whose susceptance is -0.011 siemens are all connected in parallel. The admittance vector is:

    A. $0.0044 + j0.024$.

    B. $0.035 - j0.011$.

    C. $-0.011 + j0.035$.

    D. $0.0044 + j0.046$.

14. A resistor of 100 Ω, a coil of 4.50 uH and a capacitor of 220 pF are in parallel. What is the admittance vector at 6.50 MHz?

    A. $100 + j0.00354$.

    B. $0.010 + j0.00354$.

    C. $100 - j0.0144$.

    D. $0.010 + j0.0144$.

15. The admittance for a circuit, $G + jB$, is $0.02 + j0.20$. What is the impedance, $R + jX$?

    A. $50 + j5.0$.

    B. $0.495 - j4.95$.

    C. $50 - j5.0$.

    D. $0.495 + j4.95$.

16. A resistor of 51.0 Ω, an inductor of 22.0 uH and a capacitor of 150 pF are in parallel. The frequency is 1.00 MHz. What is the complex impedance, $R + jX$?

    A. $51.0 - j14.9$.

    B. $51.0 + j14.9$.

    C. $46.2 - j14.9$.

    D. $46.2 + j14.9$.

17. A series circuit has 99.0 Ω of resistance and 88.0 Ω of inductive reactance. An ac rms voltage of 117 V is applied to this series network. What is the current?

    A. 1.18 A.

    B. 1.13 A.

    C. 0.886 A.

    D. 0.846 A.

18. What is the voltage across the reactance in the above example?

    A. 78.0 V.

    B. 55.1 V.

    C. 99.4 V.

    D. 74.4 V.

19. A parallel circuit has 10 ohms of resistance and 15 Ω of reactance. An ac rms voltage of 20 V is applied across it. What is the total current?

    A. 2.00 A.

    B. 2.40 A.

    C. 1.33 A.

    D. 0.800 A.

20. What is the current through the resistance in the above example?

    A. 2.00 A.

    B. 2.40 A.

    C. 1.33 A.

    D. 0.800 A.

17
CHAPTER

Power and resonance in ac circuits

YOU HAVE LEARNED HOW CURRENT, VOLTAGE, AND RESISTANCE BEHAVE IN ac circuits. How can all this theoretical knowledge be put to practical use?

One of the engineer's biggest challenges is the problem of efficient energy transfer. This is a major concern at radio frequencies. But audio design engineers, and even the utility companies, need to be concerned with ac circuit efficiency because it translates into energy conservation. The first two-thirds of this chapter is devoted to this subject.

Another important phenomenon, especially for the radio-frequency engineer, is *resonance*. This is an electrical analog of the reverberation you're familiar with if you've ever played a musical instrument. The last third of this chapter discusses resonance in series and parallel circuits.

What is power?

There are several different ways to define *power*. The applicable definition depends on the kind of circuit or device in use.

Energy per unit time

The most all-encompassing definition of power, and the one commonly used by physicists, is this: *Power is the rate at which energy is expended*. The standard unit of power is the watt, abbreviated W; it is equivalent to one joule per second.

This definition can be applied to motion, chemical effects, electricity, radio waves, sound, heat, light, ultraviolet, and X rays. In all cases, the energy is "used up" somehow, converted from one form into another form at a certain rate. This expression of power refers to an event that takes place at some definite place or places.

Sometimes power is given as kilowatts (kW or thousands of watts), megawatts (MW or millions of watts) or gigawatts (GW or billions of watts). It might be given as milliwatts (mW or thousandths of watts), microwatts (uW or millionths of watts), or nanowatts (nW or billionths of watts).

Volt-amperes

In dc circuits, and also in ac circuits having no reactance, power can be defined this way: *Power is the product of the voltage across a circuit or component, times the current through it.* Mathematically this is written $P = EI$. If E is in volts and I is in amperes, then P is in *volt-amperes (VA)*. This translates into watts when there is no reactance in the circuit (Fig. 17-1). The root-mean-square (rms) values for voltage and current are always used to derive the effective, or average, power.

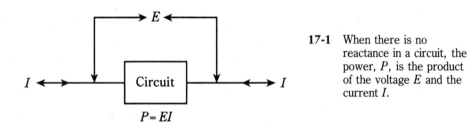

17-1   When there is no reactance in a circuit, the power, P, is the product of the voltage E and the current I.

Like joules per second, volt-amperes, also called *VA power* or *apparent power*, can take various forms. A resistor converts electrical energy into heat energy, at a rate that depends on the value of the resistance and the current through it. A light bulb converts electricity into light and heat. A radio antenna converts high-frequency ac into radio waves. A speaker converts low-frequency ac into sound waves. The power in these forms is a measure of the intensity of the heat, light, radio waves, or sound waves.

The VA power can have a meaning that the rate-of-energy-expenditure definition does not encompass. This is *reactive* or *imaginary* power, discussed shortly.

Instantaneous power

Usually, but not always, engineers think of power based on the rms, or effective, ac value. But for VA power, peak values are sometimes used instead. If the ac is a sine wave, the peak current is 1.414 times the rms current, and the peak voltage is 1.414 times the rms voltage. If the current and the voltage are exactly in phase, the product of their peak values is twice the product of their rms values.

There are instants in time when the VA power in a reactance-free, sine-wave ac circuit is twice the effective power. There are other instants in time when the VA power is zero; at still other moments, the VA power is somewhere between zero and twice the effective power level (Fig. 17-2). This constantly changing power is called *instantaneous power.*

In some situations, such as with a voice-modulated radio signal or a fast-scan television signal, the instantaneous power varies in an extremely complicated fashion. Perhaps you have seen the *modulation envelope* of such a signal displayed on an oscilloscope.

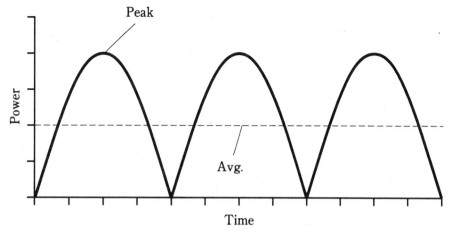

17-2 Peak versus average power for a sine wave.

Imaginary power

If an ac circuit contains reactance, things get interesting. The rate of energy expenditure is the same as the VA power in a pure resistance. But when inductance and/or capacitance exists in an ac circuit, these two definitions of power part ways. The VA power becomes greater than the power actually manifested as heat, light, radio waves, or whatever. The extra "power" is called *imaginary power*, because it exists in the reactance, and reactance can be, as you have learned, rendered in mathematically imaginary numerical form. It is also known as *reactive power*.

Inductors and capacitors store energy and then release it a fraction of a cycle later. This phenomenon, like *true power*, is expressible as the rate at which energy is changed from one form to another. But rather than any immediately usable form of power, such as radio or sound waves, imaginary power is "stashed" as a magnetic or electric field and then "dumped" back into the circuit, over and over again.

You might think of the relationship between imaginary and true power in the same way as you think of *potential* versus *kinetic* energy. A brick held out of a seventh-story window has potential energy, just as a charged-up capacitor or inductor has imaginary power.

Although the label "imaginary power" carries a connotation that it's not real or important, it's significant indeed. Imaginary power is responsible for many aspects of ac circuit behavior.

True power doesn't travel

An important semantical point should be brought up concerning true power, not only in ac circuits, but in any kind of circuit or device.

A common and usually harmless misconception about true power is that it "travels." For example, if you connect a radio transmitter to a cable that runs outdoors to an antenna, you might say you're "feeding power" through the cable to the antenna. Everybody says this, even engineers and technicians. What's moving along the cable is *imaginary power*, not *true power*. True power always involves a change in form, such as from electrical current and voltage into radio waves.

Some true power is dissipated as heat in the transmitter amplifiers and in the feed line (Fig. 17-3). The useful dissipation of true power occurs when the imaginary power, in the form of high-frequency current and voltage, gets to the antenna, where it is changed into electromagnetic waves.

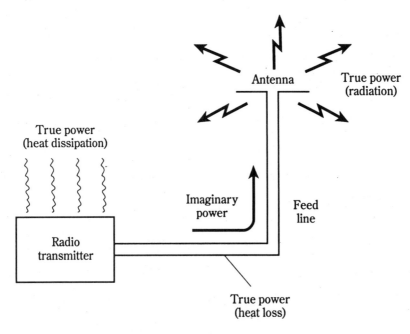

17-3   True and imaginary power in a radio antenna system.

You will often hear expressions such as "forward power" and "reflected power," or "power is fed from this amplifier to these speakers." It is all right to talk like this, but it can sometimes lead to wrong conclusions, especially concerning impedance and standing waves. Then, you need to be keenly aware of the distinction among true, imaginary and apparent power.

Reactance does not consume power

A coil or capacitor cannot dissipate power. The only thing that such a component can do is store energy and then give it back to the circuit a fraction of a cycle later. In real life, the dielectrics or wires in coils or capacitors dissipate some power as heat, but ideal components would not do this.

A capacitor, as you have learned, stores energy as an electric field. An inductor stores energy as a magnetic field.

A reactance causes ac current to shift in phase, so that it is no longer exactly in step with the voltage. In a circuit with inductive reactance, the current lags the voltage by up to 90 degrees, or one-quarter cycle. In a circuit with capacitive reactance, the current leads the voltage by up to 90 degrees.

In a resistance-reactance circuit, true power is dissipated only in the resistive components. The reactive components cause the VA power to be exaggerated compared with the true power.

Why does reactance cause this discrepancy between apparent (VA) power and true power? In a circuit that is purely resistive, the voltage and current march right along in step with each other, and therefore, they combine in the most efficient possible way (Fig. 17-4A). But in a circuit containing reactance, the voltage and current don't work together as well (Fig. 17-4B) because of their phase difference. Therefore, the actual energy expenditure, or true power, is not as great as the product of the voltage and the current.

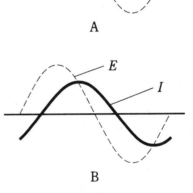

17-4 At A, current (*I*) and voltage (*E*) are in phase in a nonreactive ac circuit. At B, *I* and *E* are not in phase when reactance is present.

A

B

True power, VA power, and reactive power

In a circuit containing both resistance and reactance, the relationships among true power P_T, apparent or VA power P_{VA}, and imaginary or reactive power P_X are:

$$P_{VA}^2 = P_T^2 + P_X^2$$
$$P_T < P_{VA}$$
$$P_X < P_{VA}$$

If there is no reactance in the circuit, then:

$$P_{VA} = P_T$$
$$P_X = 0$$

Engineers often strive to eliminate, or at least minimize, the reactance in a circuit. This is particularly true for radio antenna systems, or when signals must be sent over long spans of cable. It is also important in the design of radio-frequency amplifiers. To a lesser extent, minimizing the reactance is important in audio work and in utility power transmission.

Power factor

The ratio of the true power to the VA power, P_T/P_{VA}, is called the *power factor* in an ac circuit. If there is no reactance, the ideal case, then $P_T = P_{VA}$, and the power factor (*PF*) is equal to 1. If the circuit contains all reactance and no resistance of any significance (that is, zero or infinite resistance), then $P_T = 0$, and $PF = 0$.

If you try to get a pure reactance to dissipate power, it's a little like throwing a foam-rubber ball into a gale-force wind. The ball will come right back in your face. A pure reactance cannot, and will not, dissipate power.

When a *load*, or a circuit in which you want power to be dissipated, contains some resistance and some reactance, then *PF* will be between 0 and 1. That is, $0 < PF < 1$. *PF* might be expressed as a percentage between 0 and 100, written $PF_\%$. Mathematically:

$$PF = P_T/P_{VA}$$
$$PF_\% = 100P_T/P_{VA}$$

When a load has some resistance and some reactance, a portion (but not all) of the power is dissipated as true power, and some is "rejected" by the load and sent back to the source as VA power.

Calculation of power factor

There are two ways to determine the power factor in an ac circuit that contains reactance and resistance. Either method can be used in any situation, although sometimes one scheme is more convenient than the other.

Cosine of phase angle

Recall that in a circuit having reactance and resistance, the current and the voltage are not in phase. The extent to which they differ in phase is the *phase angle*. If there is no reactance, then the phase angle is 0 degrees. If there is a pure reactance, then the phase angle is either +90 degrees or −90 degrees.

The power factor is equal to the cosine of the phase angle. You can use a calculator to find this easily.

Problem 17-1

A circuit contains no reactance, but a pure resistance of 600 Ω. What is the power factor?

Without doing any calculations, it is evident that $PF = 1$, because $P_{VA} = P_T$ in a pure resistance. So $P_T/P_{VA} = 1$. But you can also look at this by noting that the phase angle is 0 degrees, because the current is in phase with the voltage. Using your calculator, find cos 0 = 1. Therefore, *PF* = 1 = 100 percent. The vector for this case is shown in Fig. 17-5.

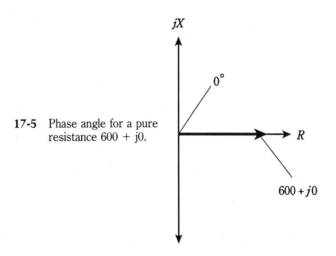

17-5 Phase angle for a pure resistance 600 + j0.

Problem 17-2

A circuit contains a pure capacitive reactance of $-40 \, \Omega$, but no resistance. What is the power factor?

Here, the phase angle is -90 degrees (Fig. 17-6). A calculator will tell you that cos $-90 = 0$. Therefore, *PF* = 0. This means that $P_T/P_{VA} = 0 = 0$ percent. That is, none of the power is true power, and all of it is imaginary or reactive power.

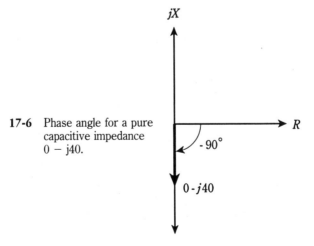

17-6 Phase angle for a pure capacitive impedance 0 − j40.

Problem 17-3

A circuit contains a resistance of 50 Ω and an inductive reactance of 50 Ω in series. What is the power factor?

The phase angle in this case is 45 degrees (Fig. 17-7). The resistance and reactance components are equal, and form two sides of a right triangle, with the complex impedance vector forming the hypotenuse. Find cos 45 = 0.707. This means that $P_T/P_{VA} = 0.707 = 70.7$ percent.

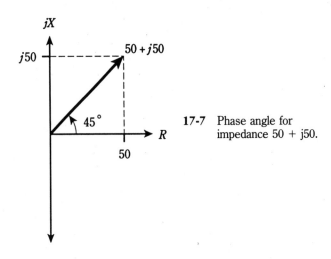

17-7 Phase angle for impedance 50 + j50.

Ratio R/Z

Another way to calculate the power factor is to find the ratio of the resistance R to the absolute-value impedance Z. In Fig. 17-7, this is visually apparent. A right triangle is formed by the resistance vector R (the base), the reactance vector jX (the height) and the absolute-value impedance Z (the hypotenuse). The cosine of the phase angle is equal to the ratio of the base length to the hypotenuse length; this represents R/Z.

Problem 17-4

A circuit has an absolute-value impedance Z of 100 Ω, with a resistance $R = 80$ Ω. What is the power factor?

Simply find the ratio: $PF = R/Z = 80/100 = 0.8 = 80$ percent. Note that it doesn't matter whether the reactance in this circuit is capacitive or inductive. In fact, you don't even have to worry about the value of the reactance here.

Problem 17-5

A circuit has an absolute-value impedance of 50 Ω, purely resistive. What is the power factor?

Here, $R = Z = 50$. Therefore, $PF = R/Z = 50/50 = 1 = 100$ percent.

Resistance and reactance

Sometimes you'll get data that tells you the resistance and reactance components in a circuit. To calculate the power factor from this, you can either find the phase angle and take its cosine, or find the absolute-value impedance and take the ratio R/Z.

Problem 17-6

A circuit has a resistance of 50 Ω and a capacitive reactance of -30 Ω. What is the power factor? Use the cosine method.

    The tangent of the phase angle is equal to X/R. Therefore, the phase angle is arctan (X/R) = arctan $(-30/50)$ = arctan (-0.60) = -31 degrees. The power factor is the cosine of this angle; PF = cos (-31) = 0.86 = 86 percent.

Problem 17-7

A circuit has a resistance of 30 Ω and an inductive reactance of 40 Ω. What is the power factor? Use the R/Z method.

    Find the absolute-value impedance: $Z^2 = R^2 + X^2 = 30^2 + 40^2 = 900 + 1600 = 2500$; therefore $Z = 50$. The power factor is therefore $PF = R/Z = 30/50 = 0.60 = 60$ percent. This problem is represented very nicely by a 3:4:5 right triangle (Fig. 17-8).

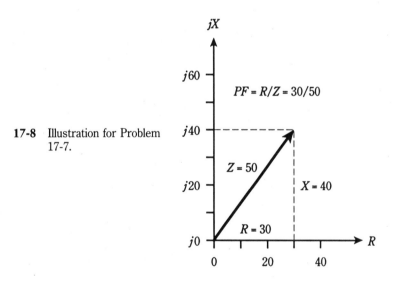

17-8 Illustration for Problem 17-7.

How much of the power is true?

The above simple formulas allow you to figure out, given the resistance, reactance, and VA power, how many watts are true or real power, and how many watts are imaginary or reactive power. This is important in radio-frequency (RF) equipment, because RF wattmeters will usually display VA power, and this reading is exaggerated when there is reactance in a circuit.

Problem 17-8

A circuit has 50 Ω of resistance and 30 Ω of inductive reactance in series. A wattmeter shows 100 watts, representing the VA power. What is the true power?

First, calculate the power factor. You might use either the phase-angle method or the R/Z method. Suppose you use the phase-angle method. Then:

$$\text{Phase angle} = \arctan (X/R)$$
$$= \arctan (30/50) = 31 \text{ degrees}$$

The power factor is the cosine of the phase angle:

$$PF = \cos 31 = 0.86 = 86 \text{ percent}$$

Remember that $PF = P_T/P_{VA}$. This means that the true power, P_T, is equal to 86 watts.

Problem 17-9

A circuit has a resistance of 1,000 Ω in parallel with a capacitance of 1,000 pF. The frequency is 100 kHz. If a wattmeter reads a VA power of 88.0 watts, what is the true power?

This problem requires several steps in calculation. First, note that the components are in *parallel*. This means that you have to find the conductance and the capacitive susceptance, and then combine these to get the admittance. Convert the frequency to megahertz: $f = 100 \text{ kHz} = 0.100 \text{ MHz}$. Convert capacitance to microfarads: $C = 1000 \text{ pF} = 0.001000 \text{ uF}$. From the previous chapter, use the equation for capacitive susceptance:

$$B_C = 6.28 fC = 6.28 \times 0.100 \times 0.001000$$
$$= 0.000628 \text{ siemens}$$

The conductance of the resistor, G, is found by taking the reciprocal of the resistance, R:

$$G = 1/R = 1/1,000 = 0.001000 \text{ siemens}$$

Although you don't need to know the actual complex admittance vector to solve this problem, note in passing that it is:

$$G + jB = 0.001000 + j0.000628$$

Now, use the formula for calculating the resistance and reactance of this circuit, in terms of the conductance and susceptance. First, find the resistance:

$$R = G/(G^2 + B^2)$$
$$= 0.001000/(0.001000^2 + 0.000628^2)$$
$$= 0.001000/0.000001394$$
$$= 717 \ \Omega$$

Then, find the reactance:

$$X = -B/(G^2 + B^2)$$
$$= -0.000628/0.000001394$$
$$= -451 \ \Omega$$

Therefore, $R = 717$ and $X = -451$.

Using the phase-angle method to solve this (the numbers are more manageable that way than they are with the R/Z method), calculate:

$$\text{Phase angle} = \arctan (X/R)$$
$$= \arctan (-451/717) = \arctan (-0.629)$$
$$= -32.2 \text{ degrees}$$

Then the power factor is:

$$PF = \cos -32.2 = 0.846 = 84.6 \text{ percent}$$

The VA power, P_{VA}, is given as 88.0 watts, and $PF = P_T/P_{VA}$. Therefore, the true power is found this way:

$$P_T/P_{VA} = 0.846$$
$$P_T/88.0 = 0.846$$
$$P_T = 0.846 \times 88.0 = 74.4 \text{ watts}$$

This is a good example of a practical problem. Although there are several steps, each requiring careful calculation, none of the steps individually is very hard. It's just a matter of using the right equations in the right order, and plugging the numbers in. You do have to be somewhat careful in manipulating plus/minus signs, and also in placing decimal points.

Power transmission

One of the most multifaceted, and important, problems facing engineers is *power transmission*.

Generators produce large voltages and currents at a power plant, say from turbines driven by falling water. The problem: getting the electricity from the plant to the homes, businesses, and other facilities that need it. This process involves the use of long wire *transmission lines*. Also needed are *transformers* to change the voltages to higher or lower values.

A radio transmitter produces a high-frequency alternating current. The problem: getting the power to be radiated by the antenna, located some distance from the transmitter. This involves the use of a radio-frequency transmission line. The most common type is coaxial cable. Two-wire line is also sometimes used. At ultra-high and microwave frequencies, another kind of transmission line, known as a *waveguide*, is often employed.

The overriding concern in any power-transmission system is minimizing the loss. Power wastage occurs almost entirely as heat in the line conductors and dielectric, and in objects near the line. Some loss can also take the form of unwanted electromagnetic radiation from a transmission line.

In an ideal transmission line, all of the power is VA power; that is, it is in the form of an alternating current in the conductors and an alternating voltage between them.

It is undesirable to have power in a transmission line exist in the form of true power. This translates either into heat loss in the line, radiation loss, or both. The place for true power dissipation is in the load, such as electrical appliances or radio antennas. Any true power in a transmission line represents power that can't be used by the load, because it doesn't show up there.

The rest of this chapter deals mainly with radio transmitting systems.

Power measurement in a transmission line

In a transmission line, power is measured by means of a voltmeter between the conductors, and an ammeter in series with one of the conductors (Fig. 17-9). Then the power, P (in watts) is equal to the product of the voltage E (in volts) and the current I (in amperes). This technique can be used in any transmission line, be it for 60-Hz utility service, or in a radio transmitting station.

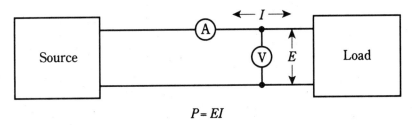

$$P = EI$$

17-9   Power measurement in a transmission line.

But is this indication of power the same as the power actually dissipated by the load at the end of the line? Not necessarily.

Recall, from the discussion of impedance, that any transmission line has a *characteristic impedance*. This value, Z_o, depends on the size of the line conductors, the spacing between the conductors, and the type of dielectric material that separates the conductors. For a coaxial cable, Z_o can be anywhere from about 50 to 150 Ω. For a parallel-wire line, it can range from about 75 Ω to 600 Ω.

If the load is a pure resistance R containing no reactance, and if $R = Z_o$, then the power indicated by the voltmeter/ammeter scheme will be the same as the true power dissipated by the load. The voltmeter and ammeter must be placed at the load end of the transmission line.

If the load is a pure resistance R, and $R < Z_o$ or $R > Z_o$, then the voltmeter and ammeter will not give an indication of the true power. Also, if there is any reactance in the load, the voltmeter/ammeter method will not be accurate.

The physics of this is rather sophisticated, and a thorough treatment of it is beyond the scope of this course. But you should remember that it is always desirable to have the load impedance be a pure resistance, a complex value of $R + j0$, where $R = Z_o$. Small discrepancies, in the form of a slightly larger or smaller resistance, or a small reactance, can sometimes be tolerated. But in very-high-frequency (VHF), ultra-high-frequency (UHF) and microwave radio transmitting systems, even a small *impedance mismatch* between the load and the line can cause excessive power losses in the line.

An impedance mismatch can usually be corrected by means of *matching transformers* and/or reactances that cancel out any load reactance. This is discussed in the next chapter.

Loss in a mismatched line

When a transmission line is terminated in a resistance $R = Z_o$, then the current and the voltage are constant all along the line, if the line is perfectly lossless. The ratio of the voltage to the current, E/I, is equal to R and also equal to Z_o.

Actually, this is an idealized case; no line is completely without loss. Therefore, as a signal goes along the line from a source to a load, the current and voltage gradually decrease. But they are always in same ratio (Fig. 17-10).

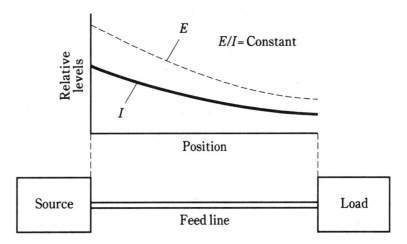

17-10   In a matched line, E/I is constant, although both E and I decrease with increasing distance from the source.

Standing waves

If the load is not matched to the line, the current and voltage vary in a complicated way along the length of the line. In some places, the current is high; in other places it is low. The maxima and minima are called *loops* and *nodes* respectively. At a current loop, the voltage is minimum (a voltage node), and at a current node, the voltage is maximum (a voltage loop). Loops and nodes make it impossible to measure power by the voltmeter/ ammeter method, because the current and voltage are not in constant proportion.

The loops and nodes, if graphed, form wavelike patterns along the length of the line. These patterns remain fixed over time. They are therefore known as *standing waves*— they just "stand there."

Standing-wave loss

At current loops, the loss in the line conductors is exaggerated. At voltage loops, the loss in the line dielectric is increased. At minima, the losses are decreased. But overall, in a mismatched line, the losses are greater than they are in a perfectly matched line.

This loss occurs at heat dissipation. It is true power. Any true power that goes into heating up a transmission line is wasted, because it cannot be dissipated in the load. The additional loss caused by standing waves, over and above the perfectly-matched line loss, is called *standing-wave loss*.

The greater the mismatch, the more severe the standing-wave loss becomes. The more loss a line has to begin with (that is, when it is perfectly matched), the more loss is caused by a given amount of mismatch. Standing-wave loss increases with frequency. It tends to be worst in long lengths of line at VHF, UHF, and microwaves.

Line overheating

A severe mismatch between the load and the transmission line can cause another problem: physical destruction of the line!

A feed line might be able to handle a kilowatt (1 kW) of power when it is perfectly matched. But if a severe mismatch exists and you try to feed 1 kW into the line, the extra current at the current loops can heat the conductors to the point where the dielectric material melts and the line shorts out.

It is also possible for the voltage at the voltage loops to cause arcing between the line conductors. This perforates and/or burns the dielectric, ruining the line.

When a line must be used with a mismatch, *derating functions* are required to determine how much power the line can safely handle. Manufacturers of prefabricated lines can supply you with this information.

Series resonance

One of the most important phenomena in ac circuits, especially in radio-frequency engineering, is the property of *resonance*. You've already learned that resonance is a condition that occurs when capacitive and inductive reactance cancel each other out. Resonant circuits and devices have a great many different applications in electricity and electronics.

Recall that capacitive reactance, X_C, and inductive reactance, X_L, can sometimes be equal in magnitude. They are always opposite in effect. In any circuit containing an inductance and capacitance, there will be a frequency at which $X_L = -X_C$. This is resonance. Sometimes $X_L = -X_C$ at just one frequency; in some special devices it can occur at many frequencies. Generally, if a circuit contains one coil and one capacitor, there will be one resonant frequency.

Refer to the schematic diagram of Fig. 17-11. You might recognize this as a series RLC circuit. At some particular frequency, $X_L = -X_C$. This is inevitable, if L and C are finite and nonzero. This is the *resonant frequency* of the circuit. It is abbreviated f_o.

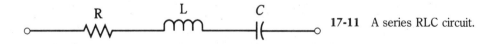

17-11   A series RLC circuit.

At f_o, the effects of capacitive reactance and inductive reactance cancel out. The result is that the circuit appears as a pure resistance, with a value very close to R. If

$R = 0$, that is, the resistor is a short circuit, then the circuit is called a *series LC circuit*, and the impedance at resonance will be extremely low. The circuit will offer practically no opposition to the flow of alternating current at the frequency f_o. This condition is *series resonance*.

Parallel resonance

Refer to the circuit diagram of Fig. 17-12. This is a parallel RLC circuit. You remember that, in this case, the resistance R is thought of as a conductance G, with $G = 1/R$. Then the circuit can be called a GLC circuit.

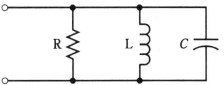

17-12   A parallel RLC circuit.

At some particular frequency f_o, the inductive susceptance B_L will exactly cancel the capacitive susceptance B_C; $B_L = -B_C$. This is inevitable for some frequency f_o, as long as the circuit contains finite, nonzero inductance and finite, nonzero capacitance.

At the frequency f_o, the susceptances cancel each other out, leaving zero susceptance. The admittance through the circuit is then very nearly equal to the conductance, G, of the resistor. If the circuit contains no resistor, but only a coil and capacitor, it is called a *parallel LC circuit*, and the admittance at resonance will be extremely low. The circuit will offer great opposition to alternating current at f_o. Engineers think more often in terms of impedance than in terms of admittance; low admittance translates into high impedance. This condition is *parallel resonance*.

Calculating resonant frequency

The formula for calculating resonant frequency f_o, in terms of the inductance L in henrys and the capacitance C in farads, is:

$$f_o = 0.159/(LC)^{1/2}$$

The ½ power is the square root.

If you know L and C in henrys and farads, and you want to find f_o, do these calculations in this order: first find the product LC; then take the square root; then divide 0.159 by this value. The result is f_o in hertz.

The formula will also work to find f_o in *mega*hertz (MHz), when L is given in *micro*henrys (uH) and C is in *micro*farads (uF). These values are far more common than hertz, henrys, and farads in electronic circuits. Just remember that millions of hertz go with million*ths* of henrys and million*ths* of farads.

This formula works for both series-resonant and parallel-resonant RLC circuits.

Problem 17-10

Find the resonant frequency of a series circuit with an inductance of 100 uH and a capacitance of 100 pF.

First, convert the capacitance to microfarads: $100\,pF = 0.000100$ uF. Then find the product $LC = 100 \times 0.000100 = 0.0100$. Take the square root of this, getting 0.100. Finally, divide 0.159 by 0.100, getting $f_o = 1.59$ MHz.

Problem 17-11

Find the resonant frequency of a parallel circuit consisting of a 33-uH coil and a 47-pF capacitor.

Again, convert the capacitance to microfarads: 47 pF $= 0.000047$ uF. Then find the product $LC = 33 \times 0.000047 = 0.00155$. Take the square root of this, getting 0.0394. Finally, divide 0.159 by 0.0394, getting $f_o = 4.04$ MHz.

There are times when you might know the resonant frequency f_o that you want, and you need to find a particular inductance or capacitance instead. The next two problems illustrate this type of situation.

Problem 17-12

A circuit must be designed to have $f_o = 9.00$ MHz. You have a 33-pF fixed capacitor available. What size coil will be needed to get the desired resonant frequency?

Use the formula $f_o = 0.159/(LC)^{1/2}$, and plug in the values. Convert the capacitance to microfarads: 33 pF $= 0.000033$ uF. Then just manipulate the numbers, using familiar rules of arithmetic, until the value of L is "ferreted out":

$$9.00 = 0.159/(L \times 0.000033)^{1/2}$$
$$9.00^2 = 0.159^2/(0.000033 \times L)$$
$$81.0 = 0.0253/(0.000033 \times L)$$
$$81.0 \times 0.000033 \times L = 0.0253$$
$$0.00267 \times L = 0.0253$$
$$L = 0.0253/0.00267 = 9.48\ uH$$

Problem 17-13

A circuit must be designed to have $f_o = 455$ kHz. A coil of 100 uH is available. What size capacitor is needed?

Convert kHz to MHz: 455 kHz $= 0.455$ MHz. Then the calculation proceeds in the same way as with the preceding problem:

$$f_o = 0.159/(LC)^{1/2}$$
$$0.455 = 0.159/(100 \times C)^{1/2}$$
$$0.455^2 = 0.159^2/(100 \times C)$$
$$0.207 = 0.0253/(100 \times C)$$
$$0.207 \times 100 \times C = 0.0253$$
$$20.7 \times C = 0.0253$$
$$C = 0.0253/20.7 = 0.00122\ uF = 1220\ pF$$

In practical circuits, variable inductors and/or variable capacitors are often placed in tuned circuits, so that small errors in the frequency can be compensated for. The most common approach is to design the circuit for a frequency slightly higher than f_o, and to use a *padder capacitor* in parallel with the main capacitor (Fig. 17-13).

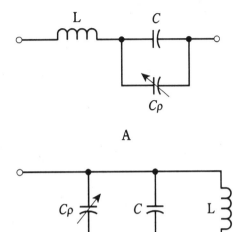

17-13 Padding capacitors (C_p) allow adjustment of resonant frequency in a series LC circuit (A) or a parallel LC circuit (B).

Resonant devices

While resonant circuits often consist of coils and capacitors in series or parallel, there are other kinds of hardware that exhibit resonance. Some of these are as follows.

Crystals

Pieces of quartz, when cut into thin wafers and subjected to voltages, will vibrate at high frequencies. Because of the physical dimensions of such a *crystal*, these vibrations occur at a precise frequency f_o, and also at whole-number multiples of f_o. These multiples, $2f_o$, $3f_o$, $4f_o$ and so on, are called *harmonics*. The frequency f_o is called the *fundamental frequency* of the crystal.

Quartz crystals can be made to act like LC circuits in electronic devices. A crystal exhibits an impedance that varies with frequency. The reactance is zero at f_o and the harmonic frequencies.

Cavities

Lengths of metal tubing, cut to specific dimensions, exhibit resonance at very-high, ultra-high, and microwave frequencies. They work in much the same way as musical instruments resonate with sound waves. But the waves are electromagnetic, rather than acoustic.

Cavities, also called *cavity resonators*, have reasonable lengths at frequencies above about 150 MHz. Below this frequency, a cavity can be made to work, but it will be long and unwieldy. Like crystals, cavities resonate at a fundamental frequency f_o, and also at harmonic frequencies.

Sections of transmission line

When a transmission line is cut to ¼ wavelength, or to any whole-number multiple of this, it behaves as a resonant circuit. The most common length for a transmission-line resonantor is a quarter wavelength. Such a piece of transmission line is called a *quarter-wave section.*

When a quarter-wave section is short-circuited at the far end, it acts like a parallel-resonant LC circuit, and has a high impedance at the resonant frequency f_o. When it is open at the far end, it acts as a series-resonant LC circuit, and has a low impedance at f_o. Thus, a quarter-wave section turns a short circuit into an open circuit and vice-versa, at a specific frequency f_o.

The length of a quarter-wave section depends on f_o. It also depends on how fast the electromagnetic energy travels along the line. This speed is specified in terms of a *velocity factor*, abbreviated v. The value of v is given as a fraction of the speed of light. Typical transmission lines have velocity factors ranging from about 0.66 to 0.95. This factor is provided by the manufacturers of prefabricated lines such as coaxial cable.

If the frequency in megahertz is f_o and the velocity factor of a line is v, then the length L_{ft} of a quarter-wave section of transmission line, in feet, is:

$$L_{ft} = 246v/f_o$$

The length in meters, L_m, is:

$$L_m = 75.0v/f_o$$

Problem 17-14

How many feet long is a quarter-wave section of transmission line at 7.05 MHz, if the velocity factor is 0.800?

Just use the formula:

$$L_{ft} = 246v/f_o$$
$$= (246 \times 0.800)/7.05$$
$$= 197/7.05 = 27.9 \text{ feet}$$

Antennas

Many types of antennas exhibit resonant properties. The simplest type of resonant antenna, and the only kind that will be mentioned here, is the center-fed, half-wavelength *dipole antenna* (Fig. 17-14).

The length L_{ft}, in feet, for a 1/2-wave dipole at a frequency of f_o MHz is given by the following formula:

$$L_{ft} = 468/f_o$$

This takes into account the fact that electromagnetic fields travel along a wire at about 95 percent of the speed of light. If L_m is specified in meters, then:

$$L_m = 143/f_o$$

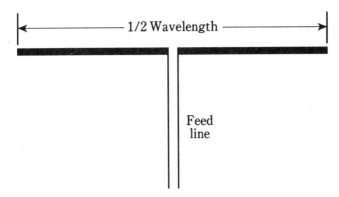

17-14 A half-wave dipole antenna.

A half-wave dipole has a purely resistive impedance of about 70 Ω at resonance. This is like a series-resonant RLC circuit with $R = 70$ Ω.

A half-wave dipole is resonant at all harmonics of its fundamental frequency f_o. The dipole is a full wavelength long at $2f_o$; it is 3/2 wavelength long at $3f_o$; it is two full wavelengths long at $4f_o$ and so on.

At f_o and odd harmonics, that is, at f_o, $3f_o$, $5f_o$ and so on, the antenna behaves like a series-resonant RLC circuit with a fairly low resistance. At even harmonics, that is, at $2f_o$, $4f_o$, $6f_o$ and so on, the antenna acts like a parallel-resonant RLC circuit with a high resistance.

But, you say, there's no resistor in the diagram of Fig. 17-14! Where does the resistance come from? This is an interesting phenomenon that all antennas have. It is called *radiation resistance*, and is a crucial factor in the design of any antenna system.

When electromagnetic energy is fed into an antenna, power flies away into space as radio waves. This is a form of true power. True power is always dissipated in a resistance. Although you don't see any resistor in Fig. 17-14, the radiation of radio waves is, in effect, power dissipation in a resistance.

For details concerning the behavior of antennas, a text on antenna engineering is recommended. This subject is vast and many-faceted. Some engineers devote their careers exclusively to antenna design and manufacture.

Quiz

Refer to the text in this chapter if necessary. A good score is 18 or more correct. Answers are in the back of the book.

1. The power in a reactance is:

   A. Radiated power.

   B. True power.

   C. Imaginary power.

   D. Apparent power.

2. Which of the following is *not* an example of true power?

   A. Power that heats a resistor.

   B. Power radiated from an antenna.

   C. Power in a capacitor.

   D. Heat loss in a feed line.

3. The apparent power in a circuit is 100 watts, and the imaginary power is 40 watts. The true power is:

   A. 92 watts.

   B. 100 watts.

   C. 140 watts.

   D. Not determinable from this information.

4. Power factor is equal to:

   A. Apparent power divided by true power.

   B. Imaginary power divided by apparent power.

   C. Imaginary power divided by true power.

   D. True power divided by apparent power.

5. A circuit has a resistance of 300 Ω and an inductance of 13.5 uH in series at 10.0 MHz. What is the power factor?

   A. 0.334.

   B. 0.999.

   C. 0.595.

   D. It can't be found from the data given.

6. A series circuit has $Z = 88.4 \ \Omega$, with $R = 50.0 \ \Omega$. What is PF?

   A. 99.9 percent.

   B. 56.6 percent.

   C. 60.5 percent.

   D. 29.5 percent.

7. A series circuit has $R = 53.5 \ \Omega$ and $X = 75.5 \ \Omega$. What is PF?

   A. 70.9 percent.

   B. 81.6 percent.

   C. 57.8 percent.

   D. 63.2 percent.

8. Phase angle is equal to:

   A. Arctan Z/R.

   B. Arctan R/Z.

C. Arctan R/X.

D. Arctan X/R.

9. A wattmeter shows 220 watts of VA power in a circuit. There is a resistance of 50 Ω in series with a capacitive reactance of -20 Ω. What is the true power?

A. 237 watts.

B. 204 watts.

C. 88.0 watts.

D. 81.6 watts.

10. A wattmeter shows 57 watts of VA power in a circuit. The resistance is known to be 50 Ω, and the true power is known to be 40 watts. What is the absolute-value impedance?

A. 50 Ω.

B. 57 Ω.

C. 71 Ω.

D. It can't be calculated from this data.

11. Which of the following is the most important consideration in a transmission line?

A. The characteristic impedance.

B. The resistance.

C. Minimizing the loss.

D. The VA power.

12. Which of the following does *not* increase the loss in a transmission line?

A. Reducing the power output of the source.

B. Increasing the degree of mismatch between the line and the load.

C. Reducing the diameter of the line conductors.

D. Raising the frequency.

13. A problem that standing waves can cause is:

A. Feed line overheating.

B. Excessive power loss.

C. Inaccuracy in power measurement.

D. All of the above.

14. A coil and capacitor are in series. The inductance is 88 mH and the capacitance is 1000 pF. What is the resonant frequency?

A. 17 kHz.

B. 540 Hz.

C. 17 MHz.

D. 540 kHz.

15. A coil and capacitor are in parallel, with $L = 10.0$ uH and $C = 10$ pF. What is f_o?

    A. 15.9 kHz.

    B. 5.04 MHz.

    C. 15.9 MHz.

    D. 50.4 MHz.

16. A series-resonant circuit is to be made for 14.1 MHz. A coil of 13.5 uH is available. What size capacitor is needed?

    A. 0.945 uF.

    B. 9.45 pF.

    C. 94.5 pF.

    D. 945 pF.

17. A parallel-resonant circuit is to be made for 21.3 MHz. A capacitor of 22.0 pF is available. What size coil is needed?

    A. 2.54 mH.

    B. 254 uH.

    C. 25.4 uH.

    D. 2.54 uH.

18. A 1/4-wave line section is made for 21.1 MHz, using cable with a velocity factor of 0.800. How many meters long is it?

    A. 11.1 m.

    B. 3.55 m.

    C. 8.87 m.

    D. 2.84 m.

19. The fourth harmonic of 800 kHz is:

    A. 200 kHz.

    B. 400 kHz.

    C. 3.20 MHz.

    D. 4.00 MHz.

20. How long is a 1/2-wave dipole for 3.60 MHz?

    A. 130 feet.

    B. 1680 feet.

    C. 39.7 feet.

    D. 515 feet.

Transformers and
impedance matching

IN ELECTRICITY AND ELECTRONICS, *TRANSFORMERS* ARE EMPLOYED IN VARIOUS ways. Transformers are used to obtain the right voltage for the operation of a circuit or system. Transformers can match impedances between a circuit and a load, or between two different circuits. Transformers can be used to provide dc isolation between electronic circuits while letting ac pass. Another application is to mate balanced and unbalanced circuits, feed systems and loads.

Principle of the transformer

When two wires are near each other, and one of them carries a fluctuating current, a current will be induced in the other wire. This effect is known as *electromagnetic induction*. All ac transformers work according to the principle of electromagnetic induction. If the first wire carries sine-wave ac of a certain frequency, then the *induced current* will be sine-wave ac of the same frequency in the second wire.

The closer the two wires are to each other, the greater the induced current will be, for a given current in the first wire. If the wires are wound into coils and placed along a common axis (Fig. 18-1), the induced current will be greater than if the wires are straight and parallel. Even more *coupling*, or efficiency of induced-current transfer, is obtained if the two coils are wound one atop the other.

The first coil is called the *primary winding*, and the second coil is known as the *secondary winding*. These are often spoken of as simply the primary and secondary.

The induced current creates a voltage across the secondary. In a *step-down* transformer, the secondary voltage is less than the primary voltage. In a *step-up* transformer, the secondary voltage is greater than the primary voltage. The primary voltage is abbreviated E_{pri}, and the secondary voltage is abbreviated E_{sec}. Unless otherwise stated, effective (rms) voltages are always specified.

The windings of a transformer have inductance because they are coils. The required

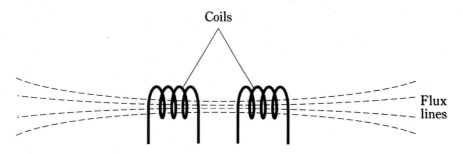

18-1   Magnetic flux between two coils of wire.

inductances of the primary and secondary depend on the frequency of operation, and also on the resistive part of the impedance in the circuit. As the frequency increases, the needed inductance decreases. At high resistive impedances, more inductance is generally needed than at low resistive impedances.

Turns ratio

The *turns ratio* in a transformer is the ratio of the number of turns in the primary, T_{pri}, to the number of turns in the secondary, T_{sec}. This ratio is written $T_{pri}:T_{sec}$ or T_{pri}/T_{sec}.

In a transformer with excellent primary-to-secondary coupling, the following relationship always holds:

$$E_{pri}/E_{sec} = T_{pri}/T_{sec}$$

That is, the primary-to-secondary voltage ratio is always equal to the primary-to-secondary turns ratio (Fig. 18-2).

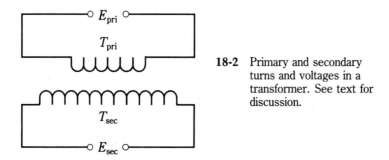

18-2   Primary and secondary turns and voltages in a transformer. See text for discussion.

Problem 18-1

A transformer has a primary-to-secondary turns ratio of exactly 9:1. The voltage at the primary is 117 V. What is the voltage at the secondary?

This is a step-down transformer. Simply plug in the numbers in the above equation and solve for E_{sec}:

$$E_{pri}/E_{sec} = T_{pri}/T_{sec}$$
$$117/E_{sec} = 9/1 = 9$$
$$1/E_{sec} = 9/117$$
$$E_{sec} = 117/9 = 13 \text{ V}$$

Problem 18-2

A transformer has a primary-to-secondary turns ratio of exactly 1:9. The voltage at the primary is 117 V. What is the voltage at the secondary?

This is a step-up transformer. Plug in numbers again:

$$117/E_{sec} = 1/9$$
$$E_{sec}/117 = 9/1 = 9$$
$$E_{sec} = 9 \times 117 = 1053 \text{ V}$$

This can be rounded off to 1050 V.

A step-down transformer always has a primary-to-secondary turns ratio greater than 1, and a step-up transformer has a primary-to-secondary turns ratio less than 1.

Sometimes the secondary-to-primary turns ratio is given. This is the reciprocal of the primary-to-secondary turns ratio, written T_{sec}/T_{pri}. In a step-down unit, $T_{sec}/T_{pri} <$ 1; in a step-up unit, $T_{sec}/T_{pri} > 1$.

When you hear someone say that such-and-such a transformer has a certain turns ratio, say 10:1, you need to be sure of which ratio is meant, T_{pri}/T_{sec} or T_{sec}/T_{pri}! If you get it wrong, you'll have the secondary voltage off by a factor of the *square* of the turns ratio. You might be thinking of 12 V when the engineer is talking about 1,200 V. One way to get rid of doubt is to ask, "Step-up or step-down?"

Transformer cores

If a ferromagnetic substance such as iron, powdered iron, or ferrite is placed within the pair of coils, the extent of coupling is increased far above that possible with an air core. But this improvement in coupling takes place with a price; some energy is invariably lost as heat in the core. Also, ferromagnetic transformer cores limit the frequency at which the transformer will work well.

The schematic symbol for an air-core transformer consists of two inductor symbols back-to-back (Fig. 18-3A). If a ferromagnetic core is used, two parallel lines are added to the schematic symbol (Fig. 18-3B).

18-3 Schematic symbols for air-core (A) and ferromagnetic-core (B) transformers.

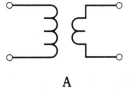
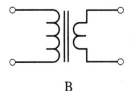

A   B

Laminated iron

In transformers for 60-Hz utility ac, and also at low audio frequencies, sheets of silicon steel, glued together in layers, are often employed as transformer cores. The silicon steel is sometimes called *transformer iron,* or simply iron.

The reason layering is used, rather than making the core from a single mass of metal, is that the magnetic fields from the coils cause currents to flow in a solid core. These *eddy currents* go in circles, heating up the core and wasting energy that would otherwise be transferred from the primary to the secondary. Eddy currents are choked off by breaking up the core into layers, so that currents cannot flow very well in circles.

Another type of loss, called *hysteresis loss,* occurs in any ferromagnetic transformer core. Hysteresis is the tendency for a core material to be "sluggish" in accepting a fluctuating magnetic field. Laminated cores exhibit high hysteresis loss above audio frequencies, and are therefore not good above a few kilohertz.

Ferrite and powdered iron

At frequencies up to several megahertz, *ferrite* works well for radio-frequency (RF) transformers. This material has high permeability and concentrates the flux efficiently. High permeability reduces the number of turns needed in the coils. But at frequencies higher than a few megahertz, ferrite begins to show loss, and is no longer effective.

For work well into the very-high-frequency (VHF) range, or up to 100 MHz or more, *powdered iron* cores work well. The permeability of powdered iron is less than that of ferrite, but at high frequencies, it is not necessary to have high magnetic permeability. In fact, at radio frequencies above a few megahertz, air core coils are often preferred, especially in transmitting amplifiers. At frequencies above several hundred megahertz, ferromagnetic cores can be dispensed with entirely.

Transformer geometry

The shape of a transformer depends on the shape of its core. There are several different core geometries commonly used with transformers.

Utility transformers

A common core for a power transformer is the *E core,* so named because it is shaped like the capital letter E. A bar, placed at the open end of the E, completes the core once the coils have been wound on the E-shaped section (Fig. 18-4A).

The primary and secondary windings can be placed on an E core in either of two ways.

The simpler winding method is to put both the primary and the secondary around the middle bar of the E (Fig. 18-4B). This is called the *shell method* of transformer winding. It provides maximum coupling between the windings. However, the shell- winding scheme results in a considerable capacitance between the primary and the secondary. This capacitance can sometimes be tolerated; sometimes it cannot. Another disadvantage of the shell geometry is that, when windings are placed one atop the other, the transformer cannot handle very much voltage.

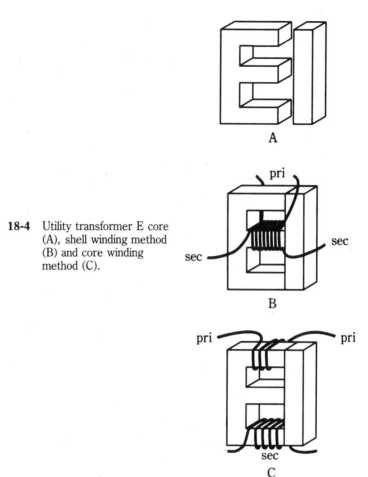

18-4 Utility transformer E core (A), shell winding method (B) and core winding method (C).

Another winding method is the *core method*. In this scheme, the primary is placed at the bottom of the E section, and the secondary is placed at the top (Fig. 18-4C). The coupling occurs via magnetic flux in the core. The capacitance between the primary and secondary is much lower with this type of winding. Also, a core-wound transformer can handle higher voltages than the shell-wound transformer. Sometimes the center part of the E is left out of the core when this winding scheme is used.

Shell-wound and core-wound transformers are almost universally employed at 60 Hz. These configurations are also common at audio frequencies.

Solenoidal core

A pair of cylindrical coils, wound around a rod-shaped piece of powdered iron or ferrite, was once a common configuration for transformers at radio frequencies. Sometimes this type of transformer is still seen, although it is most often used as a *loopstick antenna* in portable radio receivers and in radio direction-finding equipment.

The coil windings might be placed one atop the other, or they might be separated (Fig. 18-5) to reduce the capacitance between the primary and secondary.

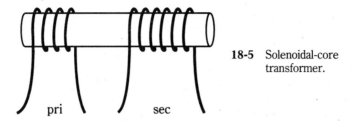

18-5   Solenoidal-core transformer.

In a loopstick antenna, the primary serves to pick up the radio signals. The secondary winding provides the best *impedance match* to the first amplifier stage, or *front end*, of the radio. The use of transformers for impedance matching is discussed later in this chapter.

Toroidal core

In recent years, the *toroidal core* has become the norm for winding radio-frequency transformers. The toroid is a donut-shaped ring of powdered iron or ferrite. The coils are wound around the donut. The primary and secondary might be wound one over the other, or they might be wound over different parts of the core (Fig. 18-6). As with other transformers, when the windings are one atop the other, there is more inter-winding capacitance than when they are separate.

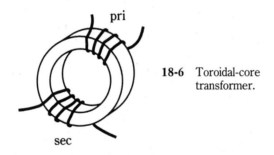

18-6   Toroidal-core transformer.

Toroids confine practically all the magnetic flux within the core material. This allows toroidal coils and transformers to be placed near other components without inductive interaction. Also, a toroidal coil or transformer can be mounted directly on a metal chassis, and the operation will not be affected (assuming the wire is insulated).

A toroidal core provides considerably more inductance per turn, for the same kind of ferromagnetic material, than a solenoidal core. It is not uncommon to see toroidal coils or transformers that have inductances of 10 mH or even 100 mH.

Pot core

Still more inductance per turn can be obtained with a *pot core*. This is a shell of ferromagnetic material that wraps around a loop-shaped coil. The core comes in two halves (Fig. 18-7). You wind the coil inside one of the halves, and then bolt the two together. The final core completely surrounds the loop, and the magnetic flux is confined to the core material.

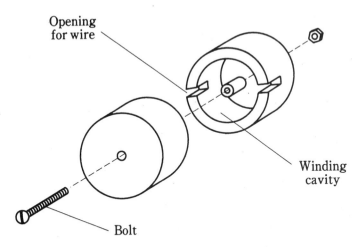

18-7 Exploded view of pot core (windings not shown).

Like the toroid, the pot core is self-shielding. There is essentially no coupling to external components. A pot core can be used to wind a single, high-inductance coil; sometimes the value can be upwards of 1 H.

In a pot-core transformer, the primary and secondary must always be wound on top of, or right next to, each other; this is unavoidable because of the geometry of the shell. Therefore, the inter-winding capacitance of a pot-core transformer is always rather high.

Pot cores are useful at the lower frequencies. They are generally not employed at higher frequencies because it isn't necessary to get that much inductance.

The autotransformer

Sometimes, it's not necessary to provide dc isolation between the primary and secondary windings of a transformer. Then an *autotransformer* can be used. This has a single, tapped winding. Its schematic symbol is shown in Fig. 18-8A for an air core, and Fig. 18-8B for a ferromagnetic core.

An autotransformer can be either a step-down or a step-up device. In Fig. 18-8, the autotransformer at A is step-down, and the one at B is step-up.

An autotransformer can have an air core, or it can be wound on any of the aforementioned types of ferromagnetic cores. You'll sometimes see this type of transformer in a radio-frequency receiver or transmitter. It works quite well in impedance- matching applications, and also in solenoidal loopsticks.

Autotransformers are occasionally, but not often, used at audio frequencies and in 60-Hz utility wiring. In utility circuits, autotransformers can step down by a large factor, but they aren't used to step up by more than a few percent.

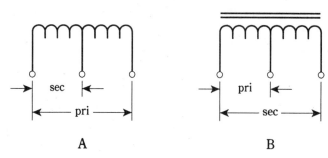

18-8 Schematic symbols for autotransformers. At A, air core; at B, ferromagnetic core. The unit at A steps down, and the one at B steps up.

Power transformers

Any transformer used in the 60-Hz utility line, intended to provide a certain rms ac voltage for the operation of electrical circuits, is a *power transformer*. Power transformers exist in sizes ranging from smaller than a tennis ball to as big as a room.

At the generating plant

The largest transformers are employed right at the place where electricity is generated. Not surprisingly, high-energy power plants have bigger transformers that develop higher voltages than low-energy, local power plants. These transformers must be able to handle not only huge voltages, but large currents as well. Their primaries and secondaries must withstand a product EI of *volt-amperes* that is equal to the power P ultimately delivered by the transmission line.

When electrical energy must be sent over long distances, extremely high voltages are used. This is because, for a given amount of power ultimately dissipated by the loads, the current is lower when the voltage is higher. Lower current translates into reduced loss in the transmission line.

Recall the formula $P = EI$, where P is the power in watts, E is the voltage in volts and I is the current in amperes. If you can make the voltage 10 times larger, for a given power level, then the current is reduced to 1/10 as much. The *ohmic losses* in the wires are proportional to the *square* of the current; remember that $P = I^2R$, where P is the power in watts, I is the current in amperes and R is the resistance in ohms.

Engineers can't do much about the wire resistance or the power consumed by the loads, but they can adjust the voltage, and thereby the current. Increasing the voltage 10 times will cut the current to 0.1 its previous value. This will render the I^2R loss $(0.1)^2 = 0.01$ (1 percent!) as much as before.

For this reason, regional power plants have massive transformers capable of generating hundreds of thousands of volts. A few can produce 1,000,000 V rms. A transmission line that carries this much voltage requires gigantic insulators, sometimes several meters long, and tall, sturdy towers.

Along the line

Extreme voltage is good for *high-tension* power transmission, but it's certainly of no use to an average consumer. The wiring in a high-tension system must be done using precautions to prevent arcing (sparking) and short circuits. Personnel must be kept at least several feet, or even tens of feet, away from the wires. Can you imagine trying to use an appliance, say a home computer, by plugging it into a 500-kV electrical outlet? A bolt of artificial lightning would dispatch you before you even got near the receptacle.

Medium-voltage power lines branch out from the major lines, and step-down transformers are used at the branch points. These lines fan out to still lower-voltage lines, and step-down transformers are employed at these points, too. Each transformer must have windings heavy enough to withstand the product $P = EI$, the amount of power delivered to all the subscribers served by that transformer, at periods of peak demand.

Sometimes, such as during a heat wave, the demand for electricity rises above the normal peak level. This loads down the circuit to the point that the voltage drops several percent. This is a *brownout*. If consumption rises further still, a dangerous current load is placed on one or more intermediate power transformers. Circuit breakers in the transformers protect them from destruction by opening the circuit. Then there is a temporary *blackout*.

Finally, at individual homes and buildings, transformers step the voltage down to either 234 V or 117 V. Usually, 234-V electricity is in three phases, each separated by 120 degrees, and each appearing at one of the three prongs in the outlet (Fig. 18-9A). This voltage is commonly employed with heavy appliances, such as the kitchen oven/stove (if they are electric), heating (if it is electric) and the laundry washer and dryer. A 117-V outlet supplies just one phase, appearing between two of the three prongs in the outlet. The third prong is for ground (Fig. 18-9B).

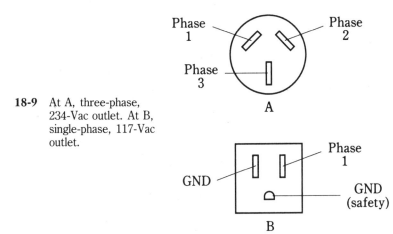

18-9   At A, three-phase, 234-Vac outlet. At B, single-phase, 117-Vac outlet.

In electronic devices

The lowest level of power transformer is found in electronic equipment such as television sets, ham radios, and home computers.

Most solid-state devices use low voltages, ranging from about 5 V up to perhaps 50 V. This equipment needs step-down power transformers in its power supplies. Solid-state equipment usually (but not always) consumes relatively little power, so the transformers are usually not very bulky. The exception is high-powered audio-frequency or radio-frequency amplifiers, whose transistors can demand more than 1,000 watts (1 kW) in some cases. At 12 V, this translates to 90 A or more.

Television sets have cathode-ray tubes that need several hundred volts. This is derived by using a step-up transformer in the power supply. Such transformers don't have to supply a lot of current, though, so they are not very big or heavy. Another type of device that needs rather high voltage is a ham-radio amplifier with vacuum tubes. Such an amplifier requires from 2 kV to 5 kV.

Any voltage higher than about 12 V should be treated with respect. *The voltages in televisions and ham radios are extremely dangerous, even after the equipment has been switched off. Do not try to service such equipment unless you are trained to do so!*

Audio-frequency transformers

Transformers for use at audio frequency (AF) are similar to those employed for 60-Hz electricity. The differences are that the frequency is somewhat higher (up to 20 kHz), and that audio signals exist in a band of frequencies (20 Hz to 20 kHz) rather than at just one frequency.

Most audio transformers look like, and are constructed like, miniature utility transformers. They have laminated E cores with primary and secondary windings wound around the cross bars, as shown in Fig. 18-4.

Audio transformers can be either the step-up or step-down type. However, rather than being made to produce a specific voltage, audio transformers are designed to match impedances.

Audio circuits, and in fact all electronic circuits that handle sine-wave or complex-wave signals, exhibit impedance at the input and output. The load has a certain impedance; a source has another impedance. Good audio design strives to minimize the reactances in the circuitry, so that the absolute-value impedance, Z, is close to the resistance R in the complex vector $R + jX$. This means that X must be zero or nearly zero.

In the following discussion of impedance-matching transformers, both at audio and at radio frequencies, assume that the reactance is zero, so that the impedance is purely resistive: $Z = R$.

Isolation transformers

One useful function of a transformer is that it can provide isolation between electronic circuits. While there is *inductive coupling* in a transformer, there is comparatively little *capacitive coupling*. The amount of capacitive coupling can be reduced by using cores that minimize the number of wire turns needed in the windings, and by keeping the windings separate from each other (rather than overlapping).

Balanced and unbalanced loads

A *balanced load* is one whose terminals can be reversed without significantly affecting circuit behavior. A plain resistor is a good example. The two-wire antenna input in a television receiver is another example of a balanced load. A *balanced transmission line* is usually a two-wire line, such as television antenna ribbon.

An *unbalanced load* is a load that must be connected a certain way; switching its leads will result in improper circuit operation. Some radio antennas are of this type. Usually, unbalanced sources and loads have one side connected to ground. The coaxial input of a television receiver is unbalanced; the shield (braid) of the cable is grounded. An *unbalanced transmission line* is usually a coaxial line, such as you find in a cable television system.

Normally, you cannot connect an unbalanced line to a balanced load, or a balanced line to an unbalanced load, and expect good performance from an electrical or electronic system. But a transformer can allow for mating between these two types of systems.

In Fig. 18-10A, a balanced-to-unbalanced transformer is shown. Note that the balanced side is center-tapped, and the tap is grounded. In Fig. 18-10B, an unbalanced-to-balanced transformer is illustrated. Again, the balanced side has a grounded center tap.

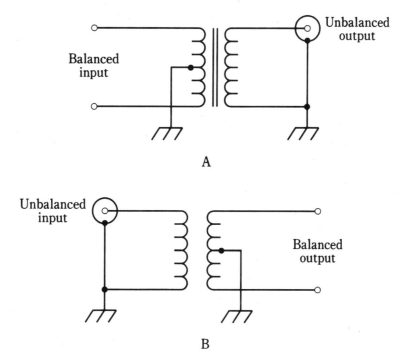

18-10   At A, a balanced-to-unbalanced transformer. At B, an unbalanced-to-balanced transformer.

The turns ratio of a balanced-to-unbalanced ("balun") or unbalanced-to-balanced ("unbal") transformer might be 1:1, but it doesn't have to be. If the impedances of the balanced and unbalanced parts of the systems are the same, then a 1:1 turns ratio is ideal. But if the impedances differ, the turns ratio should be such that the impedances are matched. This is discussed in the section on impedance transfer ratio that follows.

Transformer coupling

Transformers are sometimes used between amplifier stages in electronic equipment where a large *amplification factor* is needed. There are other methods of *coupling* from one amplifier stage to another, but transformers offer some advantages, especially in radio-frequency receivers and transmitters.

Part of the problem in getting a radio to work is that the amplifiers must operate in a stable manner. If there is too much feedback, a series of amplifiers will oscillate, and this will severely degrade the performance of the radio. Transformers that minimize the capacitance between the amplifier stages, while still transferring the desired signals, can help to prevent this oscillation.

Impedance-transfer ratio

One of the most important applications of transformers is in audio-frequency (AF) and radio-frequency (RF) electronic circuits. In these applications, transformers are generally employed to match impedances. Thus, you might hear of an *impedance step-up* transformer, or an *impedance step-down* device.

The *impedance-transfer ratio* of a transformer varies according to the square of the turns ratio, and also according to the square of the voltage-transfer ratio. Recall the formula for voltage-transfer ratio:

$$E_{pri}/E_{sec} = T_{pri}/T_{sec}$$

If the input and output, or source and load, impedances are purely resistive, and are denoted by Z_{pri} (at the primary winding) and Z_{sec} (at the secondary), then:

$$Z_{pri}/Z_{sec} = (T_{pri}/T_{sec})^2$$

and:

$$Z_{pri}/Z_{sec} = (E_{pri}/E_{sec})^2$$

The inverses of these formulas, in which the turns ratio or voltage-transfer ratio are expressed in terms of the impedance-transfer ratio, are:

$$T_{pri}/T_{sec} = (Z_{pri}/Z_{sec})^{1/2}$$

and:

$$E_{pri}/E_{sec} = (Z_{pri}/Z_{sec})^{1/2}$$

The ½ power is the same thing as the square root.

Problem 18-3

A transformer is needed to match an input impedance of 50.0 Ω, purely resistive, to an output impedance of 300 Ω, also purely resistive. What should the ratio T_{pri}/T_{sec} be?

The required transformer will have a step-up impedance ratio of $Z_{pri}/Z_{sec} = 50.0/300$ = 1:6.00. From the above formulas:

$$T_{pri}/T_{sec} = (Z_{pri}/Z_{sec})^{1/2}$$
$$= (1/6.00)^{1/2} = 0.16667^{1/2} = 0.40829$$

A couple of extra digits are included (as they show up on the calculator) to prevent the sort of error introduction you recall from earlier chapters. The decimal value 0.40829 can be changed into ratio notation by taking its reciprocal, and then writing "1:" followed by that reciprocal value:

$$0.40829 = 1:(1/0.40829) = 1:2.4492$$

This can be rounded to three significant figures, or 1:2.45. This is the primary-to-secondary turns ratio for the transformer. The secondary winding has 2.45 times as many turns as the primary winding.

Problem 18-4

A transformer has a primary-to-secondary turns ratio of 4.00:1. The load, connected to the transformer output, is a pure resistance of 37.5 Ω. What is the impedance at the primary?

The impedance-transfer ratio is equal to the square of the turns ratio. Therefore:

$$Z_{pri}/Z_{sec} = (T_{pri}/T_{sec})^2$$
$$= (4.00/1)^2 = 4.00^2 = 16.0$$

This can be written 16.0:1. The input (primary) impedance is 16.0 times the secondary impedance. We know that the secondary impedance, Z_{sec}, is 37.5 Ω. Therefore:

$$Z_{pri} = 16.0 \times Z_{sec}$$
$$= 16.0 \times 37.5 = 600$$

Anything connected to the transformer primary will "see" a purely resistive impedance of 600 Ω.

Radio-frequency transformers

In radio receivers and transmitters, transformers can be categorized generally by the method of construction used. Some have primary and secondary windings, just like utility and audio units. Others employ transmission-line sections. These are the two most common types of transformer found at radio frequencies.

Coil types

In the wound radio-frequency (RF) transformer, powdered-iron cores can be used up to quite high frequencies. Toroidal cores are most common, because they are self-shielding (all of the magnetic flux is confined within the core material). The number of turns depends on the frequency, and also on the permeability of the core.

In high-power applications, air-core coils are sometimes used, because air, although it has a low permeability, also has extremely low hysteresis loss. The disadvantage of air-core coils is that some of the magnetic flux extends outside of the coil. This affects the performance of the transformer when it must be placed in a cramped space, such as in a transmitter final-amplifier compartment.

A major advantage of coil type transformers, especially when they are wound on toroidal cores, is that they can be made to work over a wide band of frequencies, such as from 3.5 MHz to 30 MHz. These are called *broadband transformers*.

Transmission-line types

As you recall, any transmission line has a characteristic impedance, or Z_o, that depends on the line construction. This property is sometimes used to make impedance transformers out of coaxial or parallel-wire line.

Transmission-line transformers are always made from quarter-wave sections. From the previous chapter, remember the formula for the length of a quarter-wave section:

$$L_{ft} = 246v/f_o$$

where L_{ft} is the length of the section in feet, v is the velocity factor expressed as a fraction, and f_o is the frequency of operation in megahertz. If the length L_m is in meters, then:

$$L_m = 75v/f_o$$

In the last chapter, you saw how a short circuit is changed into an open circuit, and vice-versa, by a quarter-wave section of line. What happens to a pure resistive impedance at one end of such a line? What will be "seen" at the opposite end?

Let a quarter-wave section of line, with characteristic impedance Z_o, be terminated in a purely resistive impedance R_{out}. Then the input impedance is also a pure resistance R_{in}, and the following relationship holds:

$$Z_o^{\ 2} = R_{in}R_{out}$$

This is illustrated in Fig. 18-11. This formula can be broken down to solve for R_{in} in terms of R_{out}, or vice-versa:

$$R_{in} = Z_o^{\ 2}/R_{out}$$

and

$$R_{out} = Z_o^{\ 2}/R_{in}$$

These relationships hold at the frequency, f_o, for which the line is ¼ wavelength long. Neglecting line losses, these relationships will also hold at the *odd harmonics* of f_o, that is, at $3f_o$, $5f_o$, $7f_o$ and so on. At other frequencies, the line will not act as a transformer, but instead, will behave in complicated ways that are beyond the scope of this discussion.

Quarter-wave transformers are most often used in antenna systems, especially at the higher frequencies, where their dimensions become practical.

Problem 18-5

An antenna has a purely resistive impedance of 100 Ω. It is connected to a 1/4-wave section of 75-Ω coaxial cable. What will be the impedance at the input end of the section?

Use the formula from above:

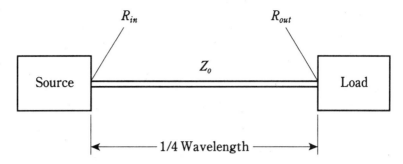

18-11 A quarter-wave matching section of transmission line. Abbreviations are discussed in the text.

$$R_{in} = Z_o^2/R_{out}$$
$$= 75^2/100 = 5625/100$$
$$= 56 \ \Omega$$

Problem 18-6

An antenna is known to have a pure resistance of 600 Ω. You want to match it to 50.0 Ω pure resistance. What is the characteristic impedance needed for a quarter-wave matching section?

Use the formula:

$$Z_o^2 = R_{in}R_{out}$$
$$Z_o^2 = 600 \times 50 = 30,000$$
$$Z_o = (30,000)^{1/2} = 173 \ \Omega$$

The challenge is to find a line that has this particular Z_o. Commercially manufactured lines come in standard Z_o values, and a perfect match might not be obtainable. In that case, the closest obtainable Z_o should be used. In this case, it would probably be 150 Ω.

If nothing is available anywhere near the characteristic impedance needed for a quarter-wave matching section, then a coil-type transformer will probably have to be used instead. A quarter-wave matching section should be made using unbalanced line if the load is unbalanced, and balanced line if the load is balanced.

The major disadvantage of quarter-wave sections is that they work only at specific frequencies. But this is often offset by the ease with which they are constructed, if radio equipment is to be used at only one frequency, or at odd-harmonic frequencies.

What about reactance?

Things are simple when there is no reactance in an ac circuit using transformers. But often, especially in radio-frequency circuits, pure resistance doesn't occur naturally. It has to be obtained by using inductors and/or capacitors to cancel the reactance out.

Reactance makes a perfect match impossible, no matter what the turns ratio or Z_o of the transformer. A small amount of reactance can be tolerated at lower radio frequencies (below about 30 MHz). A near-perfect match becomes more important at higher frequencies.

The behavior of reactance, as it is coupled through transformer windings, is too complicated for a thorough analysis here. But if you're interested in delving into it, there are plenty of good engineering texts that deal with it in all its mathematical glory.

Recall that inductive and capacitive reactances are opposite in effect, and that their magnitudes can vary. If a load presents a complex impedance $R + jX$, with X not equal to zero, it is always possible to cancel out the reactance X by adding an equal and opposite reactance $(-X)$ in the circuit. This can be done by connecting an inductor or capacitor in series with the load.

For radio communications over a wide band, adjustable impedance-matching and reactance-canceling networks can be placed between a transmitter and an antenna system. Such a device is called a *transmatch* and is popular among radio hams, who use frequencies ranging from 1.8 MHz to the microwave spectrum.

Quiz

Refer to the text in this chapter if necessary. A good score is 18 or more correct. Answers are in the back of the book.

1. In a step-up transformer:

    A. The primary impedance is greater than the secondary impedance.

    B. The secondary winding is right on top of the primary.

    C. The primary voltage is less than the secondary voltage.

    D. All of the above.

2. The capacitance between the primary and the secondary windings of a transformer can be minimized by:

    A. Placing the windings on opposite sides of a toroidal core.

    B. Winding the secondary right on top of the primary.

    C. Using the highest possible frequency.

    D. Using a center tap on the balanced winding.

3. A transformer steps a voltage down from 117 V to 6.00 V. What is its primary-to-secondary turns ratio?

    A. 1:380.

    B. 380:1.

    C. 1:19.5.

    D. 19.5:1.

4. A step-up transformer has a primary-to-secondary turns ratio of 1:5.00. If 117 V rms appears at the primary, what is the rms voltage across the secondary?

    A. 23.4 V.

B. 585 V.

C. 117 V.

D. 2.93 kV.

5. A transformer has a secondary-to-primary turns ratio of 0.167. This transformer is:

A. A step-up unit.

B. A step-down unit.

C. Neither step-up nor step-down.

D. A reversible unit.

6. Which of the following is *false*, concerning air cores versus ferromagnetic cores?

A. Air concentrates the magnetic lines of flux.

B. Air works at higher frequencies than ferromagnetics.

C. Ferromagnetics are lossier than air.

D. A ferromagnetic-core unit needs fewer turns of wire than an equivalent air-core unit.

7. Eddy currents cause:

A. An increase in efficiency.

B. An increase in coupling between windings.

C. An increase in core loss.

D. An increase in usable frequency range.

8. A transformer has 117 V rms across its primary and 234 V rms across its secondary. If this unit is reversed, assuming it can be done without damaging the windings, what will be the voltage at the output?

A. 234 V.

B. 468 V.

C. 117 V.

D. 58.5 V.

9. The shell method of transformer winding:

A. Provides maximum coupling.

B. Minimizes capacitance between windings.

C. Withstands more voltage than other winding methods.

D. Has windings far apart but along a common axis.

10. Which of these core types, in general, is best if you need a winding inductance of 1.5 H?

A. Air core.

B. Ferromagntic solenoid core.

C. Ferromagnetic toroid core.

D. Ferromagnetic pot core.

11. An advantage of a toroid core over a solenoid core is:
    A. The toroid works at higher frequencies.
    B. The toroid confines the magnetic flux.
    C. The toroid can work for dc as well as for ac.
    D. It's easier to wind the turns on a toroid.

12. High voltage is used in long-distance power transmission because:
    A. It is easier to regulate than low voltage.
    B. The I^2R losses are lower.
    C. The electromagnetic fields are stronger.
    D. Smaller transformers can be used.

13. In a household circuit, the 234-V power has:
    A. One phase.
    B. Two phases.
    C. Three phases.
    D. Four phases.

14. In a transformer, a center tap would probably be found in:
    A. The primary winding.
    B. The secondary winding.
    C. The unbalanced winding.
    D. The balanced winding.

15. An autotransformer:
    A. Works automatically.
    B. Has a center-tapped secondary.
    C. Has one tapped winding.
    D. Is useful only for impedance matching.

16. A transformer has a primary-to-secondary turns ratio of 2.00:1. The input impedance is 300 Ω resistive. What is the output impedance?
    A. 75 Ω.
    B. 150 Ω.
    C. 600 Ω.
    D. 1200 Ω.

17. A resistive input impedance of 50 Ω must be matched to a resistive output impedance of 450 Ω. The primary-to-secondary turns ratio of the transformer must be:
    A. 9.00:1.
    B. 3.00:1.

C. 1:3.00.

D. 1:9.00.

18. A quarter-wave matching section has a characteristic impedance of 75.0 Ω. The input impedance is 50.0 Ω resistive. What is the resistive output impedance?

A. 150 Ω.

B. 125 Ω.

C. 100 Ω.

D. 113 Ω.

19. A resistive impedance of 75 Ω must be matched to a resistive impedance of 300 Ω. A quarter-wave section would need:

A. Z_o = 188 Ω.

B. Z_o = 150 Ω.

C. Z_o = 225 Ω.

D. Z_o = 375 Ω.

20. If there is reactance at the output of an impedance transformer:

A. The circuit will not work.

B. There will be an impedance mismatch, no matter what the turns ratio of the transformer.

C. A center tap must be used at the secondary.

D. The turns ratio must be changed to obtain a match.

Test: Part two

DO NOT REFER TO THE TEXT WHEN TAKING THIS TEST. A GOOD SCORE IS AT LEAST 37 correct. Answers are in the back of the book. It's best to have a friend check your score the first time, so you won't memorize the answers if you want to take the test again.

1. A series circuit has a resistance of 100 Ω and a capacitive reactance of -200 Ω. The complex impedance is:

    A. $-200 + j100$.

    B. $100 + j200$.

    C. $200 - j100$.

    D. $200 + j100$.

    E. $100 - j200$.

2. Mutual inductance causes the net value of a set of coils to:

    A. Cancel out, resulting in zero inductance.

    B. Be greater than what it would be with no mutual coupling.

    C. Be less than what it would be with no mutual coupling.

    D. Double.

    E. Vary, depending on the extent and phase of mutual coupling.

3. Refer to Fig. TEST 2-1. Wave A is:

    A. Leading wave B by 90 degrees.

    B. Lagging wave B by 90 degrees.

    C. Leading wave B by 180 degrees.

    D. Lagging wave B by 135 degrees.

E. Lagging wave B by 45 degrees.

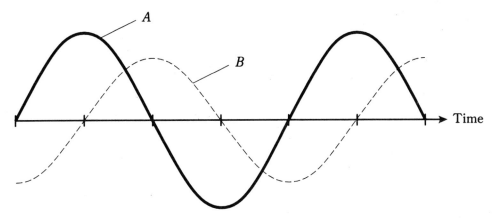

TEST 2-1   Illustration for PART TWO test question 3.

4. A sine wave has a peak value of 30.0 V. Its rms value is:

A. 21.2 V.

B. 30.0 V.

C. 42.4 V.

D. 60.0 V.

E. 90.0 V.

5. Four capacitors are connected in parallel. Their values are 100 pF each. The net capacitance is:

A. 25 pF.

B. 50 pF.

C. 100 pF.

D. 200 pF.

E. 400 pF.

6. A transformer has a primary-to-secondary turns ratio of exactly 8.88:1. The input voltage is 234 V rms. The output voltage is:

A. 2.08 kV rms.

B. 18.5 kV rms.

C. 2.97 V rms.

D. 26.4 V rms.

E. 20.8 V rms.

7. In a series RL circuit, as the resistance becomes small compared with the reactance, the angle of lag approaches:

A. 0 degrees.

   B. 45 degrees.

   C. 90 degrees.

   D. 180 degrees.

   E. 360 degrees.

8. A transmission line carries 3.50 A of ac current and 150 V ac. The *true power* in the line is:

   A. 525 W.

   B. 42.9 W.

   C. 1.84 kW.

   D. Meaningless; true power is dissipated, not transmitted.

   E. Variable, depending on standing wave effects.

9. In a parallel configuration, susceptances:

   A. Simply add up.

   B. Add like capacitances in series.

   C. Add like inductances in parallel.

   D. Must be changed to reactances before you can work with them.

   E. Cancel out.

10. A wave has a frequency of 200 kHz. How many degrees of phase change occur in a microsecond (a millionth of a second)?

   A. 180 degrees.

   B. 144 degrees.

   C. 120 degrees.

   D. 90 degrees.

   E. 72 degrees.

11. At a frequency of 2.55 MHz, a 330-pF capacitor has a reactance of:

   A. $-5.28 \ \Omega$.

   B. $-0.00528 \ \Omega$.

   C. $-189 \ \Omega$.

   D. $-18.9k \ \Omega$.

   E. $-0.000189 \ \Omega$.

12. A transformer has a step-up turns ratio of 1:3.16. The output impedance is 499 Ω purely resistive. The input impedance is:

   A. $50.0 \ \Omega$.

   B. $158 \ \Omega$.

   C. $1.58k \ \Omega$.

   D. $4.98k \ \Omega$.

   E. Not determinable from the data given.

13. A complex impedance is represented by $34 - j23$. The absolute-value impedance is:

    A. 34 Ω.

    B. 11 Ω.

    C. −23 Ω.

    D. 41 Ω.

    E. 57 Ω.

14. A coil has an inductance of 750 uH. The inductive reactance at 100 kHz is:

    A. 75.0 Ω.

    B. 75.0k Ω.

    C. 471 Ω.

    D. 47.1k Ω.

    E. 212 Ω.

15. Two waves are 180 degrees out of phase. This is a difference of:

    A. 1/8 cycle.

    B. 1/4 cycle.

    C. 1/2 cycle.

    D. A full cycle.

    E. Two full cycles.

16. If R denotes resistance and Z denotes absolute-value impedance, then R/Z is the:

    A. True power.

    B. Imaginary power.

    C. Apparent power.

    D. Absolute-value power.

    E. Power factor.

17. Two complex impedances are in series. One is $30 + j50$ and the other is $50 - j30$. The net impedance is:

    A. $80 + j80$.

    B. $20 + j20$.

    C. $20 - j20$.

    D. $-20 + j20$.

    E. $80 + j20$.

18. Two inductors, having values of 140 uH and 1.50 mH, are connected in series. The net inductance is:

    A. 141.5 uH.

    B. 1.64 uH.

    C. 0.1415 mH.

    D. 1.64 mH.

    E. 0.164 mH.

19. Which of the following types of capacitor is polarized?

    A. Mica.

    B. Paper.

    C. Electrolytic.

    D. Air variable.

    E. Ceramic.

20. A toroidal-core coil:

    A. Has lower inductance than an air-core coil with the same number of turns.

    B. Is essentially self-shielding.

    C. Works well as a loopstick antenna.

    D. Is ideal as a transmission-line transformer.

    E. Cannot be used at frequencies below about 10 MHz.

21. The efficiency of a generator:

    A. Depends on the driving power source.

    B. Is equal to output power divided by driving power.

    C. Depends on the nature of the load.

    D. Is equal to driving voltage divided by output voltage.

    E. Is equal to driving current divided by output current.

22. Admittance is:

    A. The reciprocal of reactance.

    B. The reciprocal of resistance.

    C. A measure of the opposition a circuit offers to ac.

    D. A measure of the ease with which a circuit passes ac.

    E. Another expression for absolute-value impedance.

23. The absolute-value impedance Z of a parallel RLC circuit, where R is the resistance and X is the net reactance, is found according to the formula:

    A. $Z = R + X$.

    B. $Z^2 = R^2 + X^2$.

    C. $Z^2 = RX/(R^2 + X^2)$.

    D. $Z = 1/(R^2 + X^2)$.

    E. $Z = R^2X^2/(R + X)$.

24. Complex numbers are used to represent impedance because:

    A. Reactance cannot store power.

B. Reactance isn't a real physical thing.

C. They provide a way to represent what happens in resistance-reactance circuits.

D. Engineers like to work with sophisticated mathematics.

E. No! Complex numbers aren't used to represent impedance.

25. Which of the following does *not* affect the capacitance of a capacitor?

A. The mutual surface area of the plates.

B. The dielectric constant of the material between the plates (within reason).

C. The spacing between the plates (within reason).

D. The amount of overlap between plates.

E. The frequency (within reason).

26. The zero-degree phase point in an ac sine wave is usually considered to be the instant at which the amplitude is:

A. Zero and negative-going.

B. At its negative peak.

C. Zero and positive-going.

D. At its positive peak.

E. Any value; it doesn't matter.

27. The inductance of a coil can be continuously varied by:

A. Varying the frequency.

B. Varying the net core permeability.

C. Varying the current in the coil.

D. Varying the wavelength.

E. Varying the voltage across the coil.

28. Power factor is defined as the ratio of:

A. True power to VA power.

B. True power to imaginary power.

C. Imaginary power to VA power.

D. Imaginary power to true power.

E. VA power to true power.

29. A 50-Ω feed line needs to be matched to an antenna with a purely resistive impedance of 200 Ω. A quarter-wave matching section should have:

A. $Z_o = 150$ Ω.

B. $Z_o = 250$ Ω.

C. $Z_o = 125$ Ω.

D. $Z_o = 133$ Ω.

E. $Z_o = 100$ Ω.

30. The vector $40 + j30$ represents:

    A. 40 Ω resistance and 30 uH inductance.

    B. 40 uH inductance and 30 Ω resistance.

    C. 40 Ω resistance and 30 Ω inductive reactance.

    D. 40 Ω inductive reactance and 30 Ω resistance.

    E. 40 uH inductive reactance and 30 Ω resistance.

31. In a series RC circuit where $R = 300$ Ω and $X_C = -30$ Ω:

    A. The current leads the voltage by a few degrees.

    B. The current leads the voltage by almost 90 degrees.

    C. The voltage leads the current by a few degrees.

    D. The voltage leads the current by almost 90 degrees.

    E. The voltage leads the current by 90 degrees.

32. In a step-down transformer:

    A. The primary voltage is greater than the secondary voltage.

    B. The primary impedance is less than the secondary impedance.

    C. The secondary voltage is greater than the primary voltage.

    D. The output frequency is higher than the input frequency.

    E. The output frequency is lower than the input frequency.

33. A capacitor of 470 pF is in parallel with an inductor of 4.44 uH. What is the resonant frequency?

    A. 3.49 MHz.

    B. 3.49 kHz.

    C. 13.0 MHz.

    D. 13.0 GHz.

    E. Not determinable from the data given.

34. A sine wave contains energy at:

    A. Just one frequency.

    B. A frequency and its even harmonics.

    C. A frequency and its odd harmonics.

    D. A frequency and all its harmonics.

    E. A frequency and its second harmonic only.

35. Inductive susceptance is:

    A. The reciprocal of inductance.

    B. Negative imaginary.

    C. Equal to capacitive reactance.

    D. The reciprocal of capacitive susceptance.

    E. A measure of the opposition a coil offers to ac.

36. The rate of change (derivative) of a sine wave is itself a wave, that:

    A. Is in phase with the original wave.

    B. Is 180 degrees out of phase with the original wave.

    C. Leads the original wave by 45 degrees of phase.

    D. Lags the original wave by 90 degrees of phase.

    E. Leads the original wave by 90 degrees of phase.

37. True power is equal to:

    A. VA power plus imaginary power.

    B. Imaginary power minus VA power.

    C. Vector difference of VA and reactive power.

    D. VA power; the two are the same thing.

    E. 0.707 times the VA power.

38. Three capacitors are connected in series. Their values are 47 uF, 68 uF and 100 uF. The total capacitance is:

    A. 215 uF.

    B. Between 68 uF and 100 uF.

    C. Between 47 uF and 68 uF.

    D. 22 uF.

    E. Not determinable from the data given.

39. The reactance of a section of transmission line depends on all of the following *except*:

    A. The velocity factor of the line.

    B. The length of the section.

    C. The current in the line.

    D. The frequency.

    E. The wavelength.

40. When confronted with a parallel RLC circuit and you need to find the complex impedance:

    A. Just add the resistance and reactance to get $R + jX$.

    B. Find the net conductance and susceptance, then convert to resistance and reactance, and add these to get $R + jX$.

    C. Find the net conductance and susceptance, and just add these together to get $R + jX$.

    D. Rearrange the components so they're in series, and find the complex impedance of that circuit.

    E. Subtract reactance from resistance to get $R - jX$.

41. The illustration in Fig. 2-2 of Test 2 shows a vector $R + jX$ representing:

    A. $X_C = 60 \ \Omega$ and $R = 25 \ \Omega$.

B. $X_L = 60\ \Omega$ and $R = 25\ \Omega$.

C. $X_L = 60$ uH and $R = 25\ \Omega$.

D. $C = 60$ uF and $R = 25\ \Omega$.

E. $L = 60$ uH and $R = 25\ \Omega$.

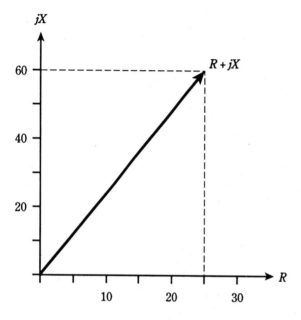

TEST 2-2   Illustration for PART TWO test question 41.

42. If two sine waves have the same frequency and the same amplitude, but they cancel out, the phase difference is:

A. 45 degrees.

B. 90 degrees.

C. 180 degrees.

D. 270 degrees.

E. 360 degrees.

43. A series circuit has a resistance of 50 Ω and a capacitive reactance of $-37\ \Omega$. The phase angle is:

A. 37 degrees.

B. 53 degrees.

C. -37 degrees.

D. -53 degrees.

E. Not determinable from the data given.

44. A 200-Ω resistor is in series with a coil and capacitor; $X_L = 200\ \Omega$ and $X_C = -100\ \Omega$. The complex impedance is:

A. $200 - j100$.

    B. $200 - j200$.

    C. $200 + j100$.

    D. $200 + j200$.

    E. Not determinable from the data given.

45. The characteristic impedance of a transmission line:

    A. Is negative imaginary.

    B. Is positive imaginary.

    C. Depends on the frequency.

    D. Depends on the construction of the line.

    E. Depends on the length of the line.

46. The period of a wave is 2×10^{-8} second. The frequency is:

    A. 2×10^8 Hz.

    B. 20 MHz.

    C. 50 kHz.

    D. 50 MHz.

    E. 500 MHz.

47. A series circuit has a resistance of 600 Ω and a capacitance of 220 pF. The phase angle is:

    A. -20 degrees.

    B. 20 degrees.

    C. -70 degrees.

    D. 70 degrees.

    E. Not determinable from the data given.

48. A capacitor with a negative temperature coefficient:

    A. Works less well as the temperature increases.

    B. Works better as the temperature increases.

    C. Heats up as its value is made larger.

    D. Cools down as its value is made larger.

    E. Has increasing capacitance as temperature goes down.

49. Three coils are connected in parallel. Each has an inductance of 300 uH. There is no mutual inductance. The net inductance is:

    A. 100 uH.

    B. 300 uH.

    C. 900 uH.

    D. 17.3 uH.

    E. 173 uH.

50. An inductor shows 100 Ω of reactance at 30.0 MHz. What is its inductance?

    A. 0.531 uH.

    B. 18.8 mH.

    C. 531 uH.

    D. 18.8 uH.

    E. It can't be found from the data given.

3 PART

Basic electronics

19
CHAPTER

Introduction to semiconductors

SINCE THE SIXTIES, WHEN THE TRANSISTOR BECAME COMMON IN CONSUMER devices, *semiconductors* have acquired a dominating role in electronics. This chapter explains what semiconducting materials actually are.

You've learned about electrical conductors, which pass current easily, and about insulators, which block the current flow. A semiconductor can sometimes act like a conductor, and at other times like an insulator in the same circuit.

The term *semiconductor* arises from the ability of these materials to conduct "part time." Their versatility lies in the fact that the conductivity can be controlled to produce effects such as amplification, rectification, oscillation, signal mixing, and switching.

The semiconductor revolution

It wasn't too long ago that vacuum tubes were the backbone of electronic equipment. Even in radio receivers and "portable" television sets, all of the amplifiers, oscillators, detectors, and other circuits required these devices. A typical vacuum tube ranged from the size of your thumb to the size of your fist.

A radio might sit on a table in the living room; if you wanted to listen to it, you would turn it on and wait for the tube filaments to warm up. I can remember this. It makes me feel like an old man to think about it.

Vacuum tubes, sometimes called "tubes" or "valves" (in England), are still used in some power amplifiers, microwave oscillators, and video display units. There are a few places where tubes work better than semiconductor devices. Tubes tolerate momentary voltage and current surges better than semiconductors. They are discussed in chapter 29.

Tubes need rather high voltages to work. Even in radio receivers, turntables, and other consumer devices, 100 V to 200 V dc was required when tubes were employed. This mandated bulky power supplies and created an electrical shock hazard.

Nowadays, a transistor about the size of a pencil eraser can perform the functions of a tube in most situations. Often, the power supply can be a couple of AA cells or a 9-V "transistor battery."

Figure 19-1 is a size comparison drawing between a typical transistor and a typical vacuum tube.

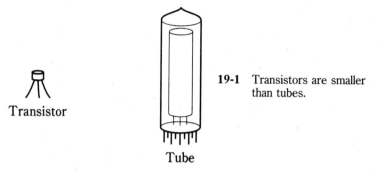

Transistor

Tube

19-1   Transistors are smaller than tubes.

Integrated circuits, hardly larger than individual transistors, can do the work of hundreds or even thousands of vacuum tubes. An excellent example of this technology is found in the personal computer, or PC. In 1950, a "PC" would have occupied a large building, required thousands of watts to operate, and probably cost well over a million dollars. Today you can buy one and carry it in a briefcase. Integrated-circuit technology is discussed in chapter 28.

Semiconductor materials

There are numerous different mixtures of elements that work as semiconductors. The two most common materials are *silicon* and a compound of gallium and arsenic known as *gallium arsenide* (often abbreviated GaAs).

In the early years of semiconductor technology, germanium formed the basis for many semiconductors; today it is seen occasionally, but not often. Other substances that work as semiconductors are selenium, cadmium compounds, indium compounds, and various metal oxides.

Many of the elements found in semiconductors can be mined from the earth. Others are "grown" as crystals under laboratory conditions.

Silicon

Silicon (chemical symbol Si) is widely used in diodes, transistors and integrated circuits. Generally, other substances, or *impurities*, must be added to silicon to give it the desired properties. The best quality silicon is obtained by growing crystals in a laboratory. The silicon is then fabricated into *wafers* or *chips*.

Gallium arsenide

Another common semiconductor is the compound gallium arsenide. Engineers and technicians call this material by its acronym-like chemical symbol, GaAs, pronounced "gas." If you hear about "gasfets" and "gas ICs," you're hearing about gallium-arsenide technology.

Gallium arsenide works better than silicon in several ways. It needs less voltage, and will function at higher frequencies because the charge carriers move faster. GaAs devices are relatively immune to the effects of ionizing radiation such as X rays and gamma rays. GaAs is used in light-emitting diodes, infrared-emitting diodes, laser diodes, visible-light and infrared detectors, ultra-high-frequency amplifying devices, and a variety of integrated circuits.

The primary disadvantage of GaAs is that it is more expensive to produce than silicon.

Selenium

Selenium has resistance that varies depending on the intensity of light that falls on it. All semiconductor materials exhibit this property, known as *photoconductivity*, to a greater or lesser degree, but selenium is especially affected. For this reason, selenium is useful for making photocells.

Selenium is also used in certain types of *rectifiers*. This is a device that converts ac to dc; you'll learn about rectification in chapters 20 and 21. The main advantage of selenium over silicon is that selenium can withstand brief *transients*, or surges of abnormally high voltage.

Germanium

Pure germanium is a poor electrical conductor. It becomes a semiconductor when impurities are added. Germanium was used extensively in the early years of semiconductor technology. Some diodes and transistors still use it.

A germanium diode has a low voltage drop (0.3 V, compared with 0.6 V for silicon and 1 V for selenium) when it conducts, and this makes it useful in some situations. But germanium is easily destroyed by heat. Extreme care must be used when soldering the leads of a germanium component.

Metal oxides

Certain metal oxides have properties that make them useful in the manufacture of semiconductor devices. When you hear about MOS (pronounced "moss") or CMOS (pronounced "seamoss") technology, you are hearing about *metal-oxide semiconductor* and *complementary metal-oxide semiconductor* devices, respectively.

One advantage of MOS and CMOS devices is that they need almost no power to function. They draw so little current that a battery in a MOS or CMOS device lasts just about as long as it would on the shelf. Another advantage is high speed. This allows operation at high frequencies, and makes it possible to perform many calculations per second.

Certain types of transistors, and many kinds of integrated circuits, make use of this technology. In integrated circuits, MOS and CMOS allows for a large number of discrete diodes and transistors on a single chip. Engineers would say that MOS/CMOS has *high component density*.

The biggest problem with MOS and CMOS is that the devices are easily damaged by static electricity. Care must be used when handling components of this type.

Doping

For a semiconductor material to have the properties needed to work in electronic components, impurities are usually added. The impurities cause the material to conduct currents in certain ways. The addition of an impurity to a semiconductor is called *doping*. Sometimes the impurity is called a *dopant*.

Donor impurities

When an impurity contains an excess of electrons, the dopant is called a *donor impurity*. Adding such a substance causes conduction mainly by means of electron flow, as in a metal like copper. The excess electrons are passed from atom to atom when a voltage exists across the material. Elements that serve as donor impurities include antimony, arsenic, bismuth and phosphorus.

A material with a donor impurity is called an *N type* semiconductor, because electrons have negative charge.

Acceptor impurities

If an impurity has a deficiency of electrons, the dopant is called an *acceptor impurity*. When a substance such as aluminum, boron, gallium, or indium is added to a semiconductor, the material conducts by means of *hole flow*. A *hole* is a missing electron; it is described in more detail shortly.

A material with an acceptor impurity is called a *P-type* semiconductor, because holes have positive charge.

Majority and minority charge carriers

Charge carriers in semiconductor materials are either electrons, which have a unit negative charge, or holes, having a unit positive charge. In any semiconductor material, some of the current is in the form of electrons passed from atom to atom in a negative-to-positive direction. Some current occurs as holes that move from atom to atom in a positive-to-negative direction.

Sometimes electrons dominate the current flow in a semiconductor; this is the case if the material has donor impurities. In substances having acceptor impurities, holes dominate. The dominating charge carriers (either electrons or holes) are the *majority carriers*. The less abundant ones are the *minority carriers*.

The ratio of majority to minority carriers can vary, depending on the nature of the semiconducting material.

Electron flow

In an N-type semiconductor, most of the current flows as electrons passed from atom to atom. But some of the current in a P-type material also takes this form. You learned about electron flow all the way back in chapter 1. It would be a good idea to turn back for a moment and review this material, because it will help you understand the concept of *hole flow*.

Hole flow

In a P-type semiconductor, most of the current flows in a way that some people find peculiar and esoteric. In a literal sense, in virtually all electronic devices, charge transfer is always the result of electron movement, no matter what the medium might be. The exceptions are particle accelerators and cloud chambers—apparatus of interest mainly to theoretical physicists.

The flow of current in a P-type material is better imagined as a flow of *electron absences*, not electrons. The behavior of P-type substances can be explained more easily this way. The absences, called "holes," move in a direction opposite that of the electrons.

Imagine a sold-out baseball stadium. Suppose that 19 of every 20 people in the stadium are randomly issued candles. Imagine it's nighttime, and the field lights are switched off. You stand at the center of the field, just behind second base. The candles are lit, and the people pass them around the stands counterclockwise. It's hard to follow the movement of individual candles, because they are so numerous. But there are dark spots: people without candles. Because they are few, they stand out. They move clockwise around the stands, against the candle movement. Of course the physical image you see is produced by candle light. But the motion you notice is that of *candle absences*.

Figure 19-2 illustrates this phenomenon. Small dots represent candles or electrons. Imagine them moving from right to left in the figure as they are passed from person to person or from atom to atom. Circles represent candle absences or holes. They "move" from left to right, contrary to the flow of the candles or the electrons, because the candles or electrons are being *passed among stationary units* (people or atoms).

This is just the way holes flow in a semiconductor material.

Behavior of a P-N junction

Simply having a semiconducting material, either P or N type, might be interesting, and a good object of science experiments. But when the two types of material are brought together, the *P-N junction* develops properties that make the semiconductor materials truly useful as electronic devices.

Figure 19-3 shows the schematic symbol for a *semiconductor diode,* formed by joining a piece of P-type material to a piece of N-type material. The N-type semiconductor is represented by the short, straight line in the symbol, and is called the *cathode.* The P-type semiconductor is represented by the arrow, and is called the *anode.*

In the diode as shown in the figure, electrons flow in the direction opposite the arrow. (Physicists consider current to flow from positive to negative, and this is in the same direction as the arrow points.) But current cannot, under most conditions, flow the other way. Electrons normally do not flow in the direction that the arrow points.

If you connect a battery and a resistor in series with the diode, you'll get a current flow if the negative terminal of the battery is connected to the cathode and the positive

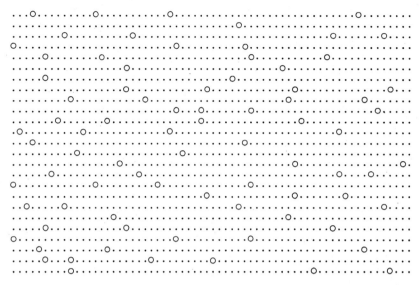

19-2 Pictorial representation of hole flow. Small dots represent electrons, moving from right to left; open circles represent holes, moving from left to right.

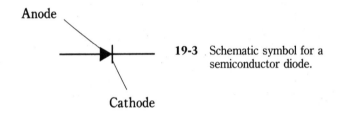

19-3 Schematic symbol for a semiconductor diode.

terminal is connected to the anode (Fig. 19-4A). No current will flow if the battery is reversed (Fig. 19-4B). The resistor is included in the circuit to prevent destruction of the diode by excessive current.

It takes a certain minimum voltage for conduction to occur. This is called the *forward breakover voltage* of the junction. Depending on the type of material, it varies from about 0.3 V to 1 V. If the voltage across the junction is not at least as great as the forward breakover value, the diode will not conduct. This effect can be of use in amplitude limiters, waveform clippers and threshold detectors.

You'll learn about the various ways diodes are used in the next chapter.

How the junction works

When the N-type material is negative with respect to the P-type, as in Fig. 19-4A, electrons flow easily from N to P. The N-type semiconductor, which already has an excess of electrons, gets even more; the P-type semiconductor, with a shortage of

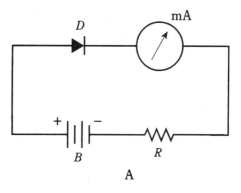

19-4 Series connection of battery B, resistor R, milliammeter mA and diode D. At A, forward bias results in current flow; at B, reverse bias results in no current.

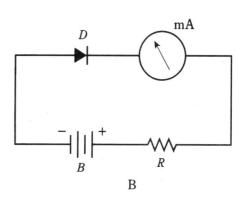

electrons, is made even more deficient. The N-type material constantly feeds electrons to the P-type in an attempt to create an electron balance, and the battery or power supply keeps robbing electrons from the P-type material. This is shown in Fig. 19-5A, and is known as *forward bias*.

When the polarity is switched so the N-type material is positive with respect to the P type, things get interesting. This is called *reverse bias*. Electrons in the N-type material

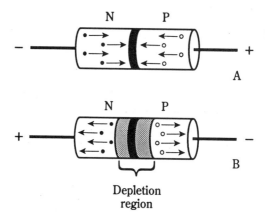

19-5 At A, forward bias of a P-N junction; at B, reverse bias. Electrons are shown as small dots, and holes are shown as open circles.

are pulled towards the positive charge, away from the junction. In the P-type material, holes are pulled toward the negative charge, also away from the junction. The electrons (in the N-type material) and holes (in the P type) are the majority charge carriers. They become depleted in the vicinity of the P-N junction (Fig. 19-5B). A shortage of majority carriers means that the semiconductor material cannot conduct well. Thus, the *depletion region* acts like an insulator.

Junction capacitance

Some P-N junctions can alternate between conduction (in forward bias) and nonconduction (in reverse bias) millions or billions of times per second. Other junctions are slower. The main limiting factor is the capacitance at the P-N junction during conditions of reverse bias. The amount of capacitance depends on several factors, including the operating voltage, the type of semiconductor material, and the cross-sectional area of the P-N junction.

By examining Fig. 19-5B, you should notice that the depletion region, sandwiched between two semiconducting sections, resembles the dielectric of a capacitor. In fact, the similarity is such that a reverse-biased P-N junction really is a capacitor. Some semiconductor components are made with this property specifically in mind.

The *junction capacitance* can be varied by changing the reverse-bias voltage, because this voltage affects the width of the depletion region. The greater the reverse voltage, the wider the depletion region gets, and the smaller the capacitance becomes.

In the next chapter, you'll learn how engineers take advantage of this effect.

Avalanche effect

The greater the reverse bias voltage, the "more determined an insulator" a P-N junction gets—to a point. If the reverse bias goes past this critical value, the voltage overcomes the ability of the junction to prevent the flow of current, and the junction conducts as if it were forward biased. This *avalanche effect* does not ruin the junction (unless the voltage is extreme); it's a temporary thing. When the voltage drops back below the critical value, the junction behaves normally again.

Some components are designed to take advantage of the avalanche effect. In other cases, avalanche effect limits the performance of a circuit.

In a device designed for voltage regulation, called a *Zener diode*, you'll hear about the *avalanche voltage* or *Zener voltage* specification. This might range from a couple of volts to well over 100 V. It's important in the design of voltage- regulating circuits in solid-state power supplies; this is discussed in the next chapter.

For *rectifier diodes* in power supplies, you'll hear about the *peak inverse voltage (PIV)* or *peak reverse voltage (PRV)* specification. It's important that rectifier diodes have PIV great enough so that avalanche effect will not occur (or even come close to happening) during any part of the ac cycle. Otherwise, the circuit efficiency will be compromised.

Quiz

Refer to the text in this chapter if necessary. A good score is at least 18 correct. Answers are in the back of the book.

1. The term "semiconductor" arises from:
    A. Resistor-like properties of metal oxides.
    B. Variable conductive properties of some materials.
    C. The fact that there's nothing better to call silicon.
    D. Insulating properties of silicon and GaAs.

2. Which of the following is *not* an advantage of semiconductor devices over vacuum tubes?
    A. Smaller size.
    B. Lower working voltage.
    C. Lighter weight.
    D. Ability to withstand high voltages.

3. The most common semiconductor among the following substances is:
    A. Germanium.
    B. Galena.
    C. Silicon.
    D. Copper.

4. GaAs is a(n):
    A. Compound.
    B. Element.
    C. Conductor.
    D. Gas.

5. A disadvantage of gallium-arsenide devices is that:
    A. The charge carriers move fast.
    B. The material does not react to ionizing radiation.
    C. It is expensive to produce.
    D. It must be used at high frequencies.

6. Selenium works especially well in:
    A. Photocells.
    B. High-frequency detectors.
    C. Radio-frequency power amplifiers.
    D. Voltage regulators.

7. Of the following, which material allows the lowest forward voltage drop in a diode?

    A. Selenium.

    B. Silicon.

    C. Copper.

    D. Germanium.

8. A CMOS integrated circuit:

    A. Can only work at low frequencies.

    B. Is susceptible to damage by static.

    C. Requires considerable power to function.

    D. Needs very high voltage.

9. The purpose of doping is to:

    A. Make the charge carriers move faster.

    B. Cause holes to flow.

    C. Give a semiconductor material certain properties.

    D. Protect devices from damage in case of transients.

10. A semiconductor material is made into N type by:

    A. Adding an acceptor impurity.

    B. Adding a donor impurity.

    C. Injecting electrons.

    D. Taking electrons away.

11. Which of the following *does not* result from adding an acceptor impurity?

    A. The material becomes P type.

    B. Current flows mainly in the form of holes.

    C. Most of the carriers have positive electric charge.

    D. The substance has an electron surplus.

12. In a P-type material, electrons are:

    A. Majority carriers.

    B. Minority carriers.

    C. Positively charged.

    D. Entirely absent.

13. Holes flow from:

    A. Minus to plus.

    B. Plus to minus.

    C. P-type to N-type material.

    D. N-type to P-type material.

14. When a P-N junction does not conduct, it is:

    A. Reverse biased.

    B. Forward biased.

    C. Biased past the breakover voltage.

    D. In a state of avalanche effect.

15. Holes flow the opposite way from electrons because:

    A. Charge carriers flow continuously.

    B. Charge carriers are passed from atom to atom.

    C. They have the same polarity.

    D. No! Holes flow in the same direction as electrons.

16. If an electron has a charge of -1 unit, a hole has:

    A. A charge of -1 unit.

    B. No charge.

    C. A charge of $+1$ unit.

    D. A charge that depends on the semiconductor type.

17. When a P-N junction is reverse-biased, the capacitance depends on all of the following *except*:

    A. The frequency.

    B. The width of the depletion region.

    C. The cross-sectional area of the junction.

    D. The type of semiconductor material.

18. If the reverse bias exceeds the avalanche voltage in a P-N junction:

    A. The junction will be destroyed.

    B. The junction will insulate; no current will flow.

    C. The junction will conduct current.

    D. The capacitance will become extremely high.

19. Avalanche voltage is routinely exceeded when a P-N junction acts as a:

    A. Current rectifier.

    B. Variable resistor.

    C. Variable capacitor.

    D. Voltage regulator.

20. An *unimportant* factor concerning the frequency at which a P-N junction will work effectively is:

    A. The type of semiconductor material.

    B. The cross-sectional area of the junction.

    C. The reverse current.

    D. The capacitance with reverse bias.

20
CHAPTER

Some uses of diodes

THE TERM *DIODE* MEANS "TWO ELEMENTS." IN THE EARLY YEARS OF ELEC-
ronics and radio, most diodes were vacuum tubes. The *cathode* element emitted
electrons, and the *anode* picked up electrons. Thus, current would flow as electrons
through the tube from the cathode to the anode, but not the other way.

Tubes had *filaments* to drive electrons from their cathodes. The filaments were
heated via a low ac voltage, but the cathodes and anodes usually wielded hundreds or even
thousands of dc volts.

Today, you'll still hear about diodes, anodes, and cathodes. But rather than large,
heavy, hot, high-voltage tubes, diodes are tiny things made from silicon or other
semiconducting materials. Some diodes can handle voltages nearly as great as their tube
counterparts. Semiconductor diodes can do just about everything that tube diodes could,
plus a few things that people in the tube era probably never imagined.

Rectification

The hallmark of a *rectifier diode* is that it passes current in only one direction. This makes
it useful for changing ac to dc. Generally speaking, when the cathode is negative with
respect to the anode, current flows; when the cathode is positive relative to the anode,
there is no current. The constraints on this behavior are the forward breakover and
avalanche voltages, as you learned about in the last chapter.

Suppose a 60-Hz ac sine wave is applied to the input of the circuit in Fig. 20-1A.
During half the cycle, the diode conducts, and during the other half, it doesn't. This cuts
off half of every cycle. Depending on which way the diode is hooked up, either the positive
half or the negative half of the ac cycle will be removed. Figure 20-1B shows the output of
the circuit at A. Remember that electrons flow from negative to positive, against the
arrow in the diode symbol.

The circuit and wave diagram of Fig. 20-1 show a *half-wave rectifier* circuit. This is

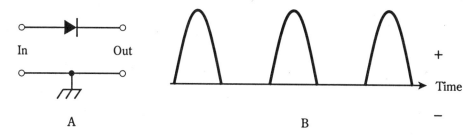

20-1   At A, half-wave rectifier. At B, output of the circuit of A with sine-wave ac input.

the simplest possible rectifier. That's its chief advantage over other, more complicated rectifier circuits. You'll learn about the different types of rectifier diodes and circuits in the next chapter.

Detection

One of the earliest diodes, existing even before vacuum tubes, was a semiconductor. Known as a *cat whisker*, this semiconductor consisted of a fine piece of wire in contact with a small piece of the mineral *galena*. This bizarre-looking thing had the ability to act as a rectifier for small radio-frequency (RF) currents. When the cat whisker was connected in a circuit like that of Fig. 20-2, the result was a receiver capable of picking up amplitude-modulated (AM) radio signals.

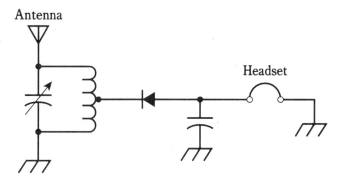

20-2   Schematic diagram of a "crystal set" radio receiver.

A cat whisker was a finicky thing. Engineers had to adjust the position of the fine wire to find the best point of contact with the galena. A tweezers and magnifying glass were invaluable in this process. A steady hand was essential.

The galena, sometimes called a "crystal," gave rise to the nickname *crystal set* for this low-sensitivity radio. You can still build a crystal set today, using a simple RF diode, a coil, a tuning capacitor, a headset and a long-wire antenna. Notice that there's no battery! The audio is provided by the received signal alone.

The diode in Fig. 20-2 acts to recover the audio from the radio signal. This is called *detection*; the circuit is a *detector*. If the detector is to be effective, the diode must be of the right type. It should have low capacitance, so that it works as a rectifier at radio frequencies, passing current in one direction but not in the other. Some modern RF diodes are actually microscopic versions of the old cat whisker, enclosed in a glass case with axial leads. You have probably seen these in electronics hobby stores.

Details about detector circuits are discussed in chapter 27. Some detectors use diodes; others do not. Modulation methods are examined in chapter 26.

Frequency multiplication

When current passes through a diode, half of the cycle is cut off, as shown in Fig. 20-1. This occurs no matter what the frequency, from 60-Hz utility current through RF, as long as the diode capacitance is not too great.

The output wave from the diode looks much different than the input wave. This condition is known as *nonlinearity*. Whenever there is nonlinearity of any kind in a circuit—that is, whenever the output waveform is shaped differently from the input waveform—there will be harmonic frequencies in the output. These are waves at integer multiples of the input frequency. (If you've forgotten what harmonics are, refer back to chapter 9.)

Often, nonlinearity is undesirable. Then engineers strive to make the circuit *linear*, so that the output waveform has exactly the same shape as the input waveform. But sometimes a circuit is needed that will produce harmonics. Then nonlinearity is introduced deliberately. Diodes are ideal for this.

A simple frequency-multiplier circuit is shown in Fig. 20-3. The output LC circuit is tuned to the desired nth harmonic frequency, nf_o, rather than to the input or fundamental frequency, f_o.

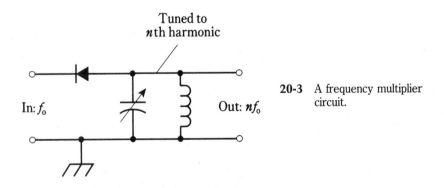

20-3   A frequency multiplier circuit.

For a diode to work as a frequency multiplier, it must be of a type that would also work well as a detector at the same frequencies. This means that the component should act like a rectifier, but not like a capacitor.

Mixing

When two waves having different frequencies are combined in a nonlinear circuit, new frequencies are produced. These new waves are at the sum and difference frequencies of the original waves. You've probably noticed this *mixing*, also called *heterodyning*, if you've ever heard two loud, sine wave tones at the same time.

Suppose there are two signals with frequencies f_1 and f_2. For mathematical convenience, assign f_2 to the wave with the higher frequency. If these signals are combined in a nonlinear circuit, new waves will result. One of them will have a frequency $f_2 - f_1$, and the other will be at $f_2 + f_1$. These are known as *beat frequencies*. The signals are called *mixing products* (Fig. 20-4).

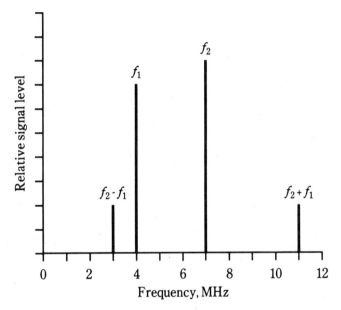

20-4   Spectral (frequency-domain) illustration of mixing. Frequency designators are discussed in the text.

Figure 20-4 is a *frequency domain* graph. Amplitude (on the vertical scale) is shown as a function of frequency (on the horizontal scale). This kind of display is what engineers see when they look at the screen of a *spectrum analyzer*. Most of the graphs you've seen so far have been *time domain* graphs, in which things are shown as a function of time. The screen of an oscilloscope normally shows things in the time domain.

How do you get the nonlinearity necessary to obtain a *mixer* circuit? There are various different schemes, but one common way is—you guessed it—to use diodes. Mixer circuits are discussed in chapter 27.

Switching

The ability of diodes to conduct with forward bias, and to insulate with reverse bias, makes them useful for switching in some electronic applications. Diodes can switch at extremely high rates, much faster than any mechanical device.

One type of diode, made for use as an RF switch, has a special semiconductor layer sandwiched in between the P-type and N-type material. This layer, called an *intrinsic semiconductor*, reduces the capacitance of the diode, so that it can work at higher frequencies than an ordinary diode. The intrinsic material is sometimes called *I type*. A diode with I-type semiconductor is called a *PIN diode* (Fig. 20-5).

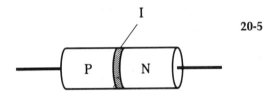

20-5 The PIN diode has a layer of intrinsic (I type) semiconductor at the P-N junction.

Direct-current bias, applied to one or more PIN diodes, allows RF currents to be effectively channeled without using complicated relays and cables. A PIN diode also makes a good RF detector, especially at frequencies above 30 MHz.

Voltage regulation

Most diodes have avalanche breakdown voltages much higher than the reverse bias ever gets. The value of the avalanche voltage depends on how a diode is manufactured. *Zener diodes* are made to have well-defined, constant avalanche voltages.

Suppose a certain Zener diode has an avalanche voltage, also called the *Zener voltage*, of 50 V. If a reverse bias is applied to the P-N junction, the diode acts as an open circuit below 50 V. When the voltage reaches 50 V, the diode starts to conduct. The more the reverse bias tries to increase, the more current flows through the P-N junction. This effectively prevents the reverse voltage from exceeding 50 V.

The current through a Zener diode, as a function of the voltage, is shown in Fig. 20-6. The Zener voltage is indicated by the abrupt rise in reverse current as the reverse bias increases. A typical Zener-diode voltage-limiting circuit is shown in Fig. 20-7.

There are other ways to get voltage regulation besides the use of Zener diodes, but Zener diodes often provide the simplest and least expensive alternative. Zener diodes are available with a wide variety of voltage and power-handling ratings. Power supplies for solid-state equipment commonly employ Zener diode regulators.

Details about power supply design are coming up in chapter 21.

Amplitude limiting

The forward breakover voltage of a germanium diode is about 0.3 V; for a silicon diode it is about 0.6 V. In the last chapter, you learned that a diode will not conduct until the forward bias voltage is at least as great as the forward breakover voltage. The "flip side"

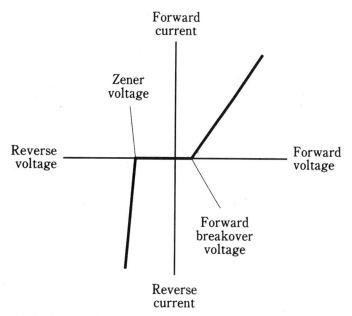

20-6 Current through a Zener diode, as a function of the bias voltage.

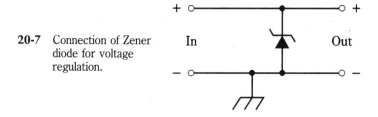

20-7 Connection of Zener diode for voltage regulation.

is that the diode will always conduct when the forward bias exceeds the breakover value. In this case, the voltage across the diode will be constant: 0.3 V for germanium and 0.6 V for silicon.

This property can be used to advantage when it is necessary to limit the amplitude of a signal, as shown in Fig. 20-8. By connecting two identical diodes back-to-back in parallel with the signal path (A), the maximum peak amplitude is limited, or *clipped,* to the forward breakover voltage of the diodes. The input and output waveforms of a clipped signal are illustrated at B. This scheme is sometimes used in radio receivers to prevent "blasting" when a strong signal comes in.

The downside of the *diode limiter* circuit is that it introduces distortion when limiting is taking place. This might not be a problem for reception of Morse code, or for signals that rarely reach the limiting voltage. But for voice signals with amplitude peaks that rise well past the limiting voltage, it can seriously degrade the audio quality, perhaps even rendering the words indecipherable.

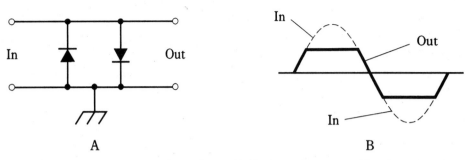

20-8   At A, two diodes can work as a limiter. At B, the peaks are cut off by the action of the diodes.

Frequency control

When a diode is reverse-biased, there is a region at the P-N junction with dielectric properties. As you know from the last chapter, this is called the depletion region, because it has a shortage of majority charge carriers. The width of this zone depends on several things, including the reverse voltage.

As long as the reverse bias is less than the avalanche voltage, the width of the depletion region can be changed by varying the bias. This results in a change in the capacitance of the junction. The capacitance, which is always quite small (on the order of picofarads), varies inversely with the square root of the reverse bias.

Some diodes are manufactured especially for use as variable capacitors. These are *varactor diodes*. Sometimes you'll hear them called *varicaps*. They are made from silicon or gallium arsenide.

A common use for a varactor diode is in a circuit called a *voltage-controlled oscillator (VCO)*. A voltage-tuned circuit, using a coil and a varactor, is shown in Fig. 20-9. This is a parallel-tuned circuit. The fixed capacitor, whose value is large compared with that of the varactor, serves to keep the coil from short-circuiting the control voltage across the varactor. Notice that the symbol for the varactor has two lines on the cathode side. This is its "signature," so that you know that it's a varactor, and not just an ordinary diode.

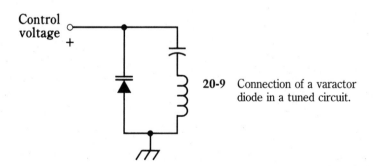

20-9   Connection of a varactor diode in a tuned circuit.

Oscillation and amplification

Under certain conditions, diodes can be made to produce microwave radio signals. There are three types of diodes that do this: *Gunn diodes, IMPATT diodes,* and *tunnel diodes.*

Gunn diodes

A Gunn diode can produce up to 1 W of RF power output, but more commonly it works at levels of about 0.1 W. Gunn diodes are usually made from gallium arsenide.

A Gunn diode oscillates because of the *Gunn effect,* named after J. Gunn of International Business Machines (IBM) who observed it in the sixties. A Gunn diode doesn't work anything like a rectifier, detector, or mixer; instead, the oscillation takes place as a result of a quirk called *negative resistance.*

Gunn-diode oscillators are often tuned using varactor diodes. A Gunn-diode oscillator, connected directly to a microwave horn antenna, is known as a *Gunnplexer.* These devices are popular with amateur-radio experimenters at frequencies of 10 GHz and above.

IMPATT diodes

The acroym *IMPATT* comes from the words *imp*act *a*valanche *t*ransit *t*ime. This, like negative resistance, is a phenomenon the details of which are rather esoteric. An *IMPATT diode* is a microwave oscillating device like a Gunn diode, except that it uses silicon rather than gallium arsenide.

An IMPATT diode can be used as an amplifier for a microwave transmitter that employs a Gunn-diode oscillator. As an oscillator, an IMPATT diode produces about the same amount of output power, at comparable frequencies, as the Gunn diode.

Tunnel diodes

Another type of diode that will oscillate at microwave frequencies is the *tunnel diode,* also known as the *Esaki diode.* It produces only a very small amount of power, but it can be used as a local oscillator in a microwave radio receiver.

Tunnel diodes work well as amplifiers in microwave receivers, because they generate very little unwanted noise. This is especially true of gallium arsenide devices.

The behavior of Gunn, IMPATT and tunnel diodes is a sophisticated topic, and is beyond the scope of this book. College-level electrical-engineering texts are good sources of information on this subject. You will want to know about how these devices work if you plan to become a microwave engineer.

Energy emission

Some semiconductor diodes emit radiant energy when a current passes through the P-N junction in a forward direction. This phenomenon occurs as electrons fall from higher to lower energy states within atoms.

LEDs and IREDs

Depending on the exact mixture of semiconductors used in manufacture, visible light of almost any color can be produced. Infrared-emitting devices also exist. The most common color for a *light-emitting diode (LED)* is bright red. An *infrared-emitting diode (IRED)* produces wavelengths too long to see.

The intensity of the light or infrared from an LED or IRED depends to some extent on the forward current. As the current rises, the brightness increases up to a certain point. If the current continues to rise, no further increase in brilliance takes place. The LED or IRED is then said to be in a state of *saturation*.

Digital displays

Because LEDs can be made in various different shapes and sizes, they are ideal for use in digital displays. You've probably seen digital clock radios that use them. They are common in car radios. They make good indicators for "on/off," "a.m./p.m.," "battery low," and other conditions.

In recent years, LED displays have been largely replaced by *liquid-crystal displays (LCDs)*. This technology has advantages over LEDs, including much lower power consumption and better visibility in direct sunlight.

Communications

Both LEDs and IREDs are useful in communications because their intensity can be modulated to carry information. When the current through the device is sufficient to produce output, but not enough to cause saturation, the LED or IRED output will follow along with rapid current changes. Voices, music, and digital signals can be conveyed over light beams in this way. Some modern telephone systems make use of modulated light, transmitted through clear fibers. This is known as *fiberoptic* technology.

Special LEDs and IREDs produce *coherent* radiation; these are called *laser diodes*. The rays from these diodes aren't the intense, parallel beams that you probably imagine when you think about lasers. A laser LED or IRED generates a cone-shaped beam of low intensity. But it can be focused, and the resulting rays have some of the same advantages found in larger lasers.

Photosensitive diodes

Virtually all P-N junctions exhibit characteristics that change when electromagnetic rays strike them. The reason that conventional diodes are not affected by these rays is that most diodes are enclosed in opaque packages.

Some photosensitive diodes have variable resistance that depends on light intensity. Others actually generate dc voltages in the presence of electromagnetic radiation.

Silicon photodiodes

A silicon diode, housed in a transparent case and constructed in such a way that visible light can strike the barrier between the P-type and N-type materials, forms a *photodiode*.

A reverse bias is applied to the device. When light falls on the junction, current flows. The current is proportional to the intensity of the light, within certain limits.

Silicon photodiodes are more sensitive at some wavelengths than at others. The greatest sensitivity is in the *near infrared* part of the spectrum, at wavelengths a little longer than visible red light.

When light of variable brightness falls on the P-N junction of a reverse-biased silicon photodiode, the output current follows the light-intensity variations. This makes silicon photodiodes useful for receiving modulated-light signals, of the kind used in fiberoptic systems.

The optoisolator

An LED or IRED and a photodiode can be combined in a single package to get a component called an *optoisolator*. This device (Fig. 20-10) actually creates a modulated-light signal and sends it over a small, clear gap to a receptor. The LED or IRED converts an electrical signal to visible light or infrared; the photodiode changes the visible light or infrared back into an electrical signal.

20-10 An optoisolator uses an LED or IRED (input) and a photodiode (output).

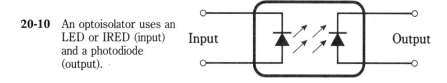

A major source of headache for engineers has always been the fact that, when a signal is electrically coupled from one circuit to another, the impedances of the two stages interact. This can lead to nonlinearity, unwanted oscillation, loss of efficiency, or other problems. Optoisolators overcome this effect, because the coupling is not done electrically. If the input impedance of the second circuit changes, the impedance that the first circuit "sees" will remain unaffected, being simply the impedance of the LED or IRED.

Photovoltaic cells

A silicon diode, with no bias voltage applied, will generate dc all by itself if enough electromagnetic radiation hits its P-N junction. This is known as the *photovoltaic effect*. It is the principle by which solar cells work.

Photovoltaic cells are made to have the greatest possible P-N junction surface area. This maximizes the amount of light that falls on the junction. A single silicon photovoltaic cell can produce about 0.6 V of dc electricity. The amount of current that it can deliver depends on the surface area of the junction. For every square inch of P-N surface area, a silicon photovoltaic cell can produce about 160 mA in direct sunlight.

Photovoltaic cells are often connected in series-parallel combinations to provide power for solid-state electronic devices like portable radios. A large assembly of solar cells is called a *solar panel*.

Solar-cell technology has advanced rapidly in the last several years. Solar power is expensive to produce, but the cost is going down—and solar cells do not pollute.

Quiz

Refer to the text in this chapter if necessary. A good score is at least 18 correct. Answers are in the back of the book.

1. When a diode is forward-biased, the anode:
    A. Is negative relative to the cathode.
    B. Is positive relative to the cathode.
    C. Is at the same voltage as the cathode.
    D. Alternates between positive and negative relative to the cathode.

2. If ac is applied to a diode, and the peak ac voltage never exceeds the avalanche voltage, then the output is:
    A. Ac with half the frequency of the input.
    B. Ac with the same frequency as the input.
    C. Ac with twice the frequency of the input.
    D. None of the above.

3. A crystal set:
    A. Can be used to transmit radio signals.
    B. Requires a battery with long life.
    C. Requires no battery.
    D. Is useful for rectifying 60-Hz ac.

4. A diode detector:
    A. Is used in power supplies.
    B. Is employed in some radio receivers.
    C. Is used commonly in high-power radio transmitters.
    D. Changes dc into ac.

5. If the output wave in a circuit has the same shape as the input wave, then:
    A. The circuit is linear.
    B. The circuit is said to be detecting.
    C. The circuit is a mixer.
    D. The circuit is a rectifier.

6. The two input frequencies of a mixer circuit are 3.522 MHz and 3.977 MHz. Which of the following frequencies might be used at the output?
    A. 455 kHz.
    B. 886 kHz.
    C. 14.00 MHz.
    D. 1.129 MHz.

7. A time-domain display might be found in:

    A. An ammeter.

    B. A spectrum analyzer.

    C. A digital voltmeter.

    D. An oscilloscope.

8. Zener voltage is also known as:

    A. Forward breakover voltage.

    B. Peak forward voltage.

    C. Avalanche voltage.

    D. Reverse bias.

9. The forward breakover voltage of a silicon diode is:

    A. About 0.3 V.

    B. About 0.6 V.

    C. About 1.0 V.

    D. Dependent on the method of manufacture.

10. A diode audio limiter circuit:

    A. Is useful for voltage regulation.

    B. Always uses Zener diodes.

    C. Rectifies the audio to reduce distortion.

    D. Can cause objectionable signal distortion.

11. The capacitance of a varactor varies with:

    A. Forward voltage.

    B. Reverse voltage.

    C. Avalanche voltage.

    D. Forward breakover voltage.

12. The purpose of the I layer in a PIN diode is to:

    A. Minimize the diode capacitance.

    B. Optimize the avalanche voltage.

    C. Reduce the forward breakover voltage.

    D. Increase the current through the diode.

13. Which of these diode types might be found in the oscillator circuit of a microwave radio transmitter?

    A. A rectifier diode.

    B. A cat whisker.

    C. An IMPATT diode.

    D. None of the above.

14. A Gunnplexer can be used as a:
    A. Communications device.
    B. Radio detector.
    C. Rectifier.
    D. Signal mixer.

15. The most likely place you would find an LED would be:
    A. In a rectifier circuit.
    B. In a mixer circuit.
    C. In a digital frequency display.
    D. In an oscillator circuit.

16. Coherent radiation is produced by a:
    A. Gunn diode.
    B. Varactor diode.
    C. Rectifier diode.
    D. Laser diode.

17. You want a circuit to be stable with a variety of amplifier impedance conditions. You might consider a coupler using:
    A. A Gunn diode.
    B. An optoisolator.
    C. A photovoltaic cell.
    D. A laser diode.

18. The power from a solar panel depends on all of the following *except*:
    A. The operating frequency of the panel.
    B. The total surface area of the panel.
    C. The number of cells in the panel.
    D. The intensity of the light.

19. Emission of energy in an IRED is caused by:
    A. High-frequency radio waves.
    B. Rectification.
    C. Electron energy-level changes.
    D. None of the above.

20. A photodiode, when not used as a photovoltaic cell, has:
    A. Reverse bias.
    B. No bias.
    C. Forward bias.
    D. Negative resistance.

21
CHAPTER

Power supplies

MOST ELECTRONIC EQUIPMENT NEEDS DIRECT CURRENT (DC) TO WORK. BATTER-
ies produce dc, but there is a limit to how much energy and how much voltage a battery
can provide. The same is true of solar panels.

The electricity from the utility company is alternating current (ac) with a frequency
of 60 Hz. In your house, most wall outlets carry an effective voltage of 117 V; some have
234 V. The energy from a wall outlet is practically unlimited, but it must be converted
from ac to dc, and tailored to just the right voltage, to be suitable for electronic equipment.

Parts of a power supply

A *power supply* provides the proper voltage and current for electronic apparatus. Most
power supplies consist of several stages, always in the same order (Fig. 21-1).

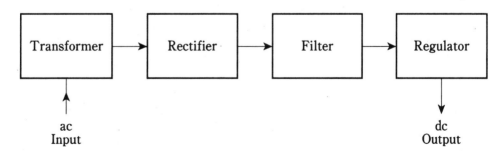

21-1   Block diagram of a power supply. Sometimes a regulator is not needed.

First, the ac encounters a transformer that steps the voltage either down or up,
depending on the exact needs of the electronic circuits.

Second, the ac is rectified, so that it becomes *pulsating dc* with a frequency of either 60 Hz or 120 Hz. This is almost always done by one or more semiconductor diodes.

Third, the pulsating dc is *filtered*, or smoothed out, so that it becomes a continuous voltage having either positive or negative polarity with respect to ground.

Finally, the dc voltage might need to be *regulated*. Some equipment is finicky, insisting on just the right amount of voltage all the time. Other devices can put up with some voltage changes.

Power supplies that provide more than a few volts must have features that protect the user (that's you!) from receiving a dangerous electrical shock. All power supplies need fuses and/or circuit breakers to minimize the fire hazard in case the equipment shorts out.

The power transformer

Power transformers can be categorized as step-down or step-up. As you remember, the output, or secondary, voltage of a step-down unit is lower than the input, or primary, voltage. The reverse is true for a step-up transformer.

Step-down

Most solid-state electronic devices, such as radios, need only a few volts. The power supplies for such equipment use step-down power transformers. The physical size of the transformer depends on the current.

Some devices need only a small current and a low voltage. The transformer in a radio receiver, for example, can be quite small physically. A ham radio transmitter or hi-fi amplifier needs much more current. This means that the secondary winding of the transformer must be of heavy-gauge wire, and the core must be bulky to contain the magnetic flux. Such a transformer is massive.

Step-up

Some circuits need high voltage. The picture tube in a TV set needs several hundred volts. Some ham radio power amplifiers use vacuum tubes working at kilovolts dc. The transformers in these appliances are step-up types. They are moderate to large in size because of the number of turns in the secondary, and also because high voltages can spark, or *arc*, between wire turns if the windings are too tight.

If a step-up transformer needs to supply only a small amount of current, it need not be big. But for ham radio transmitters and radio/TV broadcast amplifiers, the transformers are large and heavy—and expensive.

Transformer ratings

Transformers are rated according to output voltage and current. For a given unit, the *volt-ampere (VA)* capacity is often specified. This is the product of the voltage and current. A transformer with a 12-V output, capable of delivering 10 A, would have 12 V $\times$ 10 A = 120 VA of capacity.

The nature of power-supply filtering, to be discussed a bit later in this chapter, makes it necessary for the power-transformer VA rating to be greater than just the wattage needed by the load.

A high-quality, rugged power transformer, capable of providing the necessary currents and/or voltages, is crucial in any power supply. The transformer is usually the most expensive component to replace. When designing a power supply, it's wise to spend a little extra to get a reliable transformer. Engineers might call this "maintenance insurance."

The diode

Rectifier diodes are available in various sizes, intended for different purposes. Most rectifier diodes are made of silicon and are therefore known as *silicon rectifiers*. A few are fabricated from selenium, and are called *selenium rectifiers*.

Two important features of a power-supply diode are the *average forward current (I_o)* rating and the *peak inverse voltage (PIV)* rating. There are other specifications that engineers need to know when designing a specialized power supply, but in this course, you only need to be concerned about I_o and PIV.

Average forward current

Electric current produces heat. If the current through a diode is too great, the heat will destroy the P-N junction.

Generally speaking, when designing a power supply, it's wise to use diodes with an I_o rating of at least 1.5 times the expected average dc forward current. If this current is 4.0 A, the rectifier diodes should be rated at $I_o = 6.0$ A or more. Of course, it would be wasteful of money to use a 100-A diode in a circuit where the average forward current is 4.0 A. While it would work, it would be a bit like shooting a sparrow with a cannon.

Note that I_o flows through the *diodes*. The current drawn by the load is often quite different from this. Also, note that I_o is an *average* figure. The *instantaneous* forward current is another thing, and can be 15 or even 20 times I_o, depending on the nature of the power-supply filtering circuitry.

Some diodes have *heatsinks* to help carry heat away from the P-N junction. A selenium diode can be recognized by the appearance of its heatsink (Fig. 21-2).

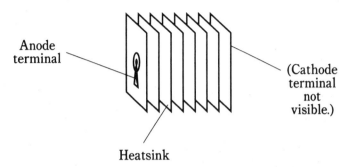

21-2 A selenium rectifier can be recognized by its heatsink.

Diodes can be connected in parallel to increase the current rating. When this is done, small-value resistors are placed in series with each diode in the set to equalize the current burden among the diodes (Fig. 21-3). Each resistor should have a voltage drop of about 1 V.

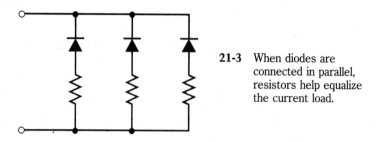

21-3 When diodes are connected in parallel, resistors help equalize the current load.

Peak inverse voltage

The PIV rating of a diode is the instantaneous inverse, or reverse-bias, voltage that it can withstand without avalanche taking place. A good power supply has diodes whose PIV ratings are significantly greater than the peak voltage of the ac at the input.

If the PIV rating is not great enough, the diode or diodes in a supply will conduct for part of the reverse cycle. This will degrade the efficiency of the supply; the reverse current will "buck" the forward current. It would be like having a team of rowers in a long boat, with one or two rowers trying to propel the boat backwards instead of forwards.

Diodes can be connected in series to get a higher PIV capacity than a single diode alone. This scheme is sometimes seen in high-voltage supplies, such as those needed for tube-type ham radio power amplifiers. High-value resistors, of about 500 Ω for each peak-inverse volt, are placed across each diode in the set to distribute the reverse bias equally among the diodes (Fig. 21-4). Also, each diode is shunted by a capacitor of 0.005 uF or 0.1 uF.

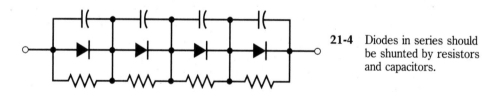

21-4 Diodes in series should be shunted by resistors and capacitors.

The half-wave rectifier

The simplest rectifier circuit uses just one diode (or a series or parallel combination) to "chop off" half of the ac input cycle. You saw this circuit in the previous chapter, diagrammed in Fig. 20-1.

In a half-wave circuit, the average output voltage is approximately 45 percent of the rms ac input voltage. But the PIV across the diode can be as much as 2.8 times the rms ac input voltage. It's a good idea to use diodes whose PIV ratings are at least 1.5 times the maximum expected PIV; therefore, with a half-wave supply, the diodes should be rated for at least 4.2 times the rms ac input voltage.

Half-wave rectification has some shortcomings. First, the output is hard to smooth out, because the waveform is so irregular. Second, the voltage output tends to drop when the supply is connected to a load. (This can be overcome to some extent by means of a good voltage regulator. Voltage regulation is discussed later in this chapter.) Third, half-wave rectification puts a disproportionate strain on the power transformer and the diodes.

Half-wave rectification is useful in supplies that don't have to deliver much current, or that don't need to be especially well regulated. The main advantage of using a half-wave circuit in these situations is that it costs a little less than full-wave or bridge circuits.

The full-wave, center-tap rectifier

A much better scheme for changing ac to dc is to use both halves of the ac cycle. Suppose you want to convert an ac wave to dc with positive polarity. Then you can allow the positive half of the ac cycle to pass unchanged, and flip the negative portion of the wave upside-down, making it positive instead. This is the principle behind *full-wave rectification*.

One common full-wave circuit uses a transformer with a center-tapped secondary, as shown in Fig. 21-5A. The center tap, a wire coming out of the exact middle of the secondary winding, is connected to common ground. This produces out-of-phase waves at the ends of the winding. These two waves can be individually half-wave rectified, cutting off the negative half of the cycle. Because the waves are 180 degrees (half a cycle) out of phase, the output of the circuit has positive pulses for both halves of the cycle (Fig. 21-5B).

In this rectifier circuit, the average dc output voltage is about 90 percent of the rms ac input voltage. The PIV across the diodes can be as much as 2.8 times the rms input voltage. Therefore, the diodes should have a PIV rating of at least 4.2 times the rms ac input.

Compare Fig. 21-5B with Fig. 20-1B from the last chapter. Can you see that the waveform of the full-wave rectifier ought to be easier to smooth out? In addition to this advantage, the *full-wave, center-tap rectifier* is kinder to the transformer and diodes than a half-wave circuit. Furthermore, if a load is applied to the output of the full-wave circuit, the voltage will drop much less than it would with a half-wave supply, because the output has more "substance."

The bridge rectifier

Another way to get full-wave rectification is the *bridge rectifier*. It is diagrammed in Fig. 21-6. The output waveform is just like that of the full-wave, center-tap circuit.

The average dc output voltage in the bridge circuit is 90 percent of the rms ac input voltage, just as is the case with center-tap rectification. The PIV across the diodes is 1.4 times the rms ac input voltage. Therefore, each diode needs to have a PIV rating of at least 2.1 times the rms ac input voltage.

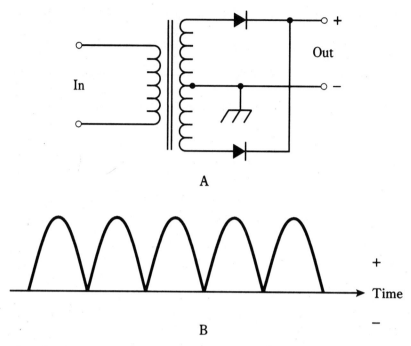

21-5   At A, schematic diagram of a full-wave, center-tap rectifier. At B, output
waveform from this rectifier.

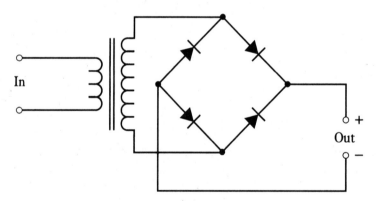

21-6   Schematic diagram of a full-wave bridge rectifier.

The bridge circuit does not need a center-tapped transformer secondary. This is its
main practical advantage. Electrically, the bridge circuit uses the entire secondary on
both halves of the wave cycle; the center-tap circuit uses one side of the secondary for
one half of the cycle, and the other side for the other half of the cycle. For this reason, the
bridge circuit makes more efficient use of the transformer.

The main disadvantage of the bridge circuit is that it needs four diodes rather than
two. This doesn't always amount to much in terms of cost, but it can be important when

a power supply must deliver a high current. Then, the extra diodes—two for each half of the cycle, rather than one—dissipate more overall heat energy. When current is used up as heat, it can't go to the load. Therefore, center-tap circuits are preferable in high-current applications.

The voltage doubler

By using diodes and capacitors connected in certain ways, a power supply can be made to deliver a multiple of the peak ac input voltage. Theoretically, large whole-number multiples are possible. But you won't often see power supplies that make use of multiplication factors larger than 2.

In practice, *voltage multipliers* are practical only when the load draws low current. Otherwise, the regulation is poor; the output voltage changes considerably with changes in the load resistance. This bugaboo gets worse and worse as the multiplication factor increases. This is why engineers don't attempt to make, say, a factor-of-16 voltage multiplier. For a good high-voltage power supply, the best approach is to use a step-up transformer, not a voltage multiplier.

A *voltage-doubler* circuit is shown in Fig. 21-7. This circuit works on the whole ac input wave cycle, and is therefore called a *full-wave voltage doubler*. Its dc output voltage, when the current drawn is low, is about twice the *peak* ac input voltage, or about 2.8 times the rms ac input voltage.

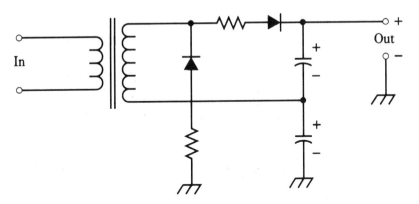

21-7   A full-wave voltage doubler.

Notice the capacitors in this circuit. The operation of any voltage multiplier is dependent on the ability of these capacitors to hold a charge, even when a load is connected to the output of the supply. Thus, the capacitors must have large values. If the intent is to get a high dc voltage from the supply, massive capacitors will be necessary.

Also, notice the resistors in series with the diodes. These have low values, similar to those needed when diodes are connected in parallel. When the supply is switched on, the capacitors draw a huge initial charging current. Without the resistors, it would be necessary to use diodes with astronomical I_o ratings. Otherwise the *surge current* would burn them out.

This circuit subjects the diodes to a PIV of 2.8 times the rms ac input voltage. Therefore, they should be rated for PIV of at least 4.2 times the rms ac input voltage.

In this circuit, each capacitor charges to the peak ac input voltage when there is no load (the output current is zero). As the load draws current, the capacitors will have trouble staying charged to the peak ac input voltage. This isn't much of a problem as long as the load is light, that is, if the current is low. But for heavy loads, the output voltage will drop, and it will not be smooth dc.

The major difference between the voltage doubler and the supplies discussed previously, besides the increased output voltage, is the fact that the dc output is filtered. The capacitors serve two purposes: to boost the voltage and to filter the output. Additional filtering might be wanted to smooth out the dc still more, but the circuit of Fig. 21-7 is a complete, if crude, power supply all by itself.

The filter

Electronic equipment doesn't like the pulsating dc that comes straight from a rectifier. The *ripple* in the waveform must be smoothed out, so that pure, battery-like dc is supplied. The *filter* does this.

Capacitors alone

The simplest filter is one or more large-value capacitors, connected in parallel with the rectifier output (Fig. 21-8). Electrolytic capacitors are almost always used. They are *polarized*; they must be hooked up in the right direction. Typical values range in the hundreds or thousands of microfarads.

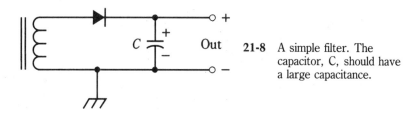

21-8 A simple filter. The capacitor, C, should have a large capacitance.

The more current drawn, the more capacitance is needed for good filtering. This is because the load resistance decreases as the current increases. The lower the load resistance, the faster the filter capacitors will discharge. Larger capacitances hold charge for a longer time with a given load.

Filter capacitors work by "trying" to keep the dc voltage at its peak level (Fig. 21-9). This is easier to do with the output of a full-wave rectifier (shown at A) as compared with a half-wave circuit (at B). The remaining waveform bumps are the ripple. With a half-wave rectifier, this ripple has the same frequency as the ac, or 60 Hz. With a full-wave supply, the ripple is 120 Hz. The capacitor gets recharged twice as often with a full-wave rectifier, as compared with a half-wave rectifier. This is why the ripple is less severe, for a given capacitance, with full-wave circuits.

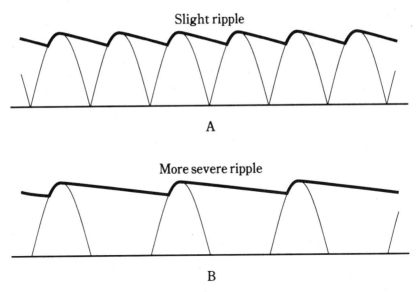

21-9 Filtered output for full-wave rectification (A) and half-wave rectification (B).

Capacitors and chokes

Another way to smooth out the dc from a rectifier is to use an extremely large inductance in series with the output. This is always done in conjunction with parallel capacitance. The inductance, called a *filter choke*, is on the order of several henrys. If the coil must carry a lot of current, it will be physically bulky.

Sometimes the capacitor is placed ahead of the choke. This circuit is a *capacitor-input filter* (Fig. 21-10A). If the coil comes ahead of the capacitor, the circuit is a *choke-input filter* (Fig. 21-10B).

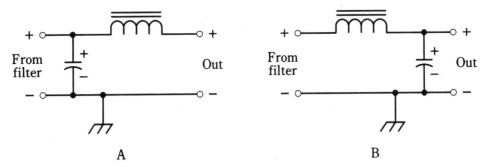

21-10 Capacitor-input (A) and choke-input (B) filtering.

Engineers might use capacitor-input filtering when the load is not expected to be very great. The output voltage is higher with a capacitor-input circuit than with a choke-input circuit. If the supply needs to deliver large or variable amounts of current, a choke-input filter is a better choice, because the output voltage is more stable.

If a supply must have a minimum of ripple, two or three capacitor/choke pairs might be *cascaded*, or connected one after the other (Fig. 21-11). Each pair is called a *section*. Multisection filters can consist of either capacitor-input or choke- input sections, but the two types are never mixed.

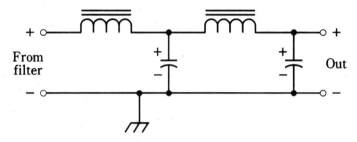

21-11   Two choke-input filter sections in cascade.

Voltage regulation

A full-wave rectifier, followed by a choke-input filter, offers fairly stable voltage under varying load conditions. But *voltage regulator* circuitry is needed for electronic devices that are finicky about the voltage they get.

Zener diodes

You learned about Zener diodes in the last chapter. If a reverse-biased Zener diode is connected across the output of a power supply, as shown back in Fig. 20-7, the diode will limit the output voltage of the supply by "brute force" as long as it has a high enough power rating.

Zener/transistor regulation

A Zener-diode voltage regulator is not very efficient if the load is heavy. When a supply must deliver high current, a power transistor is used along with the Zener diode to obtain regulation (Fig. 21-12). This greatly reduces the strain on the Zener diode, so that a lower-power (and therefore less costly) diode can be used.

Integrated circuits

In recent years, voltage regulators have become available in *integrated-circuit (IC)* form. You just connect the IC, perhaps along with some external components, at the output of the filter. This method provides the best possible regulation at low and moderate voltages. Even if the output current changes from zero to maximum, the output voltage stays exactly the same, for all practical purposes.

Regulator tubes

Occasionally, you'll find a power supply that uses a gas-filled *tube*, rather than solid-state components, to obtain regulation. The tube acts something like a very-high-power Zener

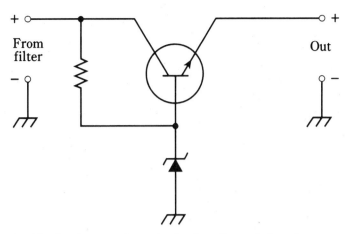

21-12 A voltage regulator circuit using a Zener diode and a power transistor.

diode. The voltage drop across a gaseous tube, designed for voltage regulation, is nearly constant. Tubes are available for regulation at moderately high voltages.

Surge current

At the instant a power supply is switched on, a sudden current surge occurs, even with no load at the output. This is because the filter capacitor(s) need an initial charge, and they draw a lot of current for a short time. The surge current is far greater than the operating current. This can destroy the rectifier diodes. The phenomenon is worst in high-voltage supplies and voltage-multiplier circuits. Diode failure can be prevented in at least four different ways.

The first method uses "brute force." You can simply use diodes with a current rating of many times the operating level. The main disadvantage is cost. High-voltage, high-current diodes can get expensive.

A second method involves connecting several units in parallel wherever a diode is called for in the circuit. This is actually a variation on the first method. The overall cost might be less. Current-equalizing resistors are necessary.

A third scheme for surge protection is to apply the input voltage little by little. A variable transformer, called a *Variac*, is useful for this. You start at zero input and turn a knob to get up to the full voltage. This can completely get rid of the current surge.

A fourth way to limit the current surge is to use an automatic switching circuit in the transformer primary. This applies a reduced ac voltage for a second or two, and then switches in the full input voltage.

Which of these methods is best? It depends on the overall cost, the operating convenience, and the whim of the design engineer.

Transient suppression

The ac on the utility line is a sine wave with a constant rms voltage near 117 V. But there are "spikes," known as *transients*, lasting microseconds or milliseconds, that attain peak values of several hundred or even several thousand volts.

Transients are caused by sudden changes in the load in a utility circuit. Lightning can also produce them. Unless they are suppressed, they can destroy the diodes in a power supply. Transients can also befuddle the operation of sensitive equipment like personal computers.

The simplest way to get rid of most transients is to place a capacitor of about 0.01 uF, rated for 600 V or more, across the transformer primary (Fig. 21-13). Commercially made *transient suppressors* are also available.

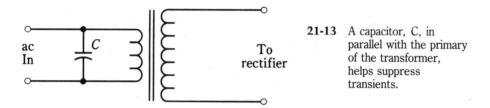

21-13   A capacitor, C, in parallel with the primary of the transformer, helps suppress transients.

In the event of a thunderstorm locally, the best way to protect equipment is to unplug it from the wall outlet. This is inconvenient, of course. But if you have a personal computer, hi-fi set, or other electronic appliance that you like a lot, it's not a bad idea.

Fuses and breakers

A *fuse* is a piece of soft wire that melts, breaking a circuit if the current exceeds a certain level. Fuses are placed in series with the transformer primary (Fig. 21-14). Any component failure, short circuit, or overload that might cause catastrophic damage (or fire!) will burn the fuse out. Fuses are easy to replace, although it's aggravating if a fuse blows and you don't have replacements on hand.

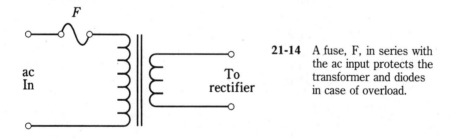

21-14   A fuse, F, in series with the ac input protects the transformer and diodes in case of overload.

If a fuse blows, it *must* be replaced with another of the same rating. If the replacement fuse is rated too low in current, it will probably blow out right away, or soon after it has been installed. If the replacement fuse is rated too high in current, it might not protect the equipment.

Fuses are available in two types: *quick-break* and *slow-blow*. You can usually recognize a slow-blow fuse by the spring inside. A quick-break fuse has only a wire or foil strip. When replacing a fuse, use the right kind. Quick-break fuses in slow-blow situations might burn out needlessly; slow-blow units in quick-break environments won't provide the proper protection.

Circuit breakers do the same thing as fuses, except that a breaker can be reset by turning off the power supply, waiting a moment, and then pressing a button or flipping a switch. Some breakers reset automatically when the equipment has been shut off for a certain length of time.

If a fuse or breaker keeps blowing out often, or if it blows immediately after you've replaced or reset it, then something is wrong with the supply or with the equipment connected to it.

Personal safety

Power supplies can be dangerous. This is especially true of high-voltage circuits, but anything over 12 V should be treated as potentially lethal.

A power supply is not necessarily safe after it has been switched off. Filter capacitors can hold the charge for a long time. In high-voltage supplies of good design, *bleeder resistors* of a high ohmic value (Fig. 21-15) are connected across each filter capacitor, so that the capacitors will discharge in a few minutes after the supply is turned off. But *don't* bet your life on components that might not be there, and that *can and do* sometimes fail.

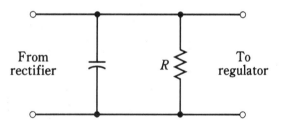

21-15 A bleeder resistor, R, allows filter capacitors to discharge when a supply is shut off.

Most manufacturers supply safety instructions and precautions with equipment carrying hazardous voltages. But *don't* assume something is safe just because dangers aren't mentioned in the instructions.

Warning　If you have any doubt about your ability to safely work with a power supply, then leave it to a professional.

In this chapter, you've had a look at power supplies from a general standpoint. Whole books have been written on the subject of power-supply engineering. If you want to design or build a power supply, you should refer to a college-level text or, better yet, a professional power-supply design manual.

Quiz

Refer to the text in this chapter if necessary. A good score is at least 18 correct. Answers are in the back of the book.

1. The output of a rectifier is:
    A. 60-Hz ac.
    B. Smooth dc.
    C. Pulsating dc.
    D. 120-Hz ac.

2. Which of the following might not be needed in a power supply?
    A. The transformer.
    B. The filter.
    C. The rectifier.
    D. All of the above are generally needed.

3. Of the following appliances, which would need the biggest transformer?
    A. A clock radio.
    B. A TV broadcast transmitter.
    C. A shortwave radio receiver.
    D. A home TV set.

4. An advantage of full-wave bridge rectification is:
    A. It uses the whole transformer secondary for the entire ac input cycle.
    B. It costs less than other rectifier types.
    C. It cuts off half of the ac wave cycle.
    D. It never needs a regulator.

5. In a supply designed to provide high power at low voltage, the best rectifier design would probably be:
    A. Half-wave.
    B. Full-wave, center-tap.
    C. Bridge.
    D. Voltage multiplier.

6. The part of a power supply immediately preceding the regulator is:
    A. The transformer.
    B. The rectifier.
    C. The filter.
    D. The ac input.

7. If a half-wave rectifier is used with 117-V rms ac (house mains), the average dc output voltage is about:

    A. 52.7 V.

    B. 105 V.

    C. 117 V.

    D. 328 V.

8. If a full-wave bridge circuit is used with a transformer whose secondary provides 50 V rms, the PIV across the diodes is about:

    A. 50 V.

    B. 70 V.

    C. 100 V.

    D. 140 V.

9. The principal disadvantage of a voltage multiplier is:

    A. Excessive current.

    B. Excessive voltage.

    C. Insufficient rectification.

    D. Poor regulation.

10. A transformer secondary provides 10 V rms to a voltage-doubler circuit. The dc output voltage is about:

    A. 14 V.

    B. 20 V.

    C. 28 V.

    D. 36 V.

11. The ripple frequency from a full-wave rectifier is:

    A. Twice that from a half-wave circuit.

    B. The same as that from a half-wave circuit.

    C. Half that from a half-wave circuit.

    D. One-fourth that from a half-wave circuit.

12. Which of the following would make the best filter for a power supply?

    A. A capacitor in series.

    B. A choke in series.

    C. A capacitor in series and a choke in parallel.

    D. A capacitor in parallel and a choke in series.

13. If you needed exceptionally good ripple filtering for a power supply, the best approach would be to:

    A. Connect several capacitors in parallel.

    B. Use a choke-input filter.

    C. Connect several chokes in series.

    D. Use two capacitor/choke sections one after the other.

14. Voltage regulation can be accomplished by a Zener diode connected in:

    A. Parallel with the filter output, forward-biased.

    B. Parallel with the filter output, reverse-biased.

    C. Series with the filter output, forward-biased.

    D. Series with the filter output, reverse-biased.

15. A current surge takes place when a power supply is first turned on, because:

    A. The transformer core is suddenly magnetized.

    B. The diodes suddenly start to conduct.

    C. The filter capacitor(s) must be initially charged.

    D. Arcing takes place in the power switch.

16. Transient suppression minimizes the chance of:

    A. Diode failure.

    B. Transformer failure.

    C. Filter capacitor failure.

    D. Poor voltage regulation.

17. If a fuse blows, and it is replaced with one having a lower current rating, there's a good chance that:

    A. The power supply will be severely damaged.

    B. The diodes will not rectify.

    C. The fuse will blow out right away.

    D. Transient suppressors won't work.

18. A fuse with a nothing but a straight wire inside is probably:

    A. A slow-blow type.

    B. A quick-break type.

    C. Of a low current rating.

    D. Of a high current rating.

19. Bleeder resistors are:

    A. Connected in parallel with filter capacitors.

    B. Of low ohmic value.

    C. Effective for transient suppression.

    D. Effective for surge suppression.

20. To service a power supply with which you are not completely familiar, you should:

    A. Install bleeder resistors.

    B. Use proper fusing.

    C. Leave it alone and have a professional work on it.

    D. Use a voltage regulator.

22
CHAPTER

The bipolar transistor

THE WORD *TRANSISTOR* IS A CONTRACTION OF "CURRENT-*TRANS*FERRING RES*IS-tor.*" This is an excellent description of what a bipolar transistor does.

Bipolar transistors have two P-N junctions connected together. This is done in either of two ways: a P-type layer sandwiched between two N-type layers, or an N type layer between two P-type layers.

Bipolar transistors, like diodes, can be made from various semiconductor substances. Silicon is probably the most common material used.

NPN versus PNP

A simplified drawing of an *NPN* transistor, and its schematic symbol, are shown in Fig. 22-1. The P-type, or center, layer is called the *base*. The thinner of the N-type semiconductors is the *emitter*, and the thicker is the *collector*. Sometimes these are labeled *B*, *E*, and *C* in schematic diagrams, although the transistor symbol alone is enough to tell you which is which.

A *PNP* bipolar transistor is just the opposite of an NPN device, having two P-type layers, one on either side of a thin, N-type layer (Fig. 22-2). The emitter layer is thinner, in most units, than the collector layer.

You can always tell whether a bipolar transistor in a diagram is NPN or PNP. With the NPN, the arrow points outward; with the PNP it points inward. The arrow is always at the emitter.

Generally, PNP and NPN transistors can do the same things in electronic circuits. The only difference is the polarities of the voltages, and the directions of the currents. In most applications, an NPN device can be replaced with a PNP device or vice-versa, and the power-supply polarity reversed, and the circuit will still work as long as the new device has the appropriate specifications.

400

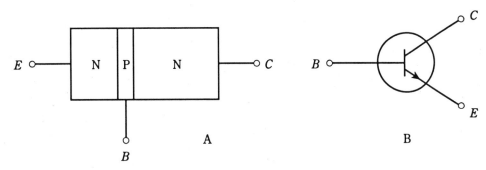

22-1 At A, pictorial diagram of an NPN transistor. At B, the schematic symbol. Electrodes are E = emitter, B = base, C = collector.

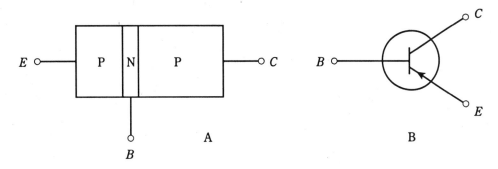

22-2 At A, pictorial diagram of a PNP transistor. At B, schematic symbol; E = emitter, B = base, C = collector.

There are many different kinds of NPN or PNP bipolar transistors. Some are used for radio-frequency amplifiers and oscillators; others are intended for audio frequencies. Some can handle high power, and others cannot, being made for weak-signal work. Some bipolar transistors are manufactured for the purpose of switching, rather than signal processing. If you look through a catalog of semiconductor components, you'll find hundreds of different bipolar transistors, each with its own unique set of specifications.

Why, you might ask, need there be two different kinds of bipolar transistor (NPN and PNP), if they do exactly the same things? Sometimes engineers need to have both kinds in one circuit. Also, there are some subtle differences in behavior between the two types. These considerations are beyond the scope of this book. But you should know that the NPN/PNP duality is not just whimsy on the part of people who want to make things complicated.

NPN biasing

You can think of a bipolar transistor as two diodes in reverse series. You can't normally connect two diodes together this way and get a good transistor, but the analogy is good for *modeling* the behavior of bipolar transistors, so that their operation is easier to understand.

A dual-diode NPN transistor model is shown in Fig. 22-3. The base is formed by the connection of the two diode anodes. The emitter is one of the cathodes, and the collector is the other.

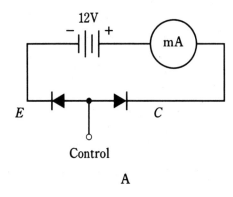

Control

A

22-3 At A, simple NPN circuit using dual-diode modeling. At B, the actual transistor circuit.

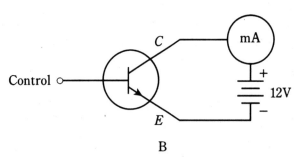

B

The normal method of biasing an NPN transistor is to have the emitter negative and the collector positive. This is shown by the connection of the battery in Fig. 22-3. Typical voltages for this battery (although it might be, and often is, a dc power supply) range from 3 V to about 50 V. Most often, 6 V, 9 V, or 12 V supplies are used.

The base is labeled "control" in the figure. This is because the flow of current through the transistor depends critically on the *base bias* voltage, E_B, relative to the *emitter-collector bias* voltage, E_C.

Zero bias

Suppose that the base isn't connected to anything, or is at the same potential as the emitter. This is *zero base bias*, sometimes simply called *zero bias*. How much current will flow through the transistor? What will the milliammeter (mA) show?

The answer is that there will be no current. The meter will register zero.

Recall the discussion of diode behavior from the previous chapter. No current flows through a P-N junction unless the forward bias is at least equal to the forward breakover voltage. (For silicon, this is about 0.6 V.) But here, the forward bias is zero. Therefore, the emitter-base current, often called simply *base current* and denoted I_B, is zero, and the emitter-base junction does not conduct. This prevents any current from flowing between the emitter and collector, unless some signal is injected at the base to change the situation. This signal would have to be of positive polarity and would need to be at least equal to the forward breakover voltage of the junction.

Reverse bias

Now imagine that another battery is connected to the base at the point marked "control," so that E_B is negative with respect to the emitter. What will happen? Will current flow through the transistor?

The answer is no. The addition of this new battery will cause the emitter-base (E-B) junction to be *reverse-biased*. It is assumed that this new battery is not of such a high voltage that avalanche breakdown takes place at the junction.

A signal might be injected to overcome the reverse-bias battery and the forward-breakover voltage of the E-B junction, but such a signal would have to be of a high, positive voltage.

Forward bias

Now suppose that E_B is made positive, starting at small voltages and gradually increasing.

If this *forward bias* is less than the forward breakover voltage, no current will flow. But as the base voltage E_B reaches breakover, the E-B junction will start to conduct.

The base-collector (B-C) junction will remain reverse-biased as long as E_B is less than the supply voltage (in this case 12 V). In practical transistor circuits, it is common for E_B to be set at a fraction of the supply voltage.

Despite the reverse bias of the B-C junction, the emitter-collector current, called *collector current* and denoted I_C, will flow once the E-B junction conducts. In a real transistor (Fig. 22-3B), the meter reading will jump when the forward- breakover voltage of the E-B junction is reached. Then even a small rise in E_B, attended by a rise in I_B, will cause a big increase in I_C. This is shown graphically in Fig. 22-4.

Saturation

If E_B continues to rise, a point will eventually be reached where I_C increases less rapidly. Ultimately, the I_C vs. E_B curve will level off. The transistor is then *saturated* or *in saturation*. It is conducting as much as it possibly can; it's "wide open."

This property of three-layer semiconductors, in which reverse-biased junctions can sometimes pass current, was first noticed in the late forties by the engineers Bardeen, Brattain, and Shockley at the Bell Laboratories. When they saw how current variations were magnified by a three-layer device of this kind, they knew they were on to something. They envisioned that the effect could be exploited to amplify weak signals, or to use small currents to switch much larger ones. They must have been excited, but they surely had no idea how much their discovery would affect the world.

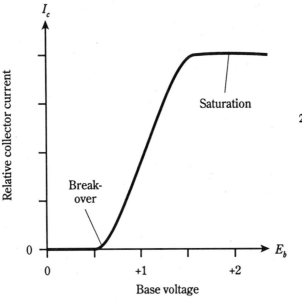

22-4 Relative collector current (I_C) as a function of base voltage (E_B) for a hypothetical silicon transistor.

PNP biasing

For a PNP transistor, the situation is just a "mirror image" of the case for an NPN device. The diodes are turned around the opposite way, the arrow points inward rather than outward in the transistor symbol, and all the polarities are reversed. The dual-diode PNP model, along with the actual bipolar transistor circuit, are shown in Fig. 22-5. In the discussion above, simply replace every occurrence of the word "positive" with the word "negative."

You need not be concerned with what actually goes on inside the semiconductor materials in NPN and PNP transistors. The important thing is the fact that either type of device can serve as a sort of "current valve." Small changes in the base voltage, E_B, cause small changes in the base current, I_B. This induces large fluctuations in the current I_C through the transistor.

In the following discussion, and in most circuits that appear later in this book, you'll see NPN transistors used almost exclusively. This doesn't mean that NPN is better than PNP; in almost every case, you can replace each NPN transistor with a PNP, reverse the polarity, and get the same results. The motivation is to save space and avoid redundancy.

Biasing for current amplification

Because a small change in the base current, I_B, results in a large collector-current (I_C) variation when the bias is just right, a transistor can operate as a *current amplifier*. It might be more technically accurate to say that it is a "current- fluctuation amplifier," because it's the magnification of current variations, not the absolute current, that's important.

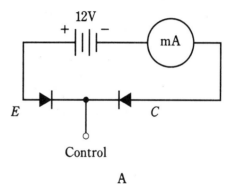

22-5   At A, simple PNP circuit
using dual-diode
modeling. At B, the
actual transistor circuit.

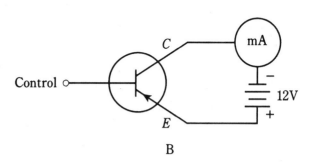

If you look at Fig. 22-4 closely, you'll see that there are some bias values at which a transistor won't give current amplification. If the E-B junction is not conducting, or if the transistor is in saturation, the curve is horizontal. A small change (to the left and right) of the base voltage, E_B, in these portions of the curve, will cause little or no up-and-down variation of I_C.

But if the transistor is biased near the middle of the straight-line part of the curve in Fig. 22-4, the transistor will work as a current amplifier.

Static current amplification

Current amplification is often called *beta* by engineers. It can range from a factor of just a few times up to hundreds of times.

One method of expressing the beta of a transistor is as the *static forward current transfer ratio*, abbreviated H_{FE}. Mathematically, this is

$$H_{FE} = I_C/I_B$$

Thus, if a base current, I_B, of 1 mA results in a collector current, I_C, of 35 mA, $H_{FE} = 35/1 = 35$. If $I_B = 0.5$ mA yields $I_C = 35$ mA, then $H_{FE} = 35/0.5 = 70$.

This definition represents the greatest current amplification possible with a given transistor.

Dynamic current amplification

Another way of specifying current amplification is as the ratio of *the difference in I_C* to *the difference in I_B*. Abbreviate the words "the difference in" by the letter *d*. Then, according to this second definition:

$$\text{Current amplification} = dI_C/dI_B$$

A graph of collector current versus base current (I_C vs I_B) for a hypothetical transistor is shown in Fig. 22-6. This graph resembles Fig. 22-4, except that current, rather than voltage, is on the horizontal scale. Three different points are shown, corresponding to various bias values.

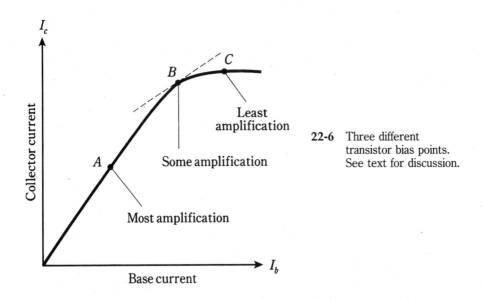

22-6   Three different transistor bias points. See text for discussion.

The ratio dI_C/dI_B is different for each of the points in this graph. Geometrically, dI_C/dI_B at a given point is the slope of a line tangent to the curve at that point. The tangent line for point B in Fig. 22-6 is a dotted, straight line; the tangent lines for points A and C lie right along the curve. The steeper the slope of the line, the greater is dI_B/dI_C.

Point A provides the highest dI_C/dI_B, as long as the input signal is small. This value is very close to H_{FE}. For small-signal amplification, point A represents a good bias level. Engineers would say that it's a good *operating point*.

At point B, dI_C/dI_B is smaller than at point A. (It might actually be less than 1.) At point C, dI_C/dI_B is practically zero. Transistors are rarely biased at these points.

Overdrive

Even when a transistor is biased for best operation (near point A in Fig. 22-6), a strong input signal can drive it to point B or beyond during part of the cycle. Then, dI_C/dI_B is reduced, as shown in Fig. 22-7. Points X and Y in the graph represent the instantaneous current extremes during the signal cycle.

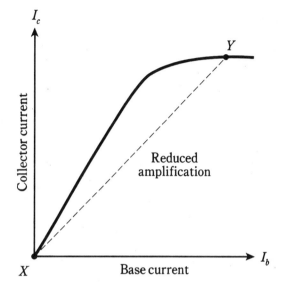

22-7 Excessive input reduces amplification.

When conditions are like those in Fig. 22-7, there will be distortion in a transistor amplifier. The output waveform will not have the same shape as the input waveform. This *nonlinearity* can sometimes be tolerated; sometimes it cannot.

The more serious trouble with *overdrive* is the fact that the transistor is in or near saturation during part of the cycle. When this happens, you're getting "no bang for the buck." The transistor is doing futile work for a portion of every wave cycle. This reduces circuit efficiency, causes excessive collector current, and can overheat the base-collector (B-C) junction. Sometimes overdrive can actually destroy a transistor.

Gain versus frequency

Another important specification for a transistor is the range of frequencies over which it can be used as an amplifier. All transistors have an amplification factor, or *gain*, that decreases as the signal frequency increases. Some devices will work well only up to a few megahertz; others can be used to several gigahertz.

Gain can be expressed in various different ways. In the above discussion, you learned a little about *current gain*, expressed as a ratio. You will also sometimes hear about *voltage gain* or *power gain* in amplifier circuits. These, too, can be expressed as ratios. For example, if the voltage gain of a circuit is 15, then the output signal voltage (rms, peak, or peak-to-peak) is 15 times the input signal voltage. If the power gain of a circuit is 25, then the output signal power is 25 times the input signal power.

There are two expressions commonly used for the gain-versus-frequency behavior of a bipolar transistor. The *gain bandwidth product*, abbreviated f_T, is the frequency at which the gain becomes equal to 1 with the emitter connected to ground. If you try to make an amplifier using a transistor at a frequency higher than its f_T, you'll fail! Thus, f_T represents an absolute upper limit of sorts.

The *alpha cutoff frequency* of a transistor is the frequency at which the gain becomes 0.707 times its value at 1 kHz. A transistor might still have considerable gain at its alpha cutoff. By looking at the alpha cutoff frequency, you can get an idea of how rapidly the transistor loses gain as the frequency goes up. Some devices "die off" faster than others.

Figure 22-8 shows the gain bandwidth product and alpha cutoff frequency for a hypothetical transistor, on a graph of gain versus frequency. Note that the scales of this graph are nonlinear; they're "scrunched up" at the higher values. This type of graph is useful for showing some functions. It is called a *log-log* graph because both scales are *logarithmic* rather than linear.

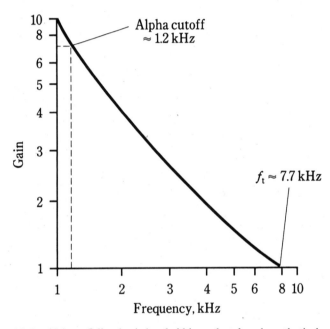

22-8   Alpha cutoff and gain bandwidth product for a hypothetical transistor.

Common emitter circuit

A transistor can be hooked up in three general ways. The emitter can be grounded for signal, the base can be grounded for signal, or the collector can be grounded for signal.

Probably the most often-used arrangement is the *common-emitter circuit*. "Common" means "grounded for the signal." The basic configuration is shown in Fig. 22-9.

A terminal can be at ground potential for the signal, and yet have a significant dc voltage. In the circuit shown, C1 looks like a dead short to the ac signal, so the emitter is at *signal ground*. But R1 causes the emitter to have a certain positive dc voltage with respect to ground (or a negative voltage, if a PNP transistor is used). The exact dc voltage at the emitter depends on the value of R1, and on the bias.

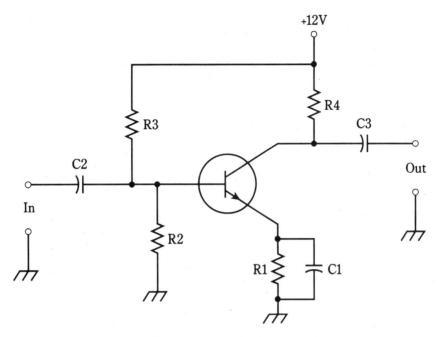

22-9   Common-emitter circuit configuration.

The bias is set by the ratio of resistances *R2* and *R3*. It can be anything from zero, or ground potential, to +12 V, the supply voltage. Normally it will be a couple of volts.

Capacitors C2 and C3 block dc to or from the input and output circuitry (whatever that might be) while letting the ac signal pass. Resistor R4 keeps the output signal from being shorted out through the power supply.

A signal voltage enters the common-emitter circuit through C2, where it causes the base current, I_B to vary. The small fluctuations in I_B cause large changes in the collector current, I_C. This current passes through R4, causing a fluctuating dc voltage to appear across this resistor. The ac part of this passes unhindered through C3 to the output.

The circuit of Fig. 22-9 is the basis for many amplifiers, from audio frequencies through ultra-high radio frequencies. The common-emitter configuration produces the largest gain of any arrangement. The output is 180 degrees out of phase with the input.

Common base circuit

As its name implies, the *common-base circuit,* shown in general form by Fig. 22-10, has the base at signal ground.

The dc bias on the transistor is the same for this circuit as for the common-emitter circuit. The difference is that the input signal is applied at the emitter, instead of at the base. This causes fluctuations in the voltage across R1, causing variations in I_B. The result of these small current fluctuations is a large change in the dc current through R4. Therefore amplification occurs.

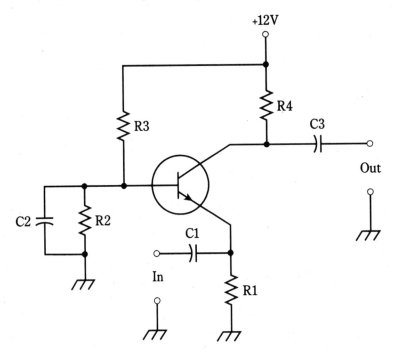

22-10   Common-base circuit configuration.

Instead of varying I_B by injecting the signal at the base, it's being done by injecting the signal at the emitter. Therefore, in the common-base arrangement, the output signal is in phase with the input, rather than out of phase.

The signal enters through C1. Resistor R1 keeps the input signal from being shorted to ground. Bias is provided by R2 and R3. Capacitor C2 keeps the base at signal ground. Resistor R4 keeps the signal from being shorted out through the power supply. The output is through C3.

The common-base circuit provides somewhat less gain than a common-emitter circuit. But it is more stable than the common-emitter configuration in some applications, especially in radio-frequency power amplifiers.

Common collector circuit

A *common-collector circuit* (Fig. 22-11) operates with the collector at signal ground. The input is applied at the base, just as it is with the common-emitter circuit.

The signal passes through C2 onto the base of the transistor. Resistors R2 and R3 provide the correct bias for the base. Resistor R4 limits the current through the transistor. Capacitor C3 keeps the collector at signal ground. A fluctuating direct current flows through R1, and a fluctuating dc voltage therefore appears across it. The ac part of this voltage passes through C1 to the output. Because the output follows the emitter current, this circuit is sometimes called an *emitter follower circuit*.

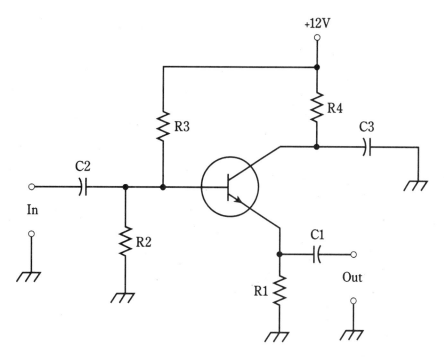

22-11 Common-collector circuit configuration. This arrangement is also known as an emitter follower.

The output of this circuit is in phase with the input. The input impedance is high, and the output impedance is low. For this reason, the common-collector circuit can be used to match high impedances to low impedances. When well designed, an emitter follower works over a wide range of frequencies, and is a low-cost alternative to a broadband impedance-matching transformer.

Quiz

Refer to the text in this chapter if necessary. A good score is at least 18 correct. Answers are in the back of the book.

1. In a PNP circuit, the collector:
    A. Has an arrow pointing inward.
    B. Is positive with respect to the emitter.
    C. Is biased at a small fraction of the base bias.
    D. Is negative with respect to the emitter.

2. In many cases, a PNP transistor can be replaced with an NPN device and the circuit will do the same thing, provided that:
    A. The supply polarity is reversed.

    B. The collector and emitter leads are interchanged.

    C. The arrow is pointing inward.

    D. No! A PNP device cannot be replaced with an NPN.

3. A bipolar transistor has:

    A. Three P-N junctions.

    B. Three semiconductor layers.

    C. Two N type layers around a P-type layer.

    D. A low avalanche voltage.

4. In the dual-diode model of an NPN transistor, the emitter corresponds to:

    A. The point where the cathodes are connected together.

    B. The point where the cathode of one diode is connected to the anode of the other.

    C. The point where the anodes are connected together.

    D. Either of the diode cathodes.

5. The current through a transistor depends on:

    A. E_C.

    B. E_B relative to E_C.

    C. I_B.

    D. More than one of the above.

6. With no signal input, a bipolar transistor would have the least I_C when:

    A. The emitter is grounded.

    B. The E-B junction is forward-biased.

    C. The E-B junction is reverse-biased.

    D. The E-B current is high.

7. When a transistor is conducting as much as it possibly can, it is said to be:

    A. In cutoff.

    B. In saturation.

    C. Forward-biased.

    D. In avalanche.

8. Refer to Fig. 22-12. The best point at which to operate a transistor as a small-signal amplifier is:

    A. A.

    B. B.

    C. C.

    D. D.

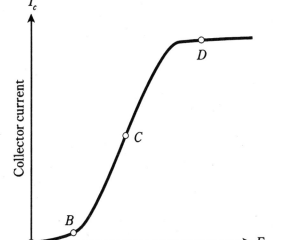

22-12 Illustration for quiz questions 8, 9, 10, and 11.

9. In Fig. 22-12, the forward-breakover point for the E-B junction is nearest to:
   A. No point on this graph.
   B. B.
   C. C.
   D. D.

10. In Fig. 22-12, saturation is nearest to point:
    A. A.
    B. B.
    C. C.
    D. D.

11. In Fig. 22-12, the greatest gain occurs at point:
    A. A.
    B. B.
    C. C.
    D. D.

12. In a common-emitter circuit, the gain bandwidth product is:
    A. The frequency at which the gain is 1.
    B. The frequency at which the gain is 0.707 times its value at 1 MHz.
    C. The frequency at which the gain is greatest.
    D. The difference between the frequency at which the gain is greatest, and the frequency at which the gain is 1.

13. The configuration most often used for matching a high input impedance to a low output impedance puts signal ground at:

    A. The emitter.

    B. The base.

    C. The collector.

    D. Any point; it doesn't matter.

14. The output is in phase with the input in a:

    A. Common-emitter circuit.

    B. Common-base circuit.

    C. Common-collector circuit.

    D. More than one of the above.

15. The greatest possible amplification is obtained in:

    A. A common-emitter circuit.

    B. A common-base circuit.

    C. A common-collector circuit.

    D. More than one of the above.

16. The input is applied to the collector in:

    A. A common-emitter circuit.

    B. A common-base circuit.

    C. A common-collector circuit.

    D. None of the above.

17. The configuration noted for its stability in radio-frequency power amplifiers is the:

    A. Common-emitter circuit.

    B. Common-base circuit.

    C. Common-collector circuit.

    D. Emitter-follower circuit.

18. In a common-base circuit, the output is taken from the:

    A. Emitter.

    B. Base.

    C. Collector.

    D. More than one of the above.

19. The input signal to a transistor amplifier results in saturation during part of the cycle. This produces:

    A. The greatest possible amplification.

    B. Reduced efficiency.

C. Avalanche effect.

D. Nonlinear output impedance.

20. The gain of a transistor in a common-emitter circuit is 100 at a frequency of 1000 Hz. The gain is 70.7 at 335 kHz. The gain drops to 1 at 210 MHz. The alpha cutoff is:

A. 1 kHz.

B. 335 kHz.

C. 210 MHz.

D. None of the above.

23
CHAPTER

The field-effect transistor

BIPOLAR TRANSISTORS BEHAVE AS THEY DO BECAUSE CURRENT VARIATIONS AT
one P-N junction produce larger current variations at another. You've seen a simplified
picture of how this happens, and how the effect can be exploited to get current
amplification.

The bipolar transistor isn't the only way that semiconductors can be combined to get
amplification effects. The other major category of transistor, besides the bipolar device,
is the *field-effect transistor* or *FET*. There are two main types of FET: the *junction FET*
(*JFET*) and the *metal-oxide FET* (*MOSFET*).

Principle of the JFET

A JFET can have any of several different forms. They all work the same way: the current
varies because of the effects of an electric field within the device.

The workings inside a JFET can be likened to the control of water flow through a
garden hose. Electrons or holes pass from the *source* (S) electrode to the *drain* (D). This
results in a *drain current*, I_D, that is generally the same as the *source current*, I_S. This is
analogous to the fact that the water comes out of a garden hose at the same rate it goes in
(assuming that there aren't any leaks in the hose).

The rate of flow of charge carriers—that is, the current—depends on the voltage at
a regulating electrode called the *gate* (G). Fluctuations in *gate voltage*, E_G, cause changes
in the current through the *channel*, I_S or I_D. Small fluctuations in the control voltage E_G
can cause large variations in the flow of charge carriers through the JFET. This translates
into voltage amplification in electronic circuits.

416

N-channel versus P-channel

A simplified drawing of an *N-channel* JFET, and its schematic symbol, are shown in Fig. 23-1. The N-type material forms the channel, or the path for charge carriers. In the N-channel device, the majority carriers are electrons. The source is at one end of the channel, and the drain is at the other. You can think of electrons as being "injected" into the source and "collected" from the drain as they pass through the channel. The drain is positive with respect to the source.

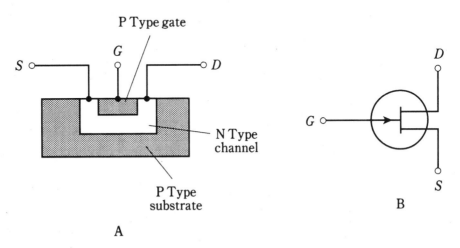

23-1 Simplified cross-sectional drawing of an N-channel JFET (at A) and its schematic symbol (at B).

In an N-channel device, the gate consists of P-type material. Another, larger section of P-type material, called the *substrate*, forms a boundary on the side of the channel opposite the gate. The JFET is formed in the substrate during manufacture by a process known as *diffusion*.

The voltage on the gate produces an electric field that interferes with the flow of charge carriers through the channel. The more negative E_G becomes, the more the electric field chokes off the current though the channel, and the smaller I_D becomes.

A *P-channel* JFET (Fig. 23-2) has a channel of P-type semiconductor. The majority charge carriers are holes. The drain is negative with respect to the source. In a sense, holes are "injected" into the source and are "collected" from the drain. The gate and the substrate are of N-type material.

In the P-channel JFET, the more positive E_G gets, the more the electric field chokes off the current through the channel, and the smaller I_D becomes.

You can recognize the N-channel device by the arrow pointing inward at the gate, and the P-channel JFET by the arrow pointing outward. Also, you can tell which is which (sometimes arrows are not included in schematic diagrams) by the power-supply polarity. A positive drain indicates an N-channel JFET, and a negative drain indicates a P-channel type.

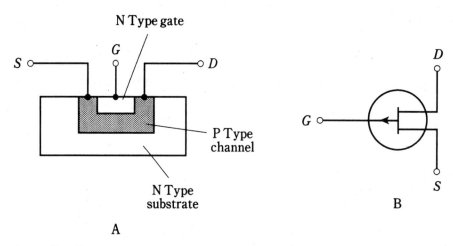

23-2 Simplified cross-sectional drawing of a P-channel JFET (at A) and its schematic symbol (at B).

In electronic circuits, N-channel and P-channel devices can do the same kinds of things. The main difference is the polarity. An N-channel device can almost always be replaced with a P-channel JFET, and the power-supply polarity reversed, and the circuit will still work if the new device has the right specifications.

Just as there are different kinds of bipolar transistors, there are various types of JFETs, each suited to a particular application. Some JFETs work well as weak-signal amplifiers and oscillators; others are made for power amplification.

Field-effect transistors have some advantages over bipolar devices. Perhaps the most important is that FETs are available that generate less internal noise than bipolar transistors. This makes them excellent for use in sensitive radio receivers at very high or ultra-high frequencies.

Field-effect transistors have high input impedances. The gate controls the flow of charge carriers by means of an electric *field*, rather than via an electric *current*.

Depletion and pinchoff

Either the N-channel or the P-channel JFET works because the voltage at the gate causes an electric field that interferes, more or less, with the flow of charge carriers along the channel. A simplified drawing of the situation for an N-channel device is shown in Fig. 23-3. For a P-channel device, just interchange polarities (minus/plus) and semiconductor types (N/P) in this discussion.

As the drain voltage E_D increases, so does the drain current I_D, up to a certain level-off value. This is true as long as the gate voltage E_G is constant, and is not too large negatively.

But as E_G becomes increasingly negative (Fig. 23-3A), a *depletion region* (solid black) begins to form in the channel. Charge carriers cannot flow in this region; they must pass through a narrowed channel. The more negative E_G becomes, the wider the depletion region gets, as shown at B. Ultimately, if the gate becomes negative enough,

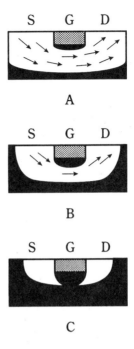

23-3 At A, depletion region (solid area) is not wide, and many charge carriers (arrows) flow. At B, depletion region is wider, channel is narrower, and fewer carriers flow. At C, channel is completely obstructed, and no carriers flow.

the depletion region will completely obstruct the flow of charge carriers. This is called *pinchoff*, and is illustrated at C.

Again, think of the garden-hose analogy. More negative gate voltages, E_G, correspond to stepping harder and harder on the hose. When pinchoff takes place, you've cut off the water flow entirely, perhaps by bearing down with all your weight on one foot! Biasing beyond pinchoff is something like loading yourself up with heavy weights as you balance on the hose, thereby shutting off the water flow with extra force.

JFET biasing

Two biasing arrangements for an N-channel JFET are shown in Fig. 23-4. These hookups are similar to the way an NPN bipolar transistor is connected, except that the source-gate (SG) junction is not forward-biased.

At A, the gate is grounded through resistor R2. The source resistor, R1, limits the current through the JFET. The drain current, I_D, flows through R3, producing a voltage across this resistor. The ac output signal passes through C2.

At B, the gate is connected to a voltage that is negative with respect to ground through potentiometer R2. Adjusting this potentiometer results in a variable negative E_G between R2 and R3. Resistor R1 limits the current through the JFET. The drain current, I_D, flows through R4, producing a voltage across it; the ac output signal passes through C2.

In both of these circuits, the drain is positive relative to ground. For a P-channel JFET, reverse the polarities in Fig. 23-4. The connections are somewhat similar to the way a PNP bipolar transistor is used, except the SG junction isn't forward-biased.

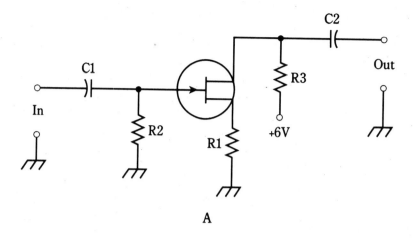

A

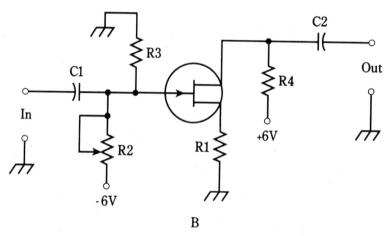

B

23-4   Two methods of biasing an N-channel JFET. At A, fixed gate bias; at B, variable gate bias.

Typical JFET power-supply voltages are comparable to those with bipolar transistors. The voltage between the source and drain, abbreviated E_D, can range from about 3 V to 50 V; most often it is 6 V to 12 V.

The biasing arrangement in Fig. 23-4A is commonly used for weak-signal amplifiers, low-level amplifiers and oscillators. The scheme at B is more often employed in power amplifiers having a substantial input signal.

Voltage amplification

The graph of Fig. 23-5 shows the drain (channel) current, I_D, as a function of the gate bias voltage, E_G, for a hypothetical N-channel JFET. The drain voltage, E_D, is assumed to be constant.

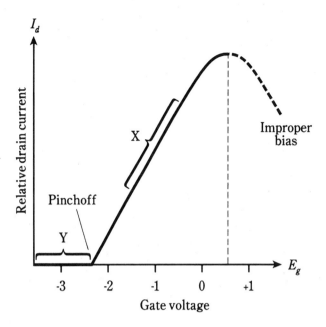

23-5 Relative drain current as a function of gate voltage for a hypothetical N-channel JFET.

When E_G is fairly large and negative, the JFET is pinched off, and no current flows through the channel. As E_G gets less negative, the channel opens up, and current begins flowing. As E_G gets still less negative, the channel gets wider and the current I_D increases. As E_G approaches the point where the SG junction is at forward breakover, the channel conducts as well as it possibly can.

If E_G becomes positive enough so that the SG junction conducts, the JFET will no longer work properly. Some of the current in the channel will then be shunted off through the gate, a situation that is never desired in a JFET. The hose will spring a leak!

The best amplification for weak signals is obtained when the gate bias, E_G, is such that the slope of the curve in Fig. 23-5 is the greatest. This is shown roughly by the range marked X in the figure. For power amplification, however, results are often best when the JFET is biased at, or even beyond, pinchoff, in the range marked Y.

The current I_D passes through the drain resistor, as shown in either diagram of Fig. 23-4. Small fluctuations in E_G cause large changes in I_D, and these variations in turn produce wide swings in the dc voltage across R3 (at A) or R4 (at B). The ac part of this voltage goes through capacitor C2, and appears at the output as a signal of much greater ac voltage than that of the input signal at the gate. That's voltage amplification.

Drain current versus drain voltage

You might expect that the current I_D, passing through the channel of a JFET, would increase linearly with increasing drain voltage E_D. But this is not, in general, what happens. Instead, the current I_D rises for awhile, and then starts to level off.

The drain current I_D (which is the same as the channel current) is often plotted as a function of drain voltage, E_D, for various values of gate voltage, E_G. The resulting set of curves is called a *family of characteristic curves* for the device. The graph of Fig. 23-6 shows a family of characteristic curves for a hypothetical N-channel JFET. Engineers make use of these graphs when deciding on the best JFET type for an electronic circuit. Also of importance is the curve of I_D vs E_G, one example of which is shown in Fig. 23-5.

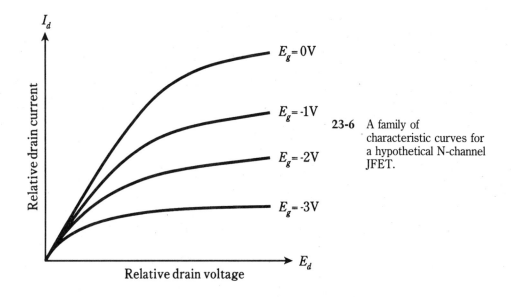

23-6 A family of characteristic curves for a hypothetical N-channel JFET.

Transconductance

Recall the discussion of *dynamic current amplification* from the last chapter. This is a measure of how well a bipolar transistor amplifies a signal. The JFET analog of this is called *dynamic mutual conductance* or *transconductance*.

Refer again to Fig. 23-5. Suppose that E_G is a certain value, with a corresponding I_D resulting. If the gate voltage changes by a small amount dE_G, then the drain current will also change by a certain increment dI_D. The transconductance is the ratio dI_D/dE_G. Geometrically, this translates to the slope of a line tangent to the curve of Fig. 23-5.

The value of dI_D/dE_G is obviously not the same everywhere along the curve. When the JFET is biased beyond pinchoff, in the region marked Y in the figure, the slope of the curve is zero. There is no drain current, even if the gate voltage changes. Only when the channel is conducting will there be a change in I_D when there is a change in E_G. The region where the transconductance, dI_D/dE_G, is the greatest is the region marked X, where the slope of the curve is steepest. This is where the most gain can be obtained from the JFET.

The MOSFET

The acronym *MOSFET* (pronounced "*moss*-fet") stands for *metal-oxide-semiconductor field-effect transistor*. A simplified cross-sectional drawing of an N-channel MOSFET, along with the schematic symbol, is shown in Fig. 23-7. The P-channel device is shown in the drawings of Fig. 23-8. The N-channel device is diffused into a substrate of P-type semiconductor material. The P-channel device is diffused into a substrate of N-type material.

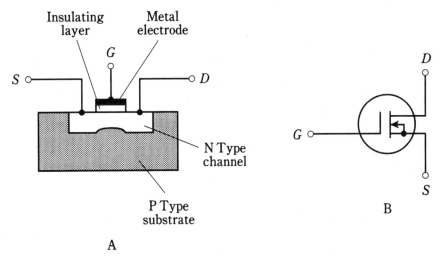

23-7   At A, simplified cross-sectional drawing of an N-channel MOSFET. At B, the schematic symbol.

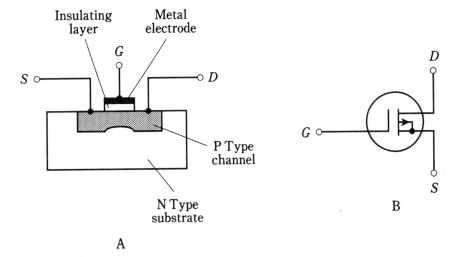

23-8   At A, simplified cross-sectional drawing of a P-channel MOSFET. At B, the schematic symbol.

Super-high input impedance

When the MOSFET was first developed, it was called an *insulated-gate FET* or *IGFET*. This is perhaps more descriptive of the device than the currently accepted name. The gate electrode is actually insulated, by a thin layer of dielectric, from the channel. As a result, the input impedance is even higher than that of a JFET; the gate-to-source resistance of a typical MOSFET is comparable to that of a capacitor! This means that a MOSFET draws essentially no current, and therefore no power, from the signal source. Some MOSFETs have input resistance exceeding a trillion (10^{12}) ohms.

The main problem

The trouble with MOSFETs is that they can be easily damaged by static electric discharges. When building or servicing circuits containing MOS devices, technicians must use special equipment to ensure that their hands don't carry static charges that might ruin the components. If a static discharge occurs through the dielectric of a MOS device, the component will be destroyed permanently. Warm and humid climates do not offer protection against the hazard. (This author's touch has dispatched several MOSFETs in Miami during the summer.)

Flexibility

In actual circuits, an N-channel JFET can sometimes be replaced directly with an N-channel MOSFET; P-channel devices can be similarly interchanged. But the characteristic curves for MOSFETs are not the same as those for JFETs. The main difference is that the SG junction in a MOSFET is not a P-N junction. Therefore, forward breakover cannot occur. An E_G of more than $+0.6$ V can be applied to an N-channel MOSFET, or an E_G more negative than -0.6 V to a P-channel device, without a current "leak" taking place.

A family of characteristic curves for a hypothetical N-channel MOSFET is shown in the graph of Fig. 23-9. The device will work with positive gate bias as well as with negative gate bias. A P-channel MOSFET behaves in a similar way, being usable with either positive or negative E_G.

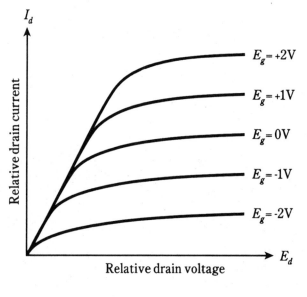

23-9   A family of characteristic curves for a hypothetical N-channel MOSFET.

Depletion mode versus enhancement mode

The JFET works by varying the width of the channel. Normally the channel is wide open; as the depletion region gets wider and wider, choking off the channel, the charge carriers are forced to pass through a narrower and narrower path. This is known as the *depletion mode* of operation for a field-effect transistor.

A MOSFET can also be made to work in the depletion mode. The drawings and schematic symbols of Figs. 23-7 and 23-8 show depletion-mode MOSFETs.

However, MOS technology also allows an entirely different means of operation. An *enhancement-mode* MOSFET normally has a pinched-off channel. It is necessary to apply a bias voltage, E_G, to the gate so that a channel will form. If $E_G = 0$ in such a MOSFET, that is, if the device is at zero bias, the drain current I_D is zero when there is no signal input.

The schematic symbols for N-channel and P-channel enhancement-mode devices are shown in Fig. 23-10. The vertical line is broken. This is how you can recognize an enhancement-mode device in circuit diagrams.

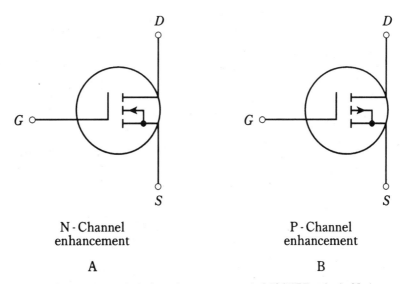

N - Channel
enhancement

A

P - Channel
enhancement

B

23-10 Schematic symbols for enhancement-mode MOSFETs. At A, N-channel; at B, P-channel.

Common source circuit

There are three different circuit hookups for FETs, just as there are for bipolar transistors. These three arrangements have the source, the gate, or the drain at signal ground.

The *common-source circuit* places the source at signal ground. The input is at the base. The general configuration is shown in Fig. 23-11. An N-channel JFET is used here, but the device could be an N-channel, depletion-mode MOSFET and the circuit diagram would be the same. For an N-channel, enhancement-mode device, an extra resistor would be necessary, running from the gate to the positive power supply terminal. For P-channel devices, the supply would provide a negative, rather than a positive, voltage.

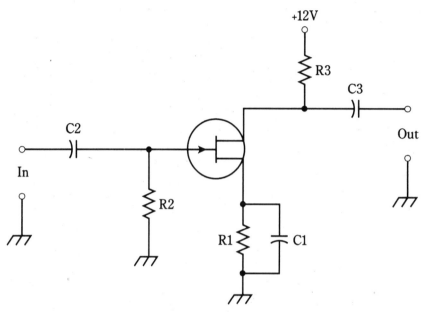

23-11   Common-source circuit configuration.

This circuit is an almost exact replica of the grounded-emitter bipolar arrangement. The only difference is the lack of a voltage-dividing network for bias on the control electrode.

Capacitor C1 and resistor R1 place the source at signal ground while elevating this electrode above ground for dc. The ac signal enters through C2; resistor R2 adjusts the input impedance and provides bias for the gate. The ac signal passes out of the circuit through C3. Resistor R3 keeps the output signal from being shorted out through the power supply.

The circuit of Fig. 23-11 is the basis for amplifiers and oscillators, especially at radio frequencies. The common-source arrangement provides the greatest gain of the three FET circuit configurations. The output is 180 degrees out of phase with the input.

Common gate circuit

The *common-gate circuit* (Fig. 23-12) has the gate at signal ground. The input is applied to the source. The illustration shows an N-channel JFET. For other types of FETs, the same considerations apply as described above for the common-source circuit. Enhancement-mode devices would require a resistor between the gate and the positive supply terminal (or the negative terminal if the MOSFET is P-channel).

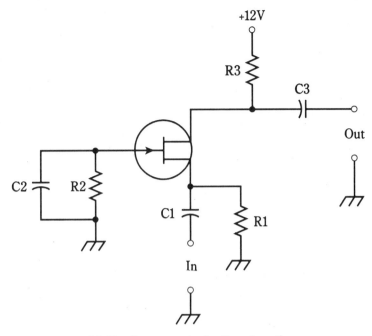

23-12   Common-gate circuit configuration.

The dc bias for the common-gate circuit is basically the same as that for the common-source arrangement. But the signal follows a different path. The ac input signal enters through C1. Resistor R1 keeps the input from being shorted to ground. Gate bias is provided by R1 and R2; capacitor C2 places the gate at signal ground. In some common-gate circuits, the gate electrode is directly grounded, and components R2 and C2 are not used. The output leaves the circuit through C3. Resistor R3 keeps the output signal from being shorted through the power supply.

The common-gate arrangement produces less gain than its common-source counterpart. But this is not all bad; a common-gate amplifier is very stable, and is not likely to break into unwanted oscillation. The output is in phase with the input.

Common drain circuit

A *common-drain circuit* is shown in Fig. 23-13. This circuit has the collector at signal ground. It is sometimes called a *source follower*.

The FET is biased in the same way as for the common-source and common-gate circuits. In the illustration, an N-channel JFET is shown, but any other kind of FET could be used, reversing the polarity for P-channel devices. Enhancement-mode MOSFETs would need a resistor between the gate and the positive supply terminal (or the negative terminal if the MOSFET is P-channel).

The input signal passes through C2 to the gate. Resistors R1 and R2 provide gate bias. Resistor R3 limits the current. Capacitor C3 keeps the drain at signal ground. Fluctuating dc (the channel current) flows through R1 as a result of the input signal; this

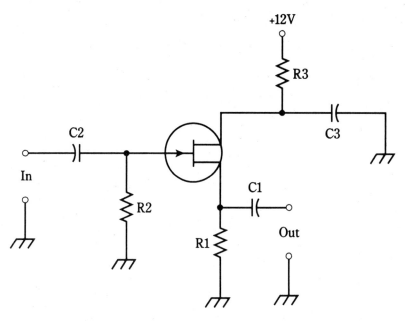

23-13   Common-drawin circuit configuration.

causes a fluctuating dc voltage to appear across the resistor. The output is taken from the source, and its ac component passes through C1.

The output of the common-drain circuit is in phase with the input. This scheme is the FET analog of the bipolar common-collector arrangement. The output impedance is rather low, making this circuit a good choice for broadband impedance matching.

Table 23-1.   Transistor circuit abbreviations.

Quantity	Abbreviations
Base-emitter voltage	E_B, V_B, E_{BE}, V_{BE}
Collector-emitter voltage	E_C, V_C, E_{CE}, V_{CE}
Collector-base voltage	E_{BC}, V_{BC}, E_{CB}, V_{CB}
Gate-source voltage	E_G, V_G, E_{GS}, V_{GS}
Drain-source voltage	E_D, V_D, E_{DS}, V_{DS}
Drain-gate voltage	E_{DG}, V_{DG}, E_{GD}, V_{GD}
Emitter current	I_E
Base current	I_B, I_{BE}, I_{EB}
Collector current	I_C, I_{CE}, I_{EC}
Source current	I_S
Gate current	I_G, I_{GS}, I_{SG}*
Drain current	I_D, I_{DS}, I_{SD}

*This is almost always insignificant.

A note about notation

In electronics, you'll encounter various different symbols that denote the same things. You might have already noticed that voltage is sometimes abbreviated by the letter E, and sometimes by the letter V. In bipolar and field-effect transistor circuits, you'll sometimes come across symbols like V_{CE} and V_{GS}; in this book they appear as E_C and E_G, respectively. Subscripts can be either uppercase or lowercase.

Remember that, although notations vary, the individual letters almost always stand for the same things. A variable might be denoted in different ways, depending on the author or engineer; but it's rare for one notation to acquire multiple meanings. The most common sets of abbreviations from this chapter and chapter 22 are shown in Table 23-1.

Wouldn't it be great if there were complete standardization in electronics? And it would be wonderful if everything were standardized in all other aspects of life, too, would it not?

Or would it?

Quiz

Refer to the text in this chapter if necessary. A good score is at least 18 correct. Answers are in the back of the book.

1. The current through the channel of a JFET is directly affected by all of the following *except*:

    A. Drain voltage.

    B. Transconductance.

    C. Gate voltage.

    D. Gate bias.

2. In an N-channel JFET, pinchoff occurs when the gate bias is:

    A. Slightly positive.

    B. Zero.

    C. Slightly negative.

    D. Very negative.

3. The current consists mainly of holes when a JFET:

    A. Has a P-type channel.

    B. Is forward-biased.

    C. Is zero-biased.

    D. Is reverse-biased.

4. A JFET might work better than a bipolar transistor in:

    A. A rectifier.

    B. A radio receiver.

    C. A filter.

    D. A transformer.

5. In a P-channel JFET:

   A. The drain is forward-biased.

   B. The gate-source junction is forward biased.

   C. The drain is negative relative to the source.

   D. The gate must be at dc ground.

6. A JFET is sometimes biased at or beyond pinchoff in:

   A. A power amplifier.

   B. A rectifier.

   C. An oscillator.

   D. A weak-signal amplifier.

7. The gate of a JFET has:

   A. Forward bias.

   B. High impedance.

   C. Low reverse resistance.

   D. Low avalanche voltage.

8. A JFET circuit essentially never has:

   A. A pinched-off channel.

   B. Holes as the majority carriers.

   C. A forward-biased P-N junction.

   D. A high input impedance.

9. When a JFET is pinched off:

   A. dI_D/dE_G is very large with no signal.

   B. dI_D/dE_G might vary considerably with no signal.

   C. dI_D/dE_G is negative with no signal.

   D. dI_D/dE_G is zero with no signal.

10. Transconductance is the ratio of:

    A. A change in drain voltage to a change in source voltage.

    B. A change in drain current to a change in gate voltage.

    C. A change in gate current to a change in source voltage.

    D. A change in drain current to a change in drain voltage.

11. Characteristic curves for JFETs generally show:

    A. Drain voltage as a function of source current.

    B. Drain current as a function of gate current.

    C. Drain current as a function of drain voltage.

    D. Drain voltage as a function of gate current.

12. A disadvantage of a MOS component is that:

    A. It is easily damaged by static electricity.

    B. It needs a high input voltage.

    C. It draws a large amount of current.

    D. It produces a great deal of electrical noise.

13. The input impedance of a MOSFET:

    A. Is lower than that of a JFET.

    B. Is lower than that of a bipolar transistor.

    C. Is between that of a bipolar transistor and a JFET.

    D. Is extremely high.

14. An advantage of MOSFETs over JFETs is that:

    A. MOSFETs can handle a wider range of gate voltages.

    B. MOSFETs deliver greater output power.

    C. MOSFETs are more rugged.

    D. MOSFETs last longer.

15. The channel in a zero-biased JFET is normally:

    A. Pinched off.

    B. Somewhat open.

    C. All the way open.

    D. Of P-type semiconductor material.

16. When an enhancement-mode MOSFET is at zero bias:

    A. The drain current is high with no signal.

    B. The drain current fluctuates with no signal.

    C. The drain current is low with no signal.

    D. The drain current is zero with no signal.

17. An enhancement-mode MOSFET can be recognized in schematic diagrams by:

    A. An arrow pointing inward.

    B. A broken vertical line inside the circle.

    C. An arrow pointing outward.

    D. A solid vertical line inside the circle.

18. In a source follower, which of the electrodes of the FET receives the input signal?

    A. None of them.

    B. The source.

    C. The gate.

    D. The drain.

19. Which of the following circuits has its output 180 degrees out of phase with its input?

    A. Common source.

    B. Common gate.

    C. Common drain.

    D. All of them.

20. Which of the following circuits generally has the greatest gain?

    A. Common source.

    B. Common gate.

    C. Common drain.

    D. It depends only on bias, not on which electrode is grounded.

24
CHAPTER

Amplifiers

IN THE PRECEDING TWO CHAPTERS, YOU SAW SCHEMATIC DIAGRAMS WITH bipolar and field-effect transistors. The main intent was to acquaint you with biasing schemes. Some of the diagrams were of basic amplifier circuits. This chapter examines amplifiers more closely, but the subject is vast. For a thorough treatment, you should consult a book devoted to amplifiers and amplification.

The decibel

The extent to which a circuit amplifies is called the *amplification factor*. This can be given as a simple number, such as 100, meaning that the output signal is 100 times as strong as the input. More often, amplification factor is specified in units called *decibels*, abbreviated *dB*.

It's important to keep in mind what is being amplified: current, voltage, or power. Circuits are designed to amplify one of these aspects of a signal, but not necessarily the others. In a given circuit, the amplification factor is not the same for all three parameters.

Perception is logarithmic

You don't perceive loudness directly. Instead, you sense it in a nonlinear way. Physicists and engineers have devised the decibel system, in which amplitude changes are expressed according to the *logarithm* of the actual value (Fig. 24-1), to define relative signal strength.

Gain is assigned positive decibel values; loss is assigned negative values. Therefore, if signal A is at +6 dB relative to signal B, then A is stronger than B; if signal A is at −14 dB relative to B, then A is weaker than B.

An amplitude change of plus or minus 1 dB is about equal to the smallest change a listener can detect if the change is expected. If the change is not expected, then the smallest difference a listener can notice is about plus or minus 3 dB.

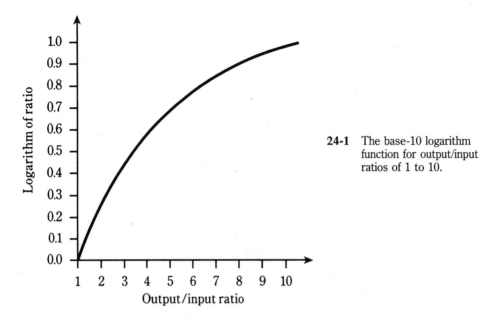

24-1 The base-10 logarithm function for output/input ratios of 1 to 10.

For voltage

Suppose there is a circuit with an rms ac input voltage of E_{in} and an rms ac output voltage of E_{out}. Then the *voltage gain* of the circuit, in decibels, is given by the formula:

$$\text{Gain (dB)} = 20 \log_{10}(E_{out}/E_{in})$$

The logarithm function is abbreviated *log*. The subscript 10 means that the *base* of the logarithm is 10. (Logarithms can have bases other than 10. This gets a little sophisticated, and it won't be discussed here.) You don't have to know the mathematical theory of logarithms to calculate them. All you need, and should buy right this minute if you don't have one, is a calculator that includes logarithm functions.

From now on, the base-10 logarithm will be called just the "logarithm," and the subscript 10 will be omitted.

Problem 24-1

A circuit has an rms ac input of 1.00 V and an rms ac output of 14.0 V. What is the gain in decibels?

First, find the ratio E_{out}/E_{in}. Because $E_{out} = 14.0$ V and $E_{in} = 1.00$ V, this ratio is 14.0/1.00, or 14.0. Then, find the logarithm of 14.0. Your calculator will tell you that log 14.0 = 1.146128036 (it adds a lot of unnecessary digits). Finally, press the various buttons to multiply this number by 20, getting (with my calculator, anyway) 22.92256071. Round off to three significant figures, because that's all you're entitled to: Gain = 22.9 dB.

Problem 24-2

A circuit has an rms ac input voltage of 24.2 V and an rms ac output voltage of 19.9 V. What is the gain in decibels?

Find the ratio E_{out}/E_{in} = 19.9/24.2 = 0.822. . . . (The three dots indicate extra digits introduced by the calculator. You can leave them in till the final roundoff.) Find the logarithm of this: log 0.822. . . = −0.0849. . . . Then multiply by 20: Gain = −1.699. . . dB, rounded off to −1.70 dB.

Negative gain translates into *loss*. A gain of −1.70 dB is equivalent to a loss of 1.70 dB. The circuit of Problem 24-2 is not an amplifier—or if it is supposed to be, it isn't working!

If a circuit has the same output voltage as input voltage, that is, if $E_{out} = E_{in}$, then the gain is 0 dB. The ratio E_{out}/E_{in} is always equal to 1 in this case, and log 1 = 0.

It's important to remember, when doing gain calculations, always to use the same units for the input and the output signal levels. Don't use millivolts for E_{out} and microvolts for E_{in}, for example. This applies to current and power also.

For current

The *current gain* of a circuit is calculated just the same way as for voltage. If I_{in} is the rms ac input current and I_{out} is the rms ac output current, then:

$$\text{Gain (dB)} = 20 \log (I_{out}/I_{in})$$

Often, a circuit that produces voltage gain will produce current loss, and vice-versa. An excellent example is a simple ac transformer.

Some circuits have gain for both the voltage and the current, although not the same decibel figures. The reason is that the output impedance is different from the input impedance, altering the ratio of voltage to current.

For power

The *power gain* of a circuit, in decibels, is calculated according to the formula:

$$\text{Gain (dB)} = 10 \log (P_{out}/P_{in})$$

where P_{out} is the output signal power and P_{in} is the input signal power.

Problem 24-3

A power amplifier has an input of 5.03 W and an output of 125 W. What is the gain in decibels?

First find the ratio P_{out}/P_{in} = 125/5.03 = 24.85. . . . Then find the logarithm: log 24.85. . . = 1.395. . . . Finally, multiply by 10 and round off: Gain = 10 × 1.395. . . = 14.0 dB.

Problem 24-4

An attenuator provides 10 dB power reduction. The input power is 94 W. What is the output power?

This problem requires you to "plug values into" the formula. An *attenuator* produces a power *loss*. When you hear that the attenuation is 10 dB, it is the same thing as a gain of −10 dB. You know P_{in} = 94 W; the unknown is P_{out}. Therefore:

$$-10 = 10 \log (P_{out}/94)$$

Solving this formula proceeds in several steps. First, divide each side by 10, getting:

$$-1 = \log (P_{out}/94)$$

Then, take the *base-10 antilogarithm*, also known as the *antilog*, of each side. The antilog function is the *inverse* of the log function; that is, it "undoes" the log function. The function antilog (x) is sometimes written as 10^x. Thus:

$$\text{antilog} (-1) = 10^{-1} = 0.1 = P_{out}/94$$

Now, multiply each side of the equation by 94, getting:

$$94 \times 0.1 = 9.4 = P_{out}$$

Therefore, the output power is 9.4 W.

Don't confuse the voltage/current and power formulas. In general, for a given output/input ratio, the dB gain for voltage or current is twice the dB gain for power. Table 24-1 gives dB gain figures for various ratios of voltage, current, and power.

Table 24-1.   Decibel gain figures for various ratios of voltage, current and power.

Ratio	Voltage or current gain	Power gain
0.000 000 001 (10^{-9})	−180 dB	−90 dB
0.000 000 01 (10^{-8})	−160 dB	−80 dB
0.000 000 1 (10^{-7})	−140 dB	−70 dB
0.000 001 (10^{-6})	−120 dB	−60 dB
0.000 01 (10^{-5})	−100 dB	−50 dB
0.000 1 (10^{-4})	−80 dB	−40 dB
0.001	−60 dB	−30 dB
0.01	−40 dB	−20 dB
0.1	−20 dB	−10 dB
0.25	−12 dB	−6 dB
0.5	−6 dB	−3 dB
1	0 dB	0 dB
2	6 dB	3 dB
4	12 dB	6 dB
10	20 dB	10 dB
100	40 dB	20 dB
1000	60 dB	30 dB
10,000 (10^4)	80 dB	40 dB
100,000 (10^5)	100 dB	50 dB
1,000,000 (10^6)	120 dB	60 dB
10,000,000 (10^7)	140 dB	70 dB
100,000,000 (10^8)	160 dB	80 dB
1,000,000,000 (10^9)	180 dB	90 dB
10,000,000,000 (10^{10})	200 dB	100 dB

Basic bipolar amplifier circuit

In the previous chapters, you saw some circuits that will work as amplifiers. The principle is the same for all electronic amplification circuits. A signal is applied at some control point, causing a much greater signal to appear at the output.

In Fig. 24-2, an NPN bipolar transistor is connected as a common-emitter amplifier. The input signal passes through C2 to the base. Resistors R2 and R3 provide bias. Resistor R1 and capacitor C1 allow for the emitter to have a dc voltage relative to ground, while being grounded for signals. Resistor R1 also limits the current through the transistor. The ac output signal goes through capacitor C3. Resistor R4 keeps the ac output signal from being short-circuited through the power supply.

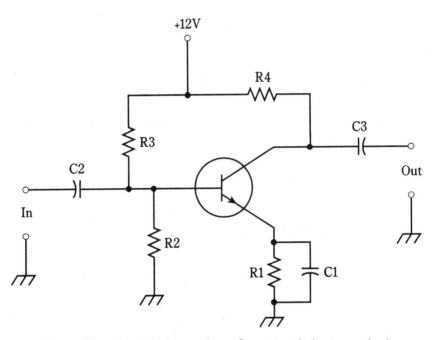

24-2 An amplifier using a bipolar transistor. Component designators and values are discussed in the text.

In this amplifier, the capacitors must have values large enough to allow the ac signal to pass with ease. But they shouldn't be much larger than the minimum necessary for this purpose. If an 0.1-uF capacitor will suffice, there's no point in using a 47-uF capacitor. That would introduce unwanted losses into the circuit, and would also make the circuit needlessly expensive to build.

The ideal capacitance values depend on the design frequency of the amplifier, and also on the impedances at the input and output. In general, as the frequency and/or circuit impedance increase, less and less capacitance is needed. At audio frequencies, say 300 Hz to 20 kHz, and at low impedance, the capacitors might be as large as 100 uF. At radio frequencies, such as 1 MHz to 50 MHz, and with high impedances, values will be only a fraction of a microfarad, down to picofarads at the highest frequencies and impedances. The exact values are determined by the design engineers, working to optimize circuit performance in the lab.

The resistor values likewise depend on the application. Typical values are R1 = 470 Ω, R2 = 4.7 KΩ, R3 = 10 KΩ, and R4 = 4.7 KΩ for a weak-signal, broadband amplifier.

If the circuit is used as a power amplifier, such as in a radio transmitter or a stereo hi-fi amplifier, the values of the resistors will be different. It might be necessary to bias the base negatively with respect to the emitter, using a second power supply with a voltage negative with respect to ground.

Basic FET amplifier circuit

In Fig. 24-3, an N-channel JFET is hooked up as a common-source amplifier. The input signal passes through C2 to the gate. Resistor R2 provides the bias. Resistor R1 and capacitor C1 give the source a dc voltage relative to ground, while grounding it for ac signals. The ac output signal goes through capacitor C3. Resistor R3 keeps the ac output signal from being short-circuited through the power supply.

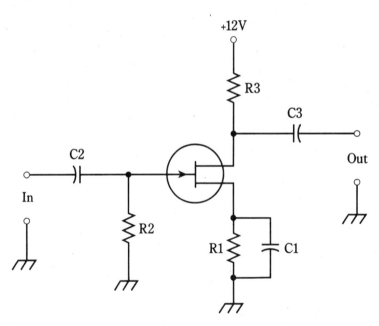

24-3   An amplifier using an FET. Component designators and values are discussed in the text.

Concerning the values of the capacitors, the same considerations apply for this amplifier, as apply in the bipolar circuit. A JFET amplifier almost always has a high input · impedance, and therefore the value of C2 will usually be small. If the device is a MOSFET, the input impedance is even higher, and C2 will be smaller yet, sometimes as little as 1 pF or less.

The resistor values depend on the application. In some instances, R1 and C1 are not used, and the source is grounded directly. If R1 is used, its value will depend on the input impedance and the bias needed for the FET. Nominal values might be R1 = 680 Ω, R2 = 10 KΩ, and R3 = 100 Ω for a weak-signal, wideband amplifier.

If the circuit is used as a power amplifier, the values of the resistors will be different. It might be necessary to bias the gate negatively with respect to the source, using a second power supply with a voltage negative relative to ground.

The class-A amplifier

With the previously mentioned component values, the amplifier circuits in Figs. 24-2 and 24-3 will operate in *class A*. Weak-signal amplifiers, such as the kind used in the first stage of a sensitive radio receiver, are always class-A. The term does not arise from inherent superiority of the design or technique (it's not like saying "grade-A eggs"). It's just a name chosen by engineers so that they know the operating conditions in the bipolar transistor or FET.

A class-A amplifier is always linear. That means that the output waveform has the same shape as (although a much greater amplitude than) the input waveform.

For class-A operation with a bipolar transistor, the bias must be such that, with no signal input, the device is near the middle of the straight-line portion of the I_C vs E_B (collector current versus base voltage) curve. This is shown for an NPN transistor in Fig. 24-4. For PNP, reverse the polarity signs.

With a JFET or MOSFET, the bias must be such that, with no signal input, the device is near the middle of the straight-line part of the I_D vs E_G (drain current versus gate

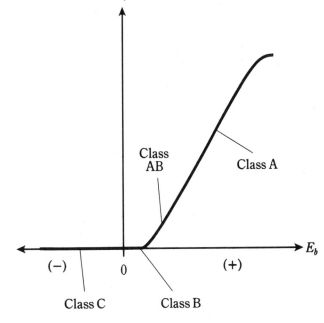

24-4 Various classes of amplifier operation for an NPN bipolar transistor.

voltage) curve. This is shown in Fig. 24-5 for an N-channel device. For P- channel, reverse the polarity signs.

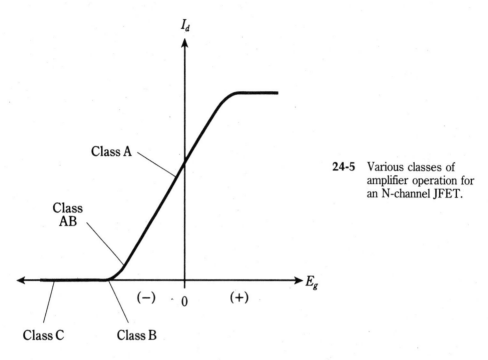

24-5   Various classes of amplifier operation for an N-channel JFET.

It is important with class-A amplifiers that the input signal not be too strong. Otherwise, during part of the cycle, the base or gate voltage will be driven outside of the straight-line part of the curve. When this occurs, the output waveshape will not be a faithful reproduction of the input waveshape; the amplifier will be nonlinear. This will cause distortion of the signal. In an audio amplifier, the output might sound "raspy" or "scratchy." In a radio-frequency amplifier, the output signal will contain a large amount of energy at harmonic frequencies. The problem of harmonics, however, can be dealt with by means of resonant circuits in the output. These circuits attenuate harmonic energy, and allow amplifiers to be biased near, at, or even past cutoff or pinchoff.

The class-AB amplifier

When a class-A amplifier is working properly, it has low distortion. But class-A operation is inefficient. (Amplifier efficiency will be discussed later in this chapter.) This is mainly because the bipolar transistor or FET draws a large current, whether there is a signal input or not. Even with zero signal, the device is working hard.

For weak-signal work, efficiency is not very important; it's gain and sensitivity that matter. In power amplifiers, efficiency is a significant consideration, and gain and sensitivity are not so important. Any power not used toward generating a strong output signal will end up as heat in the bipolar transistor or FET. If an amplifier is designed to produce high power output, inefficiency translates to a lot of heat.

When a bipolar transistor is biased close to cutoff under no-signal conditions (Fig. 24-4), or when an FET is near pinchoff (Fig. 24-5), the input signal will drive the device into the nonlinear part of the operating curve. A small collector or drain current will flow when there is no input, but it will be less than the no-signal current that flows in a class-A amplifier. This is called *class-AB operation.*

With class-AB operation, the input signal might or might not cause the device to go into cutoff or pinchoff for a small part of the cycle. Whether or not this happens depends on the actual bias point, and also on the strength of the input signal. You can visualize this by imagining the *dynamic operating point* oscillating back and forth along the curve, in either direction from the *static* (no-signal) *operating point.*

If the bipolar transistor or FET is never driven into cutoff/pinchoff during any part of the signal cycle, the amplifier is working in *class-AB$_1$*. If the device goes into cutoff/pinchoff for any part of the cycle (up to almost half), the amplifier is working in *class-AB$_2$*.

In a class-AB amplifier, the output waveshape is not identical with the input waveshape. But if the wave is *modulated,* such as in a voice radio transmitter, the waveform of the modulations will come out undistorted. Thus class-AB operation is useful in radio-frequency power amplifiers.

The class-B amplifier

When a bipolar transistor is biased exactly at cutoff, or an FET at pinchoff, under zero-input-signal conditions, an amplifier is working in *class B*. These operating points are labeled on the curves in Figs. 24-4 and 24-5.

In class-B operation, there is no collector or drain current when there is no signal. This saves energy, because the circuit is not eating up any power unless there is a signal going into it. (Class-A and class-AB amplifiers draw current even when the input is zero.) When there is an input signal, current flows in the device during exactly half of the cycle. The output waveshape is greatly different from the input waveshape in a class-B amplifier; in fact, it is half-wave rectified.

Sometimes two bipolar transistors or FETs are used in a class-B circuit, one for the positive half of the cycle and the other for the negative half. In this way, distortion is eliminated. This is called a *class-B push-pull amplifier.* A class-B push-pull circuit using two NPN bipolar transistors is illustrated in Fig. 24-6. This configuration is popular for audio-frequency power amplification. It combines the efficiency of class B with the low distortion of class A. Its main disadvantage is that it needs two center-tapped transformers, one at the input and the other at the output. This translates into two things that engineers don't like: bulk and high cost. Nonetheless, the advantages often outweigh these problems.

The class-B scheme lends itself well to radio-frequency power amplification. Although the output waveshape is distorted, resulting in harmonic energy, this problem can be overcome by a resonant LC circuit in the output. If the signal is modulated, the modulation waveform will not be distorted.

You'll sometimes hear of class-AB or class-B "linear amplifiers," especially in ham radio. The term "linear" refers to the fact that the modulation waveform is not distorted by the amplifier. The *carrier wave* is, as you've seen, affected in a nonlinear fashion, because the amplifiers are not biased in the straight-line part of the operating curve.

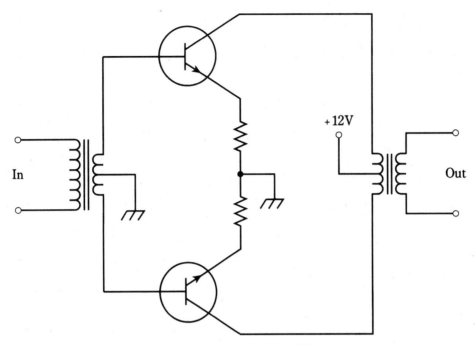

24-6   A class-B push-pull amplifier.

Class-AB$_2$ and class-B amplifiers take some power from the input signal source. Engineers say that such amplifiers require a certain amount of *drive* or *driving power* to function. Class-A and class-AB$_1$ amplifiers theoretically need no driving power, although there must be an input voltage.

The class-C amplifier

A bipolar transistor or FET can be biased past cutoff or pinchoff, and it will still work as a power amplifier (PA), provided that the drive is sufficient to overcome the bias during part of the cycle. You might think, at first, that this bias scheme couldn't possibly result in amplification. Intuitively, it seems as if this could produce a marginal signal loss, at best. But in fact, if there is significant driving power, *class-C* operation can work very well. And, it is more efficient than any of the aforementioned methods. The operating points for class C are labeled in Figs. 24-4 and 24-5.

Class-C PAs are never linear, even for amplitude modulation on a signal. Because of this, a class-C circuit is useful only for signals that are either full-on or full-off. *Continuous-wave (CW)*, also known as Morse code, and *radioteletype (RTTY)* are examples of such signals. Class-C PAs also work well with *frequency modulation (FM)* because the amplitude never changes.

A class-C PA needs a lot of drive. The gain is fairly low. It might take 300 W of radio-frequency (RF) drive to get 1 kW of RF power output. That's a gain of only a little over 5 dB. Nonetheless, the efficiency is excellent, and class-C operation is common in CW, RTTY, or FM radio transmitters.

PA efficiency

Saving energy is a noble thing. But in electronic power amplifiers, as with many other kinds of hardware, energy conservation also translates into lower cost, smaller size and weight, and longer equipment life.

Power input

Suppose you connect an ammeter or milliammeter in series with the collector or drain of an amplifier and the power supply, as shown in Fig. 24-7. While the amplifier is in operation, this meter will have a certain reading. The reading might appear constant, or it might fluctuate with changes in the input signal level.

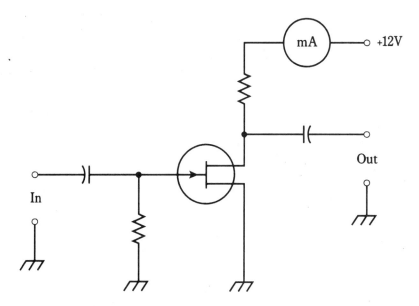

24-7   Connection of a current meter for dc power input measurement.

The *dc collector power input* to a bipolar-transistor amplifier circuit is the product of the collector current in amperes, and the collector voltage in volts. Similarly, for an FET, the *dc drain power input* is the product of the drain current and the drain voltage. These power figures can be further categorized as *average* or *peak* values. (This discussion involves only average power.)

The dc collector or drain power input can be high even when there is no signal applied to an amplifier. A class-A circuit operates this way. In fact, when a signal is applied to a class-A amplifier, the meter reading, and therefore the dc collector or drain power input, will not change compared to the value under no-signal conditions!

In class-AB$_1$ or class-AB$_2$, there is low current (and therefore low dc collector or drain power input) with zero signal, and a higher current (and therefore a higher dc power input) with signal.

In class-B and class-C, there is no current (and therefore zero dc collector or drain power input) when there is no input signal. The current, and therefore the dc power input, increases with increasing signal.

The dc collector or drain power input is usually measured in watts, the product of amperes and volts. It might be indicated in milliwatts for low-power amplifiers, or kilowatts for high-power amplifiers.

Power output

The *power output* of an amplifier must be measured by means of a specialized ac wattmeter. A dc ammeter/voltmeter combination won't work. The design of audio-frequency and radio-frequency wattmeters is a sophisticated specialty in engineering.

When there is no signal input to an amplifier, there is no signal output, and therefore the power output is zero. This is true no matter what the class of amplification. The greater the signal input, in general, the greater the power output of a power amplifier, up to a certain point.

Power output, like dc collector or drain input, is measured in watts. For very low-power circuits, it might be in milliwatts; for high-power circuits it is often given in kilowatts.

Definition of efficiency

The *efficiency* of a power amplifier is the ratio of the ac power output to the dc collector or drain power input.

For a bipolar-transistor amplifier, let P_C be the dc collector power input, and P_{out} be the ac power output. For an FET amplifier, let P_D be the dc drain power input. Then the efficiency, eff, is given by:

$$\text{eff} = P_{out}/P_C$$

for a bipolar-transistor circuit, and:

$$\text{eff} = P_{out}/P_D$$

for an FET circuit. These are ratios, and they will always be between 0 and 1.

Efficiency is often expressed as a percentage, so that the formulas become:

$$\text{eff}(\%) = 100P_{out}/P_C$$

and

$$\text{eff}(\%) = 100P_{out}/P_D$$

Problem 24-5

A bipolar-transistor amplifier has a dc collector input of 115 W and an ac power output of 65.0 W. What is the efficiency in percent?

Use the formula eff(%) = $100P_{out}/P_C$ = 100 × 65/115 = 100 × 0.565 = 56.5 percent.

Problem 24-6

An FET amplifier is 60 percent efficient. If the power output is 3.5 W, what is the dc drain power input?

"Plug in" values to the formula $eff(\%) = 100 P_{out}/P_D$. This gives:

$$60 = 100 \times 3.5/P_D$$
$$60 = 350/P_D$$
$$60/350 = 1/P_D$$
$$P_D = 350/60 = 5.8 \text{ W}$$

Efficiency versus class

Class-A amplifiers are the least efficient, in general. The efficiency ranges from 25 to 40 percent, depending on the nature of the input signal and the type of bipolar or field-effect transistor used.

If the input signal is very weak, such as might be the case in a shortwave radio receiver, the efficiency of a class-A circuit is near zero. But in that application, the circuit is not working as a power amplifier, and efficiency is not of primary importance.

Class-AB amplifiers have better efficiency. A good class-AB$_1$ amplifier might be 35 to 45 percent efficient; a class-AB$_2$ amplifier will be a little better, approaching 60 percent with the best designs.

Class-B amplifiers are typically 50 to 60 percent efficient, although some radio-frequency PA circuits work up to 65 percent or so.

Class-C amplifiers are the best of all. This author has seen well-designed class-C circuits that are 75 percent efficient.

These are not absolute figures, and you shouldn't memorize them as such. It's sufficient to know ballpark ranges, and that efficiency improves as the operating point moves towards the left on the curves shown in Figs. 24-4 and 24-5.

Drive and overdrive

Class-A power amplifiers do not, in theory, take any power from the signal source in order to produce a significant amount of output power. This is one of the advantages of class-A operation. The same is true for class-AB$_1$ amplifiers. It is only necessary that a certain voltage be present at the control electrode (the base, gate, emitter, or source).

Class-AB$_2$ amplifiers need some driving power to produce ac power output. Class-B amplifiers require more drive than class-AB$_2$, and class-C amplifiers need still more drive.

Whatever kind of PA is used in a given situation, it is important that the driving signal not be too strong. If *overdrive* takes place, there will be distortion in the output signal.

An oscilloscope can be used to determine whether or not an amplifier is being overdriven. The scope is connected to the amplifier output terminals, and the waveshape of the output signal is examined. The output waveform for a particular class of amplifier always has a characteristic shape. Overdrive is indicated by a form of distortion known as *flat topping*.

In Fig. 24-8A, the output signal waveshape for a properly operating class-B amplifier is shown. It looks like the output of a half-wave rectifier, because the bipolar transistor or FET is drawing current for exactly half (180 degrees) of the cycle.

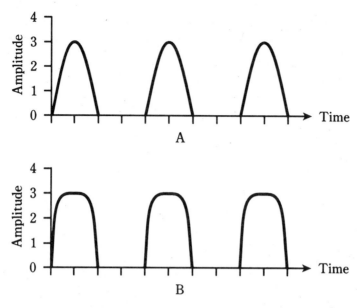

24-8 At A, waveshape at the output of a properly operating class-B amplifier. At B, distortion in the output waveshape caused by overdrive.

In Fig. 24-8B, the output of an overdriven class-B amplifier is shown. The wave is no longer a half sine wave, but instead, it shows evidence of flat topping. The peaks are blunted or truncated. The result of this is audio distortion in the modulation on a radio signal, and also an excessive amount of energy at harmonic frequencies.

The efficiency of a circuit can be degraded by overdrive. The "flat tops" of the distorted waves don't contribute anything to the strength of the signal at the desired frequency. But they do cause a higher-than-normal P_C or P_D value, which translates into a lower-than-normal efficiency P_{out}/P_C or P_{out}/P_D.

A thorough discussion of overdrive and distortion in various amplifier classes and applications would require an entire book. If you're interested in more detail, a good college or trade-school text on radio-frequency (RF) amplification is recommended.

Audio amplification

The circuits you've seen so far have been general, not application-specific. With capacitors of several microfarads, and when biased for class A, these circuits are representative of audio amplifiers. As with RF amplifiers, there isn't room enough to go into great depth about audio amplifiers in this book, but a couple of important characteristics deserve mention.

Frequency response

High-fidelity audio amplifiers, of the kind used in music systems, must have more or less constant gain from 20 Hz to 20 kHz. This is a frequency range of 1000:1. Audio amplifiers for voice communications must work from 300 Hz to 3 kHz, a 10:1 span of frequencies. In digital communications, audio amplifiers are designed to work over a narrow range of frequencies, sometimes less than 100 Hz wide.

Hi-fi amplifiers are usually equipped with resistor-capacitor (RC) networks that tailor the frequency response. These are *tone controls*, also called *bass* and *treble* controls. The simplest hi-fi amplifiers use a single knob to control the tone. More sophisticated "amps" have separate controls, one for bass and the other for treble. The most advanced hi-fi systems make use of *graphic equalizers*, having controls that affect the amplifier gain over several different frequency spans.

Gain-versus-frequency curves for three hypothetical audio amplifiers are shown in Fig. 24-9. At A, a wideband, flat curve is illustrated. This is typical of hi-fi system amplifiers. At B, a voice communications response is shown. At C, a narrowband response curve, typical of audio amplifiers in Morse code or radioteletype receivers, is illustrated.

Volume control

Audio amplifier systems usually consist of two or more *stages*. A stage is one bipolar transistor or FET (or a push-pull combination), plus peripheral resistors and capacitors. Stages are cascaded one after the other to get high gain.

In one of the stages in an audio system, a *volume control* is used. This control is usually a potentiometer that allows the gain of a stage to be adjusted without affecting its linearity. An example of a simple volume control is shown in Fig. 24-10.

In this amplifier, the gain through the transistor itself is constant. The ac output signal passes through C1 and appears across R1, a potentiometer. The *wiper* (indicated by the arrow) of the potentiometer "picks off" more or less of the ac output signal, depending on the position of the control shaft. When the shaft is fully counterclockwise, the arrow is at the bottom of the zig-zaggy line, and none of the signal passes to the output. When the shaft is fully clockwise, the arrow is at the top of the zig-zaggy line, and all of the signal passes to the output. At intermediate positions of the control shaft, various proportions of the full output signal will appear at the output. Capacitor C2 isolates the potentiometer from the dc bias of the following stage.

Volume control is usually done in a stage where the audio power level is quite low. This allows the use of a small potentiometer, rated for perhaps 1 W. If volume control were done at high audio power levels, the potentiometer would need to be able to dissipate large amounts of power, and would be needlessly expensive.

Coupling methods

In all of the amplifiers you've seen so far, with the exception of the push-pull circuit (Fig. 24-6), capacitors have been used to allow ac to pass while blocking dc. But there is another way to do this, and in some amplifier systems, it is preferred. This is the use of a transformer to couple signals from one stage to the next.

An example of *transformer coupling* is shown in Fig. 24-11. Capacitors C1 and C2 keep one end of the transformer primary and secondary at signal ground. Resistor R1

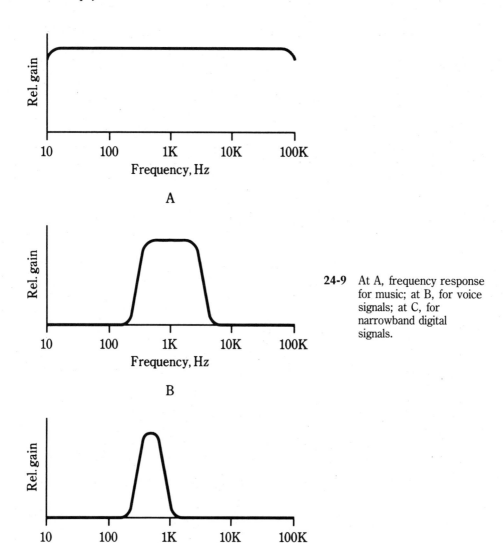

24-9 At A, frequency response for music; at B, for voice signals; at C, for narrowband digital signals.

limits the current through the first transistor, Q1. (In some cases, R1 might be eliminated.) Resistors R2 and R3 provide the proper base bias for transistor Q2.

The main disadvantage of this scheme is that it costs more than *capacitive coupling*. But transformer coupling can provide an optimum signal transfer between amplifier stages with a minimum of loss. This is because of the impedance-matching ability of transformers. Remember that the turns ratio of a transformer affects not only the input and output voltage, but the ratio of impedances. By selecting the right transformer, the output impedance of Q1 can be perfectly matched to the input impedance of Q2.

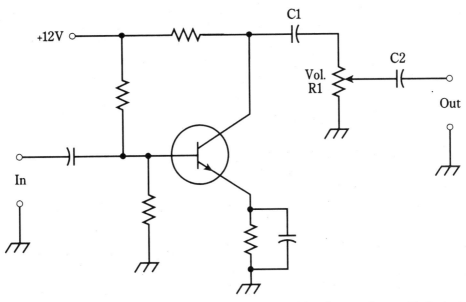

24-10 A simple volume control. Component designators and functions are discussed in the text.

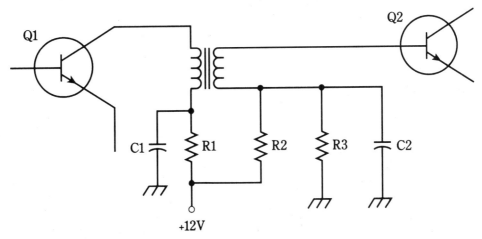

24-11 Transformer coupling. Component designators and functions are discussed in the text.

In some amplifier systems, capacitors are added across the primary and/or secondary of the transformer. This results in resonance at a frequency determined by the capacitance and the transformer winding inductance. If the set of amplifiers is intended for just one frequency (and this is often the case in RF systems), this method of coupling, called *tuned-circuit coupling*, enhances the system efficiency. But care must be taken to be sure that the amplifier chain doesn't get so efficient that it *oscillates* at the resonant frequency of the tuned circuits! You'll learn about oscillation in the next chapter.

Radio-frequency amplification

The RF spectrum begins at about 9 kHz and extends upward in frequency to well over 300 GHz, or 300,000,000,000 Hz. A complete discussion of RF amplifier design would occupy a book. Therefore, again, only a sketch of the most important characteristics can be given here.

Weak-signal versus power amplifiers

Some RF amplifiers are designed for weak-signal work. The general circuits, shown earlier in this chapter, are representative of such amplifiers, when the capacitors have values of about 1 uF or less. The higher the frequency, the smaller the values of the capacitors.

The *front end*, or first amplifying stage, of a radio receiver requires the most sensitive possible amplifier. Sensitivity is determined by two factors: *gain* and *noise figure*.

The noise figure of an amplifier is a measure of how well it can amplify the desired signal, without injecting unwanted noise. All bipolar transistors or FETs create some *white noise* because of the movement of charge carriers. In general, JFETs produce less noise than bipolar transistors. Gallium arsenide FETs, also called *GaAsFETs* (pronounced "gasfets"), are the least noisy of all.

The higher the frequency at which a weak-signal amplifier is designed, the more important the noise figure gets. This is because there is less atmospheric noise at the higher radio frequencies, as compared with the lower frequencies. At 1.8 MHz, for example, the airwaves contain much atmospheric noise, and it doesn't make a significant difference if the receiver introduces a little noise itself. But at 1.8 GHz the atmospheric noise is almost nonexistent, and receiver performance depends much more critically on the amount of internally generated noise.

Weak-signal amplifiers almost always use resonant circuits. This optimizes the amplification at the desired frequency, while helping to cut out noise on unwanted frequencies. A typical tuned GaAsFET weak-signal RF amplifier is diagrammed in Fig. 24-12. It is designed for about 10 MHz.

Broadband PAs

At RF, a PA might be either *broadband* or *tuned*.

The main advantage of a broadband PA is ease of operation, because it does not need tuning. A broadbanded amplifier is not "particular" with respect to the frequency within its design range, such as 1.5 MHz through 15 MHz. The operator need not worry about critical adjustments, nor bother to change them when changing the frequency.

One disadvantage of broadband PAs is that they are slightly less efficient than tuned PAs. This usually isn't too hard to put up with, though, considering the convenience of not having to fiddle with the tuning.

The more serious problem with broadband PAs is that they'll amplify *anything* in the design range, whether or not you want it to go over the air. If some earlier stage in a transmitter is oscillating at a frequency nowhere near the intended signal frequency, and

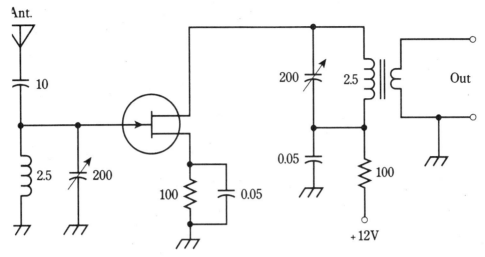

4-12 A tuned RF amplifier for use at about 10 MHz. Resistances are in ohms. Capacitances are in μF if less than 1, and in pF if more than 1. Inductances are in uH.

if this undesired "signal" falls within the design frequency range of the broadband PA, it will be amplified. The result will be unintended (and illegal!) RF emission from the radio transmitter. Such unwanted signals are called *spurious emissions*, and they occur more often than you might think.

A typical broadband PA circuit is diagrammed schematically in Fig. 24-13. The NPN bipolar transistor is a power transistor, such as General Electric D44C6. It will reliably provide about 3 W of continuous RF output from 1.5 MHz through 15 MHz. The transformers are a critical part of this circuit; they must be designed to work well over a 10:1 range of frequencies. This circuit is suitable for use on the ham radio bands at 160, 80, 75, 40, 30, and 20 meters.

Tuned PAs

A tuned RF power amplifier offers improved efficiency compared with broadband designs. Also, the tuning helps to reduce the chances of spurious signals being amplified and transmitted over the air.

Another advantage of tuned PAs is that they can work into a wide range of load impedances. In addition to a *tuning control*, or resonant circuit that adjusts the output of the amplifier to the operating frequency, there is a *loading control* that optimizes the signal transfer between the amplifier and the load (usually an antenna).

The main drawback of a tuned PA is that the adjustment takes time, and improper adjustment can result in damage to the amplifying device (bipolar transistor or FET). If the tuning and/or loading controls are out of kilter, the efficiency of the amplifier will be extremely low—sometimes practically zero—while the dc collector or drain power input is unnaturally high. Solid-state devices overheat quickly under these conditions.

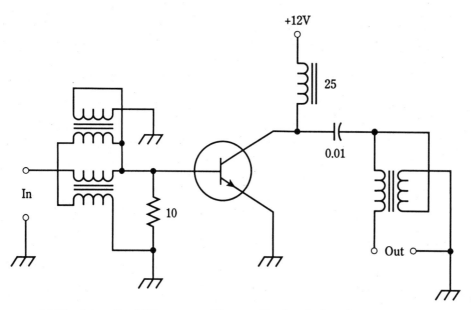

24-13 A broadband RF power amplifier, capable of producing a few watts output.

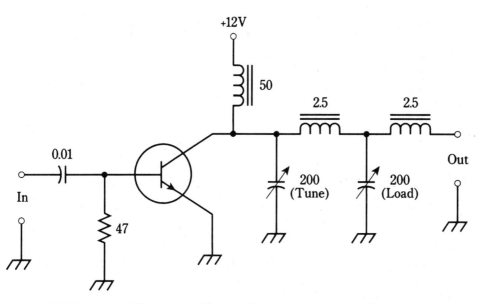

24-14 A tuned RF power amplifier, capable of producing a few watts output.

A tuned RF PA, providing 3 W output at 10 MHz or so, is shown in Fig. 24-14. The transistor is the same as for the broadband amplifier discussed above. The tuning and loading controls should be adjusted for maximum RF power output as indicated on a wattmeter in the feed line going to the load.

Quiz

Refer to the text in this chapter if necessary. A good score is at least 18 correct. Answers are in the back of the book.

1. The decibel is a unit of:
   A. Relative signal strength.
   B. Voltage.
   C. Power.
   D. Current.

2. If a circuit has a voltage-amplification factor of 20, then the voltage gain is:
   A. 13 dB.
   B. 20 dB.
   C. 26 dB.
   D. 40 dB.

3. A gain of −15 dB in a circuit means that:
   A. The output signal is stronger than the input.
   B. The input signal is stronger than the output.
   C. The input signal is 15 times as strong as the output.
   D. The output signal is 15 times as strong as the input.

4. A device has a voltage gain of 23 dB. The input voltage is 3.3 V. The output voltage is:
   A. 76 V.
   B. 47 V.
   C. 660 V.
   D. Not determinable from the data given.

5. A power gain of 44 dB is equivalent to an output/input power ratio of:
   A. 44.
   B. 160.
   C. 440.
   D. 25,000.

6. A resistor between the base of an NPN bipolar transistor and the positive supply voltage is used to:
   A. Provide proper bias.
   B. Provide a path for the input signal.
   C. Provide a path for the output signal.
   D. Limit the collector current.

7. The capacitance values in an amplifier circuit depend on:
   A. The supply voltage.
   B. The polarity.
   C. The signal strength.
   D. The signal frequency.

8. A class-A circuit would not work well as:
   A. A stereo hi-fi amplifier.
   B. A television transmitter PA.
   C. A low-level microphone preamplifier.
   D. The first stage in a radio receiver.

9. In which of the following FET amplifier types does drain current flow for 50 percent of the signal cycle?
   A. Class-A.
   B. Class-AB_1.
   C. Class-AB_2.
   D. Class-B.

10. Which of the following amplifier types produces the least distortion of the signal waveform?
    A. Class-A.
    B. Class-AB_1.
    C. Class-AB_2.
    D. Class-B.

11. Which bipolar amplifier type has some distortion in the signal wave, with collector current during most, but not all, of the cycle?
    A. Class-A.
    B. Class-AB_1.
    C. Class-AB_2.
    D. Class-B.

12. How can a class-B amplifier be made suitable for hi-fi audio applications?
    A. By increasing the bias.
    B. By using two transistors in push-pull.
    C. By using tuned circuits in the output.
    D. A class-B amplifier cannot work well for hi-fi audio.

13. How can a class-C amplifier be made linear?
    A. By reducing the bias.

B. By increasing the drive.

C. By using two transistors in push-pull.

D. A class-C amplifier cannot be made linear.

14. Which of the following amplifier classes generally needs the most driving power?

   A. Class A.

   B. Class AB_1.

   C. Class AB_2.

   D. Class B.

15. A graphic equalizer is a form of:

   A. Bias control.

   B. Gain control.

   C. Tone control.

   D. Frequency control.

16. A disadvantage of transformer coupling, as opposed to capacitive coupling, is that:

   A. Transformers can't match impedances.

   B. Transformers can't work above audio frequencies.

   C. Transformers cost more.

   D. Transformers reduce the gain.

17. A certain bipolar-transistor PA is 66 percent efficient. The output power is 33 W. The dc collector power input is:

   A. 22 W.

   B. 50 W.

   C. 2.2 kW.

   D. None of the above.

18. A broadband PA is:

   A. Generally easy to use.

   B. More efficient than a tuned PA.

   C. Less likely than a tuned PA to amplify unwanted signals.

   D. Usable only at audio frequencies.

19. A tuned PA must always be:

   A. Set to work over a wide range of frequencies.

   B. Adjusted for maximum power output.

   C. Made as efficient as possible.

   D. Operated in class C.

20. A loading control in a tuned PA:

    A. Provides an impedance match between the bipolar transistor or FET and the load.

    B. Allows broadband operation.

    C. Adjusts the resonant frequency.

    D. Controls the input impedance.

25
CHAPTER

Oscillators

SOMETIMES AMPLIFIERS WORK TOO WELL. YOU'VE PROBABLY HEARD THIS WHEN someone was getting ready to speak over a public-address system. The gain was set too high. The person began to speak; sound from the speakers got into the microphone, was amplified, went to the speakers again, and back to the microphone. A vicious cycle of *feedback* ensued. The result might have been a rumble, a howl, or a shriek. The system broke into *oscillation*. The amplifiers became temporarily useless until the gain was reduced.

Oscillation can be controlled, so that it takes place at a specific, stable, predictable frequency. An *oscillator* is a circuit that is deliberately designed to oscillate.

Uses of oscillators

Some oscillators work at audio frequencies, and others are intended to produce radio signals. Most generate sine waves, although some are built to emit square waves, sawtooth waves, or other waveshapes.

The subject of oscillators, once you understand amplifiers, is elementary. All oscillators are amplifiers with positive feedback. In this chapter, radio-frequency (RF) oscillators are discussed in some detail, and then audio oscillators are examined.

In radio communications, oscillators generate the "waves," or signals, that are ultimately sent over the air. For data to be sent, the signal from an oscillator must be *modulated*. Modulation is covered in chapter 26.

Oscillators are used in radio and TV receivers for frequency control and for *detection* and *mixing*. Detectors and mixers are discussed in chapter 27.

Audio-frequency oscillators find applications in such devices as music synthesizers, FAX modems, doorbells, beepers, sirens and alarms, and electronic toys.

Positive feedback

Feedback can be in phase or out of phase. For a circuit to oscillate, the feedback must be in phase, or *positive. Negative feedback* (out of phase) simply reduces the gain.

The output of a common-emitter or common-source amplifier is out of phase from the input. If you couple the collector to the base through a capacitor, you won't get oscillation.

The output of a common-base or common-gate amplifier is in phase with the input. But these circuits have limited gain. It's hard to make them oscillate, even with positive feedback.

Common-collector and common-drain circuits don't have enough gain to make oscillators.

But take heart: There are lots of ways to make circuits oscillate. Obtaining oscillation has never been a problem in electronics. Public-address systems do it willingly enough!

Concept of the oscillator

For a circuit to oscillate, the gain must be high, the feedback must be positive, and the *coupling* from output to input must be good. The feedback path must be easy for a signal to follow. The phase of a fed-back signal can be reversed without any trouble, so that common-emitter or common-source amplifiers can be made to oscillate.

Feedback at a single frequency

Recalling the public-address fiasco, some variation of which you've doubtless heard many times, could you know in advance whether the feedback would have a low pitch, a midrange pitch, or a high pitch? No. The oscillation was not intended, and might have started at any audio frequency.

The frequency of an oscillator is controlled by means of tuned, or resonant, circuits. These are usually inductance-capacitance (LC) or resistance-capacitance (RC) combinations. The LC scheme is common at RF; the RC method is more often used for audio oscillators.

The tuned circuit makes the feedback path easy for a signal to follow at one frequency, but hard to follow at all other frequencies (Fig. 25-1). The result is that the oscillation takes place at a predictable and stable frequency, determined by the inductance and capacitance, or by the resistance and capacitance.

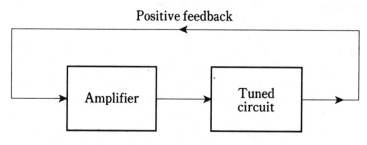

25-1   Basic concept of the oscillator.

The Armstrong oscillator

A common-emitter or common-source amplifier can be made to oscillate be coupling the output back to the input through a transformer that reverses the phase of the fed-back signal. The phase at a transformer output can be inverted by reversing the secondary terminals.

The schematic diagram of Fig. 25-2 shows a common-source amplifier whose drain circuit is coupled to the gate circuit via a transformer. In practice, getting oscillation is easy. If the circuit won't oscillate with the transformer secondary hooked up one way, you can just switch the wires.

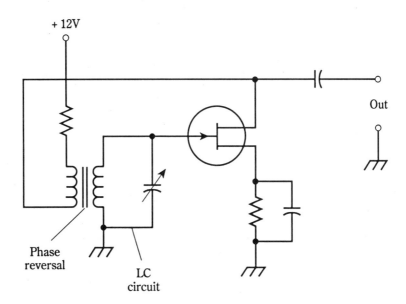

25-2   An Armstrong oscillator.

The frequency of this oscillator is controlled by means of a capacitor across either the primary or the secondary winding of the transformer. The inductance of the winding, along with the capacitance, forms a resonant circuit. The formula for determining the LC resonant frequency is in chapter 17. If you've forgotten it, now is a good time to review it.

The oscillator of Fig. 25-2 is known as an *Armstrong oscillator*. A bipolar transistor can be used in place of the JFET. It would need to be biased, using a resistive voltage-divider network, like a class-A amplifier.

The Hartley circuit

A method of obtaining controlled feedback at RF is shown in Fig. 25-3. At A, an NPN bipolar transistor is used; at B, an N-channel JFET is employed. The PNP and P-channel circuits are identical, but the power supply is negative instead of positive.

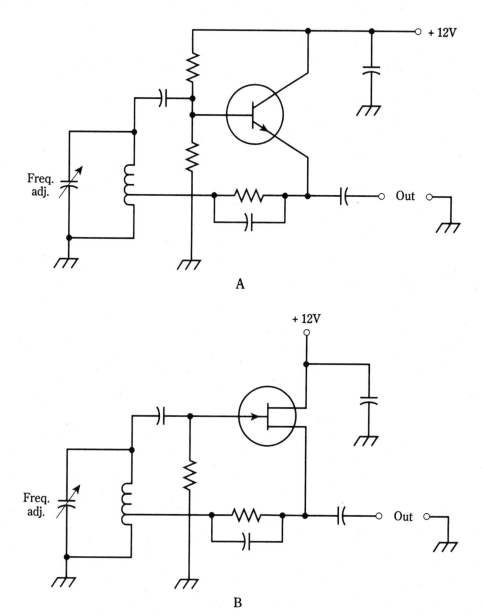

25-3   Hartley oscillators. At A, NPN bipolar transistor; at B, N-channel JFET.

The circuit uses a single coil with a tap on the windings to provide the feedback. A variable capacitor in parallel with the coil determines the oscillating frequency, and allows for frequency adjustment. This circuit is called a *Hartley oscillator.*

The Hartley oscillator uses about one-quarter of its amplifier power to produce feedback. (Remember, all oscillators are really specialized amplifiers.) The other three-quarters of the power can be used as output. Oscillators do not, in general, produce more than a fraction of a watt of power. If more power is needed, the signal can be boosted by one or more stages of amplification.

It's important to use only the minimum amount of feedback necessary to get oscillation. The amount of feedback is controlled by the position of the coil tap.

The Colpitts circuit

Another way to provide RF feedback is to tap the capacitance instead of the inductance in the tuned circuit. In Fig. 25-4, NPN bipolar (at A) and N-channel JFET (at B) *Colpitts oscillator* circuits are diagrammed.

The amount of feedback is controlled by the ratio of capacitances. The coil, rather than the capacitors, is variable in this circuit. This is a matter of convenience. It's almost impossible to find a dual variable capacitor with the right capacitance ratio between sections. Even if you find one, you cannot change the ratio of capacitances. It's easy to adjust the capacitance ratio using a pair of fixed capacitors.

Unfortunately, finding a good variable inductor might not be much easier than getting hold of a suitable dual-gang variable capacitor. A *permeability-tuned* coil can be used, but ferromagnetic cores impair the frequency stability of an RF oscillator. A *roller inductor* might be employed, but these are bulky and expensive. An inductor with several switch-selectable taps can be used, but this wouldn't allow for continuous frequency adjustment. The tradeoff is that the Colpitts circuit offers exceptional stability and reliability when properly designed.

As with the Hartley circuit, the feedback should be kept to the minimum necessary to sustain oscillation.

In these circuits, the outputs are taken from the emitter or source. Why? Shouldn't the output be taken from the collector or drain? The answer is that the output *can* be taken from the collector or drain circuit, and an oscillator will usually work just fine. But gain is not important in an oscillator; what matters is stability under varying load conditions. Stability is enhanced when the output of an oscillator is taken from the emitter or source portion of the circuit.

To prevent the output signal from being short-circuited to ground, an *RF choke (RFC)* is connected in series with the emitter or source lead in the Colpitts circuit. The choke lets dc pass while blocking ac (just the opposite of a blocking capacitor). Typical values for RF chokes range from about 100 uH at high frequencies, like 15 MHz, to 10 mH at low frequencies, such as 150 kHz.

The Clapp circuit

A variation of the Colpitts oscillator makes use of series resonance, instead of parallel resonance, in the tuned circuit. Otherwise, the circuit is basically the same as the parallel-tuned Colpitts oscillator. A schematic diagram of an N-channel JFET *Clapp oscillator* circuit is shown in Fig. 25-5. The P-channel circuit is identical, except for the power supply polarity, which is reversed.

The bipolar-transistor Clapp circuit is almost exactly the same as the circuit of Fig. 25-5, with the emitter in place of the source, the base in place of the gate, and the collector

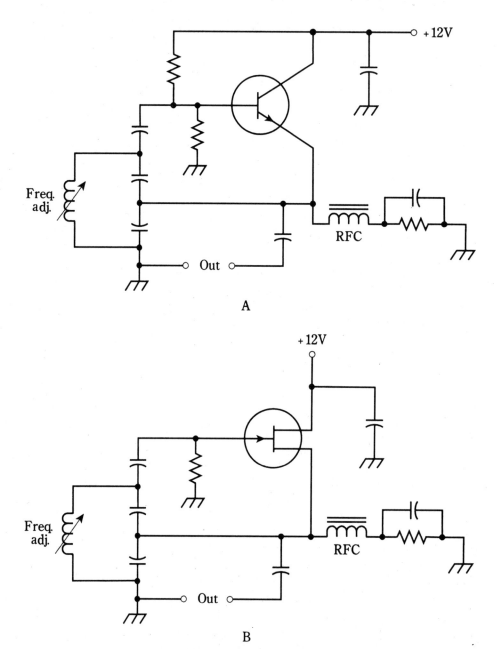

25-4   Colpitts oscillators. At A, NPN bipolar transistor; at B, N-channel JFET.

in place of the drain. The only difference, as you can probably guess by now, is the addition of a resistor between the base and the positive supply voltage (for NPN) or the negative supply voltage (for PNP).

The Clapp oscillator offers excellent stability at RF. Its frequency won't change much when high-quality components are used. The Clapp oscillator is a reliable circuit; it

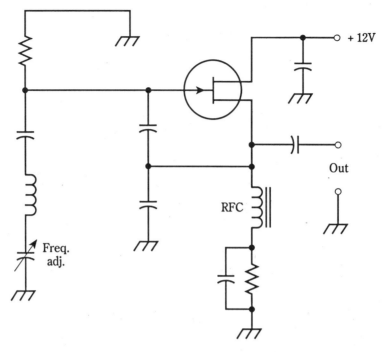

25-5 Series-tuned Colpitts oscillator using an N-channel JFET.

isn't hard to get it to oscillate. Another advantage of the Clapp circuit is that it allows the use of a variable capacitor for frequency control, while still accomplishing feedback through a capacitive voltage divider.

Stability

The term *stability* is used often by engineers when they talk about oscillators. In an oscillator, stability has two meanings: constancy of frequency and reliability of performance. Obviously, both of these considerations are important in the design of a good oscillator circuit.

Constancy of frequency

The foregoing oscillator types allow for frequency adjustment using variable capacitors or variable inductors. The component values are affected by temperature, and sometimes by humidity. When designing a *variable-frequency oscillator (VFO)*, it's crucial that the components maintain constant values, as much as possible, under all anticipated conditions.

Some types of capacitors maintain their values better than others, when the temperature goes up or down. Among the best are polystyrene capacitors. Silver-mica capacitors also work well when polystyrene units can't be found.

Inductors are most temperature-stable when they have air cores. They should be wound, when possible, from stiff wire with strips of plastic to keep the windings in place. Some air-core coils are wound on hollow cylindrical cores, made of ceramic or phenolic material. Ferromagnetic solenoidal or toroidal cores aren't very good for VFO coils, because these materials change their permeability as the temperature varies. This changes the inductance, in turn affecting the oscillator frequency.

Engineers spend much time and effort in finding components that will minimize *drift* (unwanted changes in frequency over time) in VFOs.

Reliability of performance

An oscillator should always start working as soon as power is supplied. It should keep oscillating under all normal conditions, not quitting if the load changes slightly or if the temperature rises or falls. A "finicky" oscillator is a great annoyance. The failure of a single oscillator can cause an entire receiver, transmitter or transceiver to stop working. An oscillator is sometimes called *unstable* if it has to be "coaxed" into starting, or if it quits unpredictably.

Some oscillator circuits are more reliable than others. The circuits generalized in this chapter are those that engineers have found, through trial and error over the years, to work the best.

When an oscillator is built and put to use in a radio receiver, transmitter or audio device, *debugging* is always necessary. This is a trial-and-error process of getting the flaws, or "bugs," out of the circuit. Rarely can an engineer build something straight from the drawing board and have it work just right the first time. In fact, if two oscillators are built from the same diagram, with the same component types and values in the same geometric arrangement, one circuit might work fine, and the other might be unstable. This usually happens because of differences in the quality of components that don't show up until the "acid test."

Oscillators are designed to work into a certain range of load impedances. It's important that the load impedance not be too low. (You need never be concerned that it might be too high. In general, the higher the load impedance, the better.) If the load impedance is too low, the load will try to draw power from an oscillator. Then, even a well-designed oscillator might be unstable. Oscillators aren't meant to produce powerful signals. High power can be obtained using amplification after the oscillator.

Crystal-controlled oscillators

Quartz crystals can be used in place of tuned LC circuits in RF oscillators, if it isn't necessary to change the frequency often. Crystal oscillators offer excellent frequency stability—far superior to that of LC-tuned VFOs.

There are several ways that crystals can be connected in bipolar or FET circuits to get oscillation. One common circuit is the *Pierce oscillator*. An N-channel JFET and quartz crystal are connected in a Pierce configuration as shown in the schematic diagram of Fig. 25-6.

The crystal frequency can be varied somewhat (by about 0.1 percent) by means of an inductor or capacitor in parallel with the crystal. But the frequency is determined mainly by the thickness of the crystal, and by the angle at which it is cut from the quartz rock.

Crystals change in frequency as the temperature changes. But they are far more stable than LC circuits, most of the time. Some crystal oscillators are housed in temperature-controlled chambers called *ovens*. They maintain their frequency so well that

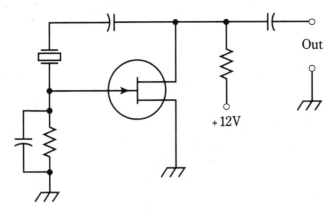

25-6 A JFET Pierce oscillator.

they are often used as *frequency standards*, against which other oscillators are calibrated. The accuracy can be within a few Hertz at working frequencies of several megahertz.

The voltage-controlled oscillator

The frequency of a VFO can be adjusted via a varactor diode in the tuned LC circuit. Recall that a varactor, also called a varicap, is a semiconductor diode that works as a variable capacitor when it is reverse-biased. The capacitance depends on the reverse-bias voltage. The greater this voltage, the lower the value of the capacitance.

The Hartley and Clapp oscillator circuits lend themselves well to varactor-diode frequency control. The varactor is placed in series or parallel with the tuning capacitor, and is isolated for dc by blocking capacitors. The schematic diagram of Fig. 25-7 shows an example of how a varactor can be connected in a tuned circuit. The resulting oscillator is called a *voltage-controlled oscillator (VCO)*.

25-7 Connection of a varactor in a tuned LC circuit.

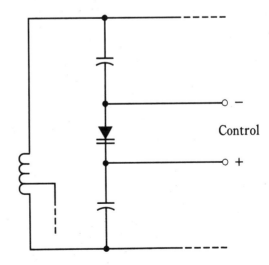

Why control the frequency of an oscillator in this way? It is commonly done in modern communications equipment; there must be a reason. In fact there are several good reasons why varactor control is better than the use of mechanically variable capacitors or inductors. But it all comes down to basically one thing: Varactors are cheaper. They're also less bulky than mechanically variable capacitors and inductors.

Nowadays, many frequency readouts are digital. You look at a numeric display instead of interpolating a dial scale. Digital control is often done by a *microcomputer*. You program the operating frequency by pressing a sequence of buttons, rather than by rotating a knob. The microcomputer might set the frequency via a synchro on the shaft of a variable capacitor or inductor. But that would be unwieldy. It would also be ridiculous, a "Rube Goldberg" contraption. A varactor can control the frequency without all that nonsense.

The PLL frequency synthesizer

One type of oscillator that combines the flexibility of a VFO with the stability of a crystal oscillator is known as a *PLL frequency synthesizer*. This scheme is extensively used in modern digital radio transmitters and receivers.

The output of a VCO is passed through a *programmable divider*, a digital circuit that divides the VCO frequency by any of hundreds or even thousands of numerical values chosen by the operator. The output frequency of the programmable divider is locked, by means of a *phase comparator*, to the signal from a crystal-controlled *reference oscillator*.

As long as the output from the programmable divider is exactly on the reference-oscillator frequency, the two signals are in phase, and the output of the phase comparator is zero volts dc. If the VCO frequency begins to drift, the output frequency of the programmable divider will drift, too (although at a different rate). But even the tiniest frequency change—a fraction of 1 Hz—causes the phase comparator to produce a dc *error voltage*. This error voltage is either positive or negative, depending on whether the VCO has drifted higher or lower in frequency. The error voltage is applied to a varactor in the VCO, causing the VCO frequency to change in a direction opposite to that of the drift. This forms a dc feedback circuit that maintains the VCO frequency at a precise multiple of the reference-oscillator frequency, that multiple having been chosen by the programmable divider. It is a *loop* circuit that *locks* the VCO onto a precise frequency, by means of *phase* sensing, hence the term *phase-locked loop (PLL)*.

The key to the stability of the PLL frequency synthesizer lies in the fact that the reference oscillator is crystal-controlled. A block diagram of such a synthesizer is shown in Fig. 25-8. When you hear that a radio receiver, transmitter or transceiver is "synthesized," it usually means that the frequency is determined by a PLL frequency synthesizer.

The stability of a synthesizer can be enhanced by using an amplified signal from the National Bureau of Standards, transmitted on shortwave by WWV at 5, 10, or 15 MHz, directly as the reference oscillator. These signals are frequency-exact to a minuscule fraction of 1 Hz, because they are controlled by atomic clocks. Most people don't need precision of this caliber, so you won't see consumer devices like ham radios and shortwave receivers with *primary-standard* PLL frequency synthesis. But it is employed by some corporations and government agencies, such as the military.

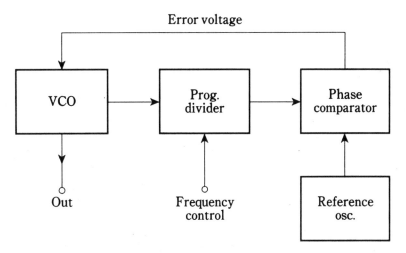

25-8 Block diagram of a PLL frequency synthesizer.

Diode oscillators

At ultra-high and microwave frequencies, certain types of diodes can be used as oscillators. These diodes, called *Gunn*, *IMPATT* and *tunnel* diodes, were discussed in chapter 20.

Audio waveforms

The above described oscillators work above the human hearing range. At audio frequencies (AF), oscillators can use RC or LC combinations to determine frequency. If LC circuits are used, the inductances must be rather large, and ferromagnetic cores are usually necessary.

All RF oscillators produce a sine-wave output. A pure sine wave represents energy at one and only one frequency. Audio oscillators, by contrast, don't necessarily concentrate all their energy at a single frequency. A pure AF sine wave, especially if it is continuous and frequency-constant, causes ear/mind fatigue. Perhaps you've experienced it.

The various musical instruments in a band or orchestra all sound different from each other, even when they play the same note (such as middle C). The reason for this is that each instrument has its own unique waveform. A clarinet sounds different than a trumpet, which in turn sounds different than a cello or piano.

Suppose you were to use an oscilloscope to look at the waveforms of musical instruments. This can be done using a high-fidelity microphone, a low-distortion amplifier and a scope. You'd see that each instrument has its own "signature." Each instrument's unique sound qualities can be reproduced using AF oscillators whose waveform outputs match those of the instrument.

The art of electronic music is a subject to which whole books have been devoted. All electronic music synthesizers use audio oscillators to generate the tones you hear.

Audio oscillators

Audio oscillators find uses in doorbells, ambulance sirens, electronic games, and those little toys that play simple musical tunes. All AF oscillators work in the same way, consisting of amplifiers with positive feedback.

A simple audio oscillator

One form of AF oscillator that is popular for general-purpose use is the *twin-T oscillator* (Fig. 25-9). The frequency is determined by the values of the resistors R and capacitors C. The output is a near-perfect sine wave. The small amount of distortion helps to alleviate the irritation produced by an absolutely pure sinusoid. This circuit uses two NPN bipolar transistors. Two JFETs could also be used, biased for class-A amplifier operation.

The multivibrator

Another audio-oscillator circuit uses two identical common-emitter or common-source amplifier circuits, hooked up so that the signal goes around and around between them.

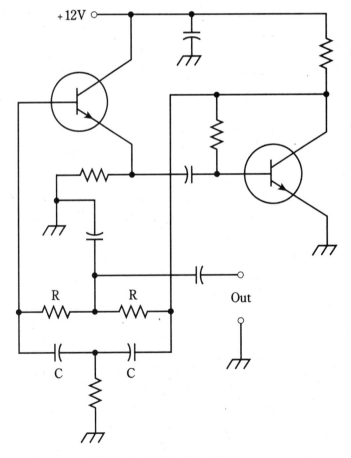

25-9   A twin-T audio oscillator.

This is sometimes called a *multivibrator* circuit, although that is technically a misnomer, the term being more appropriate to various digital signal-generating circuits.

Two N-channel JFETs are connected to form an oscillator as shown in Fig. 25-10. Each "stage" amplifies the signal in class-A, and reverses the phase by 180 degrees. Thus the signal goes through a 360-degree shift each time it gets back to any particular point. A 360-degree shift results in positive feedback, being effectively equivalent to no phase shift.

The frequency is set by means of an LC circuit. The coil uses a ferromagnetic core, because stability is not of great concern and because such a core is necessary to obtain the large inductance needed for resonance at audio frequencies. The value of L is typically from 10 mH to as much as 1 H. The capacitance is chosen according to the formula for resonant circuits, to obtain an audio tone at the frequency desired.

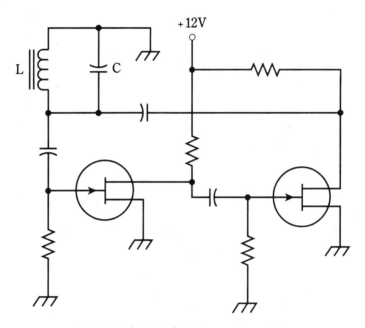

25-10 A "multivibrator" type audio oscillator.

IC oscillators

In recent years, solid-state technology has advanced to the point that whole circuits can be etched onto silicon *chips*. Such devices are called *integrated circuits (ICs)*. The *operational amplifier (op amp)* is one type of IC that is especially useful as an oscillator. Op-amp oscillators are most commonly employed as audio oscillators.

Integrated circuits are discussed in chapter 28.

Quiz

Refer to the text in this chapter if necessary. A good score is at least 18 correct. Answers are in the back of the book.

1. Negative feedback in an amplifier:
    A. Causes oscillation.
    B. Increases sensitivity.
    C. Reduces the gain.
    D. Is used in an Armstrong oscillator.

2. Oscillation requires:
    A. A common-drain or common-collector circuit.
    B. A stage with gain.
    C. A tapped coil.
    D. Negative feedback.

3. A Colpitts oscillator can be recognized by:
    A. A split capacitance in the tuned circuit.
    B. A tapped coil in the tuned circuit.
    C. A transformer for the feedback.
    D. A common-base or common-gate arrangement.

4. In an oscillator circuit, the feedback should be:
    A. As great as possible.
    B. Kept to a minimum.
    C. Just enough to sustain oscillation.
    D. Done through a transformer whose wires can be switched easily.

5. A tapped coil is used in a(n):
    A. Hartley oscillator.
    B. Colpitts oscillator.
    C. Armstrong oscillator.
    D. Clapp oscillator.

6. An RF choke:
    A. Passes RF but not dc.
    B. Passes both RF and dc.
    C. Passes dc but not RF.
    D. Blocks both dc and RF.

7. Ferromagnetic coil cores are not generally good for use in RF oscillators because:
    A. The inductances are too large.
    B. It's hard to vary the inductance of such a coil.
    C. Such coils are too bulky.
    D. Air-core coils have better thermal stability.

8. An oscillator might fail to start for any of the following reasons *except*:

    A. Low power supply voltage.

    B. Low stage gain.

    C. In-phase feedback.

    D. Very low output impedance.

9. An advantage of a crystal-controlled oscillator over a VFO is:

    A. Single-frequency operation.

    B. Ease of frequency adjustment.

    C. High output power.

    D. Low drift.

10. The frequency at which a crystal oscillator functions is determined mainly by:

    A. The values of the inductor and capacitor.

    B. The thickness of the crystal.

    C. The amount of capacitance across the crystal.

    D. The power-supply voltage.

11. The different sounds of musical instruments are primarily the result of:

    A. Differences in the waveshape.

    B. Differences in frequency.

    C. Differences in amplitude.

    D. Differences in phase.

12. A radio-frequency oscillator usually:

    A. Has an irregular waveshape.

    B. Has most or all of its energy at a single frequency.

    C. Produces a sound that depends on its waveform.

    D. Uses RC tuning.

13. A varactor diode:

    A. Is mechanically flexible.

    B. Has high power output.

    C. Can produce different waveforms.

    D. Is good for use in frequency synthesizers.

14. A frequency synthesizer has:

    A. High power output.

    B. High drift rate.

    C. Exceptional stability.

    D. Adjustable waveshape.

15. A ferromagnetic-core coil is preferred for use in the tuned circuit of an RF oscillator:

    A. That must have the best possible stability.

    B. That must have high power output.

    C. That must work at microwave frequencies.

    D. No! Air-core coils work better in RF oscillators.

16. If the load impedance for an oscillator is too high:

    A. The frequency might drift.

    B. The power output might be reduced.

    C. The oscillator might fail to start.

    D. It's not a cause for worry; it can't be too high.

17. The bipolar transistors or JFETs in a multivibrator are usually connected in:

    A. Class-B.

    B. A common-emitter or common-source arrangement.

    C. Class-C.

    D. A common-collector or common-drain arrangement.

18. The arrangement in the block diagram of Fig. 25-11 represents:

    A. A waveform analyzer.

    B. An audio oscillator.

    C. An RF oscillator.

    D. A sine-wave generator.

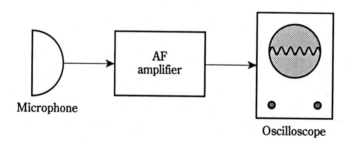

25-11 Illustration for quiz question 18.

19. Acoustic feedback in a public-address system:

    A. Is useful for generating RF sine waves.

    B. Is useful for waveform analysis.

    C. Can be used to increase the amplifier gain.

    D. Serves no useful purpose.

20. An IMPATT diode:

    A. Makes a good audio oscillator.

    B. Can be used for waveform analysis.

    C. Is used as a microwave oscillator.

    D. Allows for frequency adjustment of a VCO.

26
CHAPTER

Data transmission

TO CONVEY DATA, SOME ASPECT OF A SIGNAL MUST BE VARIED. THERE ARE several different characteristics of a signal that can be made to change in a controlled way, so that data is "imprinted" on it. *Modulation* is the process of imprinting data onto an electric current or radio wave.

Modulation can be accomplished by varying the amplitude, the frequency or the phase of a wave. Another method is to transmit a series of pulses, whose duration, amplitude, or spacing is made to change in accordance with the data to be conveyed.

The carrier wave

The "heart" of most communications signals is a sine wave, usually of a frequency well above the range of human hearing. This is called a *carrier* or *carrier wave*. The lowest carrier frequency used for radio communications is 9 kHz. The highest frequency is in the hundreds of gigahertz.

For modulation to work effectively, the carrier must have a frequency many times the highest frequency of the modulating signal. For example, if you want to modulate a radio wave with hi-fi music, which has a frequency range from a few Hertz up to 20 kHz or so, the carrier wave must have a frequency well above 20 kHz. A good rule is that the carrier must have a frequency of at least 10 times the highest modulating frequency. So for good hi-fi music transmission, a radio carrier should be at 200 kHz or higher.

This rule holds for all kinds of modulation, whether it be of the amplitude, phase, or frequency. If the rule is violated, the efficiency of transmission will be degraded, resulting in less-than-optimum data transfer.

The Morse code

The simplest, and oldest, form of modulation is on-off *keying*. Early telegraph systems used direct currents that were keyed on and off, and were sent along wires. The first radio transmitters employed spark-generated "hash" signals that were keyed using the telegraph code. The noise from the sparks, like ignition noise from a car, could be heard in crystal-set receivers several miles away. At the time of their invention, this phenomenon was deemed miraculous: a wireless telegraph!

Keying is usually accomplished at the oscillator of a *continuous-wave (CW)* radio transmitter. A block diagram of a simple CW transmitter is shown in Fig. 26-1. This is the basis for a mode of communications that is, and always has been, popular among radio amateurs and experimenters.

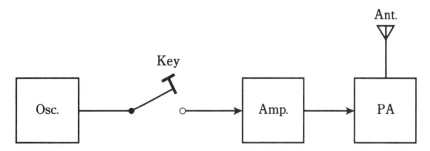

26-1   A simple CW transmitter.

While the use of Morse code might seem old-fashioned, even archaic, a CW transmitter is extremely simple to build. A human operator, listening to Morse code and writing down the characters as they are sent, is one of the most efficient data receivers ever devised. Until computers are built to have intuition, there'll always be a place for Morse code radio communications. Besides being efficient, as any "CW fanatic" radio ham will tell you, it's just plain fun to send and receive signals in Morse code.

Morse code is a form of *digital* communications. It can be broken down into *bits*, each having a length of one *dot*. A *dash* is three bits long. The space between dots and dashes, within a single character, is one bit. The space between characters in a word is three bits. The space between words is seven bits. Punctuation marks are sent as characters attached to their respective words. An *amplitude-versus-time* rendition of the Morse word "eat" is shown in Fig. 26-2.

Morse code is a rather slow way to send and receive data. Human operators typically use speeds ranging from about 5 words per minute (wpm) to 40 or 50 wpm.

Frequency-shift keying

Morse code keying is the most primitive form of *amplitude modulation (AM)*. The strength, or amplitude, of the signal is varied between two extreme conditions: full-on and full-off. There is another way to achieve two-state keying that works better with

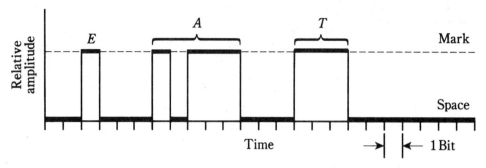

26-2 The Morse code word "eat" as sent on CW.

teleprinter machines than on-off switching. That is to shift the frequency of the carrier wave back and forth. It is called *frequency-shift keying (FSK)*.

Teleprinter codes

The Morse code is not the only digital code of its kind. There are two common teleprinter codes, used to send and receive *radioteletype (RTTY)* signals. These codes are known as the *Baudot* (pronounced "baw-*doe*") and *ASCII* (pronounced "*ask*-ee") codes. (You needn't worry about where these names come from.) A carrier wave can be keyed on and off using either of these codes, at speeds ranging from 60 wpm to over 1000 wpm. In recent years, ASCII has been replacing Baudot as the standard teleprinter code.

A special circuit, called a *terminal unit*, converts RTTY signals into electrical impulses to work a teleprinter or to display the characters on a monitor screen. The terminal unit also generates the signals necessary to send RTTY, as the operator types on the keyboard of a teleprinter *terminal*. A personal computer can be made to work as an RTTY terminal by means of *terminal emulation software*. This software is available in several different forms, and is popular among radio amateurs and electronics hobbyists.

Mark and space

The trouble with using simple on-off keying for RTTY is that noise pulses, such as thunderstorm static crashes, can be interpreted by the terminal unit as signal pulses. This causes misprints. There is no problem if a crash takes place during the full-on, or *mark*, part of the signal; but if it happens during a pause or *space* interval, the terminal unit can be fooled into thinking it's a mark pulse instead.

This problem can be helped greatly by sending a signal during the space part of the signal, but at a different frequency from the mark pulse. Then the terminal unit knows for sure that a mark is not being sent. Instead of sending "Not mark," the transmitter sends "Not mark, but space instead." The easiest way to do this is to send the mark part of the signal at one carrier frequency, and the space part at another frequency a few hundred Hertz higher or lower. This is FSK. The difference between the mark and space frequencies is called the *shift*, and is usually between 100 Hz and 1 kHz.

A *frequency-versus-time* graph of the Morse code word "eat," sent using FSK, is shown in Fig. 26-3. Normally, Baudot or ASCII, rather than Morse, is used for teleprinter operation. A block diagram of an FSK transmitter is shown in Fig. 26-4. The FSK mode, like on-off code keying, is a digital form of communications. But unlike on-off Morse keying, FSK is *frequency modulation (FM)*.

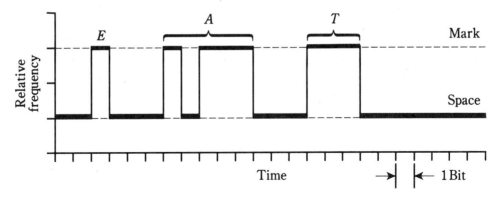

26-3   The Morse code word "eat" as sent using FSK.

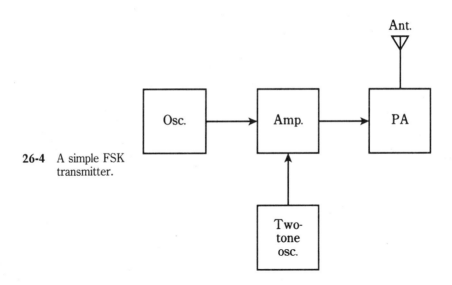

26-4   A simple FSK transmitter.

The telephone modem

Teleprinter data can be sent over the telephone lines using FSK entirely within the audio range. Two audio tones are generated, one for mark and the other for space. There are three sets of standard tone frequencies: 1200 Hz and 2200 Hz for general communications, 1070 Hz and 1270 Hz for message origination, and 2025 and 2225 Hz for answering. These represent shifts of 1000 Hz or 200 Hz.

Because this FSK takes place at audio, it is sometimes called *audio-frequency-shift keying (AFSK)*. A device that sends and receives AFSK teleprinter over the phone lines is known as a *telephone modem*. If you've used a personal computer via the phone lines, you've used a telephone modem. Perhaps you've heard the "bleep-bleep" of the tones as data is sent or received.

Amplitude modulation for voice

A voice signal is a complex waveform with frequencies mostly in the range 300 Hz to 3 kHz. Direct currents can be varied, or *modulated*, by these waveforms, thereby transmitting voice information over wires. This is how early telephones worked.

Around 1920, when CW oscillators were developed to replace spark-gap transmitters, engineers wondered, "Can a radio wave be modulated with a voice, like dc in a telephone?" If so, voices could be sent by "wireless." Radio communications via Morse code was being done over thousands of miles, but CW was slow. The idea of sending voices was fascinating, and engineers set about to find a way to do it.

A simple amplitude modulator

An amplifier was built to have variable gain. The idea was to make the gain fluctuate at voice-frequency rates—up to 3 kHz or so. *Vacuum tubes* were used as amplifiers back then, because solid-state components hadn't been invented yet. But the principle of *amplitude modulation (AM)* is the same, whether the active devices are tubes, bipolar transistors or FETs.

If bipolar transistors had been around in 1920, the first *amplitude modulator* would have resembled the circuit shown in Fig. 26-5. This circuit is simply a class-A RF amplifier, whose gain is varied in step with a voice signal coupled into the emitter circuit. The voice signal affects the instantaneous voltage between the emitter and base, varying the instantaneous bias. The result is that the instantaneous RF output increases and decreases, in a way that exactly duplicates the waveform of the voice signal.

The circuit of Fig. 26-5 will work quite well as an AM voice modulator, provided that the audio input isn't too great. If the AF is excessive, *overmodulation* will occur. This will result in a distorted signal.

The AM transmitter

Two complete AM transmitters are shown in block-diagram form in Fig. 26-6. At A, modulation is done at a low power level. This is *low-level AM*. All the amplification stages after the modulator must be linear. That means class AB or class B must be used. If a class-C PA is used, the signal will be distorted.

In some broadcast transmitters, AM is done in the final PA, as shown in Fig. 26-6B. This is *high-level AM*. The PA operates in class C; it is the modulator as well as the final amplifier. As long the PA is modulated correctly, the output will be a "clean" AM signal; RF linearity is of no concern.

The extent of modulation is expressed as a percentage, from 0 percent, representing an unmodulated carrier, to 100 percent, representing full modulation. Increasing the modulation past 100 percent will cause distortion of the signal, and will degrade, not enhance, the effectiveness of data transmission.

In an AM signal that is modulated 100 percent, only $1/3$ of the power is actually used to convey the data; the other $2/3$ is consumed by the carrier wave. For this reason, AM is rather inefficient. There are voice modulation techniques that make better use of available transmitter power. Perhaps the most widely used is *single sideband (SSB)*, which you'll learn about shortly.

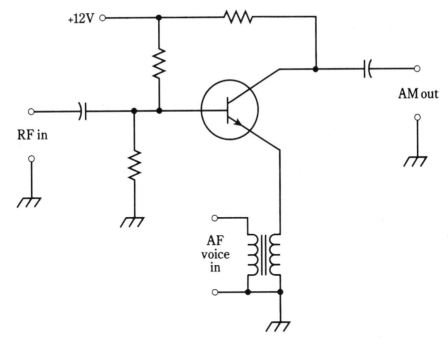

26-5 An AM modulator.

Bandwidth of an AM voice signal

Suppose you could get a graphic display of an AM signal, with frequency on the horizontal axis and amplitude on the vertical axis. This is in fact done using an instrument called a *spectrum analyzer*. In Fig. 26-7, the *spectral display* for an AM voice radio signal at 1340 kHz is illustrated.

On a spectrum analyzer, unmodulated radio carriers look like vertical lines, or "pips," of various heights depending on how strong they are. The carrier wave at 1340 kHz shows up in Fig. 26-7 as a strong pip.

The horizontal scale of the display in Fig. 26-7 is calibrated in increments of 1 kHz per division. This is an ideal scale for looking at an AM signal.

The vertical scale is calibrated in decibels below the signal level that produces 1 mW at the input terminals. Each vertical division represents 3 dB. Decibels relative to 1 mW are abbreviated *dBm* by engineers. Thus, in Fig. 26-7, the top horizontal line is 0 dBm; the first line below it is −3 dBm; the second line is −6 dBm and so on.

The audio components of the voice signal show up as *sidebands* on either side of the carrier. All of the voice energy in this example is at audio below 3 kHz. This results in sidebands within the range 1340 kHz plus or minus 3 kHz, or 1337-1343 kHz. The frequencies between 1337 and 1340 kHz are the *lower sideband (LSB)*; those from 1340 to 1343 kHz are the *upper sideband (USB)*. The *bandwidth* of the RF signal is the difference between the maximum and minimum sideband frequencies. In this case it is 1343-1337 kHz, or 6 kHz.

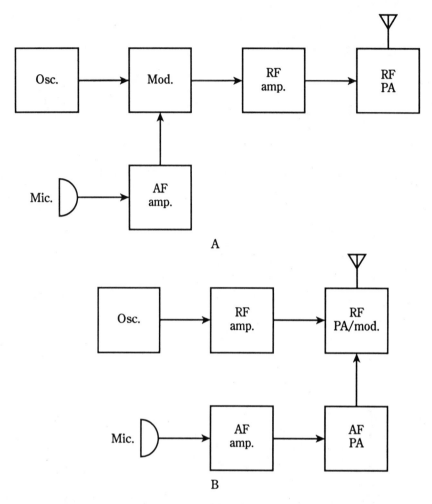

26-6 At A, transmitter using low-level AM. At B, a transmitter using high-level AM.

In an AM signal, the bandwidth is twice the highest audio modulating frequency. In the example of Fig. 26-7, the voice energy is all below 3 kHz, and the bandwidth of the complete RF signal is 6 kHz. At 3 kHz above and below the carrier, the *frequency cutoffs* are abrupt. This transmitter uses an *audio lowpass filter* that cuts out the audio above 3 kHz. Audio above 3 kHz contributes nothing to the intelligibility of a human voice. It's important to keep the bandwidth of a signal as narrow as possible, so there will be room for many signals in a given *band* of frequencies.

Single sideband

As mentioned previously, AM is not efficient. Most of the power is used up by the carrier; only 33 percent of it carries data. Besides that, the two sidebands are mirror-image duplicates. An AM signal is redundant, as well as inefficient, for voice transmission.

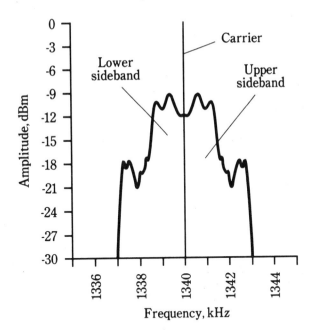

26-7 Spectral display of a typical voice AM signal.

Voice SSB

During the fifties, engineers began to work on an alternative to conventional AM. They mused, "Suppose *all* of the transmitter power could go into the voice, and *none* be taken up by the carrier? That would be a threefold effective increase in transmitter power! And what if the bandwidth could be cut to 3 kHz rather than 6 kHz, by getting rid of the sideband redundancy? That would put all the energy into half the spectrum space, so that twice as many voice signals could fit in a band." Spectrum-space conservation was getting to be a big deal. The airwaves were starting to become overcrowded.

These improvements were realized by means of circuits that cancel out, or *suppress*, the carrier in the modulator circuit, and that filter out, or phase out, one of the two sidebands. The remaining voice signal has a spectrum display that looks like the graph of Fig. 26-8. Either LSB or USB can be used, and either mode works as well as the other.

The SSB transmitter

The heart of an SSB transmitter is a *balanced modulator*. This circuit works like an ordinary AM modulator, except that the carrier wave is phased out. This leaves only the LSB and USB. One of the sidebands is removed by a *filter* that passes only the RF within a 3-kHz-wide band. A block diagram of an SSB transmitter is shown in Fig. 26-9.

High-level modulation won't work for SSB. The balanced modulator is in a low-power part of the transmitter. Therefore, the RF amplifiers after the modulator must all be linear. They usually work in class A except for the PA, which is class AB or class B. If class-C amplification is used with an SSB signal, or if any of the RF amplifiers aren't linear for any reason, the signal *envelope* (waveform) will be distorted. This will degrade the quality of the signal. It can also cause the bandwidth to exceed the nominal 3 kHz,

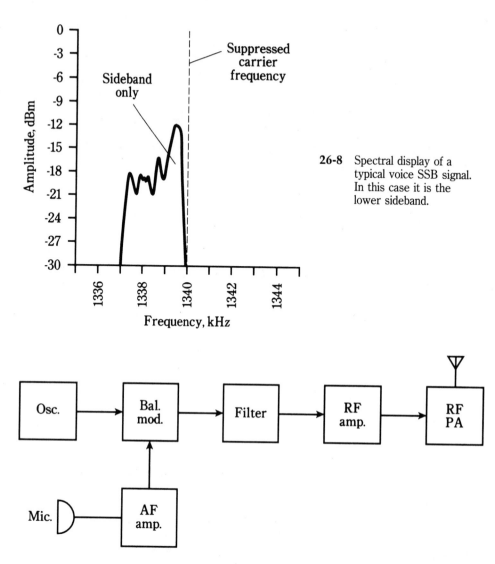

26-8 Spectral display of a typical voice SSB signal. In this case it is the lower sideband.

26-9 An SSB transmitter.

resulting in interference to other stations using the band. Engineers and technicians refer to this as *splatter*.

Frequency and phase modulation

Both AM and SSB work by varying the signal strength. This can be a disadvantage when there is *sferic noise* caused by thundershowers in the vicinity. Sferics are recognizable as "crashes" or "crackles" in an AM or SSB receiver. *Ignition noise* can also be a problem;

this sounds like a "buzz" or "whine." Sferics and ignition noise are both predominantly amplitude-modulated.

In *frequency modulation (FM)*, the amplitude of the signal remains constant, and the instantaneous frequency or phase is made to change. Because the carrier is always "full-on," class-C power amplifiers can be used. Linearity is of no concern when the signal level does not change.

Reactance modulation versus phase modulation

The most direct way to get FM is to apply the audio signal to a varactor diode in a VFO circuit. An example of this scheme, known as *reactance modulation*, is shown in Fig. 26-10. The varying voltage across the varactor causes its capacitance to change in accordance with the audio waveform. The changing capacitance results in an up-and-down swing in the frequency generated by the VFO. In the illustration, only the tuned circuit of a Hartley oscillator is shown.

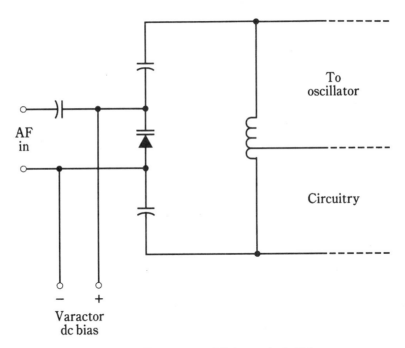

26-10   Reactance modulation to obtain FM.

Another way to get FM is to modulate the phase of the oscillator signal. This causes small fluctuations in the frequency as well, because any instantaneous phase change shows up as an instantaneous frequency change (and vice-versa). This scheme is called *phase modulation*. The circuit is more complicated than the reactance modulator. When phase modulation is used, the audio signal must be processed, adjusting the amplitude-versus-frequency response of the audio amplifiers. Otherwise the signal will sound muffled in an FM receiver.

Frequency deviation

The amount by which the carrier frequency varies will depend on the relative audio signal level, and also on the degree to which the audio is amplified before it's applied to the modulator. The *deviation* is the maximum extent to which the instantaneous carrier frequency differs from the unmodulated-carrier frequency. For most FM voice transmitters, the deviation is standardized at plus-or-minus 5.0 kHz (Fig. 26-11).

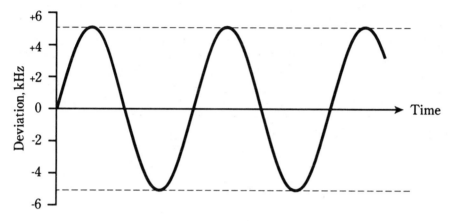

26-11   Frequency-versus-time rendition of an FM signal.

The deviation obtainable by means of direct FM is greater, for a given oscillator frequency, than the deviation that can be gotten via phase modulation. But the deviation of a signal can be increased by a *frequency multiplier*. This is an RF amplifier circuit whose output is tuned to some integral multiple of the input. A multiply-by-two circuit is called a *frequency doubler*; a multiply-by-3 circuit is a *frequency tripler*.

When an FM signal is passed through a frequency multiplier, the deviation gets multiplied along with the carrier frequency. If a modulator provides plus-or-minus 1.6 kHz deviation, the frequency can be doubled and the result will be a deviation of plus-or-minus 3.2 kHz. If the frequency is tripled, the deviation increases to plus-or-minus 4.8 kHz, which is just about the standard amount for FM voice communications.

Wideband FM

In FM hi-fi broadcasting, and in some other applications, the deviation is much greater than plus-or-minus 5.0 kHz. This is called *wideband FM*, as opposed to *narrowband FM* just discussed.

For ordinary voice communications, there's nothing to be gained by using wideband FM. The only result will be that the signal will take up an unnecessary amount of radio spectrum space. But for music, the fidelity improves as the bandwidth increases.

The deviation for an FM signal should be equal to the highest modulating audio frequency, if optimum fidelity is to be obtained. Thus, plus-or-minus 5.0 kHz is more than enough for voice (3.0 kHz would probably suffice). For music, a deviation of about plus-or-minus 15 kHz or 20 kHz is needed for excellent hi-fi reception.

The ratio of the frequency deviation to the highest modulating audio frequency is called the *modulation index*. For good fidelity, it should be at least 1:1. But it shouldn't be much more; that would waste spectrum space.

Pulse modulation

Still another method of modulation works by varying some aspect of a constant stream of signal *pulses*. Several types of *pulse modulation (PM)* are briefly described below. They are diagrammed in Fig. 26-12 as amplitude-versus-time graphs. The modulating waveforms are shown as curvy lines, and the pulses as vertical lines.

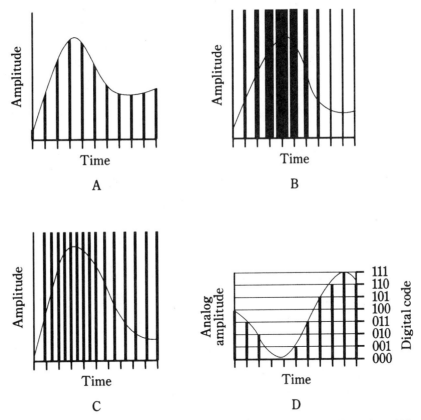

26-12 Pulse modulation. At A, pulse amplitude modulation; at B, pulse width modulation; at C, pulse interval modulation; and at D, pulse code modulation.

Pulse amplitude modulation

In *pulse amplitude modulation (PAM)*, the strength of each individual pulse varies according to the modulating waveform. In this respect, PAM is very much like ordinary amplitude modulation. An amplitude-versus-time graph of a hypothetical PAM signal is shown in Fig. 26-12A.

Normally, the pulse amplitude increases as the instantaneous modulating-signal level increases. But this can be reversed, so that higher audio levels cause the pulse amplitude to go down. Then the signal pulses are at their strongest when there is no modulation.

Either the *positive* or *negative* PAM methods provide good results. The transmitter works a little harder if the negative modulation method is used.

Pulse duration modulation

In PAM, the pulses all last for the same length of time. The effective transmitted power is varied by changing the actual peak amplitude of the pulses. Another way to change the transmitter output is to vary the duration, or width, of the pulses. This scheme is called *pulse duration modulation (PDM)* or *pulse width modulation (PWM)*, shown in Fig. 26-12B.

Normally, the pulse duration increases as the instantaneous modulating-signal level increases. But, as with PAM, this can be reversed. The transmitter must work harder to accomplish negative PDM. Regardless of whether positive or negative PDM is employed, the peak pulse amplitude remains constant.

Pulse interval modulation

Even if all the pulses have the same amplitude and the same duration, modulation can still be accomplished by varying how often they occur. In PAM and PDM, the pulses are always sent at the same time interval, such as 0.0001 second. But in *pulse interval modulation (PIM)*, also called *pulse frequency modulation (PFM)*, pulses might occur more or less frequently than the zero-modulation interval. This is shown in Fig. 26-12C. Every pulse has the same amplitude and the same duration; it is their frequency that changes.

When there is no modulation, the pulses are evenly spaced with respect to time. An increase in the instantaneous data amplitude might cause pulses to be sent more often, as is the case in Fig. 26-12C. Or, an increase in instantaneous data level might slow down the rate at which the pulses are sent. Either scheme will work equally well.

Pulse code modulation

In recent years, the transmission of data has been done more and more by *digital* means. In *digital communications*, the modulating data attains only certain defined states, rather than continuously varying in an *analog* way. Digital transmission offers better efficiency than analog transmission.

With digital modes, the *signal-to-noise ratio* is better, the bandwidth is narrower, and there are fewer errors. Teleprinter data is always sent digitally, as is Morse code. But voices and video can be sent digitally, too. The only drawback that early digital experimenters faced was somewhat degraded fidelity. But today, that has been overcome to the extent that digitized music recordings and transmissions actually sound better than the best analog reproductions.

In *pulse-code modulation (PCM)*, any of the above aspects—amplitude, duration or frequency—of a pulse train can be varied. But rather than having infinitely many possible states, there are finitely many. The greater the number of states, the better the fidelity. But the transmitting and receiving equipment must be more and more sophisticated as the number of digital states increases. An example of eight-level PAM/PCM is shown in Fig. 26-12D.

Analog-to-digital conversion

The graph of Fig. 26-12D illustrates a method of *analog-to-digital (A/D) conversion*. A voice signal, or any continuously variable signal, can be *digitized*, or converted into a string of pulses, whose amplitudes can achieve only a finite number of states.

Resolution

The number of states is always a power of 2, so that it can be represented as a binary-number code. Fidelity gets better as the exponent increases. The number of states is called the *sampling resolution*, or simply the *resolution*.

You might think that the resolution would have to be very large for good reproduction to be possible. But in fact, a resolution of $2^3 = 8$ (as shown in Fig. 26-12D) is good enough for voice transmission, and is the standard resolution for commercial digital voice circuits. A resolution of $2^4 = 16$ is adequate for compact disks used in advanced hi-fi systems!

Sampling rate

The efficiency with which a signal can be digitized depends on the frequency at which *sampling* is done. In general, the *sampling rate* must have a frequency that is at least twice the highest data frequency.

For an audio signal with components as high as 3 kHz, the minimum sampling rate for effective digitization is 6 kHz, or one sample every 167 microseconds (us). Ideally, the sampling rate should be somewhat higher; the commercial voice standard is 8 kHz, or one sample every 125 us.

For music and hi-fi digital transmission, the standard sampling rate is 44.1 kHz, or one sample every 22.7 us. This is based on a maximum audio frequency of 20 kHz, the approximate upper limit of the human hearing range.

Image transmission

The modulation techniques used for *image transmission* are similar to those employed for sending voices. Nonmoving pictures can be sent within the same bandwidth as a voice. For high-resolution, fast-scan moving images, the necessary bandwidth is greater.

A thorough discussion of image transmission is beyond the scope of this book. The basics of the three most common video communications modes are discussed here. For further detail, a text on video communications is recommended.

Facsimile

"Still" images are transmitted by *facsimile (fax)*. If data is sent slowly enough, any amount of detail can be transmitted within a voice band. This is how telephone fax works.

A high-resolution commercial fax image has upwards of 1000 lines of data. The image is *scanned* from left to right and from top to bottom like reading a book. The complete image takes several minutes to send. Many of the black-and-white photographs in the daily newspaper are sent via fax. Practically all weather satellite images are faxed.

A fax signal sounds somewhat like AFSK. But the modulation occurs over a continuously variable range of audio tones, rather than at only two frequencies.

To send an image by fax, a document or photo is wrapped around a *drum*. The drum is rotated at a slow, controlled rate. A spot of light scans from left to right; the drum moves the document or photo so that a "slice," or *line*, is scanned with each pass of the light spot. This continues, line by line, until the complete *frame*, or picture, has been scanned. The reflected light is picked up by a *photodetector*. Darker parts of the image reflect less light than whiter parts, so the current through the photodetector varies. This current modulates a carrier in one of the modes described earlier, such as AM, FM or SSB. Typically, black is sent as a 1.5-kHz tone, and white as 2.3 kHz. Gray shades produce intermediate tones.

At the receiver, the scanning rate and pattern can be duplicated, and a cathode-ray tube or special printer used to reproduce the image in black-and-white. Cathode-ray-tube reception of fax is popular among radio amateurs. Personal computers can be programmed to act as fax receivers.

Slow-scan television

One way to think of *slow-scan television (SSTV)* is to imagine "fast fax." An SSTV signal, like a fax signal, is sent within a band of frequencies as narrow as that of a human voice. And, like fax, SSTV transmission is of still pictures, not moving ones.

The big difference between SSTV and fax is that SSTV images are sent in much less time. The frame time is 8 seconds, rather than several minutes. This speed bonus comes with a tradeoff: lower *resolution,* meaning less fineness of detail. The resolution of an SSTV image is a bit less than that of an ordinary television picture.

All SSTV signals are received on cathode-ray-tube (CRT) displays. A computer can be programmed so that its monitor will act as an SSTV receiver. Converters are also available that allow SSTV signals to be viewed on a consumer type TV set.

An SSTV frame has 120 lines. The black and white frequencies are the same as for fax transmission; the darkest parts of the picture are sent at 1.5 kHz and the brightest at 2.3 kHz.

Synchronization (sync) pulses, that keep the receiving apparatus in step with the transmitter, are sent at 1.2 kHz. A *vertical sync pulse* tells the receiver it's time to begin a new frame; it lasts for 30 milliseconds (ms). A *horizontal sync pulse* tells the receiver it's time to start a new line in a frame; its duration is 5 ms. These pulses prevent "rolling" or "tearing" of the image.

Ham radio operators like to send SSTV with SSB transmitters. It's also possible to transmit SSTV using AM, FM, or PM. But these modes take up more spectrum space than SSB.

An SSTV signal, like a fax signal, can be sent over the telephone. This puts the "video phone" within reach of current technology, and the equipment isn't too expensive. Although the images don't convey movement, because the frame time is long, telephone SSTV lets you see people on the other end of the line, and also lets them see you. The bugaboo is that the other person—or you—might not want to be looked at. The camera can be switched off easily enough. And, as often as not, users of video phones prefer it that way.

Fast-scan television

Conventional television is also known as *fast-scan TV (FSTV)*. This is the TV that brings you sports events, newscasts, and all the other programming with which you're familiar.

In FSTV, the frames come at the rate of 30 per second. The human eye/brain perceives bursts of motion down to a time resolution of about 1/20 second. Therefore, in FSTV, the sequence of still images blends together to give the appearance of continuous motion.

The FSTV image has 525 lines per frame. In recent years, technological advances have been made that promise to make *high-resolution TV* widely available and affordable. This mode has more than 525 lines per frame.

The quick frame time, and the increased resolution, of FSTV make it necessary to use a much wider frequency band than is the case with fax or SSTV. A typical video FSTV signal takes up 6 MHz of spectrum space, or 2000 times the bandwidth of a fax or SSTV signal.

Fast-scan TV is almost always sent using conventional AM. Wideband FM can also be used. With AM, one of the sidebands can be filtered out, leaving just the carrier and the other sideband. This mode is called *vestigial sideband (VSB)* transmission. It cuts the bandwidth of a FSTV signal down to about 3 MHz.

Because of the large amount of spectrum space needed to send FSTV, this mode isn't practical at frequencies below about 30 MHz (10 times the bandwidth of a VSB signal). All commercial FSTV transmission is done above 50 MHz, with the great majority of channels having frequencies far higher than this. Channels 2 through 13 on your TV receiver are sometimes called the *VHF (very-high-frequency)* channels; the higher channels are called the *UHF (ultra-high-frequency)* channels.

An amplitude-versus-time graph of the waveform of a TV signal is illustrated in Fig. 26-13. This represents one line of one frame, or $1/525$ of a complete picture. The highest instantaneous signal amplitude corresponds to the blackest shade, and the lowest amplitude to the lightest shade. Thus, the FSTV signal is sent "negatively."

The reason that FSTV signals are sent "upside down" is that *retracing* (moving from the end of one line to the beginning of the next) must stay synchronized between the transmitter and receiver. This is guaranteed by a defined, strong *blanking pulse*. This pulse tells the receiver when to retrace; it also shuts off the beam while the CRT is retracing. You've probably noticed that weak TV signals have poor *contrast*. Weakened blanking pulses result in incomplete retrace blanking. But this is better than having the TV receiver completely lose track of when it should retrace!

Weak TV signals are received better when the strongest signals correspond to black, rather than to white. This was discovered, as things so often are, by experimentation.

When you tune your TV set to a vacant channel, you see "snow," or white-and-gray, fast-moving dots. If a TV signal comes on the air without modulation, the screen goes dark. Only when there is modulation do portions of the screen get light again.

Color FSTV works by sending three separate monocolor signals, corresponding to the primary colors red, blue, and green. The signals are literally black-and-red, black-and-blue, and black-and-green. These are recombined at the receiver and displayed on the screen as a fine, interwoven matrix of red, blue, and green dots. When viewed

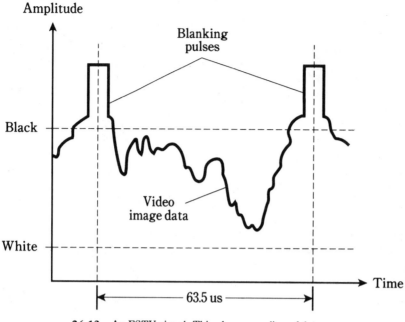

26-13 An FSTV signal. This shows one line of data.

from a distance, the dots are too small to be individually discernible. Various combinations of red, blue, and green intensities result in reproduction of all possible hues and saturations of color.

The electromagnetic field

In a radio or television transmitting antenna, electrons are moving back and forth at an extreme speed. Their velocity is constantly changing as they speed up in one direction, slow down, reverse direction, speed up again, and so on. Any change of velocity is *acceleration.*

When electrons move, a magnetic field is created. When electrons accelerate, a changing magnetic field is produced. An alternating magnetic (M) field gives rise to an alternating electric (E) field, and this generates another changing M field. The process has come full circle. Thus it repeats, the effects propagating through space at the speed of light. The E and M fields expand alternately outward from the source in spherical wavefronts. At any given point in space, the E flux is perpendicular to the M flux. The direction of wave travel is perpendicular to both the E and M flux lines (Fig. 26-14).

The E/M flux field can oscillate at any conceivable frequency, ranging from many years per cycle to quadrillions of cycles per second. The sun has a magnetic field that oscillates with 22-year a cycle. Radio waves oscillate at thousands, millions, or billions of cycles per second. Infrared, visible light, ultraviolet, and X rays oscillate at many trillions of cycles per second. All of these effects are *electromagnetic fields,* and as such, they all

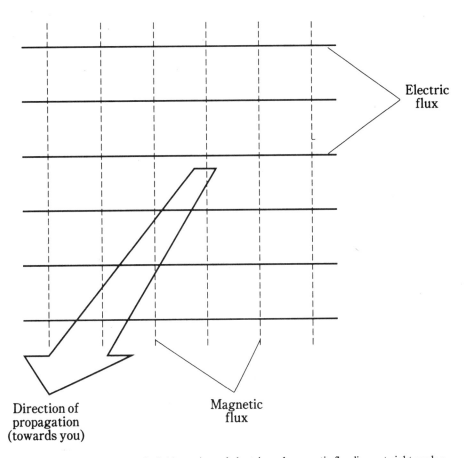

Electric flux

Direction of propagation (towards you)

Magnetic flux

26-14 An electromagnetic field consists of electric and magnetic flux lines at right angles.

have exactly the same form. The difference is in their *frequency*. The frequency of an *electromagnetic wave* is directly related to the *wavelength* in space.

Frequency versus wavelength

All electromagnetic fields have frequencies and wavelengths that are inversely related. If f_{MHz} is the frequency of a wave in megahertz, and L_{ft} is the wavelength in feet, then:

$$L_{\mathrm{ft}} = 984/f_{\mathrm{MHz}}$$

for waves in outer space or in the atmosphere of the earth. If the wavelength is given as L_{m} in meters, then:

$$L_{\mathrm{m}} = 300/f_{\mathrm{MHz}}$$

The inverses of these formulas, for finding the frequency if the wavelength is known, are:

$$f_{\mathrm{MHz}} = 984/L_{\mathrm{ft}}$$

and

$$f_{\text{MHz}} = 300/L_{\text{m}}$$

The electromagnetic spectrum

The whole range of electromagnetic frequencies or wavelengths is called the *electromagnetic spectrum*. Theoretically there is no limit to how low or high the frequency can be, nor, correspondingly, to how long or short the wavelength can be. The most common electromagnetic wavelengths range from about 10^6 m, or 1000 km, to around 10^{-12} m, or a trillionth of a meter.

Scientists use a logarithmic scale to depict the electromagnetic spectrum. A simplified rendition is shown in Fig. 26-15, labeled for wavelength in meters. To find the frequencies in megahertz, divide 300 by the wavelength shown. For frequencies in Hertz, use 300,000,000 instead of 300. For kilohertz, use 300,000; for gigahertz, use 0.300.

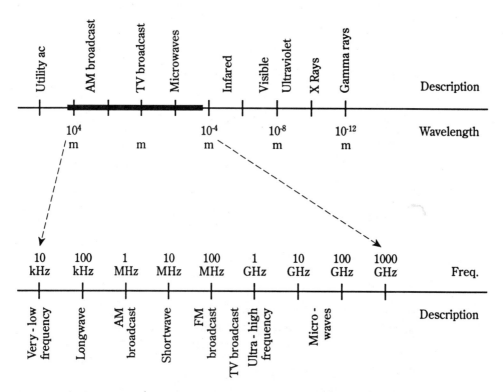

26-15   The electromagnetic spectrum, with the radio portion expanded.

Radio waves fall into a subset of this spectrum, at frequencies between approximately 9 kHz and 1000 GHz. This corresponds to wavelengths of 33 km and 0.3 mm. The *radio spectrum*, which includes television and microwaves, is "blown up" in Fig. 26- 15, and is labeled for frequency. To find wavelengths, use the conversion formulas. Just be sure you have the decimal point in the right place with respect to the "3".

Transmission media

Data can be transmitted over various different *media*. The most common are *cable*, *radio*, *satellite links*, and *fiberoptics*. Cable, radio/TV and satellite communications use the *radio-frequency spectrum*. Fiberoptics uses infrared or visible light energy.

Cable

The earliest cables were wires that carried dc. Nowadays, data-transmission cables more often carry ac at radio frequencies. One advantage of using RF is that the signals can be amplified at intervals on a long span. This greatly increases the distances over which data can be sent by cable. Another advantage of using RF is that numerous signals can be carried over a single cable, with each signal on a different frequency.

Cables can consist of pairs of wires, somewhat akin to lamp cords. But more often, *coaxial cable* is used. This has a *center conductor* surrounded by a cylindrical *shield*. The shield is grounded, and the center conductor carries the signals (Fig. 26-16). The center conductor is kept in place by an insulating *dielectric*, usually made of polyethylene. The shield keeps signals confined to the cable, and also keeps external electromagnetic fields from interfering with the signals.

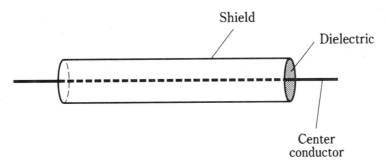

26-16   A coaxial cable has a center conductor surrounded by a cylindrical shield.

Cable signals can be modulated using any of the techniques outlined earlier in this chapter. The most familiar example is cable television.

Radio

All radio and TV signals are electromagnetic waves. The radio or TV transmitter output is coupled into an *antenna system* located at some distance from the transmitter. The energy follows a *transmission line*, also called a *feed line*, from the transmitter PA output to the antenna itself.

Most transmission lines are coaxial cables. There are other types, used in special applications. At microwaves, hollow tubes called *waveguides* are used to transfer the energy from a transmitter to the antenna. A waveguide is more efficient than coaxial cable at the shortest radio wavelengths.

Radio amateurs sometimes use *parallel-wire* transmission lines, resembling the "ribbon" cable popular for use with consumer TV receiving antennas. In a parallel-wire line, the RF currents in the two conductors are always 180 degrees out of phase, so that their electromagnetic fields cancel each other. This keeps the transmission line from radiating, guiding the EM field along toward the antenna. The energy is radiated when it reaches the antenna.

The radio frequency bands are generally categorized from *very low frequency (VLF)* through *microwaves*, according to the breakdown in Table 26-1. These waves propagate through the atmosphere, or through space, in different ways depending on their frequency. Radio signal propagation is discussed in the next chapter.

Table 26-1. Radio frequency classifications.

Classification	Abbreviation	Frequency range
Very Low Frequency	VLF	9 kHz and below
Low Frequency (Longwave)	LF	30 kHz–300 kHz
Medium Frequency	MF	300 kHz–3 MHz
High Frequency (Shortwave)	HF	3 MHz–30 MHz
Very High Frequency	VHF	30 MHz–300 MHz
Ultra High Frequency	UHF	300 MHz–3 GHz
Microwaves		3 GHz and above

Satellite links

At very high frequencies (VHF) and above, many communications circuits use satellites in *geostationary orbits* around the earth. If a satellite is directly over the equator at an altitude of 22,300 miles and orbits from west to east, it will follow the earth's rotation, thereby staying in the same spot in the sky as seen from the surface.

A single geostationary satellite is on a line of sight with about 40 percent of the earth's surface. Three such satellites, placed at 120-degree (1/3-circle) intervals around the planet, allow coverage of all populated regions. A *dish antenna* can be aimed at a geostationary satellite, and once the antenna is in place, it needn't be turned or adjusted. Perhaps you have a satellite TV system.

Fiberoptics

Beams of infrared or visible light can be modulated just as can radio-frequency carriers. The frequencies of infrared and visible light are extremely high, allowing modulation by data at rates into the gigahertz range.

A simple *modulated-light* transmitter is diagrammed schematically in Fig. 26-17. The output of the light-emitting diode (LED) is modulated by audio from the transistor. The light is guided into an *optical fiber*, made from a special mixture of very clear glass. In recent years, *fiberoptics* has begun to replace conventional cable networks on a massive scale.

Fiberoptics has several advantages over wire cables. A fiberoptic cable is cheap, and it's light in weight. It is totally immune to interference from outside electromagnetic

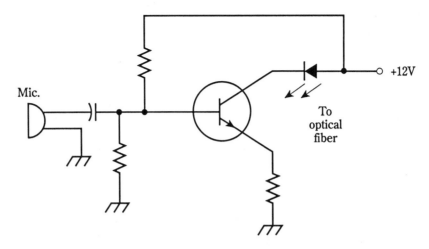

26-17   A simple circuit for voice modulating a light beam.

fields. A fiberoptic cable will not corrode as metallic wires do. Fiberoptic cables are inexpensive to maintain and easy to repair. An optical fiber can carry far more signals than a cable, because the "carrier" frequency is practically infinite. The whole radio spectrum, from VLF through microwaves, can be imprinted on a beam of light and sent through a single glass fiber no thicker than a strand of your hair.

Quiz

Refer to the text in this chapter if necessary. A good score is at least 18 correct. Answers are in the back of the book.

1. A radio wave has a frequency of 1.55 MHz. The highest modulating frequency that can be used effectively is about:

   A. 1.55 kHz.

   B. 15.5 kHz.

   C. 155 kHz.

   D. 1.55 MHz.

2. Morse code is a form of:

   A. Digital modulation.

   B. Analog modulation.

   C. Phase modulation.

   D. dc modulation.

3. An advantage of FSK over simple on-off keying for RTTY is:

   A. Better frequency stability.

   B. Higher speed capability.

C. Reduced number of misprints.

D. On-off keying is just as good as FSK.

4. The maximum AM percentage possible without distortion is:

A. 33 percent.

B. 67 percent.

C. 100 percent.

D. 150 percent.

5. If an AM signal is modulated with audio having frequencies up to 5 kHz, then the complete signal bandwidth will be:

A. 10 kHz.

B. 6 kHz.

C. 5 kHz.

D. 3 kHz.

6. An AM transmitter using a class-C PA should employ:

A. Carrier suppression.

B. High-level modulation.

C. Lower sideband.

D. Single sideband.

7. Which of the following modulation methods is used to send teleprinter data over the phone lines?

A. CW.

B. SSB.

C. AM.

D. AFSK.

8. An advantage of SSB over AM is:

A. Higher data transmission rate.

B. More effective use of transmitter power.

C. Greater bandwidth.

D. Enhanced carrier wave level.

9. An SSB suppressed carrier is at 14.335 MHz. The voice data is contained in a band from 14.335-14.338 MHz. The mode is:

A. AM.

B. LSB.

C. USB.

D. FSK.

10. A spectrum analyzer displays:
    A. Time as a function of frequency.
    B. Frequency as a function of time.
    C. Signal strength as a function of time.
    D. Signal strength as a function of frequency.

11. The deviation for voice FM signals is usually:
    A. Plus-or-minus 3 kHz.
    B. Plus-or-minus 5 kHz.
    C. Plus-or-minus 6 kHz.
    D. Plus-or-minus 10 kHz.

12. Wideband FM is preferable to narrowband FM for music transmission because:
    A. Lower frequencies are heard better.
    B. Spectrum space is conserved.
    C. The fidelity is better.
    D. No! Narrowband FM is better for music.

13. In which mode of PM does the pulse level vary?
    A. PAM.
    B. PDM.
    C. PWM.
    D. PFM.

14. In which PM mode do pulses last for varying times?
    A. PAM.
    B. PWM.
    C. PFM.
    D. PCM.

15. How many states are commonly used for the transmission of digitized voice signals?
    A. Two.
    B. Four.
    C. Six.
    D. Eight.

16. In a SSTV signal, the frame time is:
    A. $1/525$ second.
    B. $1/30$ second.
    C. $1/8$ second.
    D. 8 seconds.

17. The bandwidth of a fax signal is kept narrow by:
    A. Sending the data at a slow rate of speed.
    B. Limiting the image resolution.
    C. Limiting the range of shades sent.
    D. Using pulse modulation.

18. What is the wavelength of a 21.3-MHz signal?
    A. 46.2 m.
    B. 14.1 m.
    C. 21.0 km.
    D. 6.39 km.

19. A coaxial cable:
    A. Keeps the signal confined.
    B. Radiates efficiently.
    C. Works well as a transmitting antenna.
    D. Can pick up signals from outside.

20. An advantage of fiberoptics over cable communications is:
    A. More sensitivity to noise.
    B. Improved antenna efficiency.
    C. Higher RF output.
    D. Simpler and easier maintenance.

27
CHAPTER

Data reception

ONCE A SIGNAL HAS LEFT A TRANSMITTER, THE IMPULSES TRAVEL, OR *PROPA-gate*, in a cable, optical fiber, or space.

In cables, the signals are ac (usually) or dc (sometimes). In fiberoptic systems, the signals are infrared or visible light. The signals are confined in cables and fibers; the only important variable is the attenuation per kilometer. This depends on the ac frequency, the thickness of wire, or the clarity of optical fiber material.

In communication via electromagnetic waves, the propagation is affected by several factors.

Radio wave propagation

Here is a summary of the main things that affect EM wave communications.

Polarization

The orientation of the E flux is the *polarization* of an EM wave. If the E flux lines are parallel to the earth's surface, you have *horizontal polarization*. If the E flux lines are perpendicular to the surface, you have *vertical polarization*. Polarization can be slanted at any angle between horizontal and vertical.

The orientation of the E flux lines sometimes rotates as the wave travels through space. This is *circular polarization* if the E-field intensity remains constant. If the E-field intensity is more intense in some planes than in others, the polarization is said to be *elliptical*. Rotating polarization can be either *clockwise* or *counterclockwise*, viewed as the wavefronts approach you. This is the *sense* of polarization.

The line-of-sight wave

Electromagnetic waves follow straight lines unless something makes them bend. *Line-of-sight* propagation can take place even when the receiving antenna can't be seen from the transmitting antenna. To some extent, radio waves penetrate nonconducting objects such as trees and frame houses. The line-of-sight wave consists of two components: the *direct wave* and the *reflected wave*.

The direct wave The longest wavelengths are least affected by obstructions. At very low, low and medium frequencies, direct waves can *diffract* around things. As the frequency rises, especially above about 3 MHz, obstructions have a greater and greater blocking effect.

The reflected wave Electromagnetic waves reflect from the earth's surface and from conducting objects like wires and steel beams. The reflected wave always travels farther than the direct wave (Fig. 27-1). The two waves are usually not in phase at the receiving antenna. If they're equally strong but 180 degrees out of phase, a *dead spot* occurs. This is most common at the highest frequencies.

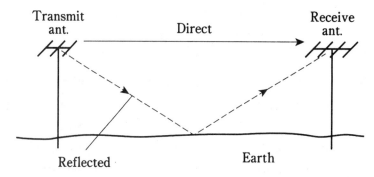

27-1 The reflected wave travels farther than the direct wave.

At VHF and UHF, an improvement in reception can result from moving the transmitting or receiving antenna just a few inches. In *mobile* VHF/UHF operation, when the transmitter and/or receiver are moving, dead spots produce "holes" or "picket fencing" in the received signal.

The surface wave

At frequencies below about 10 MHz, the earth's surface conducts ac quite well. Because of this, vertically polarized EM waves follow the surface for hundreds or even thousands of miles, with the earth helping to conduct the E flux. The lower the frequency, the lower the *ground loss* and the farther the waves travel by *surface-wave propagation*. Horizontally polarized waves do not travel well in this mode, because horizontal E flux is shorted out by the earth.

Above about 10 MHz, the earth becomes lossy, and surface-wave propagation is not useful for more than a few miles. Significant surface-wave communications are done mainly at very low, low and medium frequencies (up to 3 MHz).

Sky-wave EM propagation

The earth's *ionosphere*, at altitudes from about 35 miles to 250 miles, has a great effect on EM waves at frequencies below about 100 MHz.

The ionosphere consists of three layers, called the *D layer, E layer* and *F layer.* The layers form at different altitudes, with the D layer lowest and the F layer highest (Fig. 27-2). The D layer absorbs radio waves at frequencies below about 7 MHz. The E and F layers return radio waves to the earth by a process called *ionospheric refraction.*

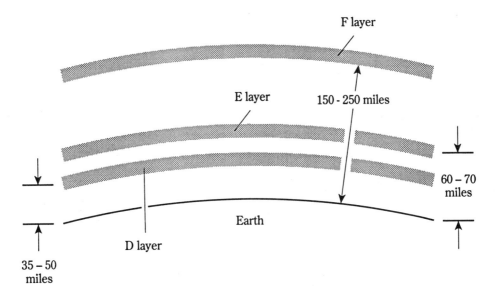

27-2   Simplified rendition of the ionosphere.

The D layer is dense during daylight and vanishes at night. This makes long-distance sky-wave propagation poor during the daytime and excellent at night on frequencies below about 7 MHz. The D layer's disappearance at dusk is responsible for the dramatic change in the behavior of the AM broadcast band (535 kHz to 1.605 MHz) that occurs after sunset.

The E layer varies in density with the 11-year sunspot cycle. The greatest ionization is during peak sunspot years (1979, 1990, 2001 etc.) and the least is during slack years. Solar flares cause the E layer to form dense regions that return radio waves to earth at frequencies as high as 100 MHz. The E layer can produce communications over distances of hundreds or even thousands of miles.

The F layer returns radio waves at all frequencies up to a certain maximum, called the *maximum usable frequency (MUF).* The MUF varies depending on the sunspot cycle and also on day-to-day solar activity. Reliable communications can usually be had up to about 7 MHz during the night and 15 MHz during the day when sunspot activity is minimum; this increases to perhaps 20 MHz at night and 50 MHz during the day when sunspot numbers are maximum.

All the ionospheric modes are called *sky-wave propagation.* This is the mode responsible for worldwide shortwave communications, and for nighttime AM broadcast reception over thousands of miles.

Tropospheric EM propagation

At frequencies above about 30 MHz, the lower atmosphere bends radio waves towards the surface (Fig. 27-3). *Tropospheric banding* or *tropo* occurs because the *index of refraction* of air, with respect to EM waves, decreases with altitude. The effect is similar to sound waves hugging the surface of a lake in the early morning, letting you hear a conversation a mile away. Tropo makes it possible to communicate for hundreds of miles when the ionosphere will not return waves to the earth.

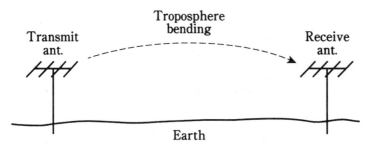

27-3   The lower atmosphere bends EM waves toward the surface.

Another type of tropospheric propagation is called *ducting*. It takes place when EM waves are trapped in a layer of cool, dense air sandwiched between two layers of warmer air. Like bending, ducting occurs almost entirely at frequencies above 30 MHz.

Still another tropospheric-propagation mode is *troposcatter*. This takes place because air molecules, dust grains and water droplets scatter some of the EM field at very high and ultra-high frequencies (above 30 MHz).

Exotic modes of EM propagation

Radio waves can bounce off the aurora (northern and southern lights). This is *auroral propagation*, and it occurs at frequencies from roughly 15 to 250 MHz. It can take place between stations separated by up to about 2,000 miles.

Meteors entering the upper atmosphere produce ionized trails that persist for several seconds up to about a minute; these ions reflect EM waves and cause *meteor-scatter propagation*. This mode allows communication for hundreds of miles at frequencies from 20 to 150 MHz.

The moon, like the earth, reflects EM fields. This makes it possible to communicate *via earth-moon-earth* (*EME*), also called *moonbounce*. High-powered transmitters, sophisticated antenna systems and sensitive receivers are needed for EME. Most EME is done by radio hams at frequencies from 50 MHz to over 2 GHz.

Receiver specifications

Any communications receiver, whether analog or digital, audio or video, must do certain basic things well.

Sensitivity

The *sensitivity* of a receiver is its ability to recover weak signals and process them into readable data. The most common way to express receiver sensitivity is to state the number of ac signal microvolts at the antenna, needed to produce a given *signal-to-noise (S/N) ratio*. Sometimes, the *signal-plus-noise-to-noise ratio*, abbreviated $S+N/N$, is given.

When you look at the specifications table for a radio receiver, you might see, for example, "better than 0.3 uV for 10 dB S/N." This means that a signal of 0.3 uV or less, at the antenna terminals, will result in a S/N ratio of 10 dB. This figure is given as an example only, and not as any sort of standard separating the good from the bad. Technological advancements are always improving the sensitivity figures for communications receivers. Besides that, a poor figure for one application, or on one frequency band, might be fine for some other intended use, or on some other frequency band.

The *front end*, or first RF amplifier stage, of a receiver is the most important stage with regard to sensitivity. Sensitivity is directly related to the gain of this stage; but the amount of noise it generates is even more significant. A good front end should produce the best possible S/N or S+N/N ratio at its output. All subsequent stages will amplify the front-end noise output as well as the front-end signal output.

Selectivity

Selectivity is the ability of a receiver to respond to a desired signal, but not to undesired ones. This generally means that a receiver must have a frequency "window" within which it is sensitive, but outside of which it is not at all sensitive. This "window" is established by a *preselector* in the early RF amplification stages of the receiver, and is honed to precision by a *narrowband* filter in a later amplifier stage.

Suppose you want to receive an LSB signal whose suppressed-carrier frequency is 3.885 MHz. The signal information is contained in a band from 3.882 to 3.885 MHz, a span of 3 kHz. The preselector makes the receiver sensitive in a range of about plus-or-minus 10 percent of the signal frequency; other frequencies are attenuated. This reduces the chance for a strong, out-of-band signal to impair the performance of the receiver. The narrowband filter responds only to the frequencies 3.882 to 3.885 MHz, so that signals in adjacent channels are rejected. The better this filter works, the better the *adjacent-channel rejection*.

Dynamic range

The signals at a receiver input vary over several orders of magnitude. If you are listening to a weak signal and a strong one comes on the same frequency, you don't want to get your eardrums damaged! But also, you want the receiver to work well if a strong signal appears at a frequency near (but not exactly on top of) the weak one. *Dynamic range* is the ability of a receiver to maintain a fairly constant output, and to keep its sensitivity, in the presence of signals ranging from the very weak to the extremely strong. Dynamic range is specified in decibels, and is typically over 100 dB.

Noise figure

All RF amplifier circuits generate some wideband noise. The less internal noise a receiver produces, the better the S/N ratio. This becomes more and more important as the frequency increases. It is paramount at very-high frequencies and above (over 30 MHz).

The *noise figure* (*NF*) of a receiver is specified in various different ways. The lower the noise figure, the better the sensitivity. You might see a specification such as "NF: 6 dB or better."

Noise figure depends on the type of active amplifier device used in the front end of a receiver. Gallium-arsenide field-effect transistors (GaAsFETs) are well known for the low levels of noise they generate, even at quite high frequencies. Other types of FETs can be used at lower frequencies. Bipolar transistors tend to be rather noisy.

Definition of detection

Detection, also called *demodulation*, is the recovery of information such as audio, a visible image (either still or moving), or printed data from a signal. The way this is done depends on the modulation mode.

Detection of AM signals

Amplitude modulation is done by varying the instantaneous strength of the carrier wave. The fluctuations follow the waveform of the voice or video data. The carrier itself is ac, of a frequency many times that of the modulating audio or video. The *modulation peaks* occur on the positive and negative swings of the carrier. This is shown in Fig. 27-4.

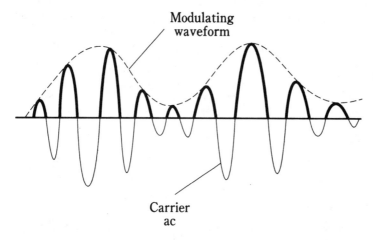

Modulating
waveform

Carrier
ac

27-4   An AM signal. The heavy lines show the output after passing
through the envelope-detector diode or transistor.

If you feed the AM carrier current directly into a headset, fax machine, or whatever, you can't recover the data because the positive and negative peaks come by in such rapid succession that they cancel each other out.

A simple way to get the modulating waveform out of an AM signal is to cut off half of the carrier wave cycle. The result (Fig. 27-4) is fluctuating, pulsating dc. The *pulsations* occur at the carrier frequency; the slower *fluctuation* is a duplication of the modulating audio or video. This process can be done by passing the signal through a diode with a small enough capacitance so that it follows the carrier frequency. A transistor, biased for class-B operation, can also be used.

The rapid pulsations are smoothed out by passing the output of the diode or transistor through a capacitor of just the right value. The capacitance should be large enough so that it holds the charge for one carrier current cycle, but not so large that it smooths out the cycles of the modulating signal.

This scheme is known as *envelope detection.* It is used extensively for reception of AM audio and video.

Detection of CW signals

If you tune in an unmodulated carrier with an envelope detector, you won't hear anything. A keyed carrier might produce barely audible thumps or clicks, but it will be impossible to read the code.

For detection of CW, it's necessary to inject a signal into the receiver a few hundred hertz from the carrier. This injected signal will *beat* against the carrier, producing a tone whose frequency is the difference between the carrier and injected- signal frequencies. The injected signal is produced by a *beat-frequency oscillator (BFO).* The beating occurs in a signal combiner or mixer.

A block diagram of a simple CW receiver is shown in Fig. 27-5. The BFO is tunable. Suppose there is a CW signal at 3.550 MHz. As the BFO approaches 3.550 MHz from below, a high-pitched tone will appear at the output. When the BFO reaches 3.549 MHz, the tone will be 3.550–3.549 MHz, or 1 kHz. This is a comfortable listening pitch for most people. The BFO setting isn't too critical; in fact, it can be changed to get a different tone pitch if you get tired of listening to one pitch. As the BFO frequency passes 3.550 MHz, the pitch will descend to a rumble, then to a swish-swish sound. As the BFO frequency continues to rise, the tone pitch will increase again, eventually rising beyond the range of human hearing.

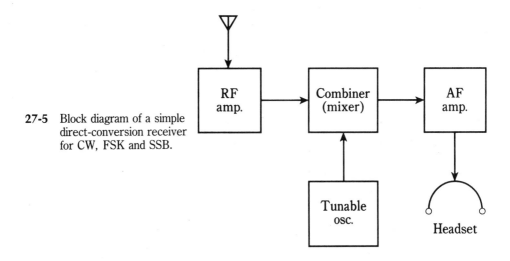

27-5 Block diagram of a simple direct-conversion receiver for CW, FSK and SSB.

This demodulation scheme is called *heterodyne detection.* A receiver that makes use of a heterodyne detector is called a *direct-conversion receiver.*

Detection of FSK signals

Frequency-shift keying can be detected using the same methods as CW detection. The carrier beats against the BFO in the mixer, producing an audio tone that alternates between two different pitches. The block diagram of Fig. 27-5 can therefore apply to FSK reception, as well as to CW reception.

With FSK, the BFO frequency is always set well above, or well below, both the mark and the space carrier frequencies. The *offset* affects the audio tone frequencies, and is set so that certain standard pitches result. There are several sets of standard tone frequencies, depending on whether the communications is amateur, commercial or military. To get the proper pitches, the BFO must be set precisely at the right frequency. There is little tolerance for error.

Detection of SSB signals

A single-sideband signal is really just an AM signal, minus the carrier and one of the sidebands. If the BFO frequency of a CW receiver is set exactly where the carrier should be, the sideband, either upper or lower, will beat against the BFO and produce audio output. In this way, the receiver of Fig. 27-5 can be used to detect SSB.

The BFO frequency is critical for good SSB reception. If it is not set exactly at the frequency of the suppressed carrier, the voice will sound bizarre, like "monkey chatter."

A more advanced method of receiving CW, FSK, and SSB makes use of a *product detector.* This is a specialized form of mixer, and is discussed a little later in this chapter.

Detection of FM signals

Frequency-modulated or phase-modulated signals can be detected in several ways. The best FM receivers respond to frequency/phase changes, but not to amplitude changes.

Slope detection

An AM receiver can be used to detect FM. This is done by setting the receiver near, but not exactly on, the FM signal.

An AM receiver has a narrowband filter with a *passband* of about 6 kHz. This gives a *selectivity curve* such as that shown in Fig. 27-6. If the FM carrier frequency is near the *skirt*, or *slope*, of the filter response, modulation will cause the signal to move in and out of the passband. This will make the receiver output vary with the modulating data.

Because this scheme takes advantage of the filter slope, it is called *slope detection.* It has two disadvantages. First, the receiver will respond to amplitude variations (because that's what it's designed for). Second, there will be nonlinearity in the received signal, producing distortion, because the slope is not a straight line.

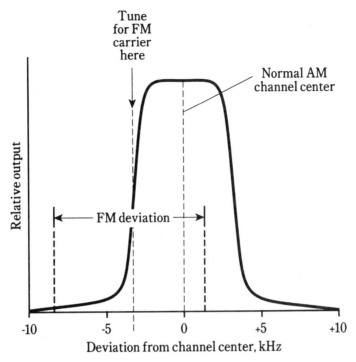

27-6 Slope detection lets an AM receiver demodulate FM.

The phase-locked loop

If an FM signal is injected into a phase-locked loop (PLL) circuit, the loop will produce an error voltage that is a duplicate of the modulating waveform. The frequency changes might be too fast for the PLL to lock onto, but the error voltage will still appear. Many modern receivers take advantage of this effect to achieve FM detection.

A circuit called a *limiter* can be placed ahead of the PLL so that the receiver does not respond to amplitude modulation. Thus, one of the major advantages of FM over AM is realized. Atmospheric noise and ignition noise cause much less disruption of a good FM receiver than to AM, CW, or SSB receivers, provided that the signal is strong enough. Weak signals tend to appear and disappear, rather than fading, in an FM receiver that employs limiting.

The discriminator

A *discriminator* produces an output voltage that depends on the instantaneous signal frequency. When the signal is at the center of the discriminator passband, the output voltage is zero. If the instantaneous signal frequency decreases, a momentary phase shift results, and the instantaneous output voltage becomes positive. If the frequency rises above center, the output becomes negative. The instantaneous voltage level (positive or negative) is directly proportional to the instantaneous frequency. Therefore, the output voltage is a duplicate of the modulating waveform.

A discriminator is sensitive to amplitude variations in the signal, but this problem can be overcome by the use of a limiter.

The ratio detector

A discriminator with a built-in limiting effect is known as a *ratio detector*. This type of FM detector was developed by RCA and is used in high-fidelity receivers and in the audio portions of TV receivers.

A simple ratio detector circuit is shown in Fig. 27-7. A transformer splits the signal into two components.

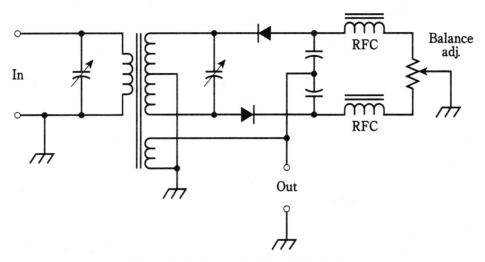

27-7   A ratio detector for demodulating FM.

A change in signal amplitude causes equal level changes in both halves of the circuit. These effects cancel because they are always 180 degrees out of phase. This nullifies amplitude variations on the signal.

A change in signal frequency causes a phase shift in the circuit. This unbalances it, so that the outputs in the two halves of the circuit become different. This produces an output in direct proportion to the instantaneous phase shift. The output signal is a duplication of the modulating waveform on the FM signal.

Detection of PM signals

Pulse modulation (PM) operates at a low *duty cycle*. The pulses are far shorter in duration than the intervals between them. A PM signal is "mostly empty space." The ratio of average signal power to peak signal power is low, often much less than 1 percent.

When the amplitude or duration of the pulses changes, the average transmitter power also changes. Stronger or longer pulses increase the effective signal amplitude;

weaker or shorter pulses result in decreased amplitude. Because of this, PM can be detected in the same way as AM.

A major advantage of PM is that it's "mostly empty space." With pulse amplitude modulation (PAM), pulse duration modulation (PDM) or pulse code modulation (PCM), the time interval is constant between pulse centers. Even at maximum modulation, the ratio of "on" time to "off" time is low. Therefore, two or more signals can be intertwined on a single carrier (Fig. 27-8). A PM receiver can pick out one of these signals and detect it, ignoring the others. This is known as *time-division multiplexing*.

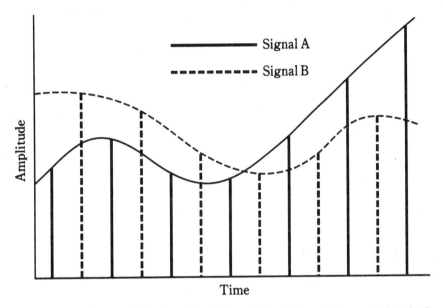

27-8 Time-division multiplexing of two different signals (A and B) on a single pulsed carrier.

A time-division-multiplex communications circuit requires that the receiver be synchronized with the transmitter. This is easy to do if the pulse frequency is constant. The receiver and transmitter can be clocked from a single, independent, primary time standard such as the broadcasts of radio station WWV. The receiver detector is blocked off during intervals between transmitter pulses and "opens up" only during "windows" lasting as long as the longest transmitter pulses.

The received data is selected by adjusting the windows to correspond with the desired pulse train. Because the duty cycle of any single signal is so low, it is possible to multiplex dozens or even hundreds of signals on one transmitted carrier.

Digital-to-analog conversion

When receiving a digital signal such as pulse code modulation (PCM), a *digital-to-analog* (*D/A*) converter is used. This reverses the process of A/D conversion at the transmitter, so that the original analog data is recovered.

You might ask, "Why convert a signal to digital form in the first place, if it's going to be changed back to analog form at the receiver anyway?" The reason is that a digital signal is inherently simpler than an analog signal, in the sense that is *less random*. Thus, a digital signal resembles noise less than an analog signal.

It's good to make a signal as different from noise as possible, in as many ways (or senses) as possible. This is because the more different a signal is from unwanted noise, the easier it is to separate the data from the noise, and the better is the realizable S/N ratio.

You might think of signal/noise separation in terms of apples, oranges, and a watermelon. It takes awhile to find an orange in a tub of apples. (You'll probably have to dump the tub). Think of the orange as an analog signal, and the apples as noise. But suppose there's a watermelon in a bushel basket with apples. You'll have no trouble at all finding the watermelon. Think of the melon as a digital signal, and the apples as noise.

Another, more interesting feature of digital communications arises when you think of a watermelon in a tub of oranges. It's as easy to separate a digital signal from a jumble of analog signals, as it is to extract a digital signal from noise. In a band occupied by thousands of analog signals, a lone digital signal can be picked out easily—far more easily than any of the analog signals.

In recent years, digitization has become commonplace not only in data communications, but in music recording and even in video recording. The main advantage of digital recording is that a selection can be recorded, re-recorded, re-re-recorded, etc., and the quality does not diminish.

Digital signal processing

A new and rapidly advancing communications technique, *digital signal processing (DSP)*, promises to revolutionize voice, digital, and image communications.

In analog modes, DSP works by converting the received voice or video signal input into digital data by means of an analog-to-digital (A/D) converter. The digital signal is processed, and is reconverted back to the original voice or video via a D/A converter (Fig. 27-9).

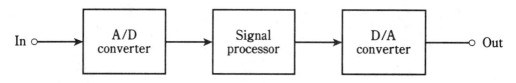

27-9 Digital signal processing.

In digital modes, A/D and D/A conversion is not necessary, but DSP can still be used to "clean up" the signal. This reduces the number of errors.

It is in the digital part of the DSP circuit that the signal enhancement takes place. Digital signals have a finite number of discrete, well-defined states. It is easier to process a signal of this kind than to process an analog signal, which has a theoretically infinite number of possible states. The DSP circuit gets rid of confusion between digital states. The result is an output that is essentially free from interference.

Digital signal processors are available from several commercial sources. They can be installed in existing communications receivers.

The benefits of DSP are improved signal-to-noise ratio, superior intelligibility and enhanced fidelity or image clarity. In addition, DSP can make a CW, AM, or SSB receiver less susceptible to interference from atmospheric noise and ignition noise. This is because a digital signal is as different from noise as a watermelon is from an apple.

The principle of signal mixing

A *mixer* is a circuit that combines two signals having different frequencies, producing a signal whose frequency is either the sum or the difference of the input frequencies. One of the signals is usually an unmodulated carrier, so that the mixer has the effect of *converting* a modulated signal at one frequency to a modulated signal at some other frequency. A mixer has two inputs and one output.

Up conversion versus down conversion

Suppose the signals to a mixer are at frequencies f1 and f2, with f2 being the higher frequency. The mixer has a nonlinear active element, such as a diode or a class-B-biased transistor. The circuit generates new signals, in addition to passing the original two signals. Outputs will appear at f1, f2, f2 + f1, and f2 − f1.

A typical mixer circuit is diagrammed in Fig. 27-10. The output has a tuned circuit that is set to either f2 + f1 or at f2 − f1. If the output is tuned to the sum of the two input frequencies, the output frequency, F_{out}, is higher than either of the input frequencies. If the output is tuned to the difference frequency, the new signal is either in between the two inputs, or else lower than both.

Let f1 represent the frequency of a signal that you want to convert via mixing. Let f2 be the signal from a *local oscillator* (*LO*). If the sum output frequency is used, you have *up conversion*. If the difference frequency is used, and if f2 is selected so that f_{out} is less than f1, you have *down conversion*.

A VLF/LF converter

Up conversion is sometimes used for reception of very-low-frequency (VLF) and low-frequency (LF) radio (9 kHz to 300 kHz). The VLF or LF input is mixed with an LO to provide an output that falls within the range of a shortwave receiver, say over the range 3.509 MHz to 3.800 MHz. A block diagram of an up converter for VLF/LF is shown in Fig. 27-11A. The LO frequency is 3.500 MHz.

This VLF/LF converter produces mirror-image output duplicates, one above 3.509 MHz and the other from 3.200 to 3.491 MHz. Does this seem like a super-broadbanded AM signal, having an overall bandwidth of 600 kHz, with USB at 3.509 to 3.800 MHz and LSB at 3.200 to 3.491 MHz? If you think so, you're right. Up conversion produces AM. When you tune from 3.509 to 3.800 MHz, thereby hearing VLF/LF signals from 9 to 300 kHz, you are actually listening to little "slices" of the USB of a wideband AM signal.

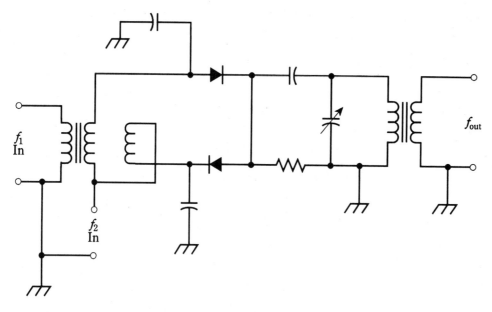

27-10   A passive mixer.

A UHF/microwave converter

Down conversion is often used to allow reception of ultra-high-frequency (UHF) and microwave signals (above 300 MHz). The UHF or microwave input is mixed with an LO to provide an output that falls within the tuning range of a shortwave VHF receiver. A block diagram of a down converter for UHF/microwave reception is shown in Fig. 27-11B.

This converter has an output that covers a huge band frequencies. In fact, a single frequency allocation at UHF or microwave might be larger than the entire frequency range of a shortwave receiver. An example is a UHF converter designed to cover 1.000 GHz to 1.100 GHz. This is a span of 100 MHz, more than three times the whole range of a shortwave radio.

To receive 1.000 to 1.100 GHz using a down converter and a shortwave receiver, the LO frequency must be switchable. Suppose you have a communications receiver that tunes in 1-MHz bands. You might choose one of these bands, say 7.000 to 8.000 MHz, and use a keypad to choose LO frequencies from 0.993 GHz to 1.092 GHz. This will produce a difference-frequency output at 7.000 to 8.000 MHz for 100 segments, each 1 MHz wide, in the desired band of reception.

If you want to hear the segment 1.023 to 1.024 GHz, you set the LO at 1.016 MHz. This produces an output range from $1023 - 1016 = 7$ MHz to $1024 - 1016 = 8$ MHz.

The product detector

For the reception of CW, FSK, and SSB signals, a *product detector* is generally used. It works according to the same basic principle as the mixer. The incoming signal combines with the signal from an unmodulated local oscillator, producing audio or video output.

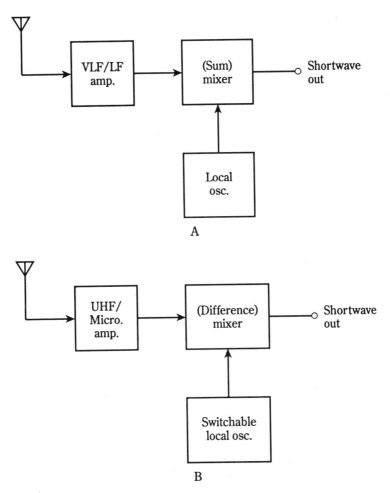

27-11 At A, an up converter that allows VLF/LF reception on a shortwave receiver. At B, a down converter that allows UHF/microwave reception on a shortwave receiver.

Product detection is similar to the heterodyne detection in a direct-conversion receiver. But product detection is done at a single frequency, rather than at a variable frequency as is the case in direct-conversion reception. The single, constant frequency is obtained by mixing the incoming signal with the output of a variable-frequency LO. The process is called *heterodyning,* and a receiver that employs this scheme is known as a *superheterodyne* or *superhet.*

Two product-detector circuits are shown in Fig. 27-12. At A, diodes are used; there is no amplification. At B, a bipolar transistor is employed; this circuit provides some gain. The essential characteristic of either circuit is the nonlinearity of the semiconductor devices. This is responsible for producing the sum and difference frequencies that you hear, or that are converted into video images.

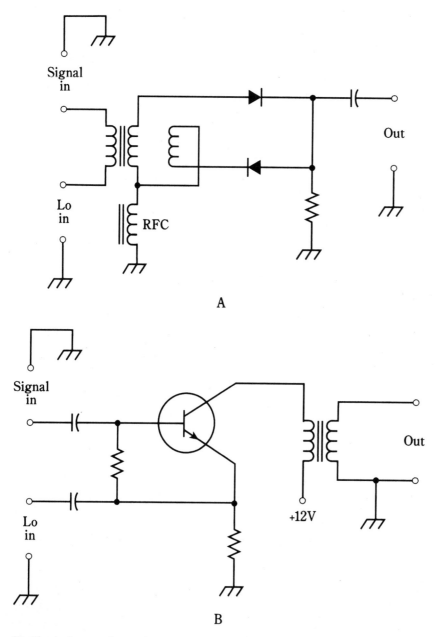

27-12   At A, a passive product detector. At B, a product detector that provides some amplification.

The superheterodyne

The superheterodyne uses one or more mixers to convert an incoming signal, regardless of its frequency, to an identically modulated signal at some other, constant frequency. The signal frequency can be heterodyned once, twice, or even three times. Thus you might hear of a *single-conversion*, *double-conversion*, or *triple-conversion* superheterodyne receiver.

A single-conversion superhet

A block diagram of a single-conversion receiver is shown in Fig. 27-13.

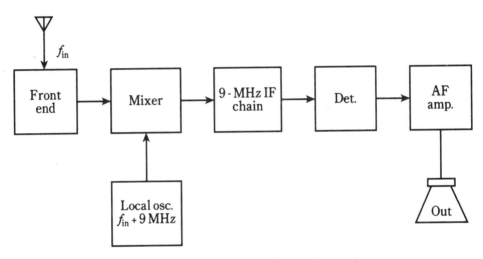

27-13 A single-conversion superheterodyne receiver.

The incoming signal first passes through a sensitive, low-noise, tunable front-end amplifier. The tuning range of this amplifier must be sufficient to cover all the desired reception frequencies f_{IN}.

The second stage is a mixer/LO combination. The LO has a variable frequency that tunes over the received-signal range plus 9.000 MHz. The LO frequency control is the *main tuning* control for the entire receiver. The LO tuning might track along with the tuning of the front end, or the front end might tune independently by means of a separate *preselector* control. The mixer output is always at 9.000 MHz, no matter what the incoming signal frequency.

The intermediate frequency

The 9.000-MHz mixer output signal is called the *intermediate frequency* (*IF*) of the superhet. This signal has the same modulation waveform, and the same bandwidth, as the incoming signal. The only difference is that it might be "upside down"; LSB would be changed to USB, or the sense of FSK would be reversed. But this is an inconsequential difference insofar as it has no effect on the quality of the received signal.

The IF is easy to process because its frequency never changes. Several *IF amplifier* stages, along with filtering, provide the best possible sensitivity and selectivity. This part of the receiver is the *IF amplifier chain* or *IF chain*.

Image rejection

In the superhet of Fig. 27-13, the LO frequency, f_{LO}, is above f_{IN}. The difference is always exactly 9.000 MHz. Thus, if $f_{IN} = 3.510$ MHz, then $f_{LO} = 3.510 + 9.000$ MHz = 12.510 MHz. And if $f_{IN} = 15.555$ MHz, then f_{LO} 15.555 + 9.000 MHz = 24.555 MHz. The IF is always obtained by *difference mixing*. The 9.000-MHz IF is the signal at f_{LO} - f_{IN}.

But there is another possible frequency that can mix with the LO signal to produce a 9.000-MHz output. That is the difference f_{IN} - f_{LO}. Suppose you have the LO set at $f_{LO} = 12.510$ MHz, so that the receiver hears signals at $f_{IN} = 3.510$ MHz. If a signal comes on at $f_{LO} + 9.000$ MHz = 12.510 + 9.000 MHz = 21.510 MHz, and if this signal gets through the front end of the receiver, then it will mix with the LO signal to produce a false signal at 21.510 - 12.510 MHz = 9.000 MHz. This is called an *image signal*.

Suppose you set the LO at $f_{LO} = 24.555$ MHz, so that the receiver hears signals at $f_{IN} = 15.555$ MHz. If a signal appears at $f_{LO} + 9.000$ MHz = 24.555 + 9.000 MHz = 33.555 MHz, then you'll hear it in the 9.000-MHz IF chain as an image signal.

Clearly, you don't want the receiver to hear image signals; you want it to receive only at the desired f_{IN}. A good preselector will provide *image rejection*. This is helped along by choosing a high IF, so that the image frequency, f_{IMAGE}, is much different from f_{IN}. That way, the tuned circuit in the front end has an easy time separating the real thing from the impostor.

Multiple conversion

Another way to improve image rejection is to convert the 9.000-MHz (or whatever frequency) *first IF* to a lower IF, known as the *second IF*. This is usually a frequency between about 50 kHz and 500 kHz. A standard second IF is 455 kHz. The resulting receiver is a double-conversion superhet.

Sometimes the second IF is converted to a still lower frequency, called the *third IF*, usually 50 kHz. This is a triple-conversion superhet.

The main advantage of using double or triple conversion, besides superior image rejection, is the fact that low IFs can be easily filtered to provide superior adjacent-channel rejection.

Low IFs aren't practical with single-conversion receivers, because f_{IMAGE} would be too close to f_{IN}, and the front end could not adequately differentiate between them. This problem would be worst at the highest frequencies f_{IN}.

Detector and audio/video amplifier

Once the IF signal has been filtered and amplified, it passes through a detector. This is generally a product detector for CW, FSK, and SSB, a ratio detector or PLL for FM, and an envelope detector for AM.

Following the detector, a DSP circuit might be used. This further enhances the S/N ratio, minimizing the number of printer or display errors, or clarifying the audio or video.

Finally, an audio/video amplifier is used to boost the level to whatever is needed to drive the speaker, teleprinter, monitor, or instrumentation at the receiver output.

A modulated-light receiver

Modulated light beams can be demodulated easily using photodiodes or photovoltaic cells, as long as the modulating frequency isn't too high. At audio frequencies, a circuit such as the one in Fig. 27-14 will provide sufficient output to drive a headset. This demonstrates the method by which fiberoptic signals are detected.

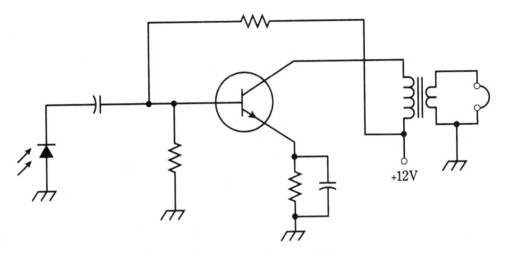

27-14 A simple modulated-light receiver.

For use with a speaker, another bipolar amplifier stage can be added following the stage shown. To increase the sensitivity of the receiver (that is, to allow reception of much fainter modulated-light beams), an FET stage can be added between the photovoltaic cell and the bipolar stage shown.

Modulated-light experiments can be done with simple apparatus, costing only a few dollars. If reasonably large lenses and/or reflectors are used, line-of-sight voice communications can be had for distances up to several hundred feet, using circuits based on the transmitter of Fig. 26-17 and the receiver of Fig. 27-14. With additional amplification and very large reflectors, the range can be increased to several miles on clear nights. This could make a great science-fair project.

Quiz

Refer to the text in this chapter if necessary. A good score is at least 18 correct. Answers are in the back of the book.

1. The reflected wave in a radio signal:

    A. Travels less distance than the direct wave.

B. Travels just as far as the direct wave.

C. Travels farther than the direct wave.

D. Might travel less far than, just as far as, or farther than the direct wave.

2. The reflected wave:

A. Arrives in phase with the direct wave.

B. Arrives out of phase with the direct wave.

C. Arrives in a variable phase compared with the direct wave.

D. Is horizontally polarized.

3. The ionospheric layer that absorbs radio waves is:

A. The D layer.

B. The E layer.

C. The F layer.

D. No layers ever absorb radio waves.

4. The highest layer of the ionosphere is:

A. The D layer.

B. The E layer.

C. The F layer.

D. Dependent on the time of day and the solar cycle.

5. Radio waves that curve earthward in the lower atmosphere are being affected by:

A. Troposcatter.

B. The D layer.

C. Ionospheric ducting.

D. Tropospheric bending.

6. Single-sideband can be demodulated by:

A. An envelope detector.

B. A diode.

C. A BFO and mixer.

D. A ratio detector.

7. A diode and capacitor can be used to detect:

A. CW.

B. AM.

C. SSB.

D. FSK.

8. The S+N/N ratio is a measure of:

A. Sensitivity.

B. Selectivity.

C. Dynamic range.

D. Adjacent-channel rejection.

9. The ability of a receiver to perform in the presence of strong signals is a consequence of its:

A. Sensitivity.

B. Noise figure.

C. Dynamic range.

D. Adjacent-channel rejection.

10. A receiver that responds to a desired signal, but not to one very nearby in frequency, has good:

A. Sensitivity.

B. Noise figure.

C. Dynamic range.

D. Adjacent-channel rejection.

11. An AM receiver can be used to demodulate FM by means of:

A. Envelope detection.

B. Product detection.

C. Slope detection.

D. Pulse detection.

12. An FM detector with built-in limiting is:

A. A ratio detector.

B. A discriminator.

C. An envelope detector.

D. A product detector.

13. Time-division multiplex is often done with:

A. AM.

B. FM.

C. FSK.

D. PM.

14. A continuously variable signal is recovered from a signal having discrete states by:

A. A ratio detector.

B. A D/A converter.

C. A product detector.

D. An envelope detector.

15. Digital modulation is superior to analog modulation in the sense that:
    A. Analog signals have discrete states, while digital ones vary continuously.
    B. Digital signals resemble noise less than analog ones.
    C. Digital signals are easier to use with FM.
    D. Digital signals have greater bandwidth.

16. A product detector would most often be used to receive:
    A. AM.
    B. CW.
    C. FM.
    D. None of the above.

17. To receive UHF signals on a shortwave receiver, you would need:
    A. A heterodyne detector.
    B. A product detector.
    C. An up converter.
    D. A down converter.

18. Image rejection in a superhet receiver is enhanced by:
    A. Front-end selectivity.
    B. A product detector.
    C. A variable LO.
    D. A sensitive IF amplifier chain.

19. A low IF is not practical with a single-conversion receiver because:
    A. Product detection cannot be used.
    B. The image frequency would be too close to the incoming-signal frequency.
    C. Sensitivity would be impaired.
    D. Adjacent-channel rejection would be poor.

20. Digital signal processing can be used to advantage with:
    A. SSB.
    B. SSTV.
    C. FSK.
    D. Any of the above.

28
CHAPTER

Integrated circuits and data storage media

THE MOST ADVANCED ELECTRONIC CIRCUITS CONTAIN HUNDREDS, THOUSANDS, or even millions of diodes, transistors, resistors, and capacitors. This is especially true of industrial and personal computers. These devices use circuits fabricated onto wafers, or *chips*, of semiconductor material, usually silicon. The chips are enclosed in little boxes or cans with pins for connection to external components.

Integrated circuits (ICs) have stimulated as much change as, and perhaps more evolution than, any other single development in the history of electronic technology.

Boxes and cans

Most ICs look like gray or black plastic boxes with protruding pins. Common configurations are the *single inline package (SIP)*, the *dual inline package (DIP)* and the *flatpack*. Another package looks like a transistor with too many leads. This is a *metal-can package*, sometimes also called a *TO package*. Examples are shown in Fig. 28-1.

Not only does an IC look like a box or can; the schematic symbols for most ICs are simple geometric shapes such as triangles or rectangles. A component designator is usually written inside the polygon, and the wires emerging from it are labeled according to their specific functions. But the circuit details are usually too complicated to be drawn in schematic diagrams. There's no point in rendering all that detail anyway; it would only confuse engineers and technicians, and make the schematic diagrams so large that they might cover a wall, or a football field, or the state of Kansas.

So then an IC is a "black box." The engineer or technician can forget about its inner workings, just as you don't dwell upon what's happening inside your personal computer as you use it. Even with the simplicity of the IC symbol, an apparatus using many ICs can have nightmarish schematic diagrams, with dozens of wires running around as parallel lines so close together that you must run a pencil along them to keep track of them.

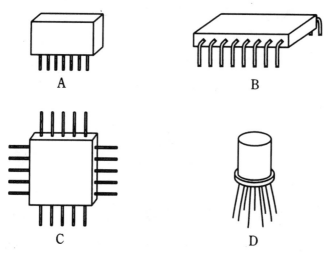

28-1 IC packages: SIP (A), DIP (B), flatpack (C) and metal can (D).

Advantages of IC technology

Integrated circuits have several advantages over individual, or *discrete*, components.

Compactness

An obvious asset of IC design is economy of space; ICs are far more compact than equivalent circuits made from individual transistors, diodes, capacitors, and resistors.

A corollary to this is the fact that far more complex circuits can be built, and kept down to a reasonable size, using ICs as compared with discrete components. Thus, you see *notebook computers*, also known as *laptops*, with capabilities more advanced than early computers that took up whole rooms.

High speed

Another corollary to the compactness of ICs is the fact that the interconnections among components are physically tiny, making high switching speeds possible. Electric currents travel fast, but not instantaneously. The less time charge carriers need to get from component X to component Y, in general, the more computations are possible within a given span of time, and the less time is needed for complex operations.

Low power requirement

Another advantage of ICs is that they use less power than equivalent discrete-component circuits. This is especially important if batteries are to be used for operation. Because ICs use so little current, they produce less heat than their discrete-component equivalents. This translates into better efficiency. It also minimizes the problems that plague equipment that gets hot with use, such as frequency drift and generation of internal noise.

Reliability

Integrated circuits fail less often, per component-hour of use, than appliances that make use of discrete components. This is mainly a result of the fact that all interconnections are sealed within the IC case, preventing corrosion or the intrusion of dust. The reduced failure rate translates into less *downtime*, or time during which the equipment is out of service for repairs.

Ease of maintenance

Integrated-circuit technology lowers maintenance costs, mainly because repair procedures are simplified when failures do occur.

Many appliances use sockets for ICs, and replacement is simply a matter of finding the faulty IC, unplugging it, and plugging in a new one. Special desoldering equipment is used with appliances having ICs soldered directly to the circuit boards.

Modular construction

Modern IC appliances use *modular construction*. In this scheme, individual ICs perform defined functions within a circuit board; the circuit board or *card*, in turn, fits into a socket and has a specific purpose. Computers, programmed with customized software, are used by repair technicians to locate the faulty card in an appliance. The whole card can be pulled and replaced, getting the appliance back to the consumer in the shortest possible time. Then the computer can be used to troubleshoot the faulty card, getting the card ready for use in the next appliance that happens to come along with a failure in the same card.

Modular construction can theoretically be used with discrete-component design. But this is rarely done. When a device is sophisticated enough to need modular construction, IC design is usually needed to keep the size within reason. Can you envision several people, or perhaps a crane, pulling and replacing a 500-pound, 10-by-15-foot card from an appliance as big as a house?

Limitations of IC technology

No technological advancement ever comes without some sacrifice or compromise. Integrated circuits aren't a ticket to electronic utopia.

Inductors impractical

While some components are easy to fabricate onto chips, other components defy the IC manufacturing process. Inductances, except for extremely low values, are one such bugaboo. Devices using ICs must generally be designed to work without inductors. Fortunately, resistance-capacitance (RC) circuits are capable of doing most things that inductance-capacitance (LC) circuits can do.

Mega-power impossible

The small size and low current consumption of ICs comes with a "flip side." This is the fact that high-power amplifiers cannot, in general, be fabricated onto semiconductor chips.

High power necessitates a certain minimum physical bulk, because such amplifiers always generate large amounts of heat.

This isn't a serious drawback. Power transistors and vacuum tubes are available to perform high-power tasks. Integrated circuits are no different than anything else: they're good at some things, and not so good at others.

Linear versus digital

A *linear IC* is used to process analog signals such as voices, music, and radio transmissions. The term "linear" arises from the fact that, in general, the amplification factor is constant as the input amplitude varies. That is, the output signal strength is a *linear function* of the input signal strength (Fig. 28-2).

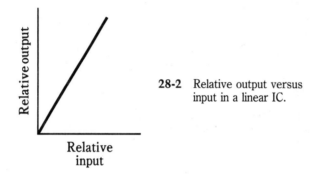

28-2 Relative output versus input in a linear IC.

A *digital IC*, also sometimes called a *digital-logic IC*, operates using just two states, called *high* and *low*. These are sometimes called *logic 1* and *logic 0*, respectively. Digital logic is discussed in detail in chapter 30. Digital ICs consist of many *logic gates* that perform operations in *binary algebra*. Several different digital IC technologies exist; these are briefly outlined later in this chapter.

Types of linear ICs

There are numerous situations in which linear integrated circuits can be used. Linear ICs include amplifiers, regulators, timers, multiplexers, and comparators.

The operational amplifier

An *operational amplifier*, or *op amp*, is a specialized form of linear IC. The op amp consists of several transistors, resistors, diodes, and capacitors, interconnected so that high gain is possible over a wide range of frequencies. An op amp might comprise an entire IC. Or, an IC might consist of two or more op amps. Thus you'll sometimes hear of *dual op amps* or *quad op amps*. Some ICs have op amps in addition to various other circuits.

An op amp has two inputs, one *noninverting* and one *inverting*, and one output. When a signal goes into the noninverting input, the output is in phase with it; when a signal goes into the inverting input, the output is 180 degrees out of phase with it. An op amp has two power supply connections, one for the emitters of the transistors (V_{ee}) and one for the collectors (V_{cc}).

The symbol for an op amp is a triangle. The inputs, output, and power supply connections are drawn as lines emerging from it, as shown in Fig. 28-3. You don't have to be concerned with the details of the circuit inside the triangle.

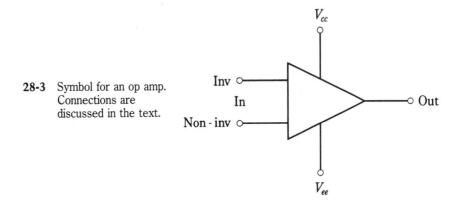

28-3 Symbol for an op amp. Connections are discussed in the text.

The gain characteristics of an op amp are determined by external resistors. Normally, a resistor is connected between the output and the inverting input. This is called the *closed-loop* configuration. The feedback is negative (out of phase), causing the gain of the op amp to be less than it would be if there were no feedback (*open loop*). A closed-loop amplifier using an op amp is shown in Fig. 28-4.

Why, you might ask, would you want to reduce the gain of an amplifier? The answer is that excessive gain can cause problems. An amplifier can be too sensitive, overloading the following circuits, generating too much noise, or producing unwanted responses.

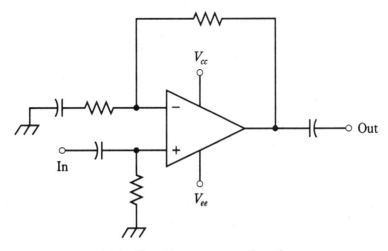

28-4 Closed-loop op-amp configuration.

Open-loop op amps, when used at low frequencies, have extremely high gain, and are prone to instability. They are also usually quite "noisy."

When resistor-capacitor (RC) combinations are used in the feedback loop of an op amp, the amplification changes with frequency. It is possible to get a *lowpass response*, a *highpass response*, a *resonant peak*, or a *resonant notch* using an op amp and various RC feedback arrangements. These four responses are shown in Fig. 28-5 as amplitude-versus-frequency graphs.

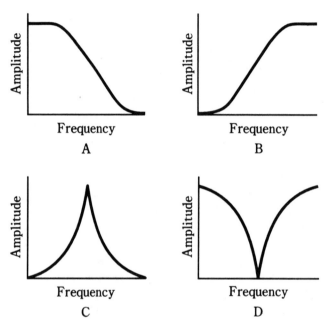

28-5   At A, lowpass response; at B, highpass response; at C, resonant peak; at D, resonant notch.

The regulator

A *voltage regulator IC* acts to control the output voltage of a power supply. This is important with precision electronic equipment. These ICs are available in various different voltage and current ratings.

Typical voltage regulator ICs have three terminals. They look somewhat like power transistors.

The timer

A *timer IC* is actually a form of oscillator. It produces a delayed output, with the delay being adjustable to suit the needs of a particular device. The delay is generated by counting the number of oscillator pulses; the length of the delay can be adjusted by means of external resistors and capacitors.

The analog multiplexer

The *analog multiplexer IC* allows several different signals to be combined in a single channel via time-division multiplexing, in a manner similar to that used with pulse modulation. (This was discussed in chapter 27.) An analog multiplexer can also be used in reverse; then it works as a *demultiplexer*. Thus, you'll sometimes hear engineers talk about *multiplexer/demultiplexer* ICs.

The comparator

Like an op amp, a *comparator* IC has two inputs. The comparator does just what its name implies: it compares the voltages at the two inputs (called A and B). If the input at A is significantly greater than the input at B, the output will be about +5 V. This is logic 1, or high. If the input at A is not greater than the input at B, the output voltage will be about +2 V. This is designated as logic 0, or low.

Voltage comparators are available for a variety of applications. Some can switch between low and high states at a rapid rate of speed; others are slower. Some have low input impedance, and others have high impedance. Some are intended for audio or low-frequency use; others are fabricated for video or high-frequency applications.

Voltage comparators are used to actuate, or *trigger*, other devices such as relays and electronic switching circuits.

Bipolar digital ICs

Digital ICs consist of gates that perform logical operations at high speeds. There are several different technologies, each with its own unique characteristics. Digital-logic technology might use bipolar transistors or metal-oxide-semiconductor (MOS) devices.

TTL

In *transistor-transistor logic (TTL)*, arrays of bipolar transistors, some with multiple emitters, operate on dc pulses. This technology has several variants, some of which date back to around 1970. The hallmark of TTL is immunity to noise pulses. A simple TTL gate is illustrated in Fig. 28-6. The transistors are either completely cut off, or else completely saturated. This is the reason why TTL is not very much affected by external noise "distractions."

ECL

Another bipolar-transistor logic form is known as *emitter-coupled logic (ECL)*. In ECL, the transistors are not operated at saturation, as they are with TTL. This increases the speed of operation of ECL compared with TTL. But noise pulses have a greater effect in ECL, because unsaturated transistors amplify as well as switch signals. The schematic of Fig. 28-7 shows a simple ECL gate.

MOS digital ICs

Several variants of MOS technology have been developed for use in digital devices. They all offer superior miniaturization and reduced power requirements as compared with bipolar digital ICs.

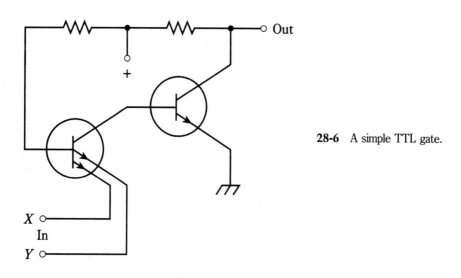

28-6 A simple TTL gate.

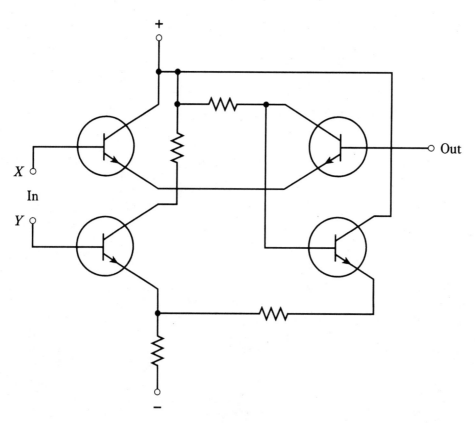

28-7 A simple ECL gate.

CMOS

Complementary-metal-oxide-semiconductor (CMOS), pronounced "seamoss" (and sometimes written that way by lay people who have heard the term but never seen it in documentation), employs both N-type and P-type silicon on a single chip. This is analogous to using N-channel and P-channel FETs in a circuit. The main advantages of CMOS technology are extremely low current drain, high operating speed, and immunity to noise.

NMOS/PMOS

N-channel MOS (NMOS) offers simplicity of design, along with high operating speed. *P-channel MOS* is similar to NMOS, but the speed is slower. An NMOS or PMOS digital IC is like a circuit that uses only N-channel FETs, or only P-channel FETs. You might think of NMOS as batting right-handed, and PMOS as batting left-handed. (Then CMOS is analogous to switch hitting.)

New trends

Research is constantly being done to find ways to increase the speed, reduce the power requirements, and improve the miniaturization of digital ICs. The motivation arises mainly from constant consumer pressure for more sophisticated and portable computers. As the technology advances, you can expect to see notebook (laptop) personal computers with more memory and higher working speed. The batteries will also last longer, because evolving technologies will consume less current.

Component density

The number of elements per chip in an IC is called the *component density*. There has been a steady increase in the number of components that can be fabricated on a single chip. Of course there is an absolute limit on the component density that can be attained; it is imposed by the atomic structure of the semiconductor material. A logic gate will never be devised that is smaller than an individual atom. Technology hasn't bumped up against that barrier yet.

MSI

In *medium-scale integration (MSI)*, there are 10 to 100 gates per chip. This allows for considerable miniaturization, but it is not a high level of component density, relatively speaking. An advantage of MSI (in a few applications) is that fairly large currents can be carried by the individual gates. Both bipolar and MOS technologies can be adapted to MSI.

LSI

In *large-scale integration (LSI)*, there are 100 to 1000 gates per semiconductor chip. This is an *order of magnitude* (a factor of 10) more dense than MSI.

Electronic wristwatches, single-chip calculators, and small microcomputers are examples of devices using LSI ICs. They can be 10 times more sophisticated than MSI devices.

VLSI

Very-large-scale integration (VLSI) devices have from 1,000 to 10,000 components per chip. This is an order of magnitude more dense than LSI. Complex microcomputers, and peripheral circuits such as memory storage ICs, are made using VLSI.

ULSI, ELSI, and OLSI

You might sometimes hear of *ultra-large-scale integration (ULSI)*. Devices of this kind have more than 10,000 elements per chip. The principal use for this technology lies in ever-larger *random-access memory (RAM)* capability for personal computing.

In the future, you can expect new names to arise to fit new technologies. But common sense will ultimately prevail over the ridiculous. While abbreviations like "ELSI," for "extremely large-scale integration," might be coined (this one happens to be an acronym, too), engineers have to be careful. What is "extremely dense" at one point in time might be "ho-hum" a few years later. No one wants to be cornered into christening a new invention with some laughable name like "OLSI" (overwhelmingly large-scale integration), just because all the other superlatives have been used up.

IC memory

Binary digital data, in the form of high and low levels (logic ones and zeros), can be stored in ICs. Data storage is generally known as *memory*. In ICs, memory can take various forms.

RAM

A *random-access memory (RAM)* stores binary data in *arrays*. The data can be *addressed* (selected) from anywhere in the matrix. Data is easily changed and stored back in RAM, in whole or in any part. A RAM is sometimes called a *read/write memory*.

An example of RAM is a word-processing computer file that you are actively working on. This paragraph, this chapter, and in fact the whole text of this book was written in semiconductor RAM before being stored on disk (another kind of RAM) and ultimately printed on the paper now before you.

There are two kinds of RAM: *dynamic RAM (DRAM)* and *static RAM (SRAM)*. A DRAM employs IC transistors and capacitors; data is stored as charges on the capacitors. The charge must be replenished frequently, or it will be lost via discharge. Replenishing is done automatically several hundred times per second. An SRAM uses a circuit called a *flip flop* to store the data. This gets rid of the need for constant replenishing of charge, but the tradeoff is that SRAM ICs require more elements to store a given amount of data.

With any RAM, the data is erased when the appliance is switched off, unless some provision is made for *memory backup*. The most common means of memory backup is the use of a cell or battery. Modern IC memories need so little current to store their data that a *backup battery* lasts as long in the circuit as it would on the shelf.

A memory that disappears when power is removed is called a *volatile memory*. If memory is retained when power is removed, it is *nonvolatile*.

ROM

By contrast to RAM, *read-only memory (ROM)* can be accessed, in whole or in any part, but not written over. A standard ROM is *programmed* at the factory. This permanent programming is known as *firmware*. But there are also ROMs that you can program and reprogram yourself.

EPROM

An *erasable programmable ROM (EPROM)* is an IC whose memory is of the read-only type, but that can be reprogrammed by a certain procedure. It is more difficult to rewrite data in an EPROM than in a RAM; the usual process for erasure involves exposure to ultraviolet. An EPROM IC can be recognized by the presence of a transparent window with a removable cover, through which the ultraviolet is focused to erase the data. The IC must be taken from the circuit in which it is used, exposed to the ultraviolet for several minutes, and then reprogrammed via a special process.

There are EPROMs that can be erased by electrical means. Such an IC is called an *EEPROM,* for *electrically erasable programmable read-only memory.* These do not have to be removed from the circuit for reprogramming.

Bubble memory

Bubble memory uses magnetic fields within ICs. The scheme is especially popular in computers, because a large amount of data can be stored in a small physical volume.

A single *bubble* is a tiny magnetic field about 0.002 millimeters across. Logic highs and lows correspond to the existence or absence, respectively, of a bubble. The IC contains a ferromagnetic film that acts as a reprogrammable permanent magnet, on which bubbles are stored (Fig. 28-8).

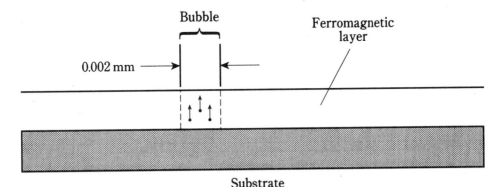

28-8 Cross-sectional view of a bubble-memory IC.

Magnetic bubbles do not disappear when power is removed from the IC. Bubbles are easily moved by electrical signals. An advantage of bubble memory is that it's a nonvolatile

RAM that doesn't need a backup battery. Another asset is that data can be moved from place to place in large chunks. This process is called *block memory transfer*.

IC memory storage capacity

It's difficult to write about the amount of data that can be stored in an IC memory, simply because the numbers keep increasing so fast. The best way to keep track of the technology is to look at the advertisements for personal computers (PCs). These use RAM ICs in practical, affordable appliances.

This book was written with word-processing software on a PC with 640 *kilobytes* (640 Kb) RAM. A kilobyte is about 1,000 *bytes*; this is about 200 words of text. A double-spaced manuscript page in pica type size (10 characters per inch) has about 200 words. Thus, the RAM in this computer can theoretically hold 640 typewritten pages. That's an average novel.

By the time you read this, PCs will probably have several *megabytes (Mb)*, perhaps even 100 Mb, of RAM. A megabyte is about 1,000 Kb or a million bytes.

How much RAM is really needed by the average PC user? Technology eventually comes up against a point of diminishing returns. In consumer word processing, this is already happening. But in other computer endeavors, such as graphics, simulations, interactive games and high-level mathematics, more RAM simply means more accurate, more realistic, faster and more powerful results. As technological capabilities increase, demand has a way of following along. (Have you ever noticed this in regards to your spending habits versus your income?) The largest RAMs are found in computers used by commerce and the government.

Trendier than fashion

New technologies for IC memory are constantly in the research-and-development (R & D) phase. Thus, you'll hear and read abbreviations and acronyms not mentioned here. So rapidly does the art advance, that manufacturers have trouble keeping their own databooks up-to-date. It's even worse than the fashion industry. In electronic technology, once something is obsolete, it never comes back. You can't store today's PROMs and hope that they'll be of any use to your grandchildren.

Magnetic media

For permanent storage of information, *magnetic media* are commonly used. These include magnetic tape and magnetic disks. They're compact (especially modern disks and diskettes) and easy to reprogram. But they are damaged by heat, and their contents can be destroyed by magnetic fields.

Audio tape

The earliest computers used *audio tape* to store data. Printed matter was recorded on the tape using audio-frequency-shift keying (AFSK). This can still be done, but disks are more often used in modern computers, especially personal computers. Nowadays, tape is used mainly for storing voice, music, and/or image data.

Audio tapes are available in *reel-to-reel* or *cassette* form. You've used them for recording voices and/or music, or at least for enjoying your favorite musical talent.

Recently, however, *compact disks (CDs)* have begun to displace magnetic tape for the recording of speech and music.

Magnetic tape has millions of iron-oxide or other ferromagnetic particles on a mylar film. In the recording process, a signal is applied to coils in the *recording head*. The recording head is an electromagnet. It polarizes the particles on the tape as the tape moves by (Fig. 28-9). The particles hold tiny magnetic fields until they are changed by rerecording. When the tape is played back, the movement of magnetic fields past the *pickup head* generates a fluctuating current in a small coil; this current replicates the original recorded signal.

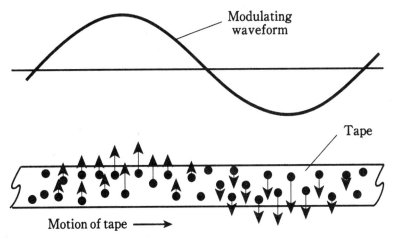

28-9   In a magnetic tape, particles are polarized by the modulating waveform.

Typical speeds for audio tape are from about one inch per second (1 ips) for dictation *microcassettes*, to 3.75 or 7.5 ips for reel-to-reel hi-fi tapes, to 15 ips for commercial hi-fi audio tapes.

There are usually at least four *tracks* on an audio tape, providing a left and right channel for each direction of movement. Some tapes have 8, 16, or 24 tracks. The tracks run parallel to the length of the tape.

Typical tape widths are ⅛ or ¼ inch for consumer use, although in commercial and military applications, wider tape is often used. The thickness varies, being specified in *mils*, where 1 mil = 0.001 inch. Typical thicknesses are 0.5 mil, 1.0 mil or 1.5 mils. The thinner a tape, the longer it will run for a given cassette or reel size. But thicker tapes have less tendency to stretch. Tape stretching causes "wow" in music and garbling of digital signals.

Video tape

Video tape is wider than audio tape, and must deal with a much larger amount of data per unit time. This is simply because a video signal contains far more information than an audio signal. A picture is, indeed, worth many words, or in this case, many magnetized particles!

The *videocassette recorder (VCR)* is used to produce video tapes. These devices have become so common that most families today own them, and use them in the same way that 8-mm film cameras and projectors were once used. The VCR literally records and plays back a fast-scan television signal.

The most common videocassette tape is called *VHS*. The tape itself comes in cassettes that measure 7-3/8 x 4 inches, and are about an inch thick. The tape is ½ inch wide.

The video tracks on a video tape run slantwise with respect to the tape, rather than lengthwise (Fig. 28-10). This facilitates a higher scanning speed as the tape moves through the VCR, than would be possible if the tracks were parallel to the edges of the tape. Each slantwise track represents one video frame. The audio and control tracks run parallel with respect to the edges of the tape.

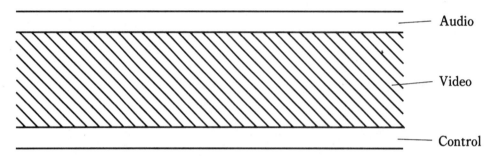

28-10   On video tape, image data is recorded diagonally; audio and control signals are recorded lengthwise.

A typical VCR can use a camera for recording home video, or can be connected to a TV set to record movies, sports events, newscasts, etc. Playback is done through the TV set. A VCR allows you to fast-forward and rewind, so that you can skip over parts of a recording or replay certain parts. Slow-motion and stop-motion features are also included in many VCRs.

Video tape, like audio tape, must be protected from magnetic fields and from excessive heat.

Magnetic disks

A major problem with any recording tape is the fact that the data is strewn out along a "highway." To get from one bit of information to some other bit, it might be necessary to traverse the entire length of the highway. The distance can be hundreds of feet!

But on a *magnetic disk*, no two bits of data are, in practice, ever separated by a distance greater than the diameter of the disk. Because of this, disks are faster than tapes for data storage and retrieval. Magnetic disks are popular for digital data. They aren't often used in analog applications, although R & D continues in the quest for a reliable, high-resolution, magnetic video disk.

Personal and commercial computers almost always use magnetic disks. They come in two forms: the *hard disk* and the *floppy disk* or *diskette*.

The hard disk in a typical PC can hold more than 40 Mb of data. This figure doubles every few years, and the trend can be expected to continue until the point of diminishing returns is reached. Hard disks are permanently installed in the computer. You can't take out and replace a hard disk. Complex software, particularly the kind involving graphics, video games and simulations, and complex mathematics, needs a lot of memory space, and as the software itself becomes more and more sophisticated, consumer demand will follow along with it. Thus you might see, before too long, a hard disk with several *gigabytes (Gb)* of capacity. A gigabyte is about 1,000 Mb or 10^9 bytes. (You are invited to figure out how many average-sized novels could be written on such a disk!)

Floppy disks and diskettes can be interchanged in seconds, so there is no limit to how much data you can sock away on them. A full-wall bookcase of floppies could have more manuscript than you'd write in 1,000,000 years. A typical PC floppy disk comes in a flexible, square package measuring 5-1/4 x 5-1/4 inches, and is about 1 mm thick. These disks usually have 1.2 Mb of capacity. Larger "floppies" are used in commercial computers. A smaller diskette is in a rigid, square package measuring 3-1/2 x 3-1/2 inches, and is slightly thicker than the floppy package. It holds (oddly) a little more data than its physically larger floppy cousin: 1.44 Mb.

It probably won't be long before the memory capacities of floppy disks and diskettes increase, so don't take these figures too seriously.

Compact disks

Speech and music are commonly stored on *compact disks (CDs)*. This medium offers several advantages over magnetic tape. Some CDs can store video data.

The main asset of CDs is their superior reproduction quality. All of the problems associated with older vinyl disks are eliminated. There is no limit to the number of times a CD can be replayed, because laser beams are used to recover the sound. The lasers bounce off tiny *pits* on the disk. Light beams, of course, do not scratch the CD; nothing mechanically rubs against the disk. A CD won't jam because it does not move while it is being replayed. The data can't be distorted by stretching, as can happen with magnetic tape.

Modern CD recording devices *digitize* the speech or music. This results in theoretically infinite reproducibility, because digital highs and lows (logic 1s and 0s) are unmistakable and do not get distorted. A *multigeneration* reproduction (say, a recording of a recording of a recording of a recording of a recording) sounds every bit as good as a *first-generation* reproduction. Of course, there's a limit to how many digital reproductions can be made without error; no system is perfect. A 500th-generation digital recording might have a few errors, where stray noise pulses have mutated 1s into 0s or vice-versa. But a 500th-generation analog reproduction would consist of indecipherable "hash."

Quiz

Refer to the text in this chapter if necessary. A good score is at least 18 correct. Answers are in the back of the book.

1. Because of the small size of ICs compared with equivalent circuits made from discrete components:

    A. More heat is generated.

B. Higher power output is possible.

C. Higher switching speeds are attainable.

D. Fewer calculations need be done in a given time.

2. Which of the following is *not* an advantage of ICs over discrete components?

A. Higher component density.

B. Ease of maintenance.

C. Greater power capability.

D. Lower current consumption.

3. In which of the following devices would you be least likely to find an integrated circuit as the main component?

A. A radio broadcast transmitter's final amplifier.

B. A notebook computer.

C. A battery-powered calculator.

D. A low-power audio amplifier.

4. Which type of component is generally not practical for fabrication in an IC?

A. Resistors.

B. Inductors.

C. Diodes.

D. Capacitors.

5. An op amp usually employs negative feedback to:

A. Maximize the gain.

B. Control the gain.

C. Allow oscillation over a wide band of frequencies.

D. No! Op amps do not employ negative feedback.

6. A channel carries several signals at once. Which type of IC might be used to select one of the signals for reception?

A. An op amp.

B. A timer.

C. A comparator.

D. A multiplexer/demultiplexer.

7. Which type of IC is used to determine whether voltage levels are the same or not?

A. An op amp.

B. A timer.

C. A comparator.

D. A multiplexer/demultiplexer.

8. Which type of digital IC is least susceptible to noise?

    A. Transistor-transistor logic.

    B. Base-coupled logic.

    C. Emitter-coupled logic.

    D. N-channel-coupled logic.

9. Which of the following is *not* an advantage of CMOS?

    A. Relative immunity to noise pulses.

    B. Low current requirements.

    C. Ability to work at high speed.

    D. Ability to handle high power levels.

10. An absolute limit on IC component density is:

    A. The current levels needed.

    B. The maximum attainable impedance.

    C. The size of the semiconductor atoms.

    D. No! There is no limit on component density.

11. In a ROM:

    A. It's easy to get data out and put it in.

    B. It's hard to get data out, but easy to put it in.

    C. It's easy to get data out, but hard to put it in.

    D. It's hard to get data out or put it in.

12. In a RAM:

    A. It's easy to get data out and put it in.

    B. It's hard to get data out, but easy to put it in.

    C. It's easy to get data out, but hard to put it in.

    D. It's hard to get data out or put it in.

13. Which of the following IC types must be physically removed from the circuit to have its memory contents changed?

    A. EEPROM.

    B. EPROM.

    C. ROM.

    D. RAM.

14. A kilobyte is:

    A. Equivalent to a novel.

    B. About 1,000 bytes.

    C. About 1,000,000 bytes.

    D. Equivalent to about one typewritten line.

15. In magnetic audio tape:
    A. The tracks are parallel to the edges.
    B. The tracks are diagonal.
    C. The tracks are perpendicular to the edges.
    D. The tracks can be oriented at any angle.

16. In magnetic video tape:
    A. The video tracks are parallel to the edges.
    B. The video tracks are diagonal.
    C. The video tracks are perpendicular to the edges.
    D. The video tracks can be oriented at any angle.

17. An advantage of magnetic disks over magnetic tape is:
    A. Disks are immune to damage by heat.
    B. Tapes are difficult to rewind.
    C. Disks allow faster data storage and retrieval.
    D. Disks are immune to external magnetic fields.

18. A typical audio recording tape thickness is:
    A. 0.001 mil.
    B. 0.01 mil.
    C. 0.1 mil.
    D. 1 mil.

19. Compact disks (CDs) are not generally used for recording:
    A. Voices.
    B. Music.
    C. In digital form.
    D. Via magnetic fields.

20. A reason CDs don't wear out with repeated playback is:
    A. The magnetic fields are strong.
    B. Nothing touches the disk.
    C. The data is analog.
    D. The magnetic particle density is high.

29
CHAPTER

Electron tubes

WHEN YOU THINK OF *ELECTRON TUBES*, DO YOU IMAGINE THE OLD DAYS OF electronics? Do you envision radios in racks taller than their operators, with strangely shaped, glowing, glass globes? Have you heard that "tubes" are totally obsolete?

Well, some tubes are still around. The most common example is the cathode-ray tube in your TV set. The final amplifier in a TV broadcast transmitter is probably a tube, too.

Vacuum versus gas-filled

There are two kinds of tubes: *vacuum tubes* and *gas-filled tubes*. Vacuum tubes are by far more common.

Vacuum tubes allow electrons to be accelerated to high speeds, resulting in a large current. This current can be made more or less intense, or focused into a beam and guided in a particular direction. The intensity and/or beam direction can be changed with extreme rapidity, making possible a variety of different useful effects.

Gas-filled tubes have a constant voltage drop, no matter what the current. This makes them useful as voltage regulators for high-voltage, high-current power supplies. Gas-filled tubes can withstand conditions that would destroy semiconductor regulating devices. Gas-filled tubes also *fluoresce*, or emit infrared, visible light and/or ultraviolet at well-defined wavelengths. This property can be put to use for decorative lighting. "Neon signs" are gas-filled electron tubes.

In any electron tube, the charge carriers are *free electrons*. This means that the electrons are not bound to atoms, but instead, fly through space in a barrage, somewhat like *photons* of visible light, or like the atomic nuclei in a particle accelerator.

The diode tube

Before the start of the twentieth century, scientists knew that electrons could carry a current through a vacuum. They also knew that hot *electrodes* would emit electrons more easily than cool ones. These phenomena were put to use in the first electron tubes, known as *diode tubes*, for the purpose of rectification.

Cathode, filament, plate

In any tube, the electron-emitting electrode is called the *cathode*. The cathode is usually heated by means of a wire *filament*, similar to the glowing element in an incandescent bulb. The electron-collecting electrode is known as the *anode* or *plate*.

Directly heated cathode

In some tubes, the filament also serves as the cathode. This is called a *directly heated cathode*. The negative supply voltage is applied directly to the filament. The *filament voltage* for most tubes is 6 V or 12 V. The schematic symbol for a diode tube with a directly heated cathode is shown at A in Fig. 29-1.

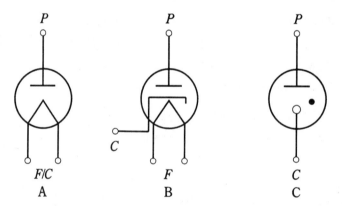

29-1 Schematic symbols for diode tubes; C = cathode, F = filament, P = plate. At A, directly heated cathode; at B, indirectly heated cathode; at C, cold cathode.

Indirectly heated cathode

In many tubes, the filament is enclosed within a cylindrical cathode, and the cathode gets hot from infrared radiation. This is known as an *indirectly heated cathode*. The cathode itself is grounded. The filament normally receives 6 V or 12 Vac. The schematic symbol for a diode tube with an indirectly heated cathode is shown at B in Fig. 29-1.

In either the directly heated or indirectly heated cathode, electrons are driven off the element by the heat of the filament. The cathode of a tube is thus somewhat analogous to the source of a field-effect transistor, or to the emitter of a bipolar transistor.

Because the electron emission in a tube depends on the filament or "heater," tubes need a certain amount of time—normally 30 seconds to a few minutes—to "warm up."

This waiting period can be an annoyance, and it seems bizarre at first to people who haven't dealt with tubes before.

Cold cathode

In a gas-filled voltage-regulator tube, the cathode might not have a filament to heat it. Such a device is called a *cold-cathode* tube. The schematic symbol for a cold-cathode tube is shown at C in Fig. 29-1. The solid dot indicates that the tube is gas-filled, rather than completely evacuated. Various different chemical elements are used in gas-filled tubes; mercury vapor is probably most common. In this type of tube, the "warmup" period is the time needed for the elemental mercury to vaporize, usually a couple of minutes.

Plate

The plate, or anode, of a tube is a cylinder concentric with the cathode and filament (Fig. 29-2). The plate is connected to the positive dc supply voltage. Tubes typically operate at about 50 V to more than 3 kVdc.

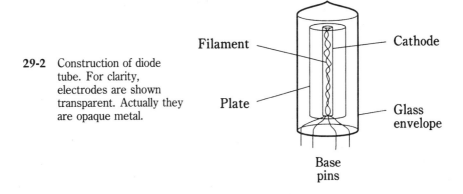

29-2 Construction of diode tube. For clarity, electrodes are shown transparent. Actually they are opaque metal.

Filament — Cathode

Plate —

Glass envelope

Base pins

Because the plate readily attracts electrons but is not a good emitter of them, and because the exact opposite is true of the cathode, a diode tube works well as a rectifier for ac. Diode tubes can also work as envelope detectors for AM, although they are no longer used for that purpose.

The triode

In a diode tube, the flow of electrons from cathode to plate depends mainly on the dc power supply voltage. The greater this voltage, the greater the current through the tube.

The flow of current can be controlled by means of an electrode between the cathode and the plate. This electrode, called the *control grid*, is a wire mesh or screen that lets electrons physically pass through. But the control grid (also called simply the *grid*) interferes with the electrons if it is provided with a voltage that is negative with respect to ground. The greater this negative *grid bias*, the more the grid impedes the flow of electrons through the tube.

A tube with a control grid, in addition to the cathode and plate, is a *triode*. This is illustrated schematically in Fig. 29-3. In this case the cathode is indirectly heated; the filament is not shown. This omission is almost standard in schematics showing tubes with indirectly heated cathodes. When the cathode is directly heated, the filament symbol serves as the cathode symbol.

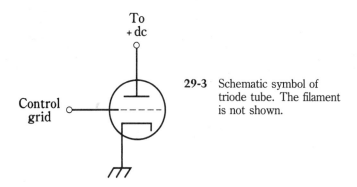

29-3 Schematic symbol of triode tube. The filament is not shown.

Now perhaps you see the resemblance between a triode tube and an N-channel FET. The cathode is analogous to the source; the grid is analogous to the gate; the plate corresponds to the drain. The only major differences, in fact, between the triode tube and the N-channel FET are that the tube is constructed differently, and the tube works with much higher voltages.

In a tube amplifier, the control grid is where the signal input is normally applied. A triode amplifier or oscillator circuit looks just like an FET circuit, except that the tube plate and grid bias voltages are far greater than the FET drain and gate voltages.

Extra grids

Many tubes have multiple grids. These extra elements allow for improved gain and stability in tube type amplifiers.

The tetrode

A second grid can be added between the control grid and the plate. This is a spiral of wire or a coarse screen, and is called the *screen grid* or *screen*. This grid normally carries a positive dc voltage, roughly ¼ to ⅓ that of the plate voltage.

The screen grid reduces the capacitance between the control grid and plate, minimizing the tendency of a tube amplifier to oscillate. The screen grid can also serve as a second control grid, allowing two signals to be injected into a tube. This tube has four elements, and is known as a *tetrode*. Its schematic symbol is shown at A in Fig. 29-4.

The pentode

The electrons in a tetrode can strike the plate with extreme speed, especially if the plate voltage is high. The electrons might bombard the plate with such force that some of them bounce back, or knock other electrons from the plate. Imagine hurling a baseball into a

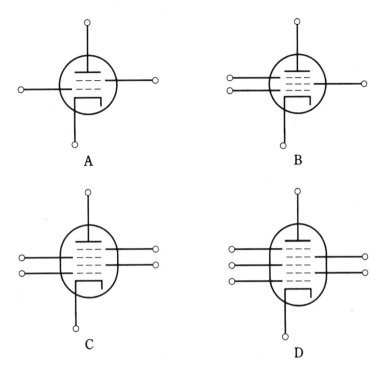

29-4   At A, a tetrode tube. At B, a pentode; at C, a hexode; at D, a heptode.

bushel basket full of other baseballs. If you throw the ball hard enough, some of the other balls might be jostled out of the basket. Many of these electrons end up "leaking out" through the screen grid, rather than going through the plate circuit of the tube. The result is diminished *plate current* and increased *screen current*. Because it's the plate current that produces the output signal for the tube, this *secondary emission* hinders the performance of the tube. If secondary emission is extensive, it can cause screen current so high that the screen grid might be damaged.

The problem of secondary emission can be dealt with by placing yet another grid, called the *suppressor grid* or *suppressor*, between the screen and the plate. The suppressor also reduces the capacitance between the control grid and the plate still more than is the case with the tetrode.

Greater gain and stability are possible with a *pentode*, or tube with five elements, than with a tetrode or triode. The schematic symbol for a pentode is shown at B in Fig. 29-4. The suppressor grid is generally grounded, or connected directly to the cathode. It therefore carries a negative charge with respect to the screen grid and the plate.

The hexode and heptode

In some older radio and TV receivers, tubes with four or five grids were sometimes used. These tubes had six and seven elements, respectively, and were called *hexode* and *heptode.* The usual function of such tubes was signal mixing. The schematic symbol for a hexode is shown at C in Fig. 29-4; the symbol for a heptode is illustrated at D.

You'll probably never hear about these devices in modern electronics, because solid-state components are used for signal mixing nowadays. You might elicit a raised eyebrow if you talk about a "pentagrid converter" or a "heptode mixer." And you can be sure an engineer is kidding you if (s)he mentions an "octode" or "nonode."

Some tubes are obsolete

Vacuum tubes aren't used in receivers anymore, except for picture tubes in TV sets. This is because, at low signal levels, solid-state components (bipolar transistors, FETs and, ICs) can do everything that tubes ever could, with greater efficiency and using lower voltages and currents.

Some electronics hobbyists like to work with antique radios. There is a certain charm in a broadcast receiver that takes up as much space as, and that weighs as much as, a small refrigerator. It brings back memories of a time when drama was broadcast on local AM stations, complete with whining heterodynes and static from summer thundershowers. The action was not rendered on high-resolution, color video, but instead was envisioned in listeners' minds as they sat around the radio on a bare, varnished hardwood floor.

Tube type, antique broadcast/shortwave receivers were about twice as bulky as necessary, because compactness was not a major concern. Such radios were 10 to 100 times bigger and heavier than their modern semiconductor counterparts. The high (and potentially lethal) voltages caused dust to accumulate via electrostatic precipitation from the air, giving the radio's "innards" a greasy, gritty film.

If the above scenarios, possibilities, and hazards appeal to you, perhaps you'd like to collect and operate radio antiques, just as some people collect and drive vintage cars. But be aware that replacement receiving tubes are hard to find. When your relic breaks, you'll need to become a spare-parts sleuth.

Radio-frequency power amplifiers

The most common application of vacuum tubes in modern technology is in RF amplifiers, especially at very high frequencies and/or power levels of more than 1 kW. Two configurations are employed: *grounded cathode* and *grounded grid.*

Grounded cathode

A simplified schematic diagram of a grounded-cathode RF power amplifier, using a pentode tube, is shown in Fig. 29-5. The output circuit is tuned to the operating frequency. The circuit can be operated in class-AB, B, or C. If the amplifier is to be linear, class C cannot be used.

The input impedance of a grounded-cathode power amplifier is moderate; the plate impedance is high. Impedance matching between the amplifier and the load (usually an antenna) is obtained by tapping the coil of the output tuned circuit, or by using a transformer. In the example of Fig. 29-5, transformer output coupling is used.

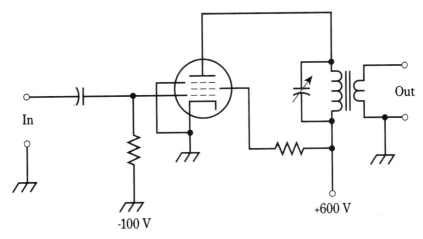

29-5   A grounded-cathode RF power amplifier.

Grounded grid

Grounded-cathode RF power amplifiers can (and sometimes do) oscillate unless they are *neutralized,* or provided with some negative feedback. The oscillation usually occurs at some frequency far removed from the operating frequency. Known as *parasitic oscillation,* it can rob the amplifier of output power at the desired frequency, as well as causing interference to communications on the frequency or frequencies of oscillation. Neutralization can be a tricky business. When the tube is replaced, the neutralizing circuit must be readjusted. The adjustment is rather critical. Improved stability, without the need for neutralization, can be obtained by grounding the grid, rather than the cathode, of the tube in an RF power amplifier.

The grounded-grid configuration requires more driving power than the grounded-cathode scheme. While a grounded-cathode amplifier might produce 1 kW of RF output for 10 W input, a grounded-grid amplifier needs about 50 W to 100 W of driving power to produce 1 kW of RF output.

A simple grounded-grid RF power amplifier is shown in Fig. 29-6. The cathode input impedance is low, and the plate output impedance is high. The output impedance is matched by the same means as with the grounded-cathode arrangement. In Fig. 29- 6, a transformer is used between the plate circuit and the load.

Plate voltages in the circuits of Figs. 29-5 and 29-6 are given only as examples. The amplifiers shown would produce perhaps 75 W to 150 W of RF power output. An amplifier with 1 kW of RF output would have a plate voltage of 2 to 5 kV, depending on the tube characteristics and the class of amplifier operation.

Typical grounded-grid linear amplifiers are operated in class-AB or B. If linearity is not important (as in CW, FSK, or FM operation), then class C provides improved efficiency, although the driving power requirement increases, so that 100 W or even 200 W of RF input is necessary to get 1 kW of RF output.

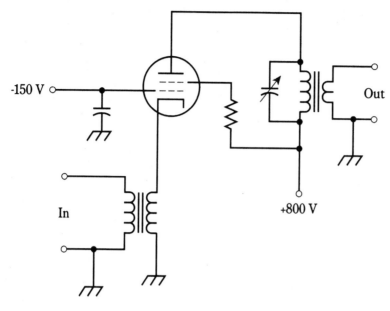

29-6 A grounded-grid RF power amplifier.

Cathode-ray tubes

Everyone encounters TV picture tubes. You've almost certainly looked at a computer monitor (even cash registers have them now), and you've probably seen an oscilloscope in some form. These video displays all use *cathode-ray tubes (CRTs)*.

The electron beam

In any CRT, an *electron gun* emits a high-intensity stream of electrons. This beam is focused and accelerated as it passes through *anodes* that carry positive charge. The anodes of a CRT work differently than the anode of a radio-frequency vacuum tube. Rather than hitting the anode, the electrons pass on until they strike a screen whose inner surface is coated with *phosphor*. The phosphor glows visibly as seen from the face of the CRT.

Unless *deflection* is used to move the electron beam around the screen of the CRT, you'll only get a brilliant spot in the center of the screen. Deflection of the beam makes various displays possible. The beam can be deflected by electrostatic fields or by magnetic fields.

Electrostatic deflection

A simplified cross-sectional drawing of an *electrostatic* CRT is shown in Fig. 29-7. The electron beam is pulled toward the positive *deflecting plates* and is repelled away from the negative plates. There are two sets of deflecting plates, one for the horizontal plane and the other for the vertical plane. The higher the voltage between the plates, the greater the intensity of the electrostatic field, and the more the electron beam is deflected.

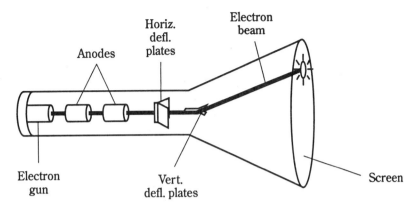

29-7 Simplified cutaway view of an electrostatic CRT.

In an oscilloscope, the horizontal deflecting plates receive a sawtooth voltage waveform. This causes the beam to *scan* or *sweep* at a precise, adjustable speed across the screen from left to right as seen from the outside. After each timed left- to-right sweep, the beam jumps instantly back to the left side of the screen for the next sweep.

The vertical deflecting plates receive the waveform to be analyzed. This waveform makes the electron beam undulate up and down. The combination of vertical and horizontal motions results in the display of the input waveform as a function of time. This is a *time-domain* display.

Oscilloscopes can be adapted to show phenomena as a function having some domain other than time. An example is a *spectrum analyzer*, showing signal amplitude as a function of frequency: a *frequency-domain* display. A radar set shows the distance and direction of signal echoes from objects in the sky or in space; these are *position-domain* or *location-domain* renditions.

Electromagnetic deflection

Coils can be used instead of charged plates for deflection of the electron beam in a CRT. When currents pass through the coils, the beam changes direction at right angles to the magnetic lines of flux. The greater the current in a deflecting coil, the stronger the magnetic field, and the greater the extent to which the beam is deflected. This is an *electromagnetic* CRT.

You can use Fig. 29-7 to imagine the workings of an electromagnetic CRT, just by thinking of coils outside the neck of the tube, rather than deflection plates inside the tube. There are two coils, one for horizontal deflection and the other for vertical deflection.

Video camera tubes

Video cameras use a form of electron tube that converts visible light into varying electric currents. The two most common types of *camera tube* are the *vidicon* and the *image orthicon.*

The vidicon

Virtually every *videocassette recorder (VCR)* makes use of a vidicon camera tube. Closed-circuit TV systems, such as the kind you see in convenience stores and banks, also employ the vidicon. The main advantage of the vidicon is its small physical bulk; it's easy to carry around.

In the vidicon, which might be almost as small as a photo camera, a lens focuses the incoming image onto a *photoconductive* screen. An electron gun generates an electron beam that scans across the screen via deflecting coils, in a manner similar to the operation of an electromagnetic cathode-ray tube. The scanning in the vidicon is exactly synchronized with the scanning in the picture tube that displays the image "seen" by the camera tube.

As the electron beam scans the photoconductive surface, the screen becomes charged. The rate of discharge in a certain region on the screen depends on the intensity of the visible light falling on that region. A simplified cutaway view of a vidicon tube is shown in Fig. 29-8.

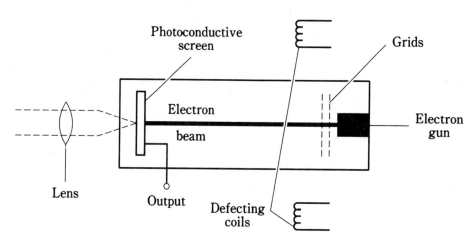

29-8   Simplified cutaway view of a vidicon camera tube.

A vidicon is sensitive, working well in dim light. But its response becomes sluggish when the level of illumination is low. This causes images to persist for a short while. You might have noticed this when using a VCR in a dimly lit environment. The motion "smears" and images blur.

The image orthicon

Another type of camera tube, also quite sensitive but having a quicker response to image changes, is the *image orthicon*. It is constructed much like the vidicon, except that there is a *target electrode* behind the photocathode (Fig. 29-9). When an electron from the photocathode strikes this target electrode, several *secondary electrons* are emitted as a result. This is a form of current amplification. The image orthicon acts as a *photomultiplier*, improving its sensitivity.

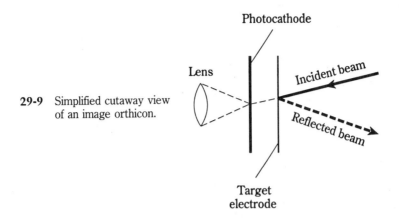

29-9 Simplified cutaway view of an image orthicon.

A fine beam of electrons, emitted from the electron gun, scans the target electrode. The secondary electrons cause some of this beam to be reflected back toward the electron gun. Areas of the target electrode with the most secondary-electron emission produce the greatest return beam intensity; regions with the least emission produce the lowest return beam intensity. The greatest return beam intensity corresponds to the brightest parts of the video image. The return beam is modulated as it scans the target electrode in a pattern just like the one in the TV receiver CRT. The return beam is picked up by a *receptor electrode*.

The main disadvantage of the image orthicon is that it produces significant noise in addition to the signal output. But when a fast response is needed (when there is a lot of action in a scene) and the light ranges from very dim to very bright, the image orthicon is the camera tube of choice. It is common in commercial broadcasting.

Traveling-wave tubes

A *traveling-wave tube* is a form of electron-beam tube that is useful at ultra-high frequencies (UHF) and microwave frequencies. There are several variations on this theme; the two most common are the *magnetron* and the *klystron*.

The magnetron

Most magnetrons contain a cathode at the center, and a surrounding plate as shown in Fig. 29-10. The plate is divided into sections, or *cavities*, by radial barriers. The output is taken from an opening in the plate, and passes into a hollow *waveguide* that serves as a transmission line for the UHF or microwave energy.

The cathode is connected to the negative terminal of a high-voltage source, and the anode is connected to the positive terminal. Therefore, electrons flow radially outward. A magnetic field is applied lengthwise through the cavities. As a result, the electrons move outward in spirals from the cathode to the anode, rather than in straight lines.

The electric field produced by the high voltage, interacting with the longitudinal magnetic field and the effects of the cavities, causes the electrons to bunch up into *clouds*. The swirling movement of the electron clouds causes a fluctuating current in the anode.

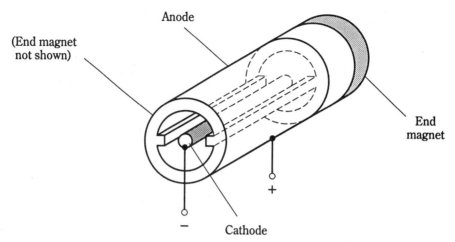

Anode

(End magnet
not shown)

End
magnet

+

−   Cathode

29-10   A magnetron. Here, only two cavities are shown for clarity; often there are several
cavities.

This is the UHF or microwave signal. The frequency depends mainly on the shape and
size of the cavities within the magnetron. Small cavities result in the highest oscillation
frequencies; larger cavities produce oscillation at relatively lower frequencies.

A magnetron can generate more than 1 kW of RF power at a frequency of 1 GHz. As
the frequency increases, the realizable power output decreases. At 10 GHz, a typical
magnetron generates about 20 W of RF power output.

Magnetrons can produce microwaves for use in cooking. The energy at these
frequencies excites the molecules in organic substances like meat, vegetables, and
grains. Your microwave oven uses a magnetron rated at about 500 W to 1 kW continuous
output power.

The klystron

A klystron is a linear-beam electron tube. It has an electron gun, one or more cavities, and
a device that modulates the electron beam. There are several different types of klystron
tube. The most common are the *multicavity* and *reflex* klystrons.

A multicavity klystron is shown in Fig. 29-11. In the first cavity, the electron beam is
velocity-modulated. This means that the speeds of the electrons are made to increase and
decrease alternately at a rapid rate. This causes the density of electrons in the beam to
change as the beam moves through subsequent cavities. The intermediate cavities
increase the magnitude of the electron-beam modulation, resulting in amplification.
Output is taken from the last cavity. Peak power levels in some multi-cavity klystrons can
exceed 1 MW (1,000,000 W).

A reflex klystron has only one cavity. A *retarding field* causes the electron beam to
periodically reverse direction. This produces a phase reversal that allows large amounts
of energy to be drawn from the electrons. The reflex klystron produces low-power UHF
and microwave signals, on the order of a few watts.

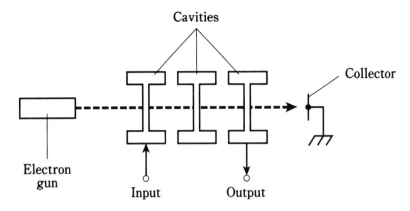

29-11   A multicavity klystron.

Quiz

Refer to the text in this chapter if necessary. A good score is at least 18 correct. Answers are in the back of the book.

1. One difference between a triode and an N-channel FET is that:

   A. Triodes work with lower voltages.

   B. Triodes are more compact.

   C. Triodes need more voltage.

   D. Triodes don't need filaments.

2. The control grid of a tube corresponds to the:

   A. Source of an FET.

   B. Collector of a bipolar transistor.

   C. Anode of a diode.

   D. Gate of an FET.

3. The intensity of the electron flow in a vacuum tube depends on all of the following *except*:

   A. The gate voltage.

   B. The power supply voltage.

   C. The grid voltage.

   D. The voltage between the cathode and the plate.

4. Which type of tube maintains constant voltage drop with changes in current?

   A. A triode.

   B. A gas-filled regulator.

   C. A tetrode.

   D. A pentagrid converter.

5. In a tube with a directly heated cathode:
   A. The filament is separate from the cathode.
   B. The grid is connected to the filament.
   C. The filament serves as the cathode.
   D. There is no filament.

6. In a tube with a cold cathode:
   A. The filament is separate from the cathode.
   B. The grid is connected to the filament.
   C. The filament serves as the cathode.
   D. There is no filament.

7. A screen grid enhances tube operation by:
   A. Decreasing the gain.
   B. Decreasing the plate voltage.
   C. Decreasing the grid-to-plate capacitance.
   D. Pulling excess electrons from the plate.

8. A tube with three grids is called a:
   A. Triode.
   B. Tetrode.
   C. Pentode.
   D. Hexode.

9. A tube type radio receiver:
   A. Is bulky and heavy.
   B. Requires low voltage.
   C. Is more sensitive than a transistorized radio.
   D. All of the above.

10. An advantage of a grounded-grid power amplifier is:
    A. Excellent sensitivity.
    B. High impedance.
    C. Low noise.
    D. Good stability.

11. A heptode tube has:
    A. Two grids.
    B. Three grids.
    C. Five grids.
    D. Seven grids.

12. The electron gun in a CRT is another name for its:

    A. Cathode.

    B. Anode.

    C. Control grid.

    D. Screen grid.

13. The electron beam in an electrostatic CRT is bent by:

    A. A magnetic field.

    B. An electric field.

    C. A fluctuating current.

    D. A constant current.

14. The horizontal displacement on an oscilloscope CRT screen is usually measured in:

    A. Frequency per unit division.

    B. Current per unit division.

    C. Time per unit division.

    D. Voltage per unit division.

15. In a time-domain oscilloscope, the waveform to be analyzed is usually applied to the:

    A. Control grid plates or coils.

    B. Anode plates or coils.

    C. Vertical deflection plates or coils.

    D. Horizontal deflection plates or coils.

16. A vidicon camera tube is noted for its:

    A. Sensitivity.

    B. Large size.

    C. Heavy weight.

    D. Rapid response.

17. In a magnetron, as the frequency is increased:

    A. The achievable power output increases.

    B. The achievable power output decreases.

    C. The output power stays the same.

    D. The output power increases and decreases alternately.

18. The paths of the electrons in a magnetron are spirals, rather than straight lines, because of:

    A. The extreme voltage used.

    B. The longitudinal magnetic flux.

    C. The bunching-up of the electrons.

    D. The shapes of the cavities.

19. A klystron is noted for its:

    A. Spiralling electrons.

    B. Low noise output.

    C. High achievable output power.

    D. Magnetic-field intensity.

20. In a multi-cavity klystron, the electrons:

    A. Have variable speed.

    B. Travel in circles.

    C. Are reflected by the cavities.

    D. Are drawn out via the cathode.

30
CHAPTER

Basic digital principles

YOU'VE SEEN HOW DIGITAL SIGNALS DIFFER FROM ANALOG SIGNALS. THE MANIpulation of digital signals is known as *digital logic*. Digital logic consists of many—sometimes innumerable—pulses racing around. It can be staggering in terms of quantity, but it is simple at heart: 1 or 0, high or low, yes or no.

Suppose a few electronics engineers from, say, 1950 could ride a time machine to the present day. They would be flabbergasted at the speed and compactness of digital logic circuitry. Some might be struck with disbelief. Others would say, "I told you so," or "I knew it would happen." And then there would inevitably be the one who would quip, "Is this all the further you've gotten?"

Well, digital-logic circuits are getting smaller, faster, and more sophisticated every day.

Numbering systems

Any number, such as 35912, can be rendered in some other number base or *modulus*. Then it's written differently. But it's always the same quantity.

The decimal system

In the decimal number system, each digit 0 through 9 represents either itself, or else itself times some power of 10. The value of the digit depends on its position or *place* in the number. In the case of the number 35912, the digit 2 is in the *units place*, and has a value of simply 2. The digit 1 is in the *10s place*; it takes the value of $1 \times 10^1 = 10$. The digit 9 is in the *100s place*, having a value of $9 \times 10^2 = 900$. The digit 5 is in the *1,000s place*, and represents $5 \times 10^3 = 5000$. The digit 3 is in the *10,000s place*, and has a value of $3 \times 10^4 = 30000$. The total value is the sum of all these: $2 + 10 + 900 + 5,000 + 30,000 = 35,912$.

A decimal number can be represented in digital form, but it requires 10 different possible states. Any decimal number has a *binary equivalent* comprised of digits that are all either 0 or 1: just two different possible states. Much simpler!

Modulo what?

You probably learned "modular arithmetic" in elementary school. If so, you ought to remember working with *modulo 8* (the *octal* system), or *modulo 16* (*hexadecimal*), or base 4, or 12, or 20. In computer electronics, octal and hexadecimal numbers are occasionally encountered, in addition to decimal numbers.

At various times, special-interest groups have tried to get the whole world to switch to some modulus besides 10. Some people think base 12 should be used (isn't a "baker's dozen" a more logical standard than the number of toes on a baby's two feet?). Others believe that base 20 makes the most sense (it's the total number of fingers and toes, including thumbs, on a typical baby). Still others argued for base 8 or base 16 because these values are "elegant": they can be repeatedly cut in half all the way down to 1.

Would you like to change the world over to, say, octal numbering because it's easy to count rhythmically by eights? Or to base 12 because then a gross would be an even 100? Lots of luck. Even the metric system hasn't caught on that well.

The Binary system

Binary numbers are in *modulo 2*. In this system, the digits are all either 1 or 0, and the places go in powers of 2. Therefore, to the immediate left of the units place is the *2's place* (2^1), then the *4's place* (2^2), then the *8's place* (2^3) and so on. In Table 30-1, powers of 2 are represented up to 2^{15}.

Table 30-1.   Powers of 2.

Exponent	Binary number value	Decimal value
0	1	1
1	10	2
2	100	4
3	1000	8
4	10000	16
5	100000	32
6	1000000	64
7	10000000	128
8	100000000	256
9	1000000000	512
10	10000000000	1024
11	100000000000	2048
12	1000000000000	4096
13	10000000000000	8192
14	100000000000000	16384
15	1000000000000000	32768

In general, the nth place from the right has the decimal value 2^n. The total decimal value, for a given binary number, is the sum of the decimal values of each of the places.

Converting decimal numbers to binary form can be done using Table 30-1. Suppose, for example, that you want to convert 35912 to binary notation. First, find the largest decimal number on the table that is no greater than the decimal number you wish to convert. In this case, it is $32768 = 2^{15}$. From this, you know that there will be 16 digits in the binary representation of this number, one for each place 2^0 through 2^{15}. Mark off 16 slots or spaces on a sheet of paper (quadrille graph paper is perfect for this), and place a digit "1" in the left-most space, representing 2^{15}.

Now, use Table 30-1 to determine which number can be added to 32768 to get the largest decimal number that doesn't exceed 35912. It happens to be $2048 = 2^{11}$. Place digits "0" in the slots for 2^{14}, 2^{13}, and 2^{12}. Then place "1" in the space for 2^{11}.

If you continue this process, you'll ultimately get the binary number 1000110001001000. This 16-digit binary number is equivalent to the decimal $35912 = 8 + 64 + 1024 + 2048 + 32768 = 2^3 + 2^6 + 2^{10} + 2^{11} + 2^{15}$. The slots for exponents 3, 6, 10, 11, and 15 each are filled with binary digit 1; the others are filled with binary digit 0.

It is possible to have fractional values in binary notation, just as it is in decimal notation. The first place to the right of the point (perhaps best called a "binary point" rather than a "decimal point") is the *1/2's place* (2^{-1}). The next place is the *1/4's place* (2^{-2}); then comes the *1/8's place* (2^{-3}), and so on. Thus, 0.001 in binary notation represents the decimal fraction ⅛. You can think of it as repeatedly dividing the size in half as you progress towards the right.

Why use such cumbersome notation?

Perhaps you've noticed that binary notation gets into long number strings. This is true. (Try writing the decimal quadrillion, or 10^{15}, in binary form!) But computers don't have trouble dealing with long strings of digits. That's what we build them for! To a computer, the important thing is to be certain of the value of each digit. This is easiest when the attainable states are as few as possible: two. High or low. Logic 1 or 0. On/off. Yes/no.

Logic signals

On a single wire or *line*, a binary digital signal is either full-on or full-off at any given moment. These are the logic high and low, and are represented by dc voltages. The high voltage is approximately +3 V to +5 V, and the low voltage is between 0 V and +2 V.

Parallel data transfer

A binary number has several binary digits, or *bits*. The number of bits depends on how big the numbers can get. For example, if you want to send binary numbers from 0000 to 1111 (which corresponds to the decimal numbers 0 through 15), you need four bits. If you want to be able to transmit numbers as high as 11111111, or the decimal 0 to 255, you need eight bits. For large decimal numbers, you'll need many bits—sometimes several dozen.

The bits can be sent along separate wires or *lines*. When this is done, the number of lines corresponds to the number of bits in the binary number. That is, in order to send binary numbers up to 1111, you need four lines; to send numbers up to 11111111, you need eight lines. This is called *parallel data transfer*, because the bits are sent along lines that are effectively in parallel with each other (Fig. 30-1). Clearly, parallel data transfer has limitations. For large numbers, it would need many lines. This can present certain logistic problems, for example in long-distance radio or wire transmission.

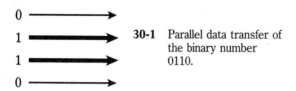

30-1 Parallel data transfer of the binary number 0110.

Serial data transfer

In radio communications using parallel data transfer, each line must correspond to a different frequency; the result is that large binary numbers need many channels. This translates into a wide signal bandwidth. The necessary bandwidth is not always easy to obtain. It is especially difficult at low and "shortwave" frequencies, where wideband signals present not only an engineering hassle, but tend to "hog" spectrum space.

Fortunately, there is another way to send binary data, and this method needs only one line or radio frequency. It is called *serial data transfer*. Instead of sending each bit on a separate line, the bits are transmitted one after the other in a defined sequence.

To send binary numbers up to 1111 serially, you need four time slots. For the serial transmission of binary numbers up to 11111111, you need eight slots. Each slot has a certain duration. Generally, all the slots are of equal length. A four-bit serial signal is shown in Fig. 30-2.

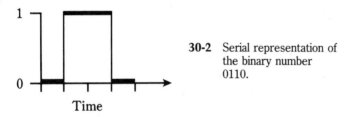

30-2 Serial representation of the binary number 0110.

Suppose the duration of a bit is 1 ms (a millisecond). The *bit rate* is then 1,000 *bits per second (bps)*. Data transmission speed is usually measured in *bauds*, where 1 baud is equal to 1 bps. If a bit lasts for 1 ms, the transmission has a speed of 1,000 bauds. In general, if the bit duration in seconds is t, then the baud rate is 1/t. Radio data-communications systems operate at speeds upwards of 10,000 bauds, or 10 *kilobauds*.

Serial data transfer is somewhat slower than parallel transfer. This is the necessary tradeoff for the convenience of a single transmission line, or for narrower radio-signal bandwidth. In electronics, as in life, you never get something for nothing.

Positive versus negative logic

Usually, logic 1 is high and logic 0 is low. This is known as *positive logic*. But these states can be reversed, and the results will be just as good. When logic 1 is low and logic 0 is high (so that a bit 1 is "off" and a bit 0 is "on"), *negative logic* is being used.

In the rest of this chapter, positive logic is the rule. This is simply to minimize confusion.

Basic logic operations

Perhaps you've taken *sentential logic,* also called *propositional logic* or *propositional calculus* (it sounds a lot more sophisticated than it really is) in high school or college. Electronic digital logic is basically the same thing.

In logic, there are three *basic operations.* These are depicted in Table 30-2.

Table 30-2. Logic operations.

Operation	Description	Symbology
NOT	Negation or complement of X	−X
		X′
AND	Conjunction of X, Y	XY
		X × Y
OR	Disjunction of X, Y	X + Y

NOT

The logic *NOT*, also called *negation* or *inversion,* involves only a single logic variable. It changes the value of a bit from 1 to 0, or from 0 to 1, as shown in the *truth table* for logic operations (Table 30-3). It is represented by a minus sign in front of a capital letter, usually from the end of the alphabet (W, X, Y, or Z). It might also be represented by a squiggly dash before the letter, an apostrophe after the letter, or a line over the letter.

Table 30-3. Truth table for logic operations.

X	Y	−X	XY	X + Y
0	0	1	0	0
0	1	1	0	1
1	0	0	0	1
1	1	0	1	1

AND

The logic *AND*, or *conjunction,* involves two variables. The resultant is high if, but only if, both variables are high. If either variable is low, or if both are low, then the resultant is low. This operation gets its name from the fact that it corresponds to the function of the word "and" in propositional calculus. It is usually represented as multiplication. See Table 30-3.

OR

The logic *OR*, or *disjunction*, also involves two variables. The resultant is high if either of the variables are high. The resultant is low if, but only if, both variables are low. This is sometimes called the *inclusive OR* operation. It corresponds to the word "or" in propositional calculus. It is represented as addition. See Table 30-3.

XOR

Occasionally you'll hear about the *exclusive OR* or *XOR* operation. The resultant of X XOR Y is high if X and Y have opposite states. The resultant X XOR Y is low if X and Y have the same state. It can be thought of as "either/or, but not both." The "X" means "exclusive." The XOR operation is shown in Table 30-4.

Table 30-4.   More logic operations.

X	Y	XOR	NAND	NOR	XNOR
0	0	0	1	1	1
0	1	1	1	0	0
1	0	1	1	0	0
1	1	0	0	0	1

NAND

The logic *NAND* is a combination of two operations, NOT and AND. It involves two variables. The AND is performed first, and the result is negated. The value of X NAND Y is low if, but only if, both X and Y are high. Otherwise, X NAND Y is high. This is shown in Table 30-4.

NOR

The logic *NOR* combines NOT and OR. It involves two variables. The OR is done first, and then the result is negated. The value of X NOR Y is high if, but only if, X and Y are low. Otherwise, X NOR Y is low (Table 30-4).

XNOR

The logic *XNOR* combines NOT and XOR. As with all the basic logic operations except NOT, it involves two variables. The XOR is done first, and then the result is negated. This is shown in Table 30-4.

Symbols for logic gates

A *logic gate* is an electronic switch that performs a logic operation. The earliest logic gates were built using vacuum tubes; later, transistors and diodes were employed. Modern logic gates are fabricated on integrated circuits (ICs), with hundreds or even thousands of gates per chip.

An *inverter* (NOT gate) has one input and one output. Other logic gates usually have two inputs and a single output. The symbols for the various logic gates are shown in Fig. 30-3.

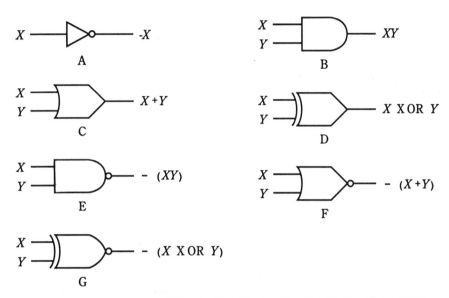

30-3 Symbols for logic gates: NOT (A), AND (B), OR (C), XOR (D), NAND (E), NOR (F), and XNOR (G).

It is possible for AND, NAND, OR, or NOR gates to have three or more inputs. For an AND gate, the output is high if and only if all of the inputs are high. For an OR gate, the output is low if and only if all the inputs are low. The multiple-input NAND and NOR gates perform the conjunction or disjunction operation first, followed by negation. Figure 30-4 shows schematic symbols for four-input AND and four-input NOR gates.

The XOR and XNOR operations are hard to define when there are three or more inputs. These fall into the category of *complex logic operations*.

Complex logic operations

The term "complex," when used to describe a logic operation, does not necessarily mean complicated. A more apt adjective might be composite. But there are cases when logic operations are indeed quite complicated.

No matter how messy a particular logic operation might appear, it can always be broken down into the elementary operations defined above.

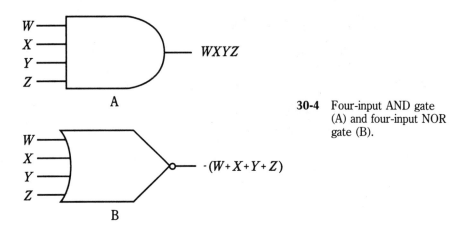

30-4 Four-input AND gate (A) and four-input NOR gate (B).

Suppose you want to find an arrangement of logic gates for some complex logic operation. This can be done in two ways. You can use either *truth tables* or *Boolean algebra*. For any complex logic operation, there might be several solutions, some requiring more gates than others. A digital design engineer has the job not only of finding an arrangement of gates for a given complex operation, but of finding the scheme that will yield the desired result with the least number of gates.

Working with truth tables

Truth tables are, in theory, infinitely versatile. It is possible to construct or break down logic operations of any complexity by using these tables, provided you have lots of paper and a fondness for column-and-row matrix drawing. Computers can also be programmed to work with truth tables, although the displays and printouts get horrible to deal with if the logic functions are messy.

Building up

You can build complex logic operations up easily by means of a truth table. An example of such a building process is shown in Table 30-5.

Table 30-5. Truth table for $-(X + Y) + XZ$.

X	Y	Z	X+Y	$-(X + Y)$	XZ	$-(X + Y) + XZ$
0	0	0	0	1	0	1
0	0	1	0	1	0	1
0	1	0	1	0	0	0
0	1	1	1	0	0	0
1	0	0	1	0	0	0
1	0	1	1	0	1	1
1	1	0	1	0	0	0
1	1	1	1	0	1	1

There are three variables X, Y, and Z. Each can be either 0 or 1 (low or high). All the possible combinations are listed by writing binary numbers XYZ upwards from 000 through 111. This forms the first column of the truth table. If there are n total variables, there will be 2^n possible combinations.

The second column lists the values of X + Y, the OR function. Some rows are duplicates of each other; you write all the resultants down anyway.

The third column lists the negations of the values in the second column; this is the NOR operation $-(X + Y)$.

The fourth column shows the values of XZ.

The fifth column is the disjunction (OR operation) of the values in the third and fourth column. It renders values for the complex logic operation $-(X + Y) + XZ$.

Breaking down

Suppose you were called upon to break down a logical operation, rather than to build it up. This is the kind of problem encountered by engineers. In the process of designing a certain digital circuit, the engineer is faced with figuring out what combination of logic gates will yield the complex operation, say, $XY + -(XZ) + YZ$.

Proceed by listing all the possible logic states for the three variables X, Y, and Z, exactly as in Table 30-5. Then find the values of XY (X AND Y), listing them in the second column. Next, find values XZ and list them in a third column. Negate these values to form a fourth column, depicting $-(XZ)$. (You might be able to perform the AND and NOT operations together in your head, skipping over the XZ column. But be careful! It's easy to make errors, and in a digital circuit, one error can be catastrophic.) Next, find values YZ and list them in a fifth column. Finally, perform the OR operation on the values in the columns for XY, $-(XZ)$ and YZ. A multiple-valued OR is 0 only if all the individual variables are 0; if any or all of the inputs are 1, then the output is 1.

This process yields Table 30-6.

Table 30-6.   Truth table for $XY + -(XZ) + YZ$.

X	Y	Z	XY	XZ	$-(XZ)$	YZ	$XY + -(XZ) + YZ$
0	0	0	0	0	1	0	1
0	0	1	0	0	1	0	1
0	1	0	0	0	1	0	1
0	1	1	0	0	1	1	1
1	0	0	0	0	1	0	1
1	0	1	0	1	0	0	0
1	1	0	1	0	1	0	1
1	1	1	1	1	0	1	1

This is a little like middle-school algebra, isn't it? The rules of algebra can, with certain adjustments, be directly applied to digital logic operations. This is called *Boolean algebra*, named after the mathematician Boole who first devised it as an alternative to truth tables.

Boolean algebra

Using Boolean representations for logic operations, some of the mathematical properties of multiplication, addition, and negation can be applied to form *Boolean equations*. The logical combinations on either side of any equation are equivalent.

In some ways, Boolean algebra differs from conventional algebra. You must use logic rules rather than "regular" rules for addition, additive inverse (negation) and multiplication. Using these rules, certain facts, called *theorems*, can be derived. Boolean theorems all take the form of equations. Some common Boolean theorems are listed in Table 30-7.

Table 30-7.   Common theorems in Boolean algebra.

$X + 0 = X$	OR identity
$X1 = X$	AND identity
$X + 1 = 1$	
$X0 = 0$	
$X + X = X$	
$XX = X$	
$-(-X) = X$	Double negation
$X + -(X) = X$	
$X(-X) = 0$	
$X + Y = Y + X$	Commutativity of OR
$XY = YX$	Commutativity of AND
$X + XY = X$	
$X(-Y) + Y = X + Y$	
$X + Y + Z = (X + Y) + Z = X + (Y + Z)$	Associativity of OR
$XYZ = (XY)Z = X(YZ)$	Associativity of AND
$X(Y + Z) = XY + XZ$	Distributivity
$-(X + Y) = (-X)(-Y)$	DeMorgan's Theorem
$-(XY) = -X + -Y$	DeMorgan's Theorem

Boolean algebra is less messy than truth tables for designing and evaluating logic circuits. Some engineers prefer truth tables because the various logic operations are easier to envision, and all the values are shown for all logic states in all parts of a digital circuit. Other engineers would rather not deal with all those 1s and 0s, nor cover whole tabletops with gigantic printouts. Boolean algebra gets around that.

For extremely complex logical circuits, computers are used as an aid in design. They're good at combinatorial derivations and optimization problems that would be uneconomical (besides tedious) if done by a salaried engineer.

The flip-flop

So far, all the logic gates discussed have outputs that depend only on the inputs. They are sometimes called *combinational logic* gates, because the output state is simply a function of the combination of input states.

A *flip-flop* is a form of *sequential logic* gate. In a sequential gate, the output state depends on both the inputs and the outputs. The term "sequential" comes from the fact that the output depends not only on the current states, but on the states immediately preceding. A flip-flop has two states, called *set* and *reset*. Usually, the set state is called logic 1, and the reset state is called logic 0.

There are several different kinds of flip-flop.

R-S

In schematic diagrams, a flip-flop is usually shown as a rectangle with two or more inputs and two outputs. If the rectangle symbol is used, the letters FF, for "flip-flop," are printed or written inside at the top.

The inputs of an *R-S flip-flop* are labeled R (reset) and S (set). The outputs are Q and $-Q$. (Often, rather than $-Q$, you will see Q', or perhaps Q with a line over it.) As their symbols imply, the two outputs are always in logically opposite states. The symbol for an R-S flip-flop is shown in Fig. 30-5.

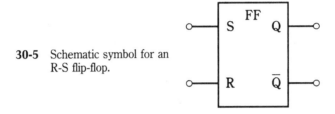

30-5 Schematic symbol for an R-S flip-flop.

In an R-S flip-flop, If R = 0 and S = 0, the output states do not change; they stay at whatever values they're already at. If R = 0 and S = 1, then Q = 1 and $-Q$ = 0. If R = 1 and S = 0, then Q = 0 and $-Q$ = 1. That is, the Q and $-Q$ outputs will attain these values, no matter what states they were at before.

But if S = 1 and R = 1, things get bizarre. The flip-flop becomes unpredictable. Because of this, engineers avoid letting logic 1s get into both inputs of an R-S flip flop. You want logic, not absurdity!

Table 30-8 is the truth table for an R-S flip-flop.

Table 30-8. Truth table for an R-S flip-flop.

R	S	Q	$-Q$
0	0	Q	$-Q$
0	1	1	0
1	0	0	1
1	1	?	?

Synchronous

An R-S flip-flop output changes state as soon as the inputs change. If the inputs change at irregular intervals, so will the outputs. For this reason, the aforementioned circuit is sometimes called an *asynchronous flip-flop*.

A *synchronous flip-flop* changes state only at certain times. The change-of-state times are determined by a circuit called a *clock*. The clock puts out a continuous *train* of pulses at regular intervals.

There are several different ways in which a synchronous flip flop can be *triggered*, or made to change state.

In *static triggering*, the outputs can change state only when the clock signal is either high or low. This type of circuit is sometimes called a *gated flip-flop*.

In *positive-edge triggering*, the outputs change state at the instant the clock signal goes from low to high, that is, while the clock pulse is positive-going. The term "edge triggering" derives from the fact that the abrupt rise or fall of a pulse looks like the edge of a cliff (Fig. 30-6). In *negative-edge triggering*, the outputs change state at the instant the clock signal goes from high to low, or when the pulse is negative-going.

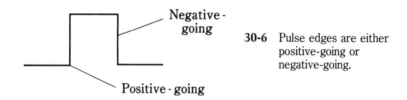

30-6 Pulse edges are either positive-going or negative-going.

Master/slave

When two flip-flops are triggered from the same clock, or if one of the outputs is used as an input, the input and output signals can sometimes get confused. A *master/slave (M/S) flip-flop* overcomes this bugaboo by storing inputs before allowing the outputs to change state.

An M/S flip-flop consists of two R-S flip-flops in series. The first flip-flop is called the *master*, and the second is called the *slave*. The master flip-flop functions when the clock output is high, and the slave acts during the next low portion of the clock output. This time delay prevents confusion between the input and output of the circuit.

J-K

The *J-K flip-flop* works just like an R-S circuit, except that it has a predictable output when the inputs are both 1. Table 30-9 shows the input and output states for this type of flip-flop. The output changes only when a triggering pulse is received.

The symbol for a J-K flip-flop is shown in Fig. 30-7. It looks just like the R-S flip-flop symbol, with the addition of a third (trigger) input, labeled T.

The counter

Some digital circuits can be programmed to change state every so-many clock pulses. Such a circuit is called a *counter*. It is also often called a *divider*, because if the output

**Table 30-9.   Truth table
for a J-K flip-flop.**

J	K	Q	$-Q$
0	0	Q	$-Q$
0	1	1	0
1	0	0	1
1	1	$-Q$	Q

30-7   Schematic symbol for a
J-K flip-flop.

changes state after every *n* input pulses, the output frequency is $1/n$ of the input frequency. Counters are built up by interconnecting many flip-flops.

A special type of counter can be instructed to divide by any desired positive integer *n*. This is a *programmable divider*. It forms the heart of a digital *frequency synthesizer*, an oscillator whose frequency can be digitally set. You learned about this circuit back in chapter 25.

The register

A *register* is a combination of flip-flops that can delay digital signals, or store them for a short time.

Shift registers

In a *shift register*, the flip-flops pass signal bits along from one to the next each time a triggering pulse is received. The most common use for a shift register is to produce a timed delay in digital data.

Any input signal to a shift register will eventually arrive at the output. The output bits occur in the same order as the input bits, except later. The length of the delay can be set by changing the clock (triggering) pulse frequency. Higher clock frequencies result in less delay, and lower frequencies result in more delay.

Storage registers

In a *storage register*, a byte, word, or group of words is held for awhile for later use. This allows the circuit to work as a small-capacity, short-term memory. Storage registers are common in digital computers.

The digital revolution

This chapter has touched upon digital basics, but cannot possibly go into depth. The scope of digital electronics is vast, and it grows and changes daily. If you are studying to be an electrical engineer, you'll take numerous courses in digital electronics.

Specialties include data communications, video communications, satellite communications, space communications, navigation, medical devices, modems, calculators, computers, video games, control systems, and so on almost *ad infinitum*. It would be easy to fill a book as large as this one with material devoted only to digital technology. Engineers devote careers to fields as specific as digital signal processing, digital CD recording, or digital robotics-control devices.

During the eighties, digital technology replaced analog circuits in everything from wristwatches to radios. Along with proliferation, costs have plummeted while sophistication has grown geometrically. These trends will continue in the twenty-first century.

Quiz

Refer to the text in this chapter if necessary. A good score is at least 18 correct. Answers are in the back of the book.

1. The value of the decimal number 23 in binary form is:
   A. 1011.
   B. 110111.
   C. 10111.
   D. 11100.

2. The binary number 110001 represents the digital number:
   A. 49.
   B. 25.
   C. 21.
   D. 13.

3. The fifth digit from the right in a binary number carries a decimal value of:
   A. 64.
   B. 32.
   C. 24.
   D. 16.

4. The largest possible decimal number that can be represented by six binary digits (bits) is:
   A. 256.
   B. 128.
   C. 64.
   D. 63.

5. Which of the following voltages could normally represent a 1 in positive logic?

   A. 0 V.

   B. +1 V.

   C. +4 V.

   D. +12 V.

6. Which of the following voltages might normally represent a 1 in negative logic?

   A. 0 V.

   B. +4 V.

   C. +6 V.

   D. +12 V.

7. If X is low, what is the state of X AND Y?

   A. There is not enough information to tell.

   B. Low.

   C. High.

   D. This logic statement makes no sense.

8. If X is high, what is the state of X NOR Y?

   A. There is not enough information to tell.

   B. Low.

   C. High.

   D. This logic statement makes no sense.

9. If X and Y are both high, what is the state of X NAND Y?

   A. There is not enough information to tell.

   B. Low.

   C. High.

   D. This logic statement makes no sense.

10. If X is high and Y is low, what is the state of X NOT Y?

    A. There is not enough information to tell.

    B. Low.

    C. High.

    D. This logic statement makes no sense.

11. A logic circuit has four inputs W, X, Y, and Z. How many possible input combinations are there?

    A. 4.

    B. 8.

    C. 16.

    D. 32.

12. Data sent along a single line, one bit after another, is called:
    A. Serial.
    B. Synchronous.
    C. Parallel.
    D. Asynchronous.

13. If $X = 1$ and $Y = 1$, then $X + YZ$ is:
    A. Always 0.
    B. 0 if $Z = 0$, and 1 if $Z = 1$.
    C. 1 if $Z = 0$, and 0 if $Z = 1$.
    D. Always 1.

14. If $X = 0$ and $Y = 1$, then $X(Y + Z)$ is:
    A. Always 0.
    B. 0 if $Z = 0$, and 1 if $Z = 1$.
    C. 1 if $Z = 0$, and 0 if $Z = 1$.
    D. Always 1.

15. An advantage of a J-K over an R-S flip-flop is that:
    A. The J-K flip-flop is faster.
    B. The J-K can attain more states.
    C. The J-K always has predictable outputs.
    D. No! An R-S flip-flop is superior to a J-K.

16. In positive-edge triggering, the change of state occurs when:
    A. The pulse level is high.
    B. The pulse level is going from high to low.
    C. The pulse level is going from low to high.
    D. The pulse level is low.

17. The inputs of an R-S flip-flop are known as:
    A. Low and high.
    B. Asynchronous.
    C. Synchronous.
    D. Set and reset.

18. When both inputs of an R-S flip-flop are 0:
    A. The outputs stay as they are.
    B. $Q = 0$ and $-Q = 1$.
    C. $Q = 1$ and $-Q = 0$.
    D. The resulting outputs can be absurd.

19. When both inputs of an R-S flip-flop are 1:

    A. The outputs stay as they are.

    B. $Q = 0$ and $-Q = 1$.

    C. $Q = 1$ and $-Q = 0$.

    D. The resulting outputs can be absurd.

20. A frequency synthesizer makes use of:

    A. An OR gate.

    B. A divider.

    C. The octal numbering system.

    D. The hexadecimal numbering system.

31
CHAPTER

Wireless technology

THE TERM *WIRELESS* WAS COINED IN THE LATE NINETEENTH CENTURY
when inventors toyed with the idea of sending and receiving telegraph messages using electromagnetic fields rather than electric currents. In the early twentieth century, the technology became known as *radio;* video and data broadcasting and
communications were added in the middle of the century, and the general term *electromagnetic communications* came into being. The word *wireless* was relegated
to history.

In the late 1980s, the term found new life. Today it refers to communications,
networking, control devices, and security systems in which signals travel without direct electrical connections.

Cellular communications

Radio transceivers can be used as telephones in a specialized communications system called *cellular.* Originally, the cellular communications network was patchy and
unreliable, and was used mainly by traveling businesspeople. Nowadays, cellular
telephone units are so common that many people regard them as necessities.

How cellular systems work

A cellular telephone set, or *cell phone,* looks and functions like a cross between a
cordless telephone and a walkie-talkie. The unit contains a radio transmitter and receiver combination called a *transceiver.* Transmission and reception take place on
different frequencies, so you can talk and listen at the same time. This capability,
which allows you to hear the other person interrupt you if he or she chooses, is
known as *full duplex.*

In an ideal cellular network, all the transceivers are always within range of at
least one *repeater.* The repeaters pick up the transmissions from the portable units
and retransmit the signals to the telephone network and to other portable units. The
region of coverage for any repeater (also known as a *base station*) is called a *cell.*

When a cell phone is in motion, say in a car on on a boat, the set goes from cell to cell in the network. This situation is shown in Fig. 31-1. The dotted line is the path of the vehicle. Base stations (dots) "hand off" access to the cell phone. Solid lines show the limits of the transmission/reception range for each base station. All the base stations are connected to the regional telephone system. This makes it possible for the user of the portable unit to place calls to, or receive calls from, anyone else in the system, whether those other subscribers have cell phones or regular phones.

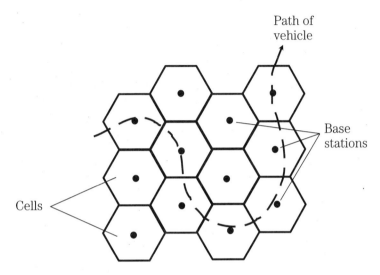

31-1 A moving vehicle is always within range of at least one base station.

To use a cellular network, you must purchase or rent a transceiver and pay a monthly fee. The fees vary, depending on the location and the amount of time per month you use the service. When using such a system, it is important to keep in mind that your conversations are not necessarily private. It is easier for unauthorized people to eavesdrop on wireless communications than to intercept wire or cable communications.

Cell phones and computers

A personal computer (PC) can be hooked up to the telephone lines for use with on-line networks such as the Internet. For some people, getting on the "information superhighway" is the main motivation for buying a computer.

You can connect a laptop or notebook computer to a cell phone with a portable *modem* that converts incoming computer data from analog to digital, and also converts outgoing data from digital to analog. This will let you get online from anywhere within range of a cellular base station. Figure 31-2 is a block diagram of this scheme.

Most commercial aircraft have telephones at each row of seats, complete with jacks into which you can plug a modem. If you plan to get online from an aircraft, you must use the phones provided by the airline, not your own cell phone, because radio

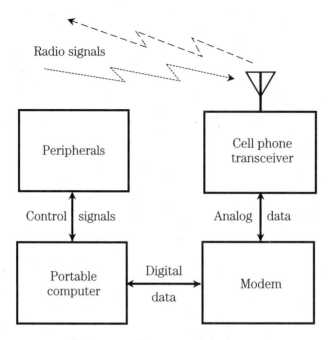

31-2 A cell phone can be equipped with a modem,
allowing portable or mobile access to online
computer networks.

transceivers can cause interference to flight instruments. You must also observe the
airline's restrictions concerning the operation of electronic equipment while in flight.
If you aren't sure what these regulations are, ask one of the flight attendants.

Satellite systems

A satellite system is, in a certain sense, a gigantic cellular network. The repeaters,
rather than being in fixed locations, are constantly moving. The "cells" are much larger,
and they change in size and shape if the satellite moves relative to the earth's surface.

Geostationary satellites

For any satellite in a circular orbit around the earth, the revolution period gets longer
as the altitude increases. At an altitude of about 22,300 miles, a satellite in a circular
orbit takes precisely one day to complete each revolution. If a satellite is placed in
such an orbit over the equator, and if it revolves in the same direction as the earth rotates, it is called a *geostationary satellite*. From the viewpoint of someone on the
earth, a geostationary satellite stays in the same spot in the sky all the time.

One geostationary satellite can cover about 40 percent of the earth's surface. A
satellite over Ecuador, for example, can link most cities in North America and South
America. Three satellites in geostationary orbits spaced 120 degrees apart (⅓ of a
circle) provide coverage over the entire civilized world. Geostationary satellites are

used in television (TV) broadcasting, telephony, and data communication; their primary uses are for gathering weather and environmental data, and for radiolocation.

In geostationary-satellite networks, earth-based stations can communicate via a single "bird" only when both stations are on a line of sight with the satellite. If two stations are nearly on opposite sides of the planet, say in Australia and Wisconsin, they must operate through two satellites to obtain a link (Fig. 31-3). In this situation, signals are relayed between the two satellites, as well as between each satellite and its respective earth-based station.

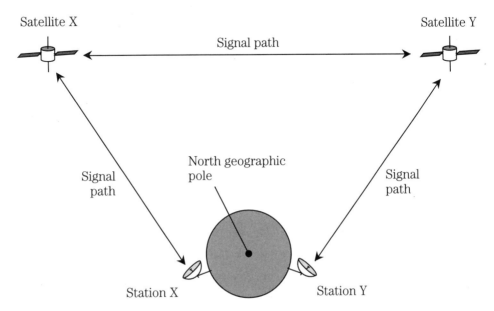

31-3 A communications link that employs two geostationary satellites.

The main problem with two-way geostationary-satellite communication is the fact that the signal path is long: at least 22,300 miles up to the satellite, and at least 22,300 miles back down to the earth. If two satellites are used in the circuit, the path is substantially longer. This doesn't cause problems in television broadcasting or in one-way data transfers, but it slows things down when computers are linked with the intention of combining their processing power. It is also noticeable in telephone conversations.

Problem 31-1

What is the minimum round-trip signal delay when a geostationary satellite is used? Assume that the satellite retransmits signals at the same instant they are received.

Radio waves travel at the speed of light (186,282 miles per second). The minimum path length to and from a geostationary satellite is 22,300 miles, for a total round-trip distance of 44,600 miles. Therefore, the delay is at least 44,600/186,282 second, or about 0.239 seconds (239 milliseconds).

The delay will be a little longer if the transmitting and receiving stations are located a great distance from each other as measured over the earth's surface. In practice, a slight additional delay might also be caused by conditions in the earth's ionosphere.

Low-earth-orbit (LEO) satellites

The earliest communications satellites orbited only a few hundred miles above the earth. They were *low-earth-orbit* (LEO) satellites. Because of their low orbits, LEO satellites took only about 90 minutes to complete one revolution. This made communication spotty and inconvenient, because a satellite was in range of any given ground station for only a few minutes at a time. Thus geostationary satellites became predominant.

Geostationary satellites have certain limitations. A geostationary orbit requires constant adjustment, because a tiny change in altitude will cause the satellite to get out-of-sync with the earth's rotation. Geostationary satellites are expensive to launch and maintain. When communicating through them, there is always a delay because of the path length. It takes rather high transmitter power and a sophisticated, precisely aimed antenna to communicate reliably.

These problems with geostationary satellites have brought about a revival of the LEO scheme. Instead of one single satellite, however, the new concept is to have a large fleet of them. Imagine dozens of LEO satellites in orbits such that, for any point on the earth, there is always at least one satellite in range. Further suppose that the satellites can relay messages throughout the fleet. Then any two points on the surface can always make, and maintain, contact through the satellites.

A LEO system employs satellites in orbits strategically spaced around the globe. The satellites are placed in *polar orbits* (routes that pass over or near the earth's geographic poles) because such orbits optimize the coverage of the system. A LEO satellite wireless communications link is easier to access and use than a geostationary-satellite link. A small, simple antenna will suffice, and it doesn't have to be aimed in any particular direction. The transmitter can reach the network using only a few watts of power. The propagation delay is much shorter than is the case with a geostationary link, usually much less than 0.1 second.

The Global Positioning System (GPS)

The *Global Positioning System* (GPS) is a network of satellites that operates on a worldwide basis. The GPS allows a user to determine his or her exact latitude, longitude, and (if applicable) altitude.

All GPS satellites transmit signals in the microwave part of the radio spectrum, where the wavelengths are a few centimeters. The signals are modulated with special timing and identification codes. A GPS receiver allows its user to find his or her position by measuring the distances to four different satellites. This is done by precisely timing the signals as they travel between the satellites and the receiver. The receiver uses a computer to process the information received from the satellites. From this information, it can give the user an indication to within a few feet (for government and industrial subscribers) or a few hundred feet (for civilians).

An increasing number of automobiles, trucks, and pleasure boats have GPS receivers installed. If you are driving your car in a remote area and get stranded, say in a blizzard, you might use the GPS to locate your position. Using a cell phone, you could call for help and inform authorities of your location. Someday, perhaps every motor vehicle and boat will be equipped with GPS and wireless communications equipment.

Acoustic transducers

An *acoustic transducer* is an electronic component that converts sound waves into some other form of energy, or vice versa. The other form of energy is usually an alternating-current (ac) electrical signal. The waveforms of the acoustical and electrical signals are identical, or nearly so.

Acoustic transducers are designed for various frequency ranges. The human hearing spectrum extends from about 20 Hz to 20 kHz, but acoustic energy can have frequencies lower than 20 Hz or higher than 20 kHz. Energy at frequencies below 20 Hz is called *infrasound*; if the frequency is above 20 kHz, it is known as *ultrasound*. In acoustic wireless devices, ultrasound is generally used, because the wavelength is short and the necessary transducers can be made very small. Also, ultrasound cannot be heard, and therefore it will not distract or annoy people.

Figure 31-4 is a simplified diagram of a *piezoelectric transducer*. This device consists of a crystal, such as quartz or ceramic material, sandwiched between two metal plates. When an acoustic wave strikes one or both of the plates, the metal vibrates. This vibration is transferred to the crystal. The piezoelectric crystal generates weak electric currents when it is subjected to mechanical stress. Therefore, an ac voltage develops between the two metal plates, with a waveform similar to that of the acoustic waves.

If an ac signal is applied to the plates, it causes the crystal to vibrate in sync with the current. The result is that the metal plates vibrate also, producing an acoustic

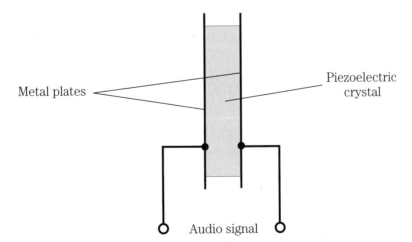

31-4 Simplified cross-sectional diagram of a piezoelectric acoustic transducer.

disturbance in the air. The piezoelectric transducer can thus act either as an *acoustic pickup* or an *acoustic emitter.* Piezoelectric transducers are common in ultrasonic applications, such as intrusion detection systems.

Figure 31-5 shows another type of acoustic transducer that can function either as a pickup or emitter. This is known as an *electrostatic transducer.* When a sound wave strikes the flexible plate, the plate vibrates. This causes rapid changes in the spacing between the flexible plate and the rigid plate, producing fluctuation in the capacitance between the plates. A direct-current (dc) voltage source is connected to the plates. As the capacitance fluctuates, the plates alternately charge and discharge a little. This causes a weak ac signal to be produced, the waveform of which is similar to that of the acoustic disturbance. The *blocking capacitor* allows this ac signal to pass to the transducer output, while keeping the dc confined to the plates.

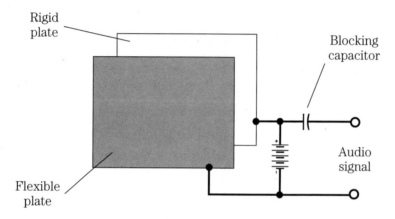

31-5 Simplified diagram of an electrostatic acoustic transducer.

If an ac signal is applied to the terminals of the transducer, it passes through the blocking capacitor and onto the metal plates. The signal creates a fluctuating voltage between the plates. The dc source, combined with the ac signal, results in an electrostatic field of constant polarity, but of rapidly varying intensity. This exerts a fluctuating force between the plates, causing the flexible plate to vibrate and generate an acoustic wave in the air.

Perhaps the most common type of acoustic transducer has a coil surrounding a permanent magnet or electromagnet. A diaphragm or cone is attached to either the coil or the magnet. When the diaphragm vibrates as a result of acoustic waves striking it, ac is produced in the coil. Conversely, if ac is applied to the coil, it produces an electromagnetic force that causes the diaphragm to vibrate. This device is called a *dynamic transducer.* Most microphones, headphones, and loudspeakers are of this type.

Acoustic transducers are used in security systems. They are also used in robotics to help mobile machines navigate in their surroundings. Acoustic transducers are employed in depth-finding apparatus commonly found on boats.

Radio-frequency transducers

The term *radio-frequency (RF) transducer* is a fancy name for an *antenna*. Antennas are so common that you probably don't think much about them. Your car radio has one. Your portable headphone radio, which you might use while jogging on a track (but never in traffic!), employs one. Cellular and cordless telephones, portable television receivers, and handheld radio transceivers use antennas. Hundreds of books have been written on the subject of antennas.

There are two basic types of RF transducer: the *receiving antenna* and the *transmitting antenna*. A receiving antenna converts electromagnetic (EM) fields in the RF range (from about 9 kHz to several hundred gigahertz, or GHz) into ac signals that are amplified by the receiving apparatus. A transmitting antenna converts strong ac signals into EM fields, which propagate through space.

There are a few significant differences between receiving antennas and transmitting antennas designed for a specific radio frequency. Generally, the *efficiency* of an antenna is very important in transmitting applications, but not so important in reception. Efficiency is the percentage of the power going into a transducer that is converted into the desired form. If the input power to a transducer is P_{in} watts and the output power is P_{out} watts, then the efficiency, *Eff,* in percent is

$$Eff = 100 \ (P_{out}/P_{in})$$

Problem 31-2

In a transmitting antenna, 75 W of RF power are delivered to the transducer, and 62 W are radiated as an EM field. What is the efficiency of the transducer?

Plug the numbers into the formula. Here, $P_{in} = 75$ and $P_{out} = 62$; therefore

$$Eff = 100 \ (62/75) = 100 \times 0.83 = 83 \text{ percent}$$

Another difference between transmitting and receiving antennas is the fact that transmitting antennas are generally larger than receiving antennas. Transmitting antennas are also more critical as to their location. While a small loop or whip antenna might work well indoors in a portable radio receiver for the frequency-modulation (FM) broadcast band, the same antenna would not function well at the broadcasting station.

Still another difference between transmitting and receiving antennas involves power-handling capability. Obviously, very little power strikes the antenna in a wireless receiver; it can be measured in fractions of a microwatt. However, a transmitter might produce kilowatts or even megawatts of output power. A small loop antenna, for example, would literally melt if it were supplied with 10 kW of RF power.

Infrared transducers

Many wireless devices transmit and receive energy at *infrared* wavelengths, rather than at radio wavelengths. Infrared energy has a frequency higher than that of radio waves, but lower than that of visible light. Infrared is sometimes called "heat radiation," although this is technically imprecise. Some wireless devices transmit and receive their signals in the visible-light range, although these are encountered much less often than infrared devices.

The most common infrared transmitting transducer is the *infrared-emitting diode (IRED)*. You learned about this type of diode in Chapter 20. A fluctuating direct current is applied to the IRED. The current causes the device to emit infrared rays; the fluctuations in the current constitute the modulation, and produce rapid variations in the intensity of the infrared emitted by the semiconductor junction. The modulation contains information, such as which channel your television set should seek, or whether the volume is to be raised or lowered. Infrared energy is not visible, but at some wavelengths it can be focused by ordinary lenses and reflected by ordinary mirrors. This makes it possible to collimate the rays (make them more or less parallel) so they can be transmitted for distances up to several hundred feet.

Infrared receiving transducers resemble photodiodes or photovoltaic cells, which were also discussed in Chapter 20. The only real difference is that the diodes are maximally sensitive in the infrared, rather than in the visible, part of the electromagnetic spectrum. The fluctuating infrared energy from the transmitter strikes the P-N junction of the receiving diode. If the receiving device is a photodiode, a current is applied to it, and this current varies rapidly in accordance with the signal waveform on the infrared beam from the transmitter. If the receiving device is a photovoltaic cell, it produces the fluctuating current all by itself, without the need for an external power supply. In either case, the current fluctuations are weak, and must be amplified before they are delivered to whatever equipment (television set, garage door, oven, security system, etc.) is controlled by the wireless system.

Infrared wireless devices work best on a line-of-sight basis; that is, when the transmitting and receiving transducers are located so the rays can travel without encountering any obstructions. You have probably noticed this when using television remote control boxes, most of which work at infrared frequencies. Sometimes enough energy will bounce off the walls or ceiling of a room to let you change the channel when the remote box can't actually "see" the television set; but the best range is obtained by making sure you and the television set can "see" each other. You cannot put the control box in your pants pocket and expect it to work, unless (of course) it uses radio waves rather than infrared. Radio and infrared control boxes are often mistaken for one another because they look very much alike to the casual observer.

Wireless local area networks

A *local area network (LAN)* is a group of computers linked together within a building, campus, or other small region. The interconnections in early LANs were made with wire cables, but increasingly, radio links are being used these days. A so-called *wireless LAN* offers flexibility because the computer users can move around without having to bother with plugging and unplugging cables. This arrangement is ideal when notebook computers (also known as laptops) are used.

The way in which a LAN is arranged is called the *LAN topology*. There are two major topologies: the *client/server wireless LAN* and the *peer-to-peer wireless LAN*.

In a client/server LAN (Fig. 31-6), there is one large, powerful central computer called a *file server*, to which all the smaller personal computers (labeled *PC*) are linked. The file server has enormous computing power, high speed, and large storage capacity, and can contain all the data for every user.

In a peer-to-peer LAN (Fig. 31-7), all of the computers are PCs with more-or-less equal computing power, speed, and storage capacity. Each user generally maintains his or her own data. This offers greater privacy and individuality than the client-server topology, but it is slower when all the users need to share the same data.

In the illustrations, only three PCs are shown in the networks. But any LAN can have as few as two or as many as several dozen PC workstations. Client/server LANs are seen mostly in large institutions. Small businesses and schools (or departments within a larger corporation or university) prefer to use peer-to-peer LANs, mainly because they are cheaper and easier to maintain.

Wireless security systems

Wireless technology lends itself well to certain security applications, especially those involving mobile vehicles such as cars, boats, and trucks. Also, a wireless link can alert you to a fire or intrusion at your home or business, when you are not physically present at the location.

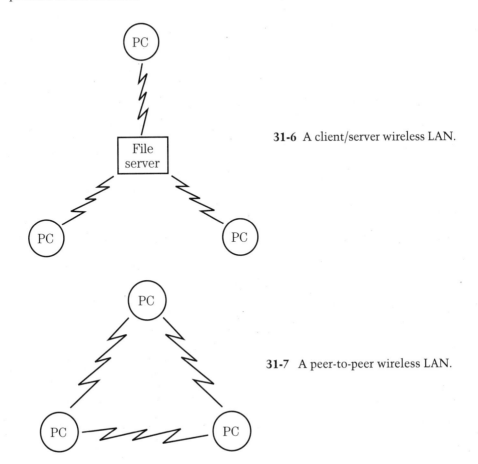

31-6 A client/server wireless LAN.

31-7 A peer-to-peer wireless LAN.

One well-known wireless security system for cars employs a device that transmits a radio signal when the vehicle is being improperly operated. The system circuits and antenna are concealed, although there might be a warning sticker on the car. If a potential thief knows your vehicle has a wireless security system, he or she will not want to spend a lot of time snooping around your car to find and disable the system. Chances are your car will be left alone in that case.

If someone manages to steal your car and it has a wireless security system installed, and if you're lucky enough to find out about the theft within a short time, you call the police. They transmit a signal that contains the vehicle identification number (VIN); this signal starts up the security transmitter in your car. Using radiolocation equipment, the police can determine where your car is. From there, mobile direction-finding radio receivers can be used to reach the car. In some cases the police have caught car thieves while they were actually driving stolen cars.

A home or business wireless security system operates like a paging device or beeper. If something happens that triggers the security system, a radio signal is transmitted that sets off a small radio receiver you carry. Such a system will ideally telephone the police and/or fire department as well. If you happen to be reasonably near your home or business, you can drive to the site.

Hobby radio

In most countries of the world, people can obtain government-issued licenses to send and receive messages via radio for nonprofessional purposes. In America, this hobby is called *amateur radio* or *ham radio*. If you want only to listen to communications and broadcasting, and not to transmit signals, you do not need a license in the United States (although you might need one in certain other countries).

Who uses amateur radio?

Anyone can use ham radio. Radio hams can communicate using any of numerous modes, including speech, Morse code, television, and radioteletype (RTTY). This last mode, RTTY, can be done in real time, or by posting messages in a manner similar to the way computer users exchange information by electronic mail (e-mail). Radio hams have set up their own radio networks, and some have "patched" into the Internet. This is known as *packet radio.*

Some radio hams chat about anything they can think of (except business matters, which are illegal to discuss via ham radio). Others like to practice emergency-communications skills, so they can be of service during crises such as hurricanes, earthquakes, or floods. Still others like to go out into the wilderness and talk to people thousands of miles away while sitting out under the stars. Amateur radio operators communicate from cars, boats, and even bicycles.

Amateur equipment and licensing

A simple ham radio station has a transceiver (transmitter/receiver), a microphone, and an antenna. A small station can fit on a desktop, and is about the size of a home computer or hi-fi stereo system. Accessories can be added until a ham "rig" is a large installation, comparable to a small commercial broadcast station.

A typical computerized ham radio station is shown in Fig. 31-8. The personal computer (PC) can be used to communicate via packet radio with other hams who own computers. The station can be equipped for online telephone (landline) services. The PC can control the antennas for the station, and can keep a log of all stations that have been contacted. Most modern transceivers can be operated by computer, either locally or by remote control over the radio or landline.

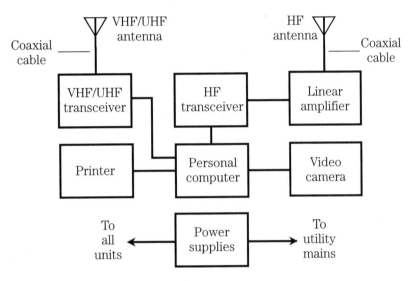

31-8 A computer-controlled amateur radio station.

A good way to learn about ham radio is to contact the headquarters of the American Radio Relay League at 225 Main Street, Newington, CT 06111.

Shortwave listening

The high-frequency (HF) portion of the radio spectrum, at frequencies between 3 and 30 MHz, is sometimes called "shortwave." The waves are actually quite long compared with microwaves (a few millimeters) or infrared (a tiny fraction of one millimeter), which are commonly used in wireless devices nowadays. In fact, 3 MHz corresponds to a wavelength of 100 meters in free space; 30 MHz corresponds to 10 meters. But in the early years of radio when the shortwave band got its name, the wavelengths between 3 and 30 MHz were short compared with the wavelengths of most broadcast and communications signals.

Anyone can build or obtain a shortwave or general-coverage radio receiver, install a modest outdoor antenna, and listen to signals from all around the world. This hobby is called *shortwave listening* or *SWLing*. In the United States, the proliferation of computers and online communications has, to some extent, overshadowed

SWLing, and many young people grow up today ignorant of a realm of broadcasting and communications that still predominates in much of the world.

There are various commercially manufactured shortwave receivers on the market today, some for low prices. An outdoor wire antenna costs practically nothing. Most electronics stores carry one or more models of shortwave receiver, along with antenna equipment for a complete installation. One problem with low-priced shortwave receivers is that they usually lack the mode flexibility, selectivity, and sensitivity necessary to engage in serious SWLing. If you are interested in this hobby and want to obtain high-end equipment, shop around in consumer electronics and amateur radio magazines. Most electronics stores and book stores carry periodicals and books for the beginner as well as the experienced SWLer. A library can also be a good source of information, especially if you are interested in older "vintage" shortwave receivers, some of which can still be found at amateur-radio conventions and flea markets.

Noise

In a wireless system, the term *noise* refers to an electromagnetic field that usually has large bandwidth; that is, it occurs over a wide range of frequencies and wavelengths. Noise does not convey information, and can be either natural or human-made.

It's bad news!

Noise never helps, and often degrades, the performance of a wireless system. It is a major concern in any device or system in which data is sent from one place to another. The higher the noise level, the stronger a signal must be if it is to be received error-free. At any given signal power level, higher noise levels translate into more errors and reduced communications range.

Figure 31-9 is a spectral display of signals and noise, with amplitude as a function of frequency. The device that generates this display is called a *spectrum analyzer.* The horizontal axis shows frequency; the vertical axis shows amplitude. The background noise level is called the *noise floor.* Signals above the noise floor appear in the display, and can be received. The strongest signals are received with the fewest errors; weak signals are subject to the most errors. Signals below the noise floor are not displayed and cannot be retrieved unless a more sophisticated receiving system is used, or the transmitter power output is increased, or both.

Minimizing noise

The noise level in any electronic system can be minimized by using components that draw the least possible current. Noise can also be kept down by lowering the temperature tremendously. Some experimentation has been done at extremely cold temperatures; this is called *cryotechnology.*

The narrower the bandwidth of the signal, the better the signal-to-noise ratio will be, if all other factors remain constant. But this improvement takes place at the expense of data speed. When the noise originates mainly in sources outside the wireless equipment (for example, atmospheric "static"), reducing the bandwidth of the

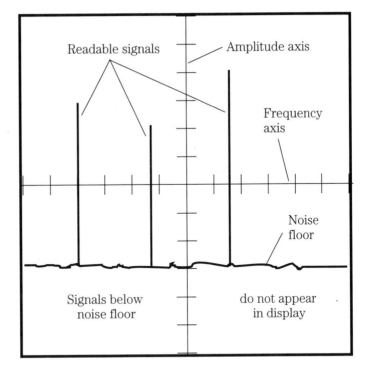

31-9 A spectrum analyzer display. The vertical axis shows amplitude, and the horizontal axis shows frequency.

receiving equipment is generally necessary. This means that, in wireless communications with high external noise levels, the data speed must usually be slower than it would be if there were little external noise.

A *noise limiter* or *noise blanker* can sometimes provide improved communications without reducing data speed. Noise limiters "chop off" high-amplitude noise peaks, and blankers in effect turn the receiver off during noise pulses. Such circuits can be effective against human-made impulse noise ("buzzing"), characterized by high-amplitude peaks of very short duration. But thermal, atmospheric, solar, and galactic noise, which are more random ("hissing" or "crashing"), are not significantly affected by limiters and blankers. The same holds true for ambient light or infrared in wireless systems that make use of optical or infrared technology.

Quiz

Refer to the text in this chapter if necessary. A good score is 18 correct. Answers are in the back of the book.

1. A network that employs one powerful central computer and several PCs is called:

    A. A wireless network.

    B. A local area network.

    C. A client/server network.

    D. A peer-to-peer network.

2. Infrared and optical wireless links work best:

    A. Over distances exceeding 1000 miles.

    B. On a line of sight.

    C. At low radio frequencies.

    D. In situations with high levels of noise.

3. Which of the following devices or systems is *not* generally considered wireless?

    A. A remote-control garage-door opener.

    B. An amateur radio station.

    C. A beeper/pager.

    D. A conventional telephone.

4. A noise blanker can improve the quality of:

    A. Radio reception.

    B. Radio transmission.

    C. Infrared reception.

    D. Infrared transmission.

5. In the United States, a license is required for:

    A. Receiving and transmitting on amateur-radio frequencies.

    B. Transmitting on amateur-radio frequencies.

    C. Using infrared wireless devices.

    D. Using a wireless automobile security system.

6. Noise in a wireless receiver can be minimized by:

    A. Raising the temperature to very high values.

    B. Reducing the temperature to very low values.

    C. Maximizing the amplification.

    D. Minimizing the relative humidity.

7. As the data speed increases in a wireless system, all other factors being equal:

    A. The signal bandwidth increases.

    B. The signal bandwidth decreases.

    C. The overall noise level increases.

    D. The overall noise level decreases.

8. The term "shortwave," in reference to radio, refers to signals having wavelengths of approximately:

    A. 10 to 100 millimeters.

    B. 100 millimeters to 1 meter.

    C. 1 to 10 meters.

    D. 10 to 100 meters.

9. In general, as the noise level in a wireless system increases:

    A. Stronger incoming signals are needed to overcome it.

    B. Weaker signals can be received.

    C. The temperature of the system rises.

    D. The bandwidth of the system increases.

10. In a spectrum analyzer, the horizontal axis shows:

    A. Voltage.

    B. Power.

    C. Frequency.

    D. Time.

11. A device consisting of a receiver and transmitter in the same box is called a:

    A. Modem.

    B. Transverter.

    C. Transceiver.

    D. Transponder.

12. The use of direction-finding equipment to pinpoint a radio transmitter is called:

    A. Radiolocation.

    B. Radionavigation.

    C. The Global Positioning System.

    D. Packet radio.

13. A LAN in which each PC stores its own data is called:

    A. A wireless LAN.

    B. A wide-area LAN.

    C. LAN topology.

    D. A peer-to-peer LAN.

14. Which of the following would be an illegal use of ham radio?

    A. Selling used cars.

    B. Talking about the weather,

    C. Talking about politics.

    D. Connecting a radio to the Internet.

15. An electromagnetic signal might be called "shortwave" if its free-space wavelength is:

    A. 55 kilometers.

    B. 55 meters.

C. 55 centimeters.

D. 55 millimeters.

16. In a cellular network, a base station is sometimes called a:

    A. Transceiver.

    B. Cell.

    C. Repeater.

    D. Cell phone.

17. An advantage of conventional "wired" telephone over cellular is:

    A. Privacy.

    B. Portability.

    C. Ease of use in a car.

    D. LAN topology.

18. An advantage of cellular over conventional "wired" telephone is:

    A. Security.

    B. Lower cost.

    C. Mobility.

    D. Data speed.

19. Infrared waves are:

    A. Longer than radio waves.

    B. Longer than visible-light waves.

    C. Shorter than visible-light waves.

    D. A misnomer; they are really heat rays.

20. The GPS might be useful for:

    A. Improving the performance of a LAN.

    B. Increasing data speed in a wireless system.

    C. Minimizing noise in a wireless system.

    D. A motorist who is lost.

32
CHAPTER

Computers and the Internet

COMPUTERS ARE DIGITAL ELECTRONIC DEVICES. IN RECENT YEARS, personal computing has boomed into a hobby all its own. Computers are used for many things. A few of these include communications, word processing, data processing, mathematical calculation, drawing, photo processing, music composition, and economic forecasting.

There are two major "lines" of personal and business desktop computers: *IBM* (International Business Machines)-*compatible* and *Macintosh* (often called "Mac")-*compatible* computers. There are a few other, less well-known systems. Regardless of the "line," however, all computers have similar components. Figure 32-1 is a block diagram showing the major parts of a typical home computer system, such as you might buy for personal or small business use.

The microprocessor and CPU

The *microprocessor* is the integrated circuit (IC), or chip, that forms the core of your computer's "brain." It coordinates all the action and does all the calculations. It is located on the *motherboard,* or main circuit board, of the main unit. This board is sometimes called the *logic board.*

Basic components

The microprocessor, together with various other circuits, comprise the *central processing unit* (CPU) of a computer. The ancillary circuits can be integrated onto the same chip as the microprocessor, but they are usually separate. The external chips contain memory and programming instructions.

You might think of the microprocessor as the computer's "conscious mind," which directs the behavior of the machine by deliberate control. The CPU, dominated by the microprocessor, represents the PC's entire mind, conscious and subconscious. All the

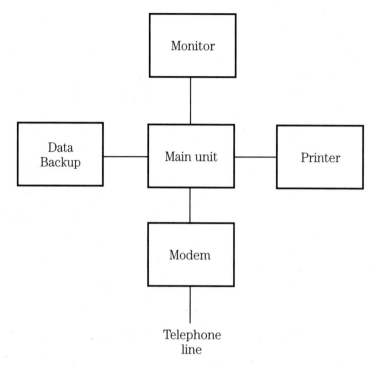

32-1 Block diagram of a simple personal computer system.

ancillary circuits, in conjunction with the CPU, create the computer's "central nervous system." Peripherals such as printers, disk drives, mice, speech recognition/synthesis apparatus, modems, and monitors are the "hands," "ears," "eyes," and "mouth" of the machine. In advanced computer systems there might be robots, vision systems, various home appliances, surveillance apparatus, medical devices, and other exotic equipment under the control of the CPU.

Always advancing

Microprocessors get more powerful every year. Physically, this translates to an increasing number of digital switching transistors per chip. The Intel '486, a primitive chip by today's standards, had more than 1,000,000 transistors. The number of digital switches that can be fabricated onto a semiconductor chip of a particular size is limited only by the structure of matter.

It's possible that someday an elementary logic unit, called a *binary digit* or *bit,* will be represented by the presence or absence of a single electron in an atom. Of course, factors will probably enter into this picture that we don't yet know about. Alternatives to digital technology, such as analog computers or neural networks, might produce new developments, creating machines with power and speed that we can only dream about today.

Bytes, kilobytes, megabytes, gigabytes

A *byte* is a unit of digital data, consisting of a string of eight bits. One byte constitutes roughly the same amount of data as one character, such as a letter, numeral, punctuation mark, space, or line-feed/return command.

Modern computers work with files that are very large in terms of bytes. Therefore, *kilobytes* (units of 2^{10} = 1024 bytes), *megabytes* (units of 2^{20} = 1,048,576 bytes), and *gigabytes* (units of 2^{30} = 1,073,741,824 bytes) are commonly mentioned by people talking about computers. The abbreviations for these units are KB, MB, and GB respectively. Alternatively you might see them abbreviated as K, M, and G.

As computer technology advances during the new century, you'll be hearing more and more about a unit of data called a *terabyte* (TB or T). This is equivalent to 2^{40} bytes, or 1,048,576 MB. And the day might come when we commonly use the term *petabyte* (PB or P), which refers to 2^{50} bytes, or 1,048,576 GB!

Personal-computer *memory* is usually specified in megabytes. The same holds true for removable data storage media such as *diskettes*. The *hard drive* in a computer generally has capacity measured in gigabytes, although an older computer might have a hard-drive capacity that is quoted as so-many-hundred megabytes. (These are 1997 figures.)

The hard drive

A hard drive, also known as a *hard disk,* is a common form of mass storage for computer data. The drive consists of several disks, called *platters,* arranged in a stack. They are made from aluminum or other rigid material, coated with a ferromagnetic substance similar to that used in audio or video tape. The platters are spaced a fraction of an inch apart. Each has two sides (top and bottom), and two *read/write heads,* one for the top and one for the bottom. The assembly is enclosed in a sealed cabinet. Fig. 32-2A is an edgewise, cutaway view of the platters and heads in a typical hard drive.

Drive action

When the computer is switched off, the hard drive mechanism locks the heads in a position away from the platters. This prevents damage to the heads and platters if the computer is moved. When the computer is powered up, the platters spin at several thousand revolutions per minute (rpm). The heads hover a few millionths of an inch above and below the platter surfaces.

When you type a command or click on an icon telling the computer to read or write data, the hard drive mechanism goes through a series of rapid, complex, precise movements. The head must be positioned over the particular spot on the platter where the data is located or is to be written. Then the head must stabilize its position and generate or detect the magnetic fields. All this takes place in a small fraction of a second.

Data arrangement and capacity

The data on a hard drive is arranged in circular *tracks*. This is not quite like the spiral groove on an old-fashioned phonograph disk. While that groove is one long path, the

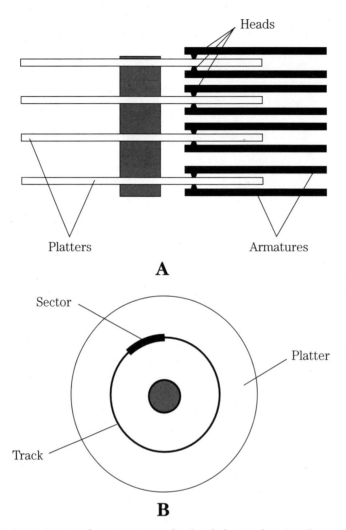

32-2 At A, edgewise view of a hard drive, showing four platters. At B, broadside view of a platter, showing one track and one sector.

tracks on a platter are individual circles. There are hundreds of tracks per radial inch of the platter surface. Each circular track is broken into a number of arcs called *sectors*. A *cylinder* is the set of equal-radius tracks on all the platters in the drive. Tracks and sectors are set up on the hard drive during the initial *formatting* process. There are also data units called *clusters*. These are units consisting of one to several sectors, depending on the arrangement of data on the platters. Figure 32-2B is a face-on view of a single hard-disk platter, showing a track and one of its constituent sectors.

The average new computer's hard drive data storage capacity roughly doubles every year. At the end of 1996, during the holiday computer buying season, a new desktop machine typically had between 1.2 GB and 4 GB of hard drive capacity.

When you buy a computer, whether it is a desktop or laptop unit, it will have a hard drive built in. The drive comes installed and formatted. All you must do is install your software, and you're ready to go. Most new computers are sold with several commonly used programs preinstalled on the hard drive.

Other forms of mass storage

There are several types of mass storage (besides the hard drive) in which data can be kept in large quantities. Computer experts categorize mass storage in two ways: access time and cost per megabyte. In general, the less the access time (that is, the faster the storage medium), the greater the cost per megabyte. The fastest mass storage media usually have the lowest capacity.

Flash memory

Flash memory is an all-electronic form of storage that is useful in high-level graphics, big-business applications, and scientific work. The capacity is comparable to that of a small hard drive, but there are no moving parts.

Because there are no mechanical components, flash memory is faster than any other mass-storage scheme. *PC cards* (also called *PCMCIA cards*) are credit-card-sized, removable components, some of which are designed to serve as removable flash memory.

Disk media

Magnetic diskettes, usually measuring 3.5 inches in diameter and enclosed in a rigid, square case about 4 millimeters thick, can be interchanged in seconds, so there is no limit to how much data you can put on them. A full-wall bookcase of diskettes could hold more work than you'd create in your lifetime.

Magneto-optical storage uses a laser-operted "guidance system" to increase the data density on a magnetic diskette. As a result, the storage capacity is many times greater than that of a typical 3.5-inch diskette.

An especially popular mass-storage medium is *compact disk, read-only memory (CD-ROM)*. You can buy CD-ROMs for various applications. They are commonly used for commercial software, and also to store reference materials such as dictionaries and telephone directories. The main asset of CD-ROM is its fairly large capacity and its long shelf life. The main drawback is that the medium cannot be erased and overwritten, unless you are willing to spend the money for a *CD-R (compact disk, recordable) drive.*

Tape media

The earliest computers used magnetic tape to store data. This is still done in some systems. You can get a *tape drive* for making an emergency backup of the data on your hard drive, or for archiving data you rarely need to use.

Magnetic tape has very high storage capacity. There are microcassettes that can hold more than 1 GB of data; standard cassettes can hold several gigabytes. But tapes are extremely slow because, unlike their disk-shaped counterparts, they are a *serial-access* storage medium. This means that the data bits are written in a string, one after another, along the entire length of the tape. The drive might have to mechanically rewind or fast-forward through a football field's length of tape to get to a particular data bit, whereas on a disk medium, the read/write head never has to travel further than the diameter of the disk to reach a given data bit.

Memory

In a computer, the term *memory* refers to integrated circuits (ICs) that store working data. The amount and speed of memory are crucial factors in determining what a computer can and cannot do.

Data flow

Figure 32-3 shows how data moves between a hard drive or diskette and the memory, controlled by the CPU. When you open a file on your hard drive or on a diskette, the data goes immediately into the memory. The CPU, under direction of the microprocessor, manipulates the data in the memory as you work on the file. Thus, the data in memory changes from moment to moment.

When you hit a key and add a character, or drag the mouse and draw a line that shows up on your display, that character or line goes into memory at the same time. If you hit the backspace key and delete a character, or drag the mouse and erase a line on the screen, it disappears from the memory too. During this time the original file on the disk stays as it was before you accessed it. No change is made to the disk data until you specifically instruct the computer to overwrite the data on the disk.

When you're done working on a file, you tell the microprocessor to close it. Then the data leaves memory and goes back to the hard drive or diskette from which it came, or to some other place, as you might direct. If you tell the computer to overwrite the file on the disk from which it came, many programs send the new data (containing the changes you have made) to unused space on the disk; the old data (as it was before you opened the file) stays in its old location. This is a safeguard, in case you decide to "undo" the changes you made.

All the data passing between the disks and the memory, and between the memory and the CPU, is in *machine language*. This consists of binary digits (bits) 0 and 1. The data passing between you and the CPU is in plain English (or whatever other language you prefer), or in some high-level programming language, having been translated by the machine into a form you can understand.

Memory capacity

The number of bytes of data that can be stored in a computer's memory is known as the *memory capacity*. The main factor that determines a computer's maximum potential memory capacity is the number of transistors that can be fabricated onto a single memory IC, or "chip." Other factors, such as microprocessor power, have a practical effect on the usable memory capacity. A gigantic memory will not be of much use if the microprocessor is slow. Nor will a fast microprocessor be of practical value if the memory capacity is too small for applications that demand high speed.

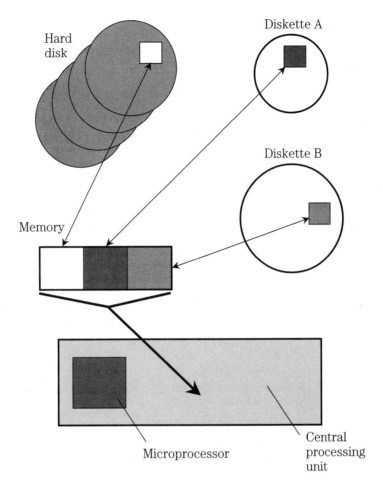

32-3 The flow of data between disk media and memory is controlled by the CPU.

The amount of memory you need depends on the applications you intend to run on your computer. Most software packages will tell you how much memory you need. They'll often quote two specifications: a minimum memory requirement and a figure for optimum performance (approximately twice the minimum requirement). If possible, you should equip your computer with enough memory for optimum performance. This year's optimum machine is next year's minimal one when it comes to popular software.

If your computer doesn't have enough memory to run a given application, you can usually add more. But this can only be done up to a certain point. Eventually, your microprocessor's power will no longer be up to running contemporary software at reasonable speed, no matter how much memory you have.

Memory volatility

In most computers, the memory is *volatile*. This means that it requires a source of power to be maintained. If you switch the computer off, or if there's a power failure, you'll lose all the data in the memory. This problem can be avoided if the memory chips are supplied with power continuously. Some machines have rechargeable memory backup batteries that keep memory data intact for short periods if there is a power failure.

In contrast to memory, the data on magnetic disks, or on optical media, will stay put when the power is removed. When you're working on a file, it's wise to store it every few minutes on the hard drive and/or diskette. That way, in the rare event a power failure does occur, you won't lose much of your work.

The display

The visual interface between you and your computer is known as the *display*. In desktop computers, an external display is often called a *monitor*. A *cathode-ray tube (CRT) monitor* resembles a television set without the tuning or volume controls. A *liquid crystal display (LCD)* is lightweight and thin, and is found in laptop (notebook) computers.

Showing detail

The *image resolution* of a computer display is important. This is the extent to which it can show detail; the better the resolution, the sharper the image. Resolution can be expressed in terms of *dot size* or *dot pitch*. This is the diameter, in millimeters (mm), of the individual elements in the display—the "smallest unbreakable pieces." A good display has a dot pitch that is a small fraction of 1 mm. The smaller the number, the higher the resolution and, all other factors being equal, the crisper the image. Image resolution is more often stated as a pair of numbers representing the number of *pixels* (picture elements) the screen shows horizontally and vertically. For a particular screen size, the greater the number of pixels the unit can display, the better the image resolution.

A good display is crucial for doing graphics work, when using the World Wide Web, in remote-control robotics, and in computer games. Besides these practical advantages, a sharp display is more pleasant to work with than a marginal one. For example, if you get used to a *Super Video Graphics Array (SVGA)*, you'll find that the older *Video Graphics Array (VGA)* display seems dull and blurry.

Choice of monitor

Some computer users can get along with one monitor until it wears out. If you use your computer only for text-based applications, you'll probably find a modest VGA monitor adequate for your needs. If, however, you find yourself scrolling left-and-right because you work with lines longer than 80 characters, you might do well with a wider screen. If you have to squint to see tiny images on screen, likewise, you should think about enlarging the image rather than wearing eyeglasses (unless you really need glasses).

If you're buying a whole new computer system, this is a good time to buy the best monitor you can afford. Many new desktop computers are sold with monitors, but some are not, allowing you to choose your own. In that case, again, you must be sure the monitor is compatible with the rest of the system. If you're buying a laptop computer, you'll have options between less expensive and more expensive displays. The best way to decide what you want is to look at several displays in the store and, if possible, use some of your favorite software on the new computer before you buy the machine.

Other display specifications

The size of a display screen is given as a diagonal measure in inches. Most screens aren't square, but have an *aspect ratio* (width-to-height) of 4:3. A 13-inch display is about 10.5 inches wide by 8 inches high, for example. Many desktop computer users opt for 17-inch monitors. Any increase in display screen area is roughly equal to the square of the increase in diagonal measure. Thus, a 17-inch display is effectively about 70 percent larger than a 13-inch unit.

Some displays can show more colors than others. The *color resolution* of a monitor is quoted as a number, such as "256 colors," "thousands of colors," or "millions of colors." Color specifications are somewhat oversimplified, because color has two distinct properties called the *hue* and the *saturation.* The hue is technically represented by the peak wavelength, but more often is described in terms of words (red, orange, yellow, green, blue, indigo, violet). Saturation is the "richness" or "intensity" of color; the higher the saturation, the more intense a given hue appears.

Interlacing (or lack thereof) is important in a desktop monitor. A noninterlaced monitor is better than an interlaced one if you're working with fast-moving graphic images. Interlacing translates into a lower *refresh rate* (number of times the entire image is renewed). A low refresh rate can cause objectionable "jitter" or jumpiness in a fast-moving image. A good refresh-rate specification is 70 Hz or more.

If you're concerned about the *extremely-low-frequency (ELF)* electromagnetic fields produced by computer monitors, you might consider buying a unit that has been designed to reduce this "radiation." Be aware, however, that there is disagreement among experts as to the actual danger posed by ELF emission.

The printer

A computer *printer* is an electromechanical device that produces *hardcopy* (text and images on paper). The most common printers are dot-matrix, thermal, inkjet, and laser.

Dot-matrix

The "old reliable" of the printing family is the *dot-matrix printer.* This type of printer is the least expensive in terms of both the purchase price and the long-term operating cost. Dot-matrix printers produce acceptable print quality for most manuscripts, reports, term papers, and theses. The mechanical parts are rugged, and maintenance requirements are minimal.

Older dot-matrix printers are quite noisy in operation; newer machines are quieter. They don't have the typeset-grade image quality of more expensive printers. Dot-matrix printers can render some simple graphic images, but the quality is fair at best, and it can take a long time to print. Dot-matrix printers can't reproduce detailed artwork or photographs with acceptable quality.

Thermal

A *thermal printer* uses temperature-sensitive dye and/or paper to create hardcopy text and images. Some thermal printers produce only black-and-white images, while others can render full color. Thermal printers are often preferred by people who use portable computers, because these printers are physically small and light.

A simple grayscale thermal printer employs a special paper that darkens when it gets hot. The print mechanism works something like that of a dot-matrix printer, but instead of the print head pressing ink onto the page, the pins in the print mechanism are heated, and they cause the paper to darken.

A color thermal printer uses thick, heat-sensitive dyes of the primary pigments: magenta (pinkish red), yellow, and cyan (bluish green). Sometimes black dye is also used, although it can be obtained by combining large, equal amounts of the primary pigments. The print head uses heat to liquefy the dye so it bleeds onto the paper. This is done for each color of pigment separately. There are three separate overlapping images produced, one for each primary pigment.

Inkjet

In an *inkjet printer,* tiny nozzles spray ink onto the paper. The machine is almost silent in operation. Inkjet printers are available in single-color and multicolor designs.

A high-end inkjet printer costs roughly twice as much as a high-end dot-matrix machine, although some monochrome inkjets are as inexpensive as dot-matrix printers. When properly used and maintained, an inkjet printer can create hard copies of exceptional quality, comparable to that of a typeset publication.

Inkjet printers require periodic replacement of the ink "well" or container. They also need the right kind of paper; some papers have fibers that carry the ink along via capillary action, causing characters and images to blur.

Laser

A *laser printer* works something like a photocopy machine. The main difference is that, while a photocopier creates a copy of a real image (the paper original), a laser printer makes a copy of a digital computer image.

When data arrives at the printer from the computer, the encoded image is stored in the printer's memory. The memory stores one page of data, and then sends it along to the laser. Some printers use a light-emitting-diode (LED) matrix rather than a single laser. These printers are called *LED printers.* For practical purposes, LED printers are equivalent to laser printers.

The laser blinks rapidly while it scans a cylindrical *drum.* The drum has special properties that cause it to attract the printing chemical (called *toner*) in some places but not others, creating a pattern that will ultimately appear on the paper. A sheet of paper is pulled past the drum, and also past an electrostatic charger. Toner from the drum is attracted to the paper. The image thus goes onto the paper, although it has not yet been permanently fused, or bonded, to the paper. The *fuser,* a hot pair of roller/squeezers, does this job, completing the printing process.

The main asset of laser printers is their excellent print and graphics quality. The image resolution of a laser printer ranges from about 300 dots per inch (DPI) to over 1200 DPI. As far as the unaided eye can tell, 600 DPI is nearly as good as a photograph. Laser printers can handle graphics and text equally well. If an image can be rendered on a photocopy machine, it can be rendered just as well on a laser printer. Another asset of laser printers is that they make almost no noise.

The laser printer is the best device for rendering high-resolution hardcopy text and graphics. If you're into desktop publishing, presentation graphics, or anything else that requires top-grade hard copy, then you'll want a laser printer.

The modem

The term *modem* is a contraction of *modulator/demodulator.* A modem interfaces a computer to a telephone line, television cable system, or radio transceiver, allowing you to communicate with other computer users or to access online networks.

External vs. internal

An *external modem* is a self-contained unit, or "box," that you can use with any computer. It might fit directly into a utility alternating-current (ac) wall outlet, or sit on your desk next to your computer. It has a cord that runs to the computer's *serial data port* (also called the *communications port*), and another cord that runs to the telephone jack. The modem will probably also have a jack into which you can plug your phone set.

An *internal modem* is a printed-circuit board, also called a *card,* that is commonly installed in a new desktop computer. A new notebook computer might have an internal modem installed, but often you must purchase the modem in the form of a *PC card* that fits into a slot on the side of the computer. This card is roughly the height and width of a credit card, but somewhat thicker.

Digital vs. analog

A computer works with binary *digital* signals which are rapidly fluctuating direct currents. In order for digital data to be conveyed over a telephone or radio circuit, the data must be converted to *analog* form. In a telephone modem or radio transceiver modem, this is done by changing the digit 1 into an audio tone, and the digit 0 into another tone with a different pitch. The result is an extremely fast back-and-forth alternation between the two tones.

If you happen to pick up a telephone extension while someone is online with the computer from another extension, you'll hear the analog signals from the two modems; it sounds like a hiss or roar. But don't make a habit of doing this. It can cause the computer to be disconnected from the online network.

In *modulation,* digital data is changed into analog data. It is a type of *digital-to-analog (D/A) conversion. Demodulation* changes the analog signals back to digital ones; this is *analog-to-digital (A/D) conversion.*

Data speed

Modems work at various speeds, usually measured in *bits per second (bps)*. You will often hear about *kilobits per second (kbps),* where 1 kbps = 1000 bps. Sometimes you'll hear about speed units called the *baud* and *kilobaud.* (A kilobaud is 1000 baud.) Baud and bps are almost the same units, but not they are not identical. People often use the term "baud" or "kilobaud" when they really mean "bps" or "kbps."

The higher the speed as specified in either bps or baud, the faster the data is sent and received through the modem. Speeds keep increasing as computer communications technology advances. In the mid-1990s, typical home computers had telephone modems that operated at 14.4 or 28.8 kbps. By the time you read this, these figures might sound prehistoric, especially if you have a television cable modem. Such devices function at speeds of several *megabits per second (Mbps)*; 1 Mbps = 1000 kbps = 1,000,000 bps.

Basic components

Figure 32-4 is a block diagram of a modem suitable for interfacing a home or business computer with the telephone line. The modulator, or D/A converter, changes the digital computer data into audio tones. The demodulator, or A/D converter, changes the incoming audio tones into digital signals for the computer. The audio tones fall within the frequency range, or band, of approximately 300 Hz to 3 kHz. This is the band needed to clearly transmit a human voice.

It's amazing how much computer data can race over a single telephone or radio circuit having such a narrow bandwidth. Even pictures can be sent and received in brilliant color and in quite good detail (high resolution). As you might imagine, color images take longer than gray-scale images to send and receive; also, the more detail an image contains, the longer it takes to be transferred at any given data speed.

The Internet

The *Internet* is a worldwide system, or *network,* of computers. It got started in the late 1960s, originally conceived as a network that could survive nuclear war. Back then it was called *ARPAnet,* named after the Advanced Research Project Agency (ARPA) of the United States federal government.

Protocol and packets

When people began to connect their computers into ARPAnet, it became clear that there must be a universal set of standards, called a *protocol,* to ensure that all the machines "speak the same language." The modern Internet is such that you can use any type of computer—IBM-compatible, Mac, or other—and take advantage of all the network's resources.

All Internet activity consists of computers "talking" to one another. This occurs in machine language. However, the situation is vastly more complicated than when data goes from one place to another within a single computer. In the Internet (often called simply *the Net*), data must often go through several different computers to get from the transmitting or *source* computer to the receiving or *destination* computer.

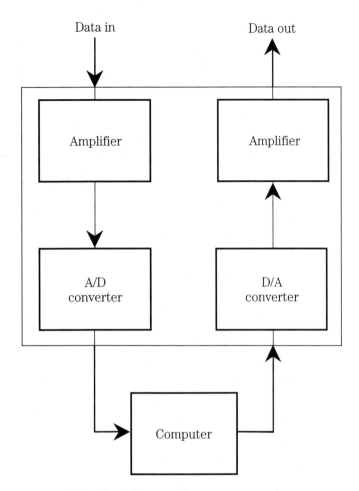

32-4 Block diagram of a computer modem.

These intermediate computers are called *nodes* or *hosts*. Millions of people are simultaneously using the Net; the most efficient route between a given source and destination can change from moment to moment. The Net is set up in such a way that signals always try to follow the most efficient route.

If you are connected to a distant computer, say a machine at the National Hurricane Center, the requests you make of it and the data it sends you are broken into small units called *packets*. Each packet coming to you has, in effect, your name written on it. But not all packets necessarily travel the same route through the network. Ultimately, all the packets are reassembled into the data you want (say, the infrared satellite image of a hurricane) even though they might not arrive in the same order they were sent.

Figure 32-5 is a simplified drawing of Net data transfer, for a hypothetical file containing five packets transferred during a period of extremely heavy usage. Nodes are shown as black dots surrounded by circles. In this example, some packets pass through more nodes, and/or over a much greater physical distance, than others. If Net traffic were very light, all the packets might follow the same route through fewer nodes. This is why it takes longer to acquire data on the Net during peak hours of use, as compared with times when there are comparatively few people connected into it. A file can't be completely reconstructed until all the packets have arrived and the destination computer has ensured that there are no errors.

E-mail and newsgroups

For many computer users, communication via Internet *electronic mail (e-mail)* and/or *newsgroups* has practically replaced the Postal Service. You can leave messages for, and receive them from, friends and relatives scattered throughout the world.

To effectively use e-mail or newsgroups, everyone must have an *Internet address*. These tend to be arcane. An example is:

stangib@aol.com

The first part of the address, before the @ symbol, is the *username*. The word after the @ sign and before the period (or "dot") represents the *domain name*. The three-letter abbreviation after the dot is the *domain type*. In this case, "com" stands for "commercial." The AOL system (America Online) is a commercial provider. Other common domain types include "net" (network), "org" (organization), "edu" (educational institution), and "gov" (government).

Internet conversations

You can carry on a "teletype" style conversation with other computer users via the Internet, but takes a bit of getting used to. This is called *Internet relay chat (IRC)*. Typing messages to, and reading from, other people in real time is more personal than letter writing, because your addressees get their messages immediately. But it's less personal than talking on the telephone, especially at first, because you can neither hear nor make any vocal inflections.

It is possible to digitize voice signals and transfer them via the Internet. This has given rise to hardware and software schemes that claim to provide virtually toll-free long-distance telephone communications. As of this writing, this is similar to amateur radio in terms of reliability and quality of connection. When Net traffic is light, such connections can be almost as good as those provided by the telephone companies. When Net traffic is heavy, the quality is marginal to poor.

Getting information

One of the most important features of Internet is the fact that it can get you in touch with thousands of sources of information. Data is transferred among computers by means of a *file transfer protocol (FTP)*. When you use FTP, the files at the remote computer become available to you, exactly as if they were stored in your own computer. When using FTP, you should be aware of the time at the remote location, and avoid, if possible, accessing files during the peak hours at the remote computer. Peak hours usually correspond to working hours, or approximately 8:00 A.M. to 5:00 P.M.

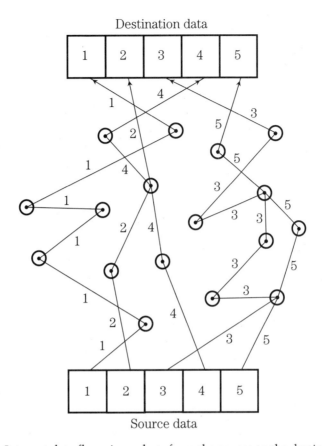

Destination data

Source data

32-5 Internet data flows in packets from the source to the destination.

local time, Monday through Friday. You'll have to take time differences into account if you're not in the same time zone as the remote computer.

The *World Wide Web* (also called *WWW* or *the Web*) is one of the most powerful information servers you will find online. Its outstanding feature is *hypertext,* a scheme of cross-referencing. Certain words or phrases are highlighted and/or underlined. When you select one of these words or phrases in a *Web page* or *Website* (a document containing text and graphics, and sometimes also other types of files), you will be linked to another document dealing with the same or a related subject. This site will probably also contain numerous links; you might find yourself "surfing" the Web for hours going from site to site. The word "surfing" derives from the similarity of this activity to television "channel surfing." (This author will vouch for the fact that it bears absolutely no resemblance to the type of surfing that takes place at the beach.)

The Web is fastest during the wee hours of the morning in the U.S., when the fewest people are on the Net. When Net traffic is heavy, Web documents can take a long time to appear. In some instances you won't be able to get a page at all; you'll sit there staring at a blank display. Then you'll know why some people refer to the Web as the "World Wide Wait." Just remember that the Web is part of the Net. The fewer people using the Net, the faster the Web will be.

Getting connected

If you're not an Internet user but would like to get access, try calling the computer science department of the nearest trade school, college, or university. Many, if not most, academic institutions have Internet access, and some will let you in for a reasonable charge. If you aren't near a school that can provide you with Internet service, you can get connected through a commercial provider. You'll have to pay a fee, generally by the month. You might also have to pay for any hours you use per month past a certain maximum.

Quiz

Refer to the text in this chapter if necessary. A good score is 18 correct. Answers are in the back of the book.

1. One megabyte is the same amount of data as:
    A. 1024 bytes.
    B. 1024 K.
    C. 1024 GB.
    D. 1/1024 K.

2. The Web would probably work fastest for a user in New York at:
    A. 2:00 A.M. local time on a Tuesday.
    B. 4:00 P.M. local time on a Wednesday.
    C. 12:30 P.M. local time on a Thursday.
    D. Any of the above times; it does not matter.

3. Image resolution can be specified in terms of:
    A. Megahertz.
    B. Color intensity.
    C. Wavelength.
    D. Dot pitch.

4. A cluster is a unit of:
    A. Frequency on a hard drive.
    B. Data on the Internet.
    C. Data on a hard drive.
    D. Bandwidth on the Internet.

5. An example of a mass-storage device is a:
    A. Hard drive.
    B. Microprocessor.

    C. Modem.

    D. Read-write head.

6. The character string *sgibilisco@noaa.gov* might represent:

    A. A Web site.

    B. The location of data in memory.

    C. An e-mail address.

    D. A computer's serial number.

7. Bits per second (bps) is a unit of:

    A. Computer memory.

    B. Mass storage.

    C. Image resolution.

    D. Data speed.

8. A platter is a:

    A. Part of a hard drive.

    B. Unit of memory.

    C. Element of a digital image.

    D. Semiconductor chip.

9. Protocol ensures that:

    A. A hard drive runs smoothly.

    B. A monitor reproduces color accurately.

    C. A printer generates a clear image.

    D. Computers can exchange data.

10. A packet is:

    A. A computer memory module.

    B. A unit of 2^{10} bytes.

    C. A piece of a file sent over the Net.

    D. A picture element in a computer monitor.

11. A motherboard contains:

    A. A microprocessor.

    B. An external modem.

    C. A diskette drive.

    D. A display screen.

12. Cross-referencing among Web pages can be accomplished by means of:

    A. Digital signal processing.

    B. A modem.

C. Internet Relay Chat (IRC).

D. Hypertext.

13. The abbreviation FTP stands for:

A. Fast Text Packet.

B. File Transfer Protocol.

C. Frequency/Time Processing.

D. Federal Trade Program.

14. A modem contains:

A. An internal hard drive.

B. A microprocessor.

C. An A/D converter.

D. A printer interface.

15. An asset of a dot-matrix printer is:

A. Low operating cost.

B. Excellent image detail.

C. Resemblance to a photocopy machine.

D. Compatibility with most modems.

16. Which of the following types of mass storage provide the fastest access time?

A. Magnetic tape.

B. CD-ROM.

C. Flash memory.

D. Hard drive.

17. Which of the following is a serial-access medium?

A. Computer memory.

B. Magnetic tape.

C. A hard drive.

D. CD-ROM.

18. When computer data is sent over long-distance telephone circuits, the digital highs and lows are generally represented by:

A. Audio tones.

B. A series of clicks.

C. Positive and negative direct currents.

D. Pixels.

19. For animated graphics involving fast motion, you should ideally use:

A. A laser printer.

B. A hard drive.

C. Hypertext.

D. A noninterlaced monitor.

20. A thermal printer might be the best type of printer for:

    A. Someone who travels a lot.

    B. Someone who works with animated graphics.

    C. Someone who needs to print huge text documents.

    D. Someone who needs top-quality printouts.

Test: Part three

Do not refer to the text when taking this test. A good score is at least 37 correct. Answers are in the back of the book. It's best to have a friend check your score the first time, so you won't memorize the answers if you want to take the test again.

1. In a junction FET, the control electrode is usually the:
    A. Source.
    B. Emitter.
    C. Drain.
    D. Base.
    E. Gate.

2 A diode can be used as a frequency multiplier because of its:
    A. Junction capacitance.
    B. Nonlinearity.
    C. Avalanche voltage.
    D. Forward breakover.
    E. Charge-carrier concentration.

3. Which of the following is an example of wireless technology?
    A. The Global Positioning System.
    B. A field-effect transistor.
    C. A junction diode.
    D. A carbon-composition resistor.
    E. Digital-to-analog conversion.

4. A very brief high-voltage spike on an ac power line is called:
    A. A surge.
    B. An arc.
    C. A transient.

    D. An avalanche.

    E. A clamp.

5. Which of the following is not characteristic of an oscillator?

    A. Negative feedback.

    B. Good output-to-input coupling.

    C. Reasonably high transistor gain.

    D. Alternating-current output.

    E. Usefulness as a signal generator.

6. Which layer of the ionosphere absorbs radio signals below about 7 MHz during the daylight hours?

    A. The F layer.

    B. The E layer.

    C. The D layer.

    B. The C layer.

    E. The B layer.

7. The beta of a bipolar transistor is the:

    A. Current amplification factor.

    B. Voltage amplification factor.

    C. Power amplification factor.

    D. Highest frequency at which it will amplify.

    E. Lowest frequency at which it will amplify.

8. Which type of component is impractical to fabricate on a silicon chip?

    A. A capacitor.

    B. A transistor.

    C. A diode.

    D. An inductor.

    E. Any component can be fabricated on a silicon chip.

9. The extent to which an oscillator maintains a constant frequency is called its:

    A. Sensitivity.

    B. Drift ratio.

    C. Gain.

    D. Selectivity.

    E. Stability.

10. A Zener diode would most likely be used in:

    A. A mixer.

    B. A voltage-controlled oscillator.

   C. A detector.

   D. A power supply regulating circuit.

   E. An oscillator.

11. When the bias in an FET stops the flow of current, the condition is called:

   A. Forward breakover.

   B. Cutoff.

   C. Reverse bias.

   D. Pinchoff.

   E. Avalanche.

12. A vacuum tube might be found in:

   A. The front end of a radio receiver.

   B. A high-power radio-frequency linear amplifier.

   C. A low-level audio amplifier.

   D. A digital computer.

   E. Antique radios only.

13. The "brain" of a computer is the:

   A. Hard drive.

   B. Internet.

   C. Microprocessor.

   D. CD-ROM.

   E. Monitor.

14. A disadvantage of a half-wave rectifier is the fact that:

   A. The voltage is excessive.

   B. The current output is low.

   C. The output is hard to filter.

   D. It needs many diodes.

   E. The transformer must have a center tap.

15. In a wireless system, noise always:

   A. Improves the bandwidth.

   B. Increases data transfer rate.

   C. Reduces the transmitted signal power.

   D. Degrades performance.

   E. Makes communication or data transfer impossible.

16. An amplifier has a dc collector power input of 300 W, and is 75.0 percent efficient. The signal output power is:

   A. 400 W.

B. 300 W.

C. 225 W.

D. Variable, depending on the bias.

E. Impossible to determine from this data.

17. When both N and P type semiconductors are found in an IC, the technology is known as:

A. Transistor-transistor logic.

B. CMOS.

C. Bipolar logic.

D. NPNP.

E. PNPN.

18. A common-base circuit is commonly employed as:

A. A microwave oscillator.

B. A lowpass filter.

C. A noise generator.

D. A phase-locked loop.

E. A radio-frequency power amplifier.

19. Which of the following devices always uses an IC as one of its main active components?

A. A radio-frequency power amplifier.

B. A digital computer.

C. A low-level audio amplifier.

D. A power transformer.

E. An impedance matching network.

20. Which type of amplifier circuit provides the greatest efficiency?

A. Class A.

B. Class AB.

C. Class B.

D. Class C.

E. Any of the above can be as efficient as any other.

21. ASCII is a form of:

A. Video modulation.

B. Diode.

C. Data transmission code.

D. Voice modulation.

E. AM detector.

22. The most stable type of oscillator circuit uses:
    A. A tapped coil.
    B. A split capacitor.
    C. Negative feedback.
    D. A common-base arrangement.
    E. A quartz crystal.

23. If the source-gate junction in an FET conducts:
    A. It is a sign of improper bias.
    B. The device will work in class C.
    C. The device will oscillate.
    D. The device will work in class A.
    E. The circuit will have good stability.

24. *Image resolution* is an important specification in:
    A. Oscillators.
    B. Computer monitors.
    C. Receiver sensitivity.
    D. Rectifier diodes.
    E. Superheterodyne receivers.

25. Signal-plus-noise-to-noise ratio (S+N/N) is often specified when stating a receiver's:
    A. Selectivity.
    B. Stability.
    C. Modulation coefficient.
    D. Sensitivity.
    E. Polarization.

26. In a reverse-biased semiconductor diode, the capacitance depends on:
    A. The width of the depletion region.
    B. The reverse current.
    C. The P:N ratio.
    D. The gate bias.
    E. The avalanche voltage.

27. In a video recording tape, the image signal tracks are:
    A. Parallel with the edges of the tape.
    B. Perpendicular to the edges of the tape.
    C. Diagonal with respect to the edges of the tape.
    D. Oriented at various angles.
    E. Interwoven with the audio signal tracks.

28. A simple power supply filter can be built with:

    A. A capacitor in series with the dc output.

    B. An inductor in parallel with the dc output.

    C. A rectifier in parallel with the dc output.

    D. A resistor in series and an inductor in parallel with the dc output.

    E. A capacitor in parallel with the dc output.

29. Which of the following bipolar-transistor circuits can provide the most amplification?

    A. Common emitter.

    B. Common base.

    C. Common collector.

    D. Common gate.

    E. Common drain.

30. A *platter* is a part of a device known as:

    A. A videotape recorder.

    B. A ceramic filter.

    C. Tape memory.

    D. A hard drive.

    E. A voltage-controlled oscillator.

31. An example of a device that commonly oscillates is:

    A. A rectifier diode.

    B. A weak-signal diode.

    C. A Gunn diode.

    D. A Zener diode.

    E. None of the above.

32. In a PNP bipolar transistor:

    A. The collector is positive relative to the emitter.

    B. The collector is at the same voltage as the emitter.

    C. The collector is negative relative to the emitter.

    D. The collector might be either positive or negative relative to the emitter.

    E. The collector must be at ground potential.

33. In a cathode-ray tube (CRT), the term *electrostatic deflection* means that:

    A. The device is not working correctly.

    B. Static electricity distorts the image.

    C. The beam is bent by an electric field.

    D. The beam is bent by coils carrying a variable current.

    E. You can get a shock if you touch the screen.

34. Which type of modulation consists of one voice sideband, with a suppressed carrier?
    A. AM.
    B. SSB.
    C. FM.
    D. RTTY.
    E. PCM.

35. A type of electron tube useful for generating microwave energy is:
    A. A triode.
    B. An oscillotron.
    C. A cathode-ray tube.
    D. A videotron.
    E. A magnetron.

36. In an AND gate, the output is high:
    A. If any input is high.
    B. Only when all inputs are low.
    C. If any input is low.
    D. Only when all inputs are high.
    E. Only when the inputs have opposite logic states.

37. A voltage-controlled oscillator makes use of:
    A. A varactor diode.
    B. A Zener diode.
    C. Negative feedback.
    D. A split capacitance.
    E. Adjustable gate or base bias.

38. A device that converts ultrasound into electrical signals by taking advantage of the properties of a crystalline or ceramic substance is:
    A. An ultrasonic wave converter.
    B. A crystal converter.
    C. A ceramic converter.
    D. A sound electrifier.
    E. A piezoelectric transducer.

39. An amplifier has an output signal voltage that is 35 times the input signal voltage. This is a gain of:
    A. 15 dB.
    B. 31 dB.
    C. 35 dB.

D. 350 dB.

E. 700 dB.

40. In an exclusive OR gate, the output is high:

    A. If any input is high.

    B. Only when all inputs are low.

    C. If any input is low.

    D. Only when all inputs are high.

    E. Only when the inputs have opposite logic states.

41. A ratio detector is a circuit for demodulating:

    A. AM.

    B. PM.

    C. FM.

    D. SSB.

    E. AFSK.

42. In a radio-frequency power amplifier using a vacuum tube, stability is enhanced by grounding the:

    A. Cathode.

    B. Plate.

    C. Filament.

    D. Control grid.

    E. Screen grid.

43. The World Wide Web is:

    A. Part of the Internet.

    B. A company that manufactures communications equipment.

    C. An agreement between nations concerning data protocols.

    D. Another name for the Global Positioning System.

    E. A satellite television network.

44. Boolean algebra is:

    A. Just like ordinary algebra.

    B. A useful tool in digital logic circuit design.

    C. Used to calculate the value of an unknown.

    D. Used with negative logic only.

    E. Used with positive logic only.

45. A voltage-doubler power supply is best for use in:

    A. Circuits that need low current at high voltage.

    B. Low-voltage devices.

    C. High-current appliances.

    D. All kinds of electronics equipment.

    E. Broadcast transmitter power amplifiers.

46. An optoisolator consists of:

    A. Two Zener diodes back-to-back.

    B. An LED and a photodiode.

    C. Two NPN transistors in series.

    D. An NPN transistor followed by a PNP transistor.

    E. A PNP transistor followed by an NPN transistor.

47. When a semiconductor is reverse-biased with a large enough voltage, it will conduct. This is because of:

    A. Bias effect.

    B. Avalanche effect.

    C. Forward breakover.

    D. An excess of charge carriers.

    E. Conduction effect.

48. Synchronizing pulses in a video signal:

    A. Keep the brightness constant.

    B. Keep the contrast constant.

    C. Keep the image from "tearing" or "rolling."

    D. Ensure that the colors are right.

    E. Keep the image in good focus.

49. In an enhancement-mode MOSFET:

    A. The channel conducts fully with zero gate bias.

    B. The channel conducts partially with zero gate bias.

    C. The channel conducts ac but not dc.

    D. The channel conducts dc but not ac.

    E. The channel does not conduct with zero gate bias.

50. In a step-up power transformer:

    A. The primary voltage is more than the secondary voltage.

    B. The secondary voltage is more than the primary voltage.

    C. The primary and secondary voltages are the same.

    D. The secondary must be center-tapped.

    E. The primary must be center-tapped.

Final exam

Do not refer to the text when taking this exam. A good score is at least 75 correct. Answers are in the back of the book. It's best to have a friend check your score the first time, so you won't memorize the answers if you want to take the test again.

1. As the frequency of ac increases in a coil, the reactance:
   A. Gets larger negatively.
   B. Gets smaller negatively.
   C. Stays the same.
   D. Gets larger positively.
   E. Gets smaller positively.

2. A beat-frequency oscillator is useful for:
   A. Generating FM.
   B. Detecting FM.
   C. Generating CW.
   D. Detecting CW.
   E. Any of the above.

3. A Colpitts circuit is a form of:
   A. Amplifier.
   B. Detector.
   C. Modulator.
   D. Oscillator.
   E. Rectifier.

4. The high component density of integrated circuits acts to:
   A. Maximize the power output.
   B. Increase the current drain.
   C. Increase the voltage requirements.

D. Increase the operating speed.

E. Reduce the frequency range.

5. A utility meter generally measures:

   A. Watts.

   B. Amperes.

   C. Kilowatt hours.

   D. Kilowatts.

   E. Volt hours.

6. High voltages are better than low voltages for long-distance electric power transmission because:

   A. The lines can better withstand lightning strokes.

   B. The magnetic fields are weaker.

   C. The electric fields are weaker.

   D. The I^2R losses are lower.

   E. No! Low voltages are better.

7. A component that discharges filter capacitors after a power supply has been switched off is called:

   A. A discharge inductor.

   B. A rectifier diode.

   C. A bleeder resistor.

   D. A filter discharger.

   E. A voltage divider.

8. An advantage of a laser printer over a dot-matrix printer for computer applications is:

   A. Lower cost.

   B. Superior image quality.

   C. Greater bandwidth.

   D. Lower resolution.

   E. Better voltage regulation.

9. The output electrode of a bipolar transistor is usually the:

   A. Grid.

   B. Gate.

   C. Base.

   D. Source.

   E. Collector.

10. The schematic symbol for an op amp is:

   A. A triangle.

B. A circle.

C. A circle with a line through it.

D. A rectangle.

E. A D-shaped figure.

11. For a given value of capacitance, as the frequency goes down, the reactance:

A. Approaches zero from the negative side.

B. Gets more and more negative.

C. Approaches zero from the positive side.

D. Gets more and more positive.

E. Stays the same.

12. In a 360-Hz wave, a degree of phase represents:

A. 1.00 second.

B. 1/360 second.

C. 360 milliseconds.

D. 7.72 microseconds.

E. A quantity you can't calculate from this data.

13. A device that converts visible light into dc is:

A. A phototransistor.

B. A photovoltaic cell.

C. An optovoltaic cell.

D. An optocell.

E. An optoisolator.

14. A communications circuit in which either operator can interrupt the other at any time is called:

A. Full duplex.

B. Full simplex.

C. A wireless system.

D. A local area network.

E. A transceiver system.

15. A circuit has a battery of 3.0 V and a bulb with a resistance of 12.0 ohms. The current through the bulb is:

A. 36 A.

B. 4.0 A.

C. 250 mA.

D. 40 mA.

E. 36 mA.

16. A small 9-V battery might be used to provide power to:

    A. An electronic calculator.

    B. A personal computer.

    C. A radio transmitter.

    D. An electric iron.

    E. Any of the above.

17. In an AM voice signal, the audio information is:

    A. Exactly at the carrier frequency.

    B. Contained in sidebands.

    C. At harmonics of the carrier frequency.

    D. Rectified before being impressed onto the carrier.

    E. Detected before being impressed onto the carrier.

18. The oscillating frequency of a quartz crystal can be varied slightly by:

    A. Changing the bias on the transistor.

    B. Changing the voltage across the varactor.

    C. Reversing the power supply polarity.

    D. Placing a small variable capacitor across the crystal.

    E. No! The frequency of a crystal cannot be changed at all.

19. When several resistances are connected in series:

    A. The current is the same through each one.

    B. The voltage is the same across each one.

    C. Both A and B are true.

    D. Neither A nor B is true.

    E. They must all have the same value.

20. In a power supply, resistors are sometimes connected in series with the diodes in order to:

    A. Increase the current output.

    B. Protect the diodes against surge currents.

    C. Help the diodes discharge.

    D. Bleed charge off of the filter capacitors.

    E. Regulate the output voltage.

21. In a purely resistive impedance, there is:

    A. A net capacitance.

    B. A net inductance.

    C. Zero resistance.

D. Zero reactance.

E. Zero conductance.

22. Two 400-µH inductors are connected in series. There is no mutual inductance. The total inductance is:

    A. 100 µH.

    B. 200 µH.

    C. 400 µH.

    D. 800 µH.

    E. 1.6 mH.

23. The current-carrying part of a field-effect transistor, analogous to a garden hose in some ways, is called the:

    A. Source.

    B. Gate.

    C. Drain.

    D. Source-drain junction.

    E. Channel.

24. In a rectifier diode, current flows for approximately how much of the ac cycle?

    A. 360 degrees.

    B. 270 degrees.

    C. 180 degrees.

    D. 90 degrees.

    E. 45 degrees.

25. A millivolt is:

    A. 10^3 V.

    B. 10^{-3} V.

    C. 10^6 V.

    D. 10^{-6} V.

    E. 10^{-9} V.

26. The reciprocal of reactance is called:

    A. Impedance.

    B. Conductance.

    C. Resistance.

    D. Admittance.

    E. Susceptance.

27. Another name for joules per second is:

    A. Volts.

B. Amperes.

C. Ohms.

D. Gilberts.

E. Watts.

28. In a pure inductance:

   A. Current lags voltage by 90 degrees.

   B. Current lags voltage by less than 90 degrees.

   C. Current and voltage are in phase.

   D. Current leads voltage by 90 degrees.

   E. Current leads voltage by 180 degrees.

29. A form of microwave electron tube is:

   A. A cavity resonator.

   B. A triode.

   C. A klystron.

   D. A cathode-ray tube.

   E. None of the above.

30. Magnetic lines of flux are generally:

   A. Parallel with the flow of electric current.

   B. In the plane of the electric current.

   C. At right angles to the flow of current.

   D. At a 45 degree angle to the flow of current.

   E. Impossible to predict as to their direction.

31. A class-A amplifier conducts during how much of the input cycle?

   A. Less than 90 degrees.

   B. 90–180 degrees.

   C. 180–270 degrees.

   D. 270–360 degrees.

   E. 360 degrees.

32. An advantage of parallel data transfer over serial transfer is:

   A. Higher speed.

   B. Narrower bandwidth.

   C. Lower frequency.

   D. Higher power.

   E. Better signal-to-noise ratio.

33. One way to keep interelectrode capacitance to a minimum is to:

   A. Use only electrolytic capacitors.

B. Avoid the use of sheet metal.

C. Use batteries as the source of power.

D. Keep wire leads short.

E. Use air-core transformers.

34. One advantage of ac (compared with dc) as a source of utility power is that:

    A. It can be used at safer voltage levels.

    B. It can be easily stepped up or down in voltage.

    C. There is lower transmission line loss.

    D. The waveshape can be varied easily.

    E. Higher voltages can be used.

35. An element of matter:

    A. Is a good conductor.

    B. Has a unique number of protons.

    C. Is an insulator.

    D. Makes a good electronic component.

    E. Can be fabricated onto an IC chip.

36. A transformer quadruples the ac voltage. The primary-to-secondary impedance ratio is therefore:

    A. 1:16.

    B. 1:4.

    C. 1:1.

    D. 4:1.

    E. 16:1.

37. Frequency multiplication is possible with a semiconductor diode because the
diode is:

    A. Amplifying.

    B. Detecting.

    C. Forward-biased.

    D. Reverse-biased.

    E. Nonlinear.

38. An ammeter measures:

    A. Current.

    B. Voltage.

    C. Resistance.

    D. Power.

    E. Energy.

39. A good type of resistor to use in a radio amplifier is:

    A. Carbon composition.

    B. Wirewound.

    C. Silver mica.

    D. Reactive.

    E. Tantalum.

40. Of the following, which does *not* denote a type of capacitor?

    A. Silver-mica.

    B. Air variable.

    C. Ferrite.

    D. Electrolytic.

    E. Tantalum.

41. A logic circuit has an output 0 when the input is 1, and vice versa. This is:

    A. A NOR gate.

    B. An AND gate.

    C. A NOT gate.

    D. An XOR gate.

    E. An XNOR gate.

42. A type of FM detector that has its limiter built in is:

    A. A balanced modulator.

    B. A beat-frequency oscillator.

    C. An envelope detector.

    D. A product detector.

    E. A ratio detector.

43. Which of the following materials is/are commonly used as a semiconductor?

    A. Silver-mica.

    B. Ferrite.

    C. Gallium arsenide.

    D. Tantalum.

    E. All of the above.

44. In an NPN bipolar transistor circuit:

    A. The dc collector voltage is negative.

    B. The output is taken from the base.

    C. The dc collector voltage is positive.

    D. The output is taken from the drain.

    E. Negative feedback must be used.

45. A simple power supply filter can be made using:

    A. A capacitor in parallel with the rectifier output.

    B. A resistor in parallel with the rectifier output.

    C. An inductor in parallel with the rectifier output.

    D. A capacitor in series with the rectifier output.

    E. A resistor in series with the rectifier output.

46. If an ac impedance is a pure resistance, then:

    A. The reactance is capacitive.

    B. The reactance is inductive.

    C. The resistance is zero.

    D. The reactance is zero.

    E. No! An ac impedance cannot be a pure resistance.

47. Three resistances are in parallel, with values of 100, 200, and 300 ohms. The current through the 200-ohm resistor is 500 mA. What is the voltage across the whole combination?

    A. There isn't enough information to figure it out.

    B. 400 V.

    C. 400 mV.

    D. 100 V.

    E. 100 mV.

48. As the frequency of an alternating current decreases toward zero (dc), the reactance of a capacitor:

    A. Remains constant.

    B. Becomes small positively.

    C. Becomes small negatively.

    D. Becomes large positively.

    E. Becomes large negatively.

49. The rate of change in a quantity is called the:

    A. Effective value.

    B. Instantaneous value.

    C. Average value.

    D. Peak value.

    E. Derivative.

50. The axis of the geomagnetic field:

    A. Corresponds exactly with the rotational axis.

    B. Is slanted with respect to the rotational axis.

    C. Is perpendicular to the rotational axis.

D. Runs parallel to lines of latitude.

E. Is circular in shape.

51. One of the main shortcomings of MOSFETs is that they:

A. Are easily damaged by static electricity.

B. Require high voltages.

C. Consume large amounts of current.

D. Have very low gain.

E. Do not have good sensitivity.

52. Resistivity of wire can be specified in:

A. Ohms.

B. Ohmmeters.

C. Ohms per meter.

D. Amperes per ohm.

E. Ohms per ampere.

53. A complementary-metal-oxide-semiconductor (CMOS) IC:

A. Employs diodes and NPN transistors on a single chip.

B. Employs N-channel and P-channel FETs on a single chip.

C. Uses two chips connected together in a special way.

D. Uses resistors and PNP transistors on a single chip.

E. Consists of metal oxide sandwiched between two layers of P-type material.

54. A piano sounds different than a saxophone, even if the notes are at the same frequency, because of a difference in:

A. Bias.

B. Waveform.

C. Voltage.

D. Current.

E. The way you imagine it.

55. Inductances in parallel, assuming there is no mutual inductance, add up like:

A. Resistances in series.

B. Resistances in parallel.

C. Capacitances in parallel.

D. Batteries in parallel.

E. No other type of electrical component.

56. A reactance modulator produces:

A. CW.

B. AM.

C. SSB.

D. FM.

E. PCM.

57. Antenna efficiency is:

A. Usually more important in transmitting systems than in receiving systems.

B. The difference between the input and output power.

C. A direct function of the noise in a transmitting system.

D. The proportion of input power that gets converted into heat.

E. Highest when the transmitter final amplifier is maximally efficient.

58. In a parallel-resonant LC circuit, the impedance is:

A. Low and reactive.

B. High and reactive.

C. Low and resistive.

D. High and resistive.

E. Any of the above.

59. In a resistance/inductance (RL) series circuit:

A. Current lags voltage by 180 degrees.

B. Current lags voltage by 90 degrees.

C. Current lags voltage by less than 90 degrees.

D. Current and voltage are in phase.

E. Current leads voltage by 90 degrees.

60. In three-phase ac, the difference in phase between any two waves is:

A. 30 degrees.

B. 45 degrees.

C. 60 degrees.

D. 90 degrees.

E. 120 degrees.

61. Electrostatic forces can be measured to directly indicate:

A. Power.

B. Frequency.

C. Current.

D. Resistance.

E. Voltage.

62. A circuit has a complex impedance of 9 + j12. The absolute-value impedance is:

A. 15 ohms.

B. 9 ohms.

C. 12 ohms.

D. 21 ohms.

E. Impossible to calculate from this data.

63. Three resistors, each of 30 ohms, are connected in parallel. The net resistance is:

A. 90 ohms.

B. 60 ohms.

C. 33 ohms.

D. 10 ohms.

E. Impossible to determine from the data given.

64. The logical statement X + Y = Y + X depicts:

A. The distributive property.

B. The associative property.

C. The commutative property.

D. The de Morgan theorem.

E. The behavior of a NOR gate.

65. A cell that can be recharged, and therefore used again and again, is called:

A. A secondary cell.

B. A multiple-use cell.

C. A primary cell.

D. A tertiary cell.

E. A battery.

66. A resistor has a positive temperature coefficient of 1.00 percent per degree C. If its value is 100 ohms at 20 degrees C, what is its value at 25 degrees C?

A. 100 ohms.

B. 105 ohms.

C. 95 ohms.

D. 125 ohms.

E. It can't be calculated from this data.

67. A memory that can be easily accessed, but not written over, is called:

A. RAM.

B. PRAM.

C. CMOS.

D. ROM.

E. CROM.

68. The capacitance between two parallel sheets of metal is:

A. Directly proportional to the distance between them.

B. Inversely proportional to the distance between them.

C. Not dependent on the distance between them.

D. Inversely proportional to their surface area.

E. Negligible unless the sheets are both gigantic.

69. The forward base bias in a transistor is increased until the collector current levels off. This condition is:

A. Cutoff.

B. Saturation.

C. Pinchoff.

D. Forward breakover.

E. Avalanche.

70. An advantage of a LEO communications satellite system over a geostationary communications satellite is the fact that:

A. The bandwidth is greater.

B. The image resolution is superior.

C. The satellites never change their position in the sky.

D. Large, high-gain antennas are not required.

E. The satellites orbit at higher altitudes.

71. A coil has 20 mH of inductance. What is the inductive reactance?

A. 20 ohms.

B. 0.05 ohms.

C. 50 ohms.

D. 20k ohms.

E. There isn't enough information to know.

72. What is an advantage of digital signal processing (DSP)?

A. Improved signal-to-noise ratio.

B. Enhanced fidelity.

C. Improved intelligibility.

D. Relative immunity to atmospheric noise.

E. All of the foregoing.

73. A dc voltage-divider network is made using:

A. Inductors.

B. Resistors.

C. Capacitors.

D. Bipolar transistors.

E. FETs.

74. The electron volt is a unit of:

    A. Voltage.

    B. Current.

    C. Power.

    D. Electric field strength.

    E. Energy.

75. A transformer has a primary-to-secondary turns ratio of 10:1. The input is 120 V rms ac. The output is:

    A. 12 kV rms ac.

    B. 1.2 kV rms ac.

    C. 120 V rms ac.

    D. 12 V rms ac.

    E. 1.2 V rms ac.

76. Wave X leads wave Y by 270 degrees. This would be better expressed by saying that:

    A. Wave X lags wave Y by 90 degrees.

    B. Wave X lags wave Y by 180 degrees.

    C. Wave X lags wave Y by 270 degrees.

    D. Waves X and Y are out of phase.

    E. Waves X and Y are in phase.

77. Which type of amplifier circuit has the transistor or FET biased exactly at cut-off or pinchoff when there is no signal input?

    A. Class-A.

    B. Class-AB.

    C. Class-B.

    D. Class-C.

    E. Class-D.

78. Weather satellites generally send images by:

    A. Pulse width modulation.

    B. Pulse amplitude modulation.

    C. Facsimile.

    D. Single sideband.

    E. Fast-scan television.

79. An audio oscillator that uses two amplifiers in cascade, with positive feedback from the output of the second stage to the input of the first stage, is known as a:

    A. Colpitts circuit.

    B. Hartley circuit.

C. Multivibrator.

D. VCO.

E. Clapp circuit.

80. The main factor that limits the frequency at which a P-N junction will rectify is the:

A. PIV rating.

B. Junction capacitance.

C. Junction resistance.

D. Junction inductance.

E. Reverse-bias current.

81. The henry is a:

A. Very small unit.

B. Unit of capacitive reactance.

C. Measure of transistor gain.

D. Unit of phase.

E. Very large unit.

82. A diode that can be used as a variable capacitance is a:

A. GaAsFET.

B. Silicon rectifier.

C. Point-contact diode.

D. Varactor.

E. Germanium detector.

83. Elements can join together to form:

A. Ions.

B. Isotopes.

C. Nuclei.

D. Compounds.

E. Majority carriers.

84. The rms value for an ac wave is also sometimes called the:

A. Absolute value.

B. Direct-current value.

C. Effective value.

D. Equivalent value.

E. Reactive value.

85. The gigabyte is a unit commonly used as a measure of:

A. Data access time.

B. Data frequency.

C. Data transfer speed.

D. Data storage capacity.

E. Data communications accuracy.

86. In a parallel combination of light bulbs, if one bulb shorts out:

A. The circuit had better have a fuse!

B. The other bulbs will brighten or burn out.

C. The other bulbs will dim, but stay lit.

D. The current drawn from the source will decrease.

E. None of the above.

87. A common lab multimeter cannot measure:

A. Current.

B. Frequency.

C. Voltage.

D. Resistance.

E. It can measure any of the above.

88. In a P-channel JFET:

A. The drain is positive with respect to the source.

B. The gate must be grounded.

C. The majority carriers are holes.

D. The source receives the input signal.

E. All of the above are true.

89. If you place a bar of iron inside a cylindrical coil of wire, and then run dc through the wire, you have:

A. A rheostat.

B. A permanent magnet.

C. A flux meter.

D. An electric generator.

E. An electromagnet.

90. Admittance is a quantity expressing:

A. Opposition to dc.

B. Opposition to audio signals.

C. Ease with which a circuit passes ac.

D. The ratio of capacitance to inductance.

E. The ratio of reactance to resistance.

91. In a common-emitter bipolar-transistor circuit:

A. The collector is at signal ground.

   B. The output is taken from the base.

   C. The emitter is at signal ground.

   D. The bases of two transistors are connected together.

   E. The output is taken from the emitter.

92. In a certain resistance-capacitance (RC) circuit, the current leads the voltage by 45 degrees. The resistance is 50 ohms. The capacitive reactance is:

   A. 25 ohms.

   B. –25 ohms.

   C. 50 ohms.

   D. –50 ohms.

   E. Impossible to determine from this information.

93. The VA power is equal to the true power only when:

   A. A circuit has no resistance.

   B. A circuit has no impedance.

   C. A circuit has no reactance.

   D. The complex impedance is high.

   E. The phase angle is 45 degrees.

94. The ac in the utility mains is usually:

   A. Of variable frequency.

   B. At radio frequencies.

   C. Sinusoidal.

   D. At a potential of 12 V.

   E. A square wave.

95. The standard unit of inductance is the:

   A. Farad.

   B. Henry.

   C. Gilbert.

   D. Gauss.

   E. Tesla.

96. The output of a circuit is 20 V and the input is 5.0 V. This is a gain of:

   A. 4 dB.

   B. 6 dB.

   C. 12 dB.

   D. –4 dB.

   E. –6 dB.

97. An example of a device that converts electrical energy into visible radiant energy is:

    A. A photocell.

    B. A phototransistor.

    C. A photovoltaic cell.

    D. A light-emitting diode.

    E. A speaker.

98. What is the purpose of a bleeder resistor in a power supply?

    A. To regulate the current.

    B. To regulate the voltage.

    C. To protect the diodes against voltage transients.

    D. To protect the diodes against current surges.

    E. None of the above.

99. A resistor of 100 ohms carries 333 mA dc. The power dissipated by that resistor is:

    A. 300 mW.

    B. 3.33 W.

    C. 33.3 W.

    D. 3.33 W.

    E. 11.1 W.

100. The data in a volatile computer memory:

    A. Is stored on magnetic disks.

    B. Consists of analog waveforms.

    C. Vanishes if power is removed.

    D. Must pass through a modem before it can be understood by the CPU.

    E. Cannot be used by a microprocessor.

A
APPENDIX

Answers to quiz, test, and exam questions

Chapter 1

1. B	2. C	3. D	4. D
5. D	6. A	7. B	8. C
9. A	10. B	11. C	12. A
13. D	14. C	15. C	16. A
17. C	18. B	19. C	20. D

Chapter 2

1. B	2. C	3. D	4. B
5. C	6. A	7. C	8. C
9. A	10. D	11. D	12. A
13. B	14. B	15. C	16. D
17. B	18. C	19. A	20. C

Chapter 3

1. B	2. A	3. C	4. C
5. A	6. D	7. C	8. A
9. D	10. C	11. A	12. B
13. A	14. C	15. D	16. B
17. D	18. A	19. C	20. B

Chapter 4

1. A	2. C	3. D	4. A
5. B	6. D	7. C	8. A
9. C	10. A	11. D	12. B
13. D	14. B	15. A	16. C
17. C	18. A	19. D	20. B

Chapter 5

1. B	2. D	3. C	4. C
5. A	6. B	7. D	8. A
9. B	10. D	11. B	12. C
13. B	14. C	15. A	16. C
17. B	18. C	19. B	20. C

Chapter 6

1. C	2. A	3. A	4. C
5. D	6. B	7. B	8. C
9. B	10. C	11. D	12. C
13. A	14. C	15. A	16. A
17. B	18. C	19. B	20. A

Chapter 7

1. C	2. C	3. B	4. C
5. D	6. B	7. A	8. B
9. A	10. D	11. B	12. A
13. B	14. C	15. D	16. C
17. C	18. C	19. A	20. D

Chapter 8

1. C	2. A	3. D	4. B
5. C	6. B	7. B	8. C
9. A	10. C	11. C	12. D
13. B	14. A	15. D	16. D
17. C	18. B	19. B	20. A

Test: Part One

1. A	2. A	3. C	4. B
5. A	6. B	7. A	8. E
9. E	10. C	11. C	12. C
13. B	14. C	15. D	16. B
17. A	18. A	19. E	20. E
21. D	22. D	23. D	24. C
25. A	26. C	27. D	28. A
29. C	30. C	31. C	32. D
33. E	34. D	35. E	36. B
37. D	38. D	39. C	40. A
41. A	42. D	43. C	44. E
45. C	46. C	47. D	48. A
49. E	50. C		

Chapter 9

1. C	2. C	3. A	4. C
5. D	6. D	7. A	8. B
9. C	10. B	11. D	12. B
13. B	14. D	15. D	16. A
17. C	18. C	19. D	20. A

Chapter 10

1. B	2. A	3. D	4. B
5. B	6. C	7. D	8. C
9. A	10. D	11. B	12. D
13. A	14. A	15. D	16. B
17. A	18. C	19. A	20. D

Chapter 11

1. D	2. A	3. A	4. A
5. C	6. D	7. B	8. A
9. D	10. D	11. B	12. C
13. B	14. A	15. B	16. C
17. D	18. A	19. B	20. B

Chapter 12

1. C	2. A	3. B	4. B
5. D	6. B	7. D	8. C
9. A	10. B	11. B	12. D
13. C	14. A	15. C	16. B
17. D	18. C	19. D	20. C

Chapter 13

1. C	2. A	3. B	4. D
5. C	6. A	7. B	8. C
9. D	10. C	11. D	12. D
13. A	14. B	15. D	16. B
17. D	18. C	19. B	20. A

Chapter 14

1. B	2. D	3. A	4. C
5. B	6. A	7. B	8. A
9. A	10. D	11. C	12. B
13. B	14. B	15. C	16. B
17. C	18. D	19. A	20. C

Chapter 15

1. B	2. B	3. A	4. D
5. B	6. C	7. A	8. C
9. D	10. A	11. A	12. A
13. C	14. A	15. D	16. B
17. D	18. C	19. C	20. B

Chapter 16

1. B	2. A	3. D	4. D
5. D	6. B	7. C	8. C
9. A	10. C	11. B	12. B
13. A	14. B	15. B	16. D
17. C	18. A	19. B	20. A

Chapter 17

1. C	2. C	3. A	4. D
5. A	6. B	7. C	8. D
9. B	10. C	11. C	12. A
13. D	14. A	15. C	16. B
17. D	18. D	19. C	20. A

Chapter 18

1. C	2. A	3. D	4. B
5. B	6. A	7. C	8. D
9. A	10. D	11. B	12. B
13. C	14. D	15. C	16. A
17. C	18. D	19. B	20. B

Test: Part Two

1. E	2. E	3. A	4. A
5. E	6. D	7. C	8. D
9. A	10. E	11. C	12. A
13. D	14. C	15. C	16. E
17. E	18. D	19. C	20. B
21. B	22. D	23. C	24. C
25. E	26. C	27. B	28. A
29. E	30. C	31. A	32. A
33. A	34. A	35. B	36. E
37. C	38. D	39. C	40. B
41. B	42. C	43. C	44. C
45. D	46. D	47. E	48. E
49. A	50. A		

Chapter 19

1. B	2. D	3. C	4. A
5. C	6. A	7. D	8. B
9. C	10. B	11. D	12. B
13. B	14. A	15. B	16. C
17. A	18. C	19. D	20. C

Chapter 20

1. B	2. D	3. C	4. B
5. A	6. A	7. D	8. C
9. B	10. D	11. B	12. A
13. C	14. A	15. C	16. D
17. B	18. A	19. C	20. A

Chapter 21

1. C	2. D	3. B	4. A
5. B	6. C	7. A	8. B
9. D	10. C	11. A	12. D
13. D	14. B	15. C	16. A
17. C	18. B	19. A	20. C

Chapter 22

1. D	2. A	3. B	4. D
5. D	6. C	7. B	8. C
9. B	10. D	11. C	12. A
13. C	14. D	15. A	16. D
17. B	18. C	19. B	20. B

Chapter 23

1. B	2. D	3. A	4. B
5. C	6. A	7. B	8. C
9. D	10. B	11. C	12. A
13. D	14. A	15. C	16. D
17. B	18. C	19. A	20. A

Chapter 24

1. A	2. C	3. B	4. B
5. D	6. A	7. D	8. B
9. D	10. A	11. C	12. B
13. D	14. D	15. C	16. C
17. B	18. A	19. B	20. A

Chapter 25

1. C	2. B	3. A	4. C
5. A	6. C	7. D	8. C
9. D	10. B	11. A	12. B
13. D	14. C	15. D	16. D
17. B	18. A	19. D	20. C

Chapter 26

1. C	2. A	3. C	4. C
5. A	6. B	7. D	8. B
9. C	10. D	11. B	12. C
13. A	14. B	15. D	16. D
17. A	18. B	19. A	20. D

Chapter 27

1. C	2. C	3. A	4. C
5. D	6. C	7. B	8. A
9. C	10. D	11. C	12. A
13. D	14. B	15. B	16. B
17. D	18. A	19. B	20. D

Chapter 28

1. C	2. C	3. A	4. B
5. B	6. D	7. C	8. A
9. D	10. C	11. C	12. A
13. B	14. B	15. A	16. B
17. C	18. D	19. D	20. B

Chapter 29

1. C	2. D	3. A	4. B
5. C	6. D	7. C	8. C
9. A	10. D	11. C	12. A
13. B	14. C	15. C	16. A
17. B	18. B	19. C	20. A

Chapter 30

1. C	2. A	3. D	4. D
5. C	6. A	7. B	8. B
9. B	10. D	11. C	12. A
13. D	14. A	15. C	16. C
17. D	18. A	19. D	20. B

Chapter 31

1. C	2. B	3. D	4. A
5. B	6. B	7. A	8. D
9. A	10. C	11. C	12. A
13. D	14. A	15. B	16. C
17. A	18. C	19. B	20. D

Chapter 32

1. B	2. A	3. D	4. C
5. A	6. C	7. D	8. A
9. D	10. C	11. A	12. D
13. B	14. C	15. A	16. C
17. B	18. A	19. D	20. A

Test: Part Three

1. E	2. B	3. A	4. C
5. A	6. C	7. A	8. D
9. E	10. D	11. D	12. B
13. C	14. C	15. D	16. C
17. B	18. E	19. B	20. D
21. C	22. E	23. A	24. B
25. D	26. A	27. C	28. E
29. A	30. D	31. C	32. C
33. C	34. B	35. E	36. D
37. A	38. E	39. B	40. E
41. C	42. D	43. A	44. B
45. A	46. B	47. B	48. C
49. E	50. B		

Final Exam

1. D	2. D	3. D	4. D
5. C	6. D	7. C	8. B
9. E	10. A	11. B	12. D
13. B	14. A	15. C	16. A
17. B	18. D	19. A	20. B
21. D	22. D	23. E	24. C
25. B	26. E	27. E	28. A
29. C	30. C	31. E	32. A
33. D	34. B	35. B	36. A
37. E	38. A	39. A	40. C
41. C	42. E	43. C	44. C
45. A	46. D	47. D	48. E
49. E	50. B	51. A	52. C
53. B	54. B	55. B	56. D
57. A	58. D	59. C	60. E
61. E	62. A	63. D	64. C
65. A	66. B	67. D	68. B
69. B	70. D	71. E	72. E
73. B	74. E	75. D	76. A
77. C	78. C	79. C	80. B
81. E	82. D	83. D	84. C
85. D	86. A	87. B	88. C
89. E	90. C	91. C	92. D
93. C	94. C	95. B	96. C
97. D	98. E	99. E	100. C

B
APPENDIX

Schematic symbols

ammeter		capacitor (feedthrough)	
amplifier (operational)		capacitor (fixed, nonpolarized)	
AND gate		capacitor (fixed, polarized)	
antenna (balanced, dipole)		capacitor (ganged, variable)	
antenna (general)		capacitor (single variable)	
antenna (loop, shielded)		capacitor (split-rotor, variable)	
antenna (loop, unshielded)		capacitor (split-stator, variable)	
antenna (unbalanced)		cathode (directly heated)	
antenna (whip)		cathode (indirectly heated)	
attenuator (or resistor, fixed)		cathode (cold)	
attenuator (or resistor, variable)		cavity resonator	
		cell	
battery		circuit breaker	

coaxial cable		headphone (single)	
coaxial cable (grounded shield)	or	headphone (stereo)	
crystal (piezoelectric)		inductor (air-core)	
delay line	or	inductor (bifilar)	
diode (field effect)		inductor (iron-core)	
diode (general)		inductor (tapped)	
diode (Gunn)		inductor (variable)	or
diode (light-emitting)			
diode (photosensitive)		integrated circuit	
diode (photovoltaic)		inverter or inverting amplifier	
diode (pin)		jack (coaxial or phono)	
diode (Schottky)		jack (phone, two-conductor)	
diode (tunnel)		jack (phone, two-conductor interrupting)	
diode (varactor)	or	jack (phone, three-conductor)	
diode (zener)		jack (phono)	
directional coupler (or wattmeter)	or	key (telegraph)	
exclusive-OR gate		lamp (incandescent)	
		lamp (neon)	or
female contact (general)		male contact (general)	
ferrite bead	or	meter (general)	
fuse	or or	microammeter	
galvanometer	or	microphone	or
ground (chassis)	or	microphone (directional)	
ground (earth)		milliammeter	
handset		NAND gate	
		negative voltage connection	

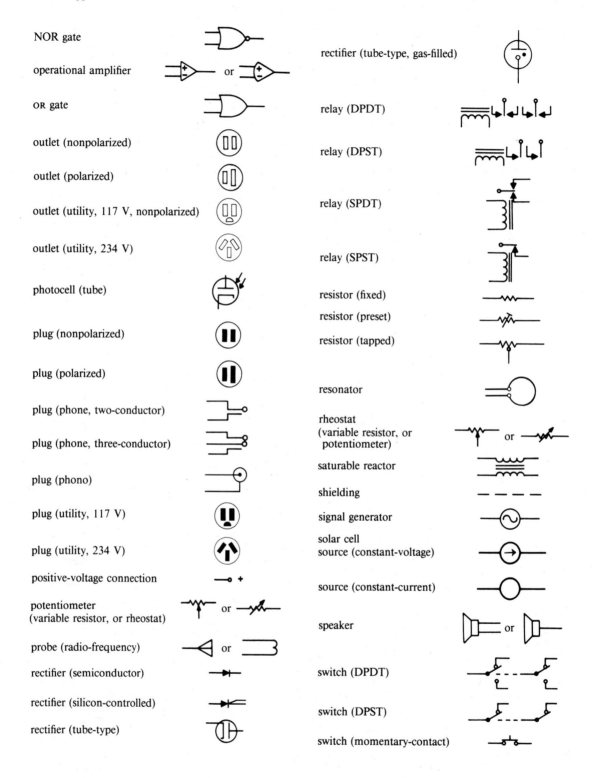

NOR gate

operational amplifier or

OR gate

outlet (nonpolarized)

outlet (polarized)

outlet (utility, 117 V, nonpolarized)

outlet (utility, 234 V)

photocell (tube)

plug (nonpolarized)

plug (polarized)

plug (phone, two-conductor)

plug (phone, three-conductor)

plug (phono)

plug (utility, 117 V)

plug (utility, 234 V)

positive-voltage connection +

potentiometer
(variable resistor, or rheostat) or

probe (radio-frequency) or

rectifier (semiconductor)

rectifier (silicon-controlled)

rectifier (tube-type)

rectifier (tube-type, gas-filled)

relay (DPDT)

relay (DPST)

relay (SPDT)

relay (SPST)

resistor (fixed)

resistor (preset)

resistor (tapped)

resonator

rheostat
(variable resistor, or
potentiometer) or

saturable reactor

shielding

signal generator

solar cell
source (constant-voltage)

source (constant-current)

speaker or

switch (DPDT)

switch (DPST)

switch (momentary-contact)

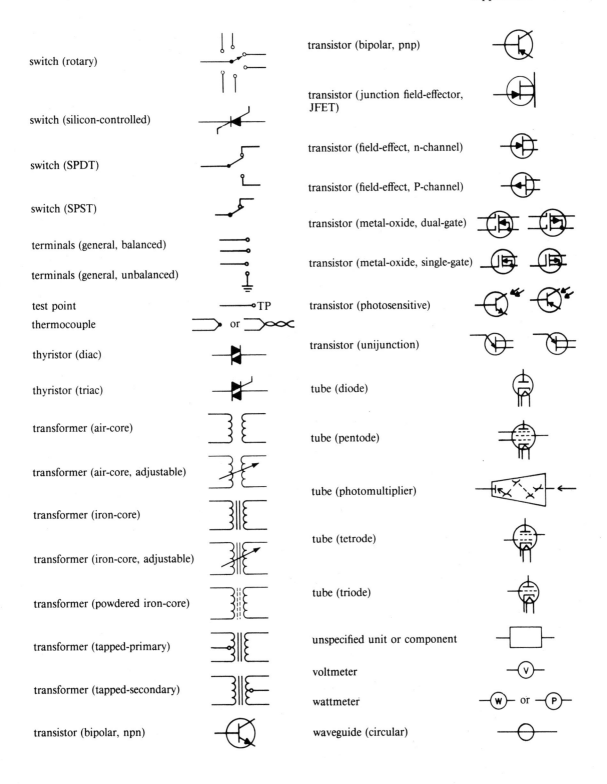

switch (rotary)	transistor (bipolar, pnp)
switch (silicon-controlled)	transistor (junction field-effector, JFET)
switch (SPDT)	transistor (field-effect, n-channel)
switch (SPST)	transistor (field-effect, P-channel)
terminals (general, balanced)	transistor (metal-oxide, dual-gate)
terminals (general, unbalanced)	transistor (metal-oxide, single-gate)
test point	transistor (photosensitive)
thermocouple	
thyristor (diac)	transistor (unijunction)
thyristor (triac)	tube (diode)
transformer (air-core)	tube (pentode)
transformer (air-core, adjustable)	tube (photomultiplier)
transformer (iron-core)	tube (tetrode)
transformer (iron-core, adjustable)	tube (triode)
transformer (powdered iron-core)	unspecified unit or component
transformer (tapped-primary)	voltmeter
transformer (tapped-secondary)	wattmeter
transistor (bipolar, npn)	waveguide (circular)

waveguide (rectangular)

waveguide (flexible)

waveguide (twisted)

wires (crossing, connected)   or   or

wires (crossing, not connected)   or

Suggested additional reading and reference

Crowhurst, Norman H. *Basic Electronics Course*. 2d ed.; McGraw-Hill, 1987.

Evans, Alvis. *Basic Digital Electronics*. Master Publishing, Inc., 1996.

Gibilisco, Stan. *The Illustrated Dictionary of Electronics*. 7th ed.; McGraw-Hill, 1997.

Gibilisco, Stan. *TAB Encyclopedia of Electronics for Technicians and Hobbyists*. McGraw-Hill, 1997.

Gibilisco, Stan, and Sclater, Neil, *Encyclopedia of Electronics*. 2d ed.; McGraw-Hill, 1990.

Graf, Rudolf F. *Encyclopedia of Electronic Circuits.*, vols. 1, 2, 3, 4. McGraw-Hill, 1985, 1988, 1990, 1992.

Horn, Delton T. *Basic Electronics Theory*. 4th ed.; McGraw-Hill, 1994.

McWharter, Gene and Alvis Evans. *Basic Electronics*. Master Publishing, Inc., 1994.

Slone, G. Randy. *The TAB Electronics Guide to Understanding Electricity and Electronics*. McGraw-Hill, 1996.

Warring, R. H. *Understanding Electronics*. 3d ed.; McGraw-Hill, 1989.

Index

About the Author

Stan Gibilisco is an electronics and science author specializing in reference works. His *Encyclopedia of Electronics* was named a "Best Reference of the 1980s" by the American Library Association. A former radio engineer and technical writer in electronics, he has also written *The Illustrated Dictionary of Electronics*, now in its seventh edition, and *The TAB Encyclopedia of Electronics for Technicians and Hobbyists*, among many other books.